The Moths and Butterflies of Great Britain and Ireland

Volume 4, Part 2

Volume 4, Part 2

Gelechiidae

Editors: A. Maitland Emmet
John R. Langmaid

Associate Editors: K. P. Bland
D. S. Fletcher, B. H. Harley
G. S. Robinson, B. Skinner
W. G. Tremewan

Artists: Richard Lewington
Michael J. Roberts

THE MOTHS AND BUTTERFLIES OF GREAT BRITAIN AND IRELAND

Harley Books

Harley Books (B. H. & A. Harley Ltd.),
Martins, Great Horkesley,
Colchester, Essex CO6 4AH, England

Text set in Plantin by Alison and Peter Guy
Text printed by St Edmundsbury Press Ltd,
 Bury St Edmunds, Suffolk
Colour reproduced by Hilo Colour Printers Ltd,
 Colchester, Essex, and printed by
 Westerham Press Ltd, Edenbridge, Kent
Bound by Woolnough Bookbinding Ltd,
 Irthlingborough, Northants

*The Moths and Butterflies of Great Britain
and Ireland Volume 4, Part 2*
© Harley Books, 2002

British Library Cataloguing-in-Publication Data applied for

ISBN 0 946589 67 4   Volume 4, Part 2 (hardback)
ISBN 0 946589 73 9   (paperback)

Volume 4, Part 2: Contents

Preface

Much of the Preface to Volume 4, Part 1, applies equally to this volume. The systematic order of genera follows that of *Log book of British Lepidoptera* by J. D. Bradley (2000), which itself incorporates changes from *The Lepidoptera of Europe. A distributional Checklist* by Karsholt & Razowski (1996). We have made every effort to comply with the most recent revisions, and the *Key to genera based on male genitalia* has been modified to take account of them.

The principal changes from earlier checklists are as follows:

The species *paripunctella* Thunberg has been transferred from *Teleiodes* Sattler to *Pseudotelphusa* Janse.

Five species formerly in the genus *Teleiodes* have been placed in the genus *Carpatolechia* Căpuşe, which is new to the British list.

The genus *Mesophleps* Hübner has been moved from Dichomeridinae to Anacampsinae.

The species *alacella* Zeller has been placed in *Dichomeris* Hübner and the genus *Acanthophila* Heinemann reduced to synonymy.

The genus *Thiotricha* Meyrick has been transposed from Anacampsinae to Pexicopiinae.

The order of species in the genus *Scrobipalpa* Janse has been altered to comply with affinities in life history at the request of the authors. The arrangement of species within the genus *Syncopacma* Meyrick has also been changed to bring closely similar species together and now follows Karsholt & Razowksi (*loc.cit.*).

In this Part of Volume 4, with the exception of those for *Bryotropha* species which are drawn by T. Rutten, all the genitalia figures are drawn by M. J. Roberts, most from specially made preparations but those of the male *Gelechia cuneatella* Douglas and the female *Monochroa niphognatha* (Gozmány) and *Coleotechnites piceaella* (Kearfott) are after Gustav Elsner's figures in *Die Palpenmotten mitteleuropas* (Elsner, Huemer & Tokár, 1999). We are most grateful to the authors for permitting us to use them. The colour plates and other diagnostic structural drawings have again been superbly executed by Richard Lewington. As with Part 1, a contribution towards the cost of the colour artwork was generously made by the John Spedan Lewis Foundation. We much appreciate their support.

February 2001   A. M. Emmet
J. R. Langmaid

Following the death of the Senior Editor in March, and while final editing and the early stages of production were in progress, the opportunity was taken to incorporate details of one or two important new discoveries, to update the distribution maps and to take account of some recent publications, now included in the References.

November 2001   J. R. Langmaid

SYSTEMATIC SECTION

Scheme of Classification

The Plan of the Work, as it was originally conceived, was based on the scheme of classification detailed by Kloet & Hincks (1972) and, with some modifications, this is printed below. However, although species continue to be described in the volumes to which they were originally assigned, re-arrangement within each volume is possible, and the scheme has been continually modified to take account of recent research. For example, that for the Hesperioidea and Papilionoidea was revised in accordance with the recommendations of Ackery (1984) and for other families in published volumes as recommended in Bradley & Fletcher (1986). The scheme for families covered in the present volume is discussed in the Preface to Part 1 (p. 9) and largely follows Bradley (2000). Current research has led to extensive revisions of classification, and one of the most recent of these was printed in *MBGBI* **3** (p. 38). The scheme for families yet to be treated will be broadly in accordance with this new plan, though further changes may in time be necessary. The effect of recent changes on families in volumes already published is clearly shown in that new arrangement; some have been reduced to subfamily status, and some new families have been erected.

ZEUGLOPTERA

Micropterigoidea
  Micropterigidae   1

DACNONYPHA

Eriocranioidea
  Eriocraniidae   1

EXOPORIA

Hepialoidae
  Hepialidae   1

MONOTRYSIA

Nepticuloidea
  Nepticulidae   1
  Opostegidae   1

Tischeroidea
  Tischeriidae   1

Incurvarioidea
  Incurvariidae   1
  Heliozelidae   1

DITRYSIA

Cossoidea
  Cossidae   2

Zygaenoidea
  Zygaenidae   2
  Limacodidae   2

Tineoidea
  Psychidae   2
  Tineidae   2
  Ochsenheimeriidae   2
  Lyonetiidae   2

Sesiodea
  Sesiidae   2
  Choreutidae   2

Gracillarioidea
  Roeslerstammiidae   3
  Bucculatricidae   2
  Douglasiidae   2
  Gracillariidae   2

Yponomeutoidea
  Yponomeutidae   3
  Glyphipterigidae
    Orthoteliinae   3
    Glyphipteriginae   2
  Heliodinidae   2
  Epermeniidae   3
  Schreckensteiniidae   3

Gelechioidea
  Oecophoridae   4(1)
  Coleophoridae   3
  Elachistidae   3
  Ethmiidae   4(1)
  Gelechiidae   **4(2)**
  Autostichidae   4(1)
  Blastobasidae   4(1)
  Batrachedridae   4(1)
  Agonoxenidae   4(1)
  Momphidae   4(1)
  Cosmopterigidae   4(1)
  Scythrididae   4(1)

Tortricoidea
  Tortricidae   5

Alucitoidea
  Alucitidae   6

Pterophoroidea   6
  Pterophoridae   6

Pyraloidea   6
  Pyralidae   6

Hesperioidea
  Hesperiidae   7(1)

Papilionoidea
  Papilionidae   7(1)
  Pieridae   7(1)
  Lycaenidae   7(1)
  Nymphalidae   7(1)

Bombycoidea
  Lasiocampidae   7(2)
  Saturniidae   7(2)
  Endromidae   7(2)

Drepanoidea
  Drepanidae   7(2)
  Thyatiridae   7(2)

Geometroidea
  Geometridae   8

Sphingoidea
  Sphingidae   9

Noctuoidea
  Notodontidae   9
  Thaumetopoiedae   9
  Lymantriidae   9
  Arctiidae   9
  Ctenuchidae   9
  Nolidae   9
  Noctuidae
    Noctuinae   9
    Hadeninae   9
    Cuculliinae to
    Hypeninae   10
  Agaristidae   10

GELECHIIDAE
K. P. Bland, M. F. V. Corley, A. M. Emmet, R. J. Heckford, P. Huemer, J. R. Langmaid, S. M. Palmer, M. S. Parsons, L. M. Pitkin, T. Rutten, K. Sattler, A. N. B. Simpson, P. H. Sterling*

Introduction

A large family of small to medium-sized Microlepidoptera; there are 160 species represented in the British Isles. Over 600 species are recorded in Europe and over 4000 worldwide. New species to the British list have been a regular feature for many years and further additions may be expected.

Classification

In the latter part of the nineteenth century and the beginning of the twentieth, the Gelechiidae included the Oecophoridae and the Blastobasidae, but these were excluded by Meyrick ([1928]). Gozmány (1963) separated the Symmocidae (detritus-feeders with termen of hindwing rounded). The modern Gelechiidae

appear to be a monophyletic group sharing the typical gelechiid hindwing shape, absence of vein CuP (1c) in the forewing, and a radial retinaculum on the female forewing (cubital in other gelechioid families).

There is reasonable agreement between authors on the grouping of genera within the family, although the placement of a few genera remains contentious. There is less agreement about the number of subfamilies (from three to nine) and tribes.

The subfamilies are distinguished by such characters as the presence or absence of an antennal pecten, the structure of abdominal sternite 2, the position of origin of the ductus seminalis, the form of abdominal segment 8 in the male and other characters of the male genitalia. In this work six subfamilies are recognized, following Bradley (2000), but subfamily descriptions are not given.

Imago The moths rest with wings folded flat or partially rolled around abdomen; body held with front end raised, slightly in some genera (e.g. *Scrobipalpa* Janse), but at about 40° to substrate in others (e.g. *Metzneria* Zeller); antennae lying above forewings, characteristically curved.

Head smooth-scaled; ocelli sometimes present; antenna filiform or shortly serrate, about three-quarters length of forewing, scape occasionally with pecten, which may be reduced to a single bristle; segment 2 of labial palpus usually strongly curved upwards so that segment 3 stands in front of frons, segment 2 frequently with projecting scales, segment 3 smooth-scaled; maxillary palpus minute, folded across base of haustellum; haustellum long, scaled nearly throughout. Forewing broad or narrow, veins R_5 (7) to costa, CuP (1c) absent, CuA_2 (2) and A_1+A_2 (1b) reach the margin close together. Hindwing with concave termen and produced apex. Hind tibia with erect hair-like scales dorsally. Abdomen without dorsal spines.

Male genitalia (figures 1–20) show such diversity between genera that it is not possible to characterize typical gelechiid genitalia. Asymmetry occurs in a few genera, unique structures may occur, and normally present structures may be absent. As a result it is not always possible to be certain of the homologies of particular structures.

The female genitalia (figures 21–46) have been neglected by many authors. In some genera this is because they show little differentiation between species, especially when the signum is absent. In *Syncopacma* Meyrick the difficulty of determining females with certainty led to their not being studied (Wolff, 1958). Nevertheless, in a few cases they provide easier separation of species than is possible with male genitalia (e.g.

*The authorship within the family is divided as follows:

K. P. Bland – The genera *Psamathocrita, Aristotelia, Xystophora, Stenolechia, Parachronistis, Recurvaria, Coleotechnites, Exoteleia, Athrips, Xenolechia, Pseudotelphusa, Chionodes, Sophronia, Anacampsis, Acompsia* less *A. schmidtiellus* (Heyden), *Platyedra, Pexicopia, Thiotricha.*

M. F. V. Corley – Family introduction; Key to genera based on male genitalia; (with P. Huemer) the genus *Caryocolum* (Description of imago).

A. M. Emmet – The genera *Metzneria, Isophrictis, Apodia, Monochroa, Chrysoesthia, Ptocheuusa, Altenia, Teleiodes, Carpatolechia, Teleiopsis, Nothris* less *N. congressariella* (Bruand), *Anarsia* less *A. lineatella* Zeller; *Sitotroga*; British distribution for most gelechiid species.

R. J. Heckford – The genera or species *Eulamprotes, Argolamprotes, Scrobipalpula, Gnorimoschema, Phthorimaea, Aproaerema, Syncopacma, Nothris congressariella, Anarsia lineatella.*

P. Huemer (with M. F. V. Corley) – The genus *Caryocolum* (introductory material and Life history).

J. R. Langmaid – The genera or species *Psoricoptera, Hypatima, Mesophleps, Acompsia schmidtiellus, Dichomeris.*

S. M. Palmer – The genera *Prolita, Aroga, Neofriseria, Neofaculta.*

M. S. Parsons (with K. Sattler) – The genus *Scrobipalpa.*

L. M. Pitkin – The genus *Mirificarma.*

T. Rutten – The genus *Bryotropha.*

K. Sattler (with M. S. Parsons) – The genus *Scrobipalpa.*

A. N. B. Simpson – The genus *Gelechia.*

P. H. Sterling – The genera *Brachmia, Helcystogramma.*

the *Bryotropha affinis* (Haworth) group). As in the male genitalia there can be unique structures.

In most genera the male genitalia are strongly three-dimensional. This makes it difficult to mount the dissected genitalia in the usual ventral side upwards position. Lateral mounting is possible, but the shape of the uncus and many of the complex features on the ventral side tend to be difficult to see, and drawing or photographs of genitalia in this position may be incomprehensible. In some genera (e.g. *Bryotropha* Heinemann) a compromise ventro-lateral position is a feasible option, but is not satisfactory for all genera. Pitkin (1986) has developed a technique in which the separation on one side of the vinculum and tegumen allows the unrolling of the ventral half of the armature so that all structures can be seen. This has the added advantage of diminishing the total depth of the mount which facilitates photographing the genitalia.

Larva There are no obvious macroscopic features characterizing the family. There is a full complement of prolegs except in *Metzneria*, *Isophrictis* Meyrick, *Apodia* Heinemann, *Ptocheuusa* Heinemann and *Psamathocrita* Meyrick where they are vestigial or absent. There is a great range of larval pabula, with British species feeding on lichens, mosses, pteridophytes and gymnosperms as well as monocotyledonous and herbaceous and woody dicotyledonous flowering plants. In all, at least 35 families of plants are used, the most important being Asteraceae (Compositae) (16 species belonging to eight genera), Fabaceae (Leguminosae) (15 species in eight genera), Caryophyllaceae (15 species in three genera), Rosaceae (11 species in eight genera) and mosses (at least ten species in four genera). Most species feed on a single genus or family, with only a very few species feeding on members of two or three different plant families. Some gelechiid genera specialize exclusively or nearly exclusively on the members of a single plant family, e.g. *Caryocolum* Gregor & Povolný on Caryophyllaceae, *Syncopacma* on Fabaceae, *Metzneria* (except *M. littorella* (Douglas)) and some related smaller genera on Asteraceae. Different species specialize on various parts of the plants, including roots, stems (in which some species cause galls), leaves, flowers, fruits and seeds. All feed concealed, those on leaves making a variety of spinnings. A number of species are initially leaf-miners, usually becoming external feeders living in spinnings, but a few mine leaves throughout their larval life. In most cases the larvae can change leaves. *Thiotricha* Meyrick forms a case from which it feeds on seeds. For a few species the biology is still unknown or is known only in continental Europe but not confirmed in the British Isles.

In warmer parts of the world some species are important crop pests. Among them are three species on the British list: *Phthorimaea operculella* (Zeller), *Sitotroga cerealella* (Olivier) and *Anarsia lineatella* Zeller but none of these can be considered to be a problem in the British Isles at present.

Pupa. First four abdominal segments fixed. Concealed, almost always in a cocoon, the pupal case remaining in the cocoon at eclosion.

The winter is usually passed as egg, larva or pupa, but for some species the overwintering stage remains unknown. Some larvae spin a cocoon in which to overwinter, but this is not necessarily used for pupation even if the larva does not feed again in the spring. Three species hibernate as adults. Most British species are univoltine, but a few are bivoltine.

Most species fly at dusk or by night, when they may come to light. A few are predominately diurnal. These usually have forewings with a metallic gloss or with metallic markings. During the day most species remain concealed, but some can be found on tree-trunks or fences, and others rest on flowers or seedheads. When disturbed they are more likely to run or drop to the ground and disappear into the herbage than to fly.

Checklist of the British and Irish Gelechiidae

ANOMOLOGINAE

METZNERIA Zeller, 1839
 CLEODORA Stephens, 1834, *nec* Péron &
  Lesueur, 1810
 PARASIA Duponchel, [1846]
723 *littorella* (Douglas, 1850)
 quinquepunctella (Herrich-Schäffer, 1854)
724 *lappella* (Linnaeus, 1758)
 silacea sensu Haworth, 1828, *nec silacella* Hübner,
  1796
 silacella sensu Stephens, 1834, *nec* Hübner, 1796
725 *aestivella* (Zeller, 1839)
 carlinella (Stainton, 1851)
726 *metzneriella* (Stainton, 1851)
 ?*ochroleucella* (Stephens, 1834)
 paucipunctella sensu Douglas, 1850, *nec* Zeller, 1839
727 *neuropterella* (Zeller, 1839)
727a *aprilella* (Herrich-Schäffer, 1854)
 sanguinolentella de Joannis, 1910

ISOPHRICTIS Meyrick, 1917
729 *striatella* ([Denis & Schiffermüller], 1775)
 tanacetella (Schrank, 1802)

APODIA Heinemann, 1870
730 *bifractella* (Duponchel, [1843])
 ?*martinii* Petry, 1911

EULAMPROTES Bradley, 1971
 LAMPROTES Heinemann, 1870, *nec* R. L., 1817
 ARGYRITIS Heinemann, 1870, *nec* Hübner,
  [1821]
731 *atrella* ([Denis & Schiffermüller], 1775)
 ?*umbriferella* (Herrich-Schäffer, 1854)
731a *immaculatella* (Douglas, 1850)
 phaeella Heckford & Langmaid, 1988
732 *unicolorella* (Duponchel, [1843])
733 *wilkella* (Linnaeus, 1758)
 pictella (Zeller, 1839)
 tarquiniella (Stainton, 1862)

ARGOLAMPROTES Benander, 1945
734 *micella* ([Denis & Schiffermüller], 1775)
 asterella (Treitschke, 1833)

MONOCHROA Heinemann, 1870
 PALTODORA Meyrick, 1894
728 *cytisella* (Curtis, 1837)
 fuscipennis (Humphreys & Westwood, 1845)
 walkeriella (Douglas, 1850)
 clinosema (Meyrick, 1935)

735 *tenebrella* (Hübner, [1817])
 fuscocuprea (Haworth, 1828)
 subcuprella (Stephens, 1834)
 tenebrosella (Zeller, 1839)
736 *lucidella* (Stephens, 1834)
737 *palustrellus* (Douglas, 1850)
738 *tetragonella* (Stainton, 1885)
739 *conspersella* (Herrich-Schäffer, 1854)
 quaestionella (Herrich-Schäffer, 1854)
 morosa (Mühlig, 1864)
740 *hornigi* (Staudinger, 1883)
740a *niphognatha* (Gozmány, 1953)
741 *suffusella* (Douglas, 1850)
 oblitella (Doubleday, 1859)
742 *lutulentella* (Zeller, 1839)
743 *elongella* (Heinemann, 1870)
 servella sensu auctt., *nec* Zeller, 1839
 micrometra (Meyrick, 1935)
744 *arundinetella* (Stainton, [1857])
744a *moyses* Uffen, 1991
745 *divisella* (Douglas, 1850)

CHRYSOESTHIA Hübner, [1825]
 MICROSETIA Stephens, 1829
746 *drurella* (Fabricius, 1775)
 hermannella sensu auctt., *nec* Fabricius, 1781
 schaefferella sensu Donovan, 1796, *nec* Linnaeus,
  1758
 zinckeella Hübner, [1813]
747 *sexguttella* (Thunberg, 1794)
 stipella sensu Hübner, 1796, *nec* Linnaeus, 1758
 knockella sensu Haworth, 1828, *nec* Fabricius, 1794
 miscella sensu Haworth, 1828, *nec* [Denis &
  Schiffermüller], 1775
 aurofasciella (Stephens, 1834)
 naeviferella (Duponchel, [1843])

PTOCHEUUSA Heinemann, 1870
748 *paupella* (Zeller, 1847)
 inulella (Curtis, 1850)
 inopella sensu auctt., *nec* Zeller, 1839

PSAMATHOCRITA Meyrick, 1925
750 *osseella* (Stainton, [1860])
750a *argentella* Pierce & Metcalfe, 1942

ARISTOTELIA Hübner, [1825]
 ERGATIS Heinemann, 1870, *nec* Blackwall, 1841
 EUCATOPTUS Walsingham, 1897
751 *subdecurtella* (Stainton, [1858])
752 *ericinella* (Zeller, 1839)
 micella sensu Hübner, 1796, *nec* [Denis &
  Schiffermüller], 1775
753 *brizella* (Treitschke, 1833)

XYSTOPHORA Wocke, 1876
 DORYPHORA Heinemann, 1870, *nec* Illiger,
  1807
 DORYPHORELLA Cockerell, 1888
754 *pulveratella* (Herrich-Schäffer, 1854)
 intaminatella (Stainton, 1854)

BRYOTROPHA Heinemann, 1870
 MNIOPHAGA Pierce & Daltry, 1938
 ADELPHOTROPHA Gozmány, 1955
777 *basaltinella* (Zeller, 1839)
777a *dryadella* (Zeller, 1850)
778 *umbrosella* (Zeller, 1839)
(781) *mundella* (Douglas, 1850)
  portlandicella (Richardson, 1890)
  anacampsoidella Hering, 1924
779 *affinis* (Haworth, 1828)
 tegulella (Herrich-Schäffer, 1854)
 affinella (Doubleday, 1859)
780 *similis* (Stainton, 1854)
 thuleella (Zeller, 1857)
 similella (Doubleday, 1859)
 confinis (Stainton, 1871)
 stolidella (Morris, 1872)
 obscurecinerea (Nolcken, 1872)
 fuliginosella Snellen, 1882
782 *senectella* (Zeller, 1839)
 obscurella Heinemann, 1870
783 *boreella* (Douglas, 1851)
784 *galbanella* (Zeller, 1839)
 angustella (Heinemann, 1870)
 ilmatariella (Hoffmann, 1893)
785 [*figulella* (Staudinger, 1859)]
786 *desertella* (Douglas, 1850)
 decrepidella (Herrich-Schäffer, 1854)
 decrepitella Heinemann, 1870
787 *terrella* ([Denis & Schiffermüller], 1775)
 inulella (Hübner, [1805])
 pauperella (Hübner, [1825])
 listeri sensu Haworth, 1828, *nec* Linnaeus, 1758
 latella (Herrich-Schäffer, 1854)
 lutescens (Constant, 1865)
 suspectella (Heinemann, 1870)
788 *politella* (Stainton, 1851)
 expolitella (Doubleday, 1859)
789 *domestica* (Haworth, 1828)
 domesticella (Doubleday, 1859)

GELECHIINAE

STENOLECHIA Meyrick, 1894
 POECILIA Heinemann, 1870, *nec* Schneider, 1801
 GIBBOSA Omelko, 1988
755 *gemmella* (Linnaeus, 1758)
 nivea (Haworth, 1828)

PARACHRONISTIS Meyrick, 1925
756 *albiceps* (Zeller, 1839)
 aleella sensu Stephens, 1834, *nec* Fabricius, 1794
 albicapitella (Doubleday, 1859)

RECURVARIA Haworth, 1828
 LITA Kollar, 1832
 TELEA Stephens, 1834, *nec* Hübner, [1819]
 APHANAULA Meyrick, 1895
 HINNEBERGIA Spuler, 1910
 MICROLECHIA Turati, 1924
757 *nanella* ([Denis & Schiffermüller], 1775)
 pumilella ([Denis & Schiffermüller], 1775)
 nana (Haworth, 1828)
758 *leucatella* (Clerck, 1759)
 leucatea (Haworth, 1828)

COLEOTECHNITES Chambers, 1880
 EVAGORA Clemens, 1860, *nec* Péron & Lesueur,
  1810
 EIDOTHEA Chambers, 1873, *nec* Risso, 1826
 EUCORDYLEA Dietz, 1900
 PULICALVARIA Freeman, 1963
759 *piceaella* (Kearfott, 1903)
 niger (Kearfott, 1903)
 obscurella (Kearfott, 1907)

EXOTELEIA Wallengren, 1881
 PARALECHIA Busck, [1903]
 HERINGIA Spuler, 1910, *nec* Rondani, 1856
 HERINGIOLA Strand, 1917
760 *dodecella* (Linnaeus, 1758)
 dodecea (Haworth, 1828)
 annulicornis (Stephens, 1834)

ATHRIPS Billberg, 1820
 RHYNCHOPACHA Staudinger, 1871
 EPITHECTIS Meyrick, 1895
 LEOBATUS Walsingham, 1904
 ZIMINIOLA Gerasimov, 1930
 CREMONA Busck, 1934
761 *tetrapunctella* (Thunberg, 1794)
 nigricostella sensu Douglas, 1852, *nec* Duponchel,
  1842
 lathyri (Stainton, 1865)
 lathyrella (Doubleday, 1866)
 subocellea sensu Pierce & Metcalfe, 1935, *nec*
  Stephens, 1834
761a *rancidella* (Herrich-Schäffer, 1854)
762 *mouffetella* (Linnaeus, 1758)
 punctifera (Haworth, 1828)

XENOLECHIA Meyrick, 1895
763 *aethiops* (Humphreys & Westwood, 1845)
 aethiopella (Doubleday, 1859)

PSEUDOTELPHUSA Janse, 1958
SATTLERIA Căpușe, 1968, *nec* Povolný, 1965
KLAUSSATTLERIA Căpușe, 1968
764 *scalella* (Scopoli, 1763)
aleella (Fabricius, 1794)
alternella Hübner, 1796, *nec* [Denis &
Schiffermüller], 1775
773 *paripunctella* (Thunberg, 1794)
triparella (Zeller, 1839)
dodecea sensu Haworth, 1828, nec *dodecella*
Linnaeus, 1758

ALTENIA Sattler, 1960
766 *scriptella* (Hübner, 1796)

TELEIODES Sattler, 1960
TELEIA Heinemann, 1870, *nec* Hübner, [1825]
TELPHUSA sensu auctt., *nec* Chambers, 1872
765 *vulgella* ([Denis & Schiffermüller], 1775)
aspera (Haworth, 1828)
769 *wagae* (Nowicki, 1860)
marsata Piskunov, 1973
774 *luculella* (Hübner, 1813)
subrosea (Haworth, 1828)
luctuella sensu Stephens, 1834, *nec* Hübner, 1793
774a *flavimaculella* (Herrich-Schäffer, 1854)
775 *sequax* (Haworth, 1828)
sequacella (Doubleday, 1859)

CARPATOLECHIA Căpușe, 1964
767 *decorella* (Haworth, 1812)
humeralis (Zeller, 1839)
lyellella (Humphreys & Westwood, 1845)
768 *notatella* (Hübner, [1813])
770 *proximella* (Hübner, 1796)
771 *alburnella* (Zeller, 1839)
772 *fugitivella* (Zeller, 1839)

TELEIOPSIS Sattler, 1960
776 *diffinis* (Haworth, 1828)
diffinella (Doubleday, 1859)

CHIONODES Hübner, [1825]
790 *fumatella* (Douglas, 1850)
celerella (Stainton, 1851)
oppletella (Herrich-Schäffer, 1854)
791 *distinctella* (Zeller, 1839)

PSORICOPTERA Stainton, 1854
859 *gibbosella* (Zeller, 1839)

MIRIFICARMA Gozmány, 1955
HELINA Guenée, 1849, *nec* Robineau-Desvoidy,
1830
792 *mulinella* (Zeller, 1839)
interruptella sensu auctt., *nec* Goeze, 1783
793 *lentiginosella* (Zeller, 1839)

PROLITA Leraut, 1993
LITA Treitschke, 1833, *nec* Kollar, 1832
794 *sexpunctella* (Fabricius, 1794)
virgella (Thunberg, 1794)
longicornis (Curtis, 1827)
longicornella (Doubleday, 1859)
795 *solutella* (Zeller, 1839)
fumosella (Douglas, 1852)

AROGA Busck, 1914
796 *velocella* (Zeller, 1839)

NEOFRISERIA Sattler, 1960
798 *peliella* (Treitschke, 1835)
799 *singula* (Staudinger, 1876)
suppeliella (Walsingham, 1896)
peliella sensu Meyrick, 1895, *nec* Treitschke, 1835

GELECHIA Hübner, [1825]
800 *rhombella* ([Denis & Schiffermüller], 1775)
rhombea (Haworth, 1828)
801 *scotinella* Herrich-Schäffer, 1854
801a *senticetella* (Staudinger, 1859)
802 *sabinellus* (Zeller, 1839)
802a *sororculella* (Hübner, [1817])
803 *muscosella* Zeller, 1839
804 *cuneatella* Douglas, 1852
805 *hippophaella* (Schrank, 1802)
basalis Stainton, 1854
806 *nigra* (Haworth, 1828)
cautella Zeller, 1839
807 *turpella* ([Denis & Schiffermüller], 1775)
pinguinella (Treitschke, 1832)

SCROBIPALPA Janse, 1951
ILSEOPSIS Povolný, 1965
822 *acuminatella* (Sircom, 1850)
?*pulliginella* (Sircom, 1850)
?*cirsiella* (Stainton, 1851)
?*gracilella* (Stainton, 1871)
814a *pauperella* (Heinemann, 1870)
klimeschi sensu auctt., *nec* Povolný, 1967
821 *murinella* (Duponchel, 1843)
excelsa (Frey, 1880)
810 *suaedella* (Richardson, 1893)
813 *salinella* (Zeller, 1847)
salicorniae (E. Hering, 1889)
812 *instabilella* (Douglas, 1846)
815 *nitentella* (Fuchs, 1902)
seminella (Pierce & Metcalfe, 1935)
816 *obsoletella* (Fischer von Röslerstamm, [1841])
818 *atriplicella* (Fischer von Röslerstamm, [1841])
atrella (Thunberg, 1788), *nec* [Denis &
Schiffermüller], 1775
814 *ocellatella* (Boyd, 1858)

811 *samadensis* (Pfaffenzeller, 1870)
  plantaginella (Stainton, 1883)
820 *artemisiella* (Treitschke, 1833)
811a *stangei* (E. Hering, 1889)
817 *clintoni* Povolný, 1968
819 *costella* (Humphreys & Westwood, 1845)
819a *hyoscyamella* (Stainton, 1869)

SCROBIPALPULA Povolný, 1964
823 *diffluella* (Frey, 1870)
  psilella sensu Povolný & Bradley, 1965, *nec*
    Herrich-Schäffer, 1854
823a *tussilaginis* (Stainton, 1867)
  tussilaginella (Heinemann, 1870)

GNORIMOSCHEMA Busck, 1900
824 *streliciella* (Herrich-Schäffer, 1854)
  strelitziella (Heinemann, 1870)

PHTHORIMAEA Meyrick, 1902
825 *operculella* (Zeller, 1873)
  terrella Walker, 1864, *nec* [Denis &
    Schiffermüller], 1775
  tabacella Ragonot, 1878
  sedata Butler, 1880

CARYOCOLUM Gregor & Povolný, 1954
826 *vicinella* (Douglas, 1851)
  inflatella (Chrétien, 1901)
  leucomelanella sensu auctt. *nec* Zeller, 1839
827 *alsinella* (Zeller, 1868)
  albifrontella (Heinemann, 1870)
  tristella (Heinemann, 1870)
  semidecandriella (Tutt, 1887)
  semidecandrella (Threlfall, 1887)
828 *viscariella* (Stainton, 1855)
829 *marmorea* (Haworth, 1828)
  manniella (Zeller, 1839)
  pulchra (Wollaston, 1858)
  marmorella (Doubleday, 1859)
830 *fraternella* (Douglas, 1851)
  intermediella (Hodgkinson, 1897)
831 *proxima* (Haworth, 1828)
  maculiferella (Douglas, 1851)
  horticolla (Peyerimhoff, 1871)
832 *blandella* (Douglas, 1852)
  maculea sensu Haworth, 1828, *nec maculella*
    Fabricius, 1794
  maculella sensu Stephens, 1834, *nec* Fabricius, 1794
  nivella sensu Stephens, 1834, *nec* Fabricius, 1794
832a *blandelloides* Karsholt, 1981
833 *junctella* (Douglas, 1851)
  aganocarpa (Meyrick, 1935)
834 *tricolorella* (Haworth, 1812)
  contigua (Haworth, 1828)

835 *blandulella* (Tutt, 1887)
836 *kroesmanniella* (Herrich-Schäffer, 1854)
  huebneri sensu auctt., *nec* Haworth, 1828
837 *huebneri* (Haworth, 1828)
  hubnerella (Doubleday, 1859)
  knaggsiella (Stainton, 1866)

ANACAMPSINAE

SOPHRONIA Hübner, [1825]
841 *semicostella* (Hübner, [1813])
  marginella sensu Thunberg, 1794, *nec* [Denis &
    Schiffermüller], 1775
  parenthesellus sensu Haworth, 1828, *nec* Linnaeus,
    1761
842 *humerella* ([Denis & Schiffermüller], 1775)

APROAEREMA Durrant, 1897
843 *anthyllidella* (Hübner, [1813])
  nigritella sensu Douglas, 1851, *nec* Zeller,
    1847
  sparsiciliella (Barrett, 1891)
  aureliana Căpuşe, 1964

SYNCOPACMA Meyrick, 1925
  HARPAGUS Stephens, 1834, *nec* Vigors,
    1824
  STOMOPTERYX sensu auctt., *nec* Heinemann,
    1870
845 *sangiella* (Stainton, 1863)
  coronillella sensu auctt. (partim), *nec* Treitschke,
    1833
844 *larseniella* Gozmány, 1957
  ligulella sensu Pierce & Metcalfe, 1935, *nec* [Denis
    & Schiffermüller], 1775
849 *cinctella* (Clerck, 1759)
  vorticella (Scopoli, 1763)
  ligulella ([Denis & Schiffermüller], 1775)
  ?*albistrigella* (Stephens, 1834)
  ?*sircomella* (Stainton, 1854)
847 *taeniolella* (Zeller, 1839)
846 *vinella* (Bankes, 1898)
  coronillella sensu auctt. (partim), *nec* Treitschke,
    1833
  cincticulella sensu auctt., *nec* Bruand, 1850
848 *albipalpella* (Herrich-Schäffer, 1854)
  leucopalpella (Herrich-Schäffer, 1853)
    (unavailable)
848a *suecicella* (Wolff, 1958)
850 *polychromella* (Rebel, 1902)

ANACAMPSIS Curtis, 1827
  TACHYPTILIA Heinemann, 1870
852 *temerella* (Lienig & Zeller, 1846)
  pernigrella (Douglas, 1850)

853 *populella* (Clerck, 1759)
    populi (Haworth, 1828)
    blattariae Haworth, 1828 (partim)
    laticinctella Stephens, 1834
    tremulella (Duponchel, 1839)
    subsp. *fuscatella* (Bentinck, 1934)
    subsp. *ambronella* (Meder, 1934)
854 *blattariella* (Hübner, 1796)
    blattariae Haworth, 1828 (partim)
    betulinella Vári, 1941

MESOPHLEPS Hübner, [1825]
860 *silacella* (Hübner, 1796)
    silacea (Haworth, 1828)

CHELARIINAE

NEOFACULTA Gozmány, 1955
797 *ericetella* (Geyer, [1832])
    betulea sensu Haworth, 1828, nec *betulella* Hübner,
    [1825]
    lanceolella (Stephens, 1834)

NOTHRIS Hübner, [1825]
838 *verbascella* ([Denis & Schiffermüller], 1775)
839 *congressariella* (Bruand, 1858)
    declaratella Staudinger, 1859

ANARSIA Zeller, 1839
856 *spartiella* (Schrank, 1802)
    robertsonella (Curtis, 1837)
    genistae Stainton, 1854
    genistella Doubleday, 1859
857 *lineatella* Zeller, 1839
    pruniella Clemens, 1860

HYPATIMA Hübner, [1825]
CHELARIA Haworth, 1828
858 *rhomboidella* (Linnaeus, 1758)
    conscriptella (Hübner, [1805])
    hubnerella (Donovan, 1806)
    conscripta (Haworth, 1828)

DICHOMERIDINAE

ACOMPSIA Hübner, [1825]
BRACHYCROSSATA Heinemann, 1870
TELEPHILA Meyrick, 1923
855 *cinerella* (Clerck, 1759)
    cinerea (Haworth, 1828)
861 *schmidtiellus* (Heyden, 1848)
    durdhamellus (Stainton, 1849)

DICHOMERIS Hübner, 1818
ACANTHOPHILA Heinemann, 1870
862 *marginella* (Fabricius, 1781)
863 *juniperella* (Linnaeus, 1761)
864 *ustalella* (Fabricius, 1794)

865 *derasella* ([Denis & Schiffermüller], 1775)
    fasciella (Hübner, 1796)
851 *alacella* (Zeller, 1839)

BRACHMIA Hübner, [1825]
866 *blandella* (Fabricius, 1798)
    gerronella (Zeller, 1850)
867 *inornatella* (Douglas, 1850)

HELCYSTOGRAMMA Zeller, 1877
868 *rufescens* (Haworth, 1828)
    rufescentella (Doubleday, 1859)
869 *lutatella* (Herrich-Schäffer, 1854)

PEXICOPIINAE

PLATYEDRA Meyrick, 1895
808 *subcinerea* (Haworth, 1828)
    vilella (Zeller, 1847)

PEXICOPIA Common, 1958
809 *malvella* (Hübner, [1805])
    lutarea (Haworth, 1828)

SITOTROGA Heinemann, 1870
749 *cerealella* (Olivier, 1789)
    hordei Beckman, 1815

THIOTRICHA Meyrick, 1886
THISTRICHA Meyrick, 1885
REUTTIA Hofmann, 1898
840 *subocellea* (Stephens, 1834)
    internella (Lienig & Zeller, 1846)
    subocellella (Doubleday, 1859)
    lathyri sensu Pierce & Metcalfe, 1935, *nec* Stainton,
    1865

Key to genera of the Gelechiidae based on male genitalia (figures 1–20, pp. 21–40)

The terms used are illustrated in *MBGBI* I, p. 21, except for:
Armature – Blanket term for the whole vinculum-tegumen-valvae complex, but excluding the aedeagus.
Harpe – A process from the inner face of the valva near its base.
See also figures 53 (p. 104) and 57a,b (p. 164).

That part of the couplet which the detailed figures illustrate is designated by the bold capital letters **A – Z** or **AA – II**

1	Abdominal segment 8 divided into separate dorsal and ventral plates **A**	2
–	Abdominal segment 8 forming complete ring around genitalia	26
2(1)	Valvae and vinculum asymmetrical **B**	
	.... *Coleotechnites* (p. 122)	
–	Valvae and vinculum symmetrical	3
3(2)	Gnathos absent, or weak and easily overlooked	4
–	Gnathos present and easily discernible	9
4(3)	Uncus deeply divided into two lobes	5
–	Uncus entire, or at most shallowly notched or excavate	6
5(4)	Uncus lobes convergent, inner margin not excavate **C** .... *Altenia* (p. 131)	
–	Uncus lobes parallel, excavate on inner margin at apex **D** .... *Xenolechia* (p. 127)	
6(4)	Uncus pointed	7
–	Uncus rounded or truncate	
	...*Teleiodes* (part) (p. 132), *Carpatolechia* (part) (p. 137)	
7(6)	Valva ending in a stout spine **E** .... *Aroga* (p. 150)	
–	Valva not ending in a stout spine	8
8(7)	Digitate lobes present on posterior margin of vinculum, but not on juxta .... *Pseudotelphusa* (p. 128)	
–	Digitate lobes present on juxta, but not on vinculum **F** .... *Carpatolechia* (part) (p. 137)	
9(3)	Gnathos two-lobed, the lobes separate or partly joined **G** .... *Gelechia* (p. 153)	
–	Gnathos simple	10
10(9)	Uncus four-lobed .... *Psoricoptera* (p. 144)	
–	Uncus entire or two-lobed	11
11(10)	Valva, except at base, tubular, tapering, sometimes filiform, without bristles	12
–	Valva flat or tubular, with bristles	16
12(11)	Base of valva swollen **H**	13
–	Base of valva not swollen **I** .... *Chionodes* (p. 142)	

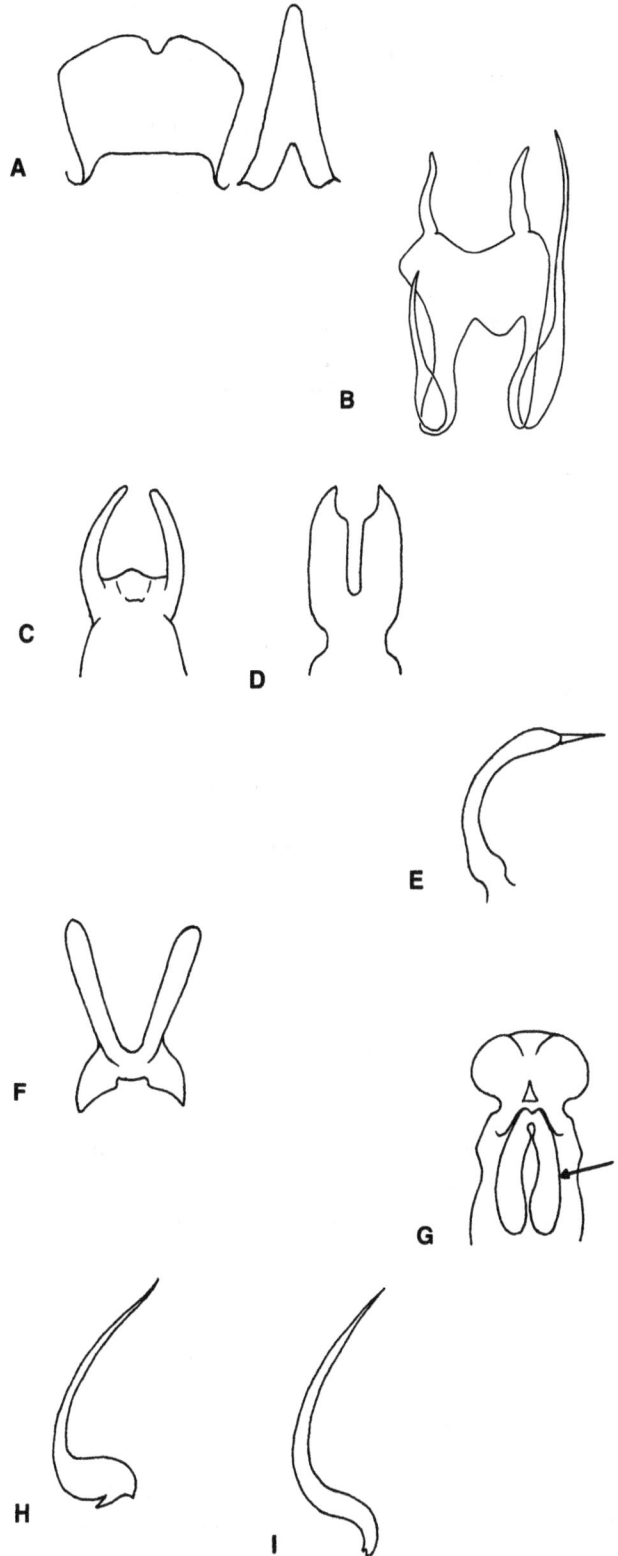

13(12) Uncus as wide as long, widest close to apex **J** 14
–   Uncus longer than wide, or widest near base **K** 15

14(13) Aedeagus straight .... *Recurvaria* (p. 120)
–   Aedeagus curved below middle or near apex ....
       .... *Stenolechia* (p. 118)

15(13) Aedeagus straight ....
       ...*Teleiodes* (part) (p. 132), *Carpatolechia* (part) (p. 137)
–   Aedeagus curved below middle or near apex ....
       .... *Exoteleia* (p. 123)

16(11) With stout filament arising adjacent to anterior end
       of saccus and directed posteriorly parallel to aedeagus
       L ....*Mirificarma* (p. 145)
–   Without such filament 17

17(16) Valva consisting of three distinct lobes and long,
       curved, sclerotized harpe **M** .... *Neofriseria* (p. 151)
–   Valva otherwise 18

18(17) Posterior margin of vinculum with broad, centrally
       placed process **N** 19
–   Posterior margin of vinculum with digitate paired
       central or lateral processes, or processes absent
       20

19(18) Uncus distally tapered, gnathos straight ....
       .... *Teleiopsis* (p. 141)
–   Uncus broad, gnathos hook-like .. *Sophronia* (p. 204)

20(18) Costal apex of valva produced in short process **O** ....
       .... *Parachronistis* (p. 119)
–   Costal apex of valva not produced 21

21(20) Uncus distally set densely with scales ....
       .... *Prolita* (p. 147)
–   Uncus at most with isolated setae 22

22(21) Valva considerably exceeding uncus; saccus broad,
       rounded .... *Phthorimaea* (p. 188)
–   Valva not exceeding uncus, or saccus narrow,
       pointed 23

23(22) Sacculus absent; gnathos wide at apex **P** ....
       .... *Scrobipalpula* (p. 184)
–   Sacculus present; gnathos narrow at apex 24

24(23) Sacculus strongly hooked, open sides of hooks facing
       away from one another **Q** ... *Gnorimoschema* (p. 187)
–   Sacculus not hooked, or open side of hooks facing
       one another 25

25(24) Gnathos hook-like .... *Scrobipalpa* (p. 161)
–   Gnathos not hook-like .... *Caryocolum* (p. 189)

26(1) Uncus well developed, divided into two lobes 27
–   Uncus entire, or rudimentary, or absent 30

27(26) Uncus with divergent lobes **R** ...... *Pexicopia* (p. 240)
–   Uncus with lobes not divergent 28

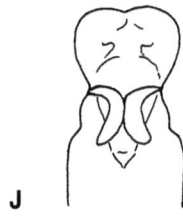

J

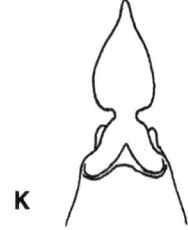

K

M

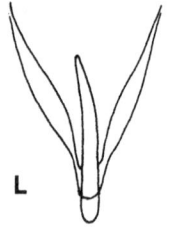

L

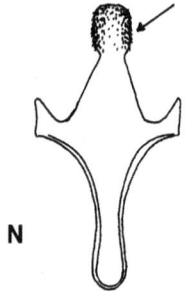

N

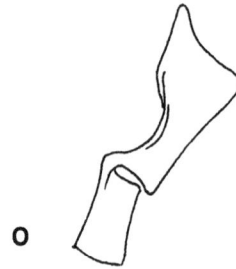

O

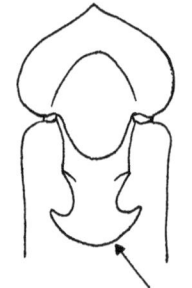

P

Q

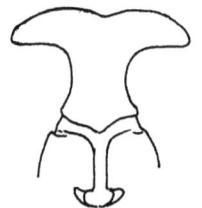

R

28(27) Gnathos absent .... *Ptocheuusa* (p. 96)
–   Gnathos present 29

29(28) Valva expanded in middle, ending in stout hook
      S .... *Sitotroga* (p. 241)
–   Valva parallel-sided, not ending in hook ....
      .... *Aproaerema* (p. 206)

30(26) Valvae asymmetrical 31
–   Valvae symmetrical 32

31(30) Right valva only with straight appendage **T** ....
      .... *Thiotricha* (p. 242)
–   Each valva with curved appendage **U** ....
      .... *Anarsia* (p. 224)

32(30) Gnathos absent, or weak and easily overlooked 33
–   Gnathos present, conspicuous 40

33(32) Tegumen short, uncus weakly developed or vestigial,
      together shorter than valvae 34
–   Tegumen and uncus well developed, uncus equalling
      or exceeding valvae 38

34(33) Uncus short and wide, scarcely differentiated from
      tegumen 35
–   Uncus elongate, slender **V** .... *Eulamprotes* (p. 75)

35(34) Valva constricted in middle, expanded towards apex
      W .... *Metzneria* (p. 67)
–   Valva not expanded towards apex 36

36(35) Aedeagus slender, not or scarcely tapering ....
      .... *Apodia* (p. 74)
–   Aedeagus stout, tapering 37

37(36) Sacculus nearly circular **X** .... *Isophrictis* (p. 72)
–   Sacculus semicircular, ovate, lanceolate or elliptic,
      often curved **Y** .... *Monochroa* (p. 81)

38(33) Uncus pointed .... *Psamathocrita* (p. 97)
–   Uncus rounded 39

39(38) Sacculus almost as long as valva ....
      .... *Argolamprotes* (p. 79)
–   Sacculus much shorter than valva ....
      .... *Chrysoesthia* (p. 93)

40(32) Gnathos forming pair of hooks ... *Mesophleps* (p. 219)
–   Gnathos simple 41

41(40) Valva with sharp median tooth on ventral side **Z** ......
      .... *Xystophora* (p. 102)
–   Valva without sharp median tooth 42

42(41) Uncus with dense brush of distally directed setae
      AA .... *Athrips* (p. 124)
–   Uncus without such setae 43

43(42) Apex of aedeagus reflexed, or with spiral process or
      flagellum 44
–   Apex of aedeagus simple, or ending with cornutus or
      small hook 46

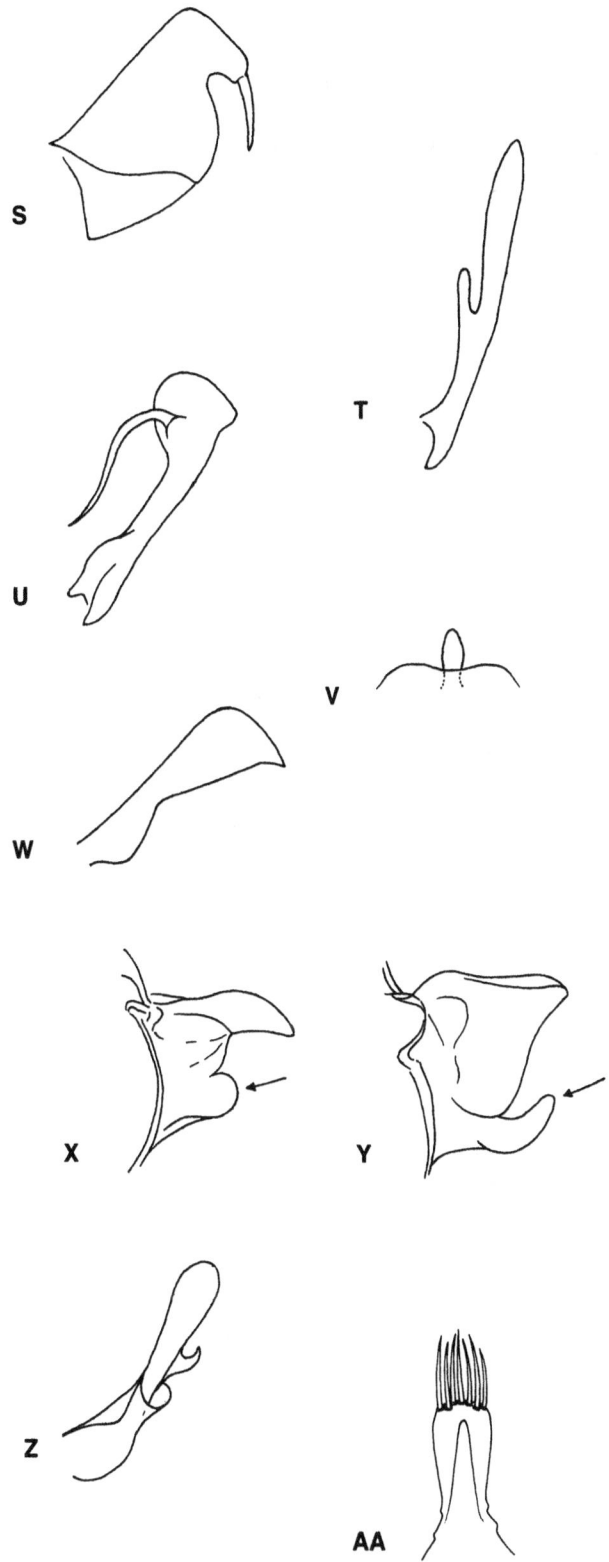

S

T

U

V

W

X

Y

Z

AA

44(43) Aedeagus rather stout, tapering to a spiral process
BB .... *Neofaculta* (p. 220)
– Aedeagus slender apart from bulbous base, ending in
flagellum or reflexed apex 45

45(44) Aedeagus curved throughout, apex reflexed **CC** ......
.... *Bryotropha* (part) (p. 103)
– Aedeagus not curved throughout, but may be
abruptly curved near apex, which bears flagellum
(sometimes broken off in preparation) **DD** ....
.... *Nothris* (p. 221)

46(43) Uncus pointed .... *Brachmia* (p. 234)
– Uncus rounded or truncate 47

47(46) Socii large, stalked, with abundant long bristles **EE** ....
.... *Platyedra* (p. 239)
– Socii absent, or inconspicuous and with few bristles
.... 48

48(47) Posterior margin of vinculum with enlarged asymmetrical lobes **FF** .... *Dichomeris* (part) (p. 229)
– Posterior margin of vinculum symmetrical, lobes
absent or not greatly enlarged 49

49(48) Aedeagus with long cornuti, the longest more than
half its total length .... *Dichomeris* (part) (p. 229)
– Aedeagus with cornuti less than half its total length,
or absent 50

50(49) Valva club-shaped, width near distal end more than
twice width near base 51
– Width of distal end of valva less than twice width
near base 52

51(50) Valva gradually widened distally **GG** ....
....*Acompsia* (p. 227)
– Valva abruptly expanded at apex **HH** ....
.... *Hypatima* (p. 226)

52(50) Uncus ventrally with group of black pegs **II** 53
– Uncus without such pegs 54

53(52) Aedeagus curved, as long as or longer than armature
.... *Anacampsis* (p. 216)
– Aedeagus straight, shorter than armature ....
.... *Syncopacma* (p. 207)

54(52) Sacculus present 55
– Sacculus absent .... *Bryotropha* (part) (p. 103)

55(54) Aedeagus with small hook at apex ....
.... *Helcystogramma* (p. 236)
– Aedeagus without hook .... *Aristotelia* (p. 99)

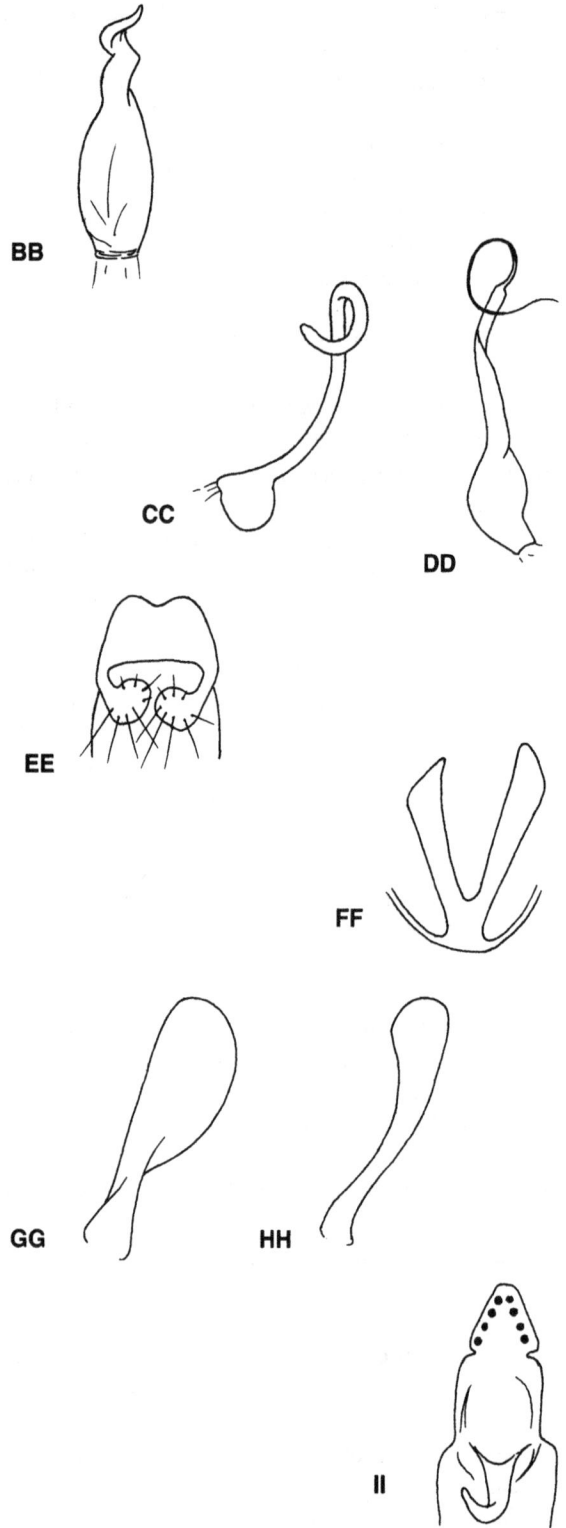

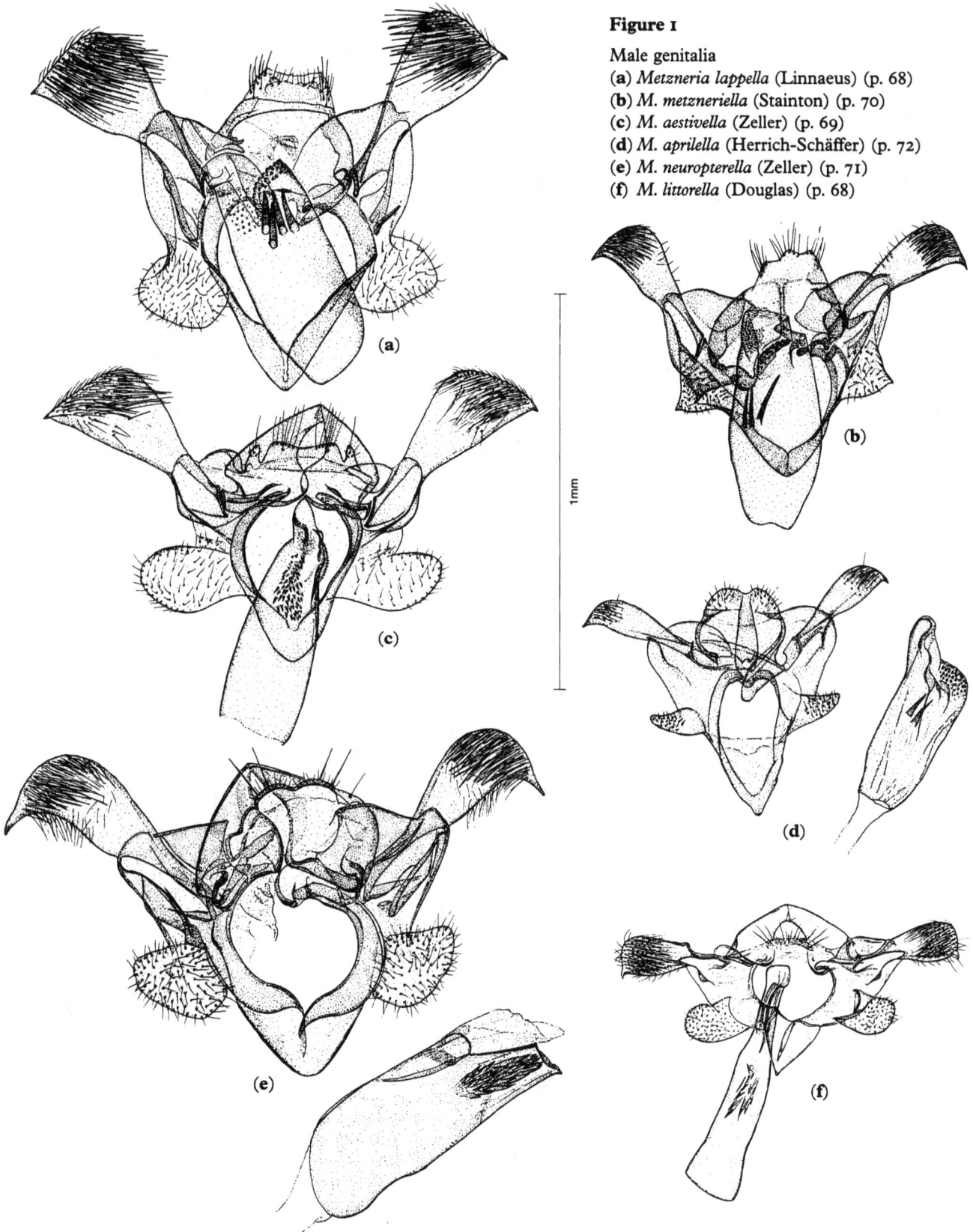

Figure 1

Male genitalia
(a) *Metzneria lappella* (Linnaeus) (p. 68)
(b) *M. metzneriella* (Stainton) (p. 70)
(c) *M. aestivella* (Zeller) (p. 69)
(d) *M. aprilella* (Herrich-Schäffer) (p. 72)
(e) *M. neuropterella* (Zeller) (p. 71)
(f) *M. littorella* (Douglas) (p. 68)

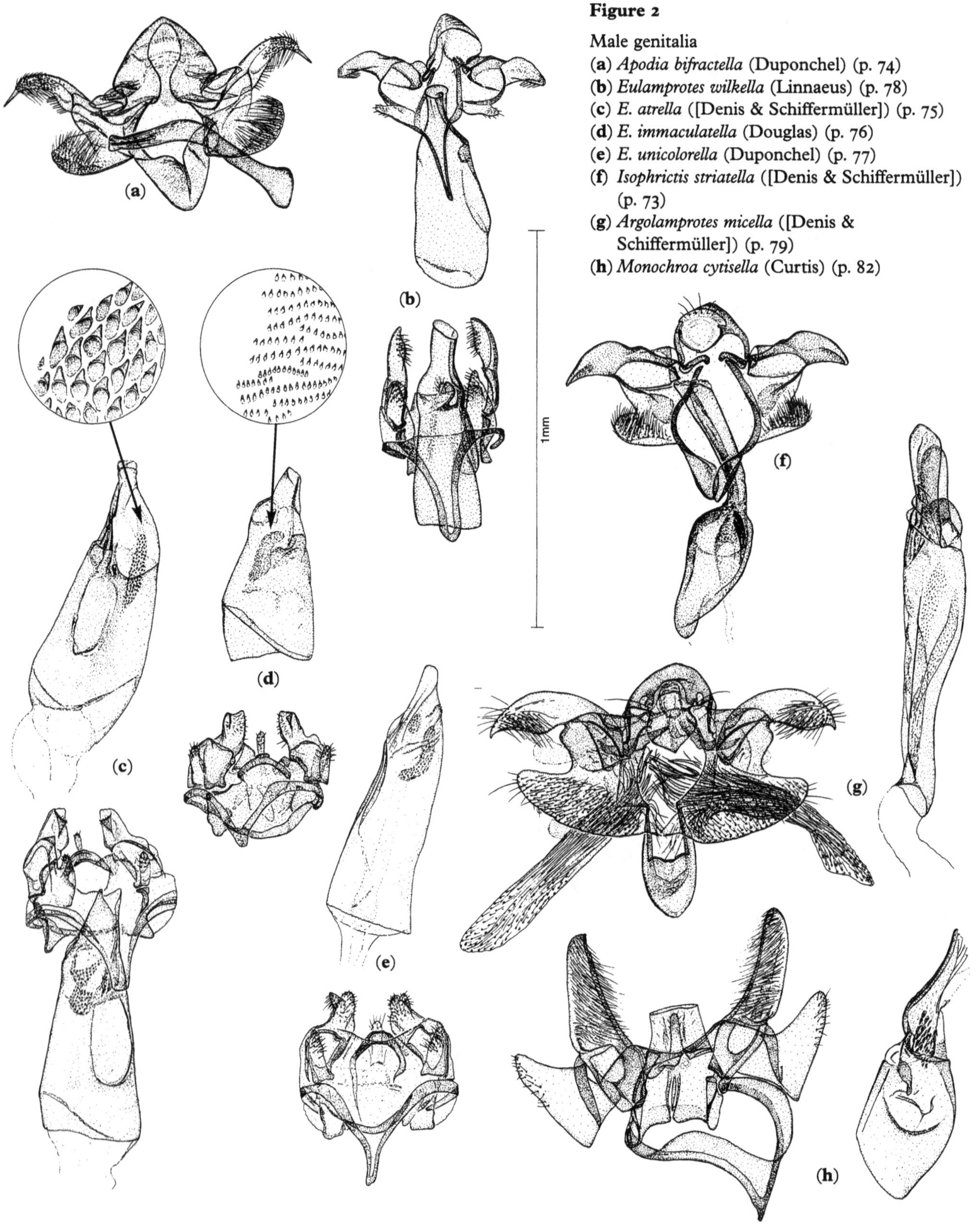

Figure 2

Male genitalia
(**a**) *Apodia bifractella* (Duponchel) (p. 74)
(**b**) *Eulamprotes wilkella* (Linnaeus) (p. 78)
(**c**) *E. atrella* ([Denis & Schiffermüller]) (p. 75)
(**d**) *E. immaculatella* (Douglas) (p. 76)
(**e**) *E. unicolorella* (Duponchel) (p. 77)
(**f**) *Isophrictis striatella* ([Denis & Schiffermüller]) (p. 73)
(**g**) *Argolamprotes micella* ([Denis & Schiffermüller]) (p. 79)
(**h**) *Monochroa cytisella* (Curtis) (p. 82)

Figure 3

Male genitalia
(**a**) *Monochroa tenebrella*
  (Hübner) (p. 83)
(**b**) *M. palustrellus* (Douglas)
  (p. 84)
(**c**) *M. lucidella* (Stephens)
  (p. 84)

(**d**) *M. conspersella* (Herrich-
  Schäffer) (p. 86)
(**e**) *M. suffusella* (Douglas)
  (p. 88)
(**f**) *M. tetragonella* (Stainton)
  (p. 85)
(**g**) *M. hornigi* (Staudinger)
  (p. 86)

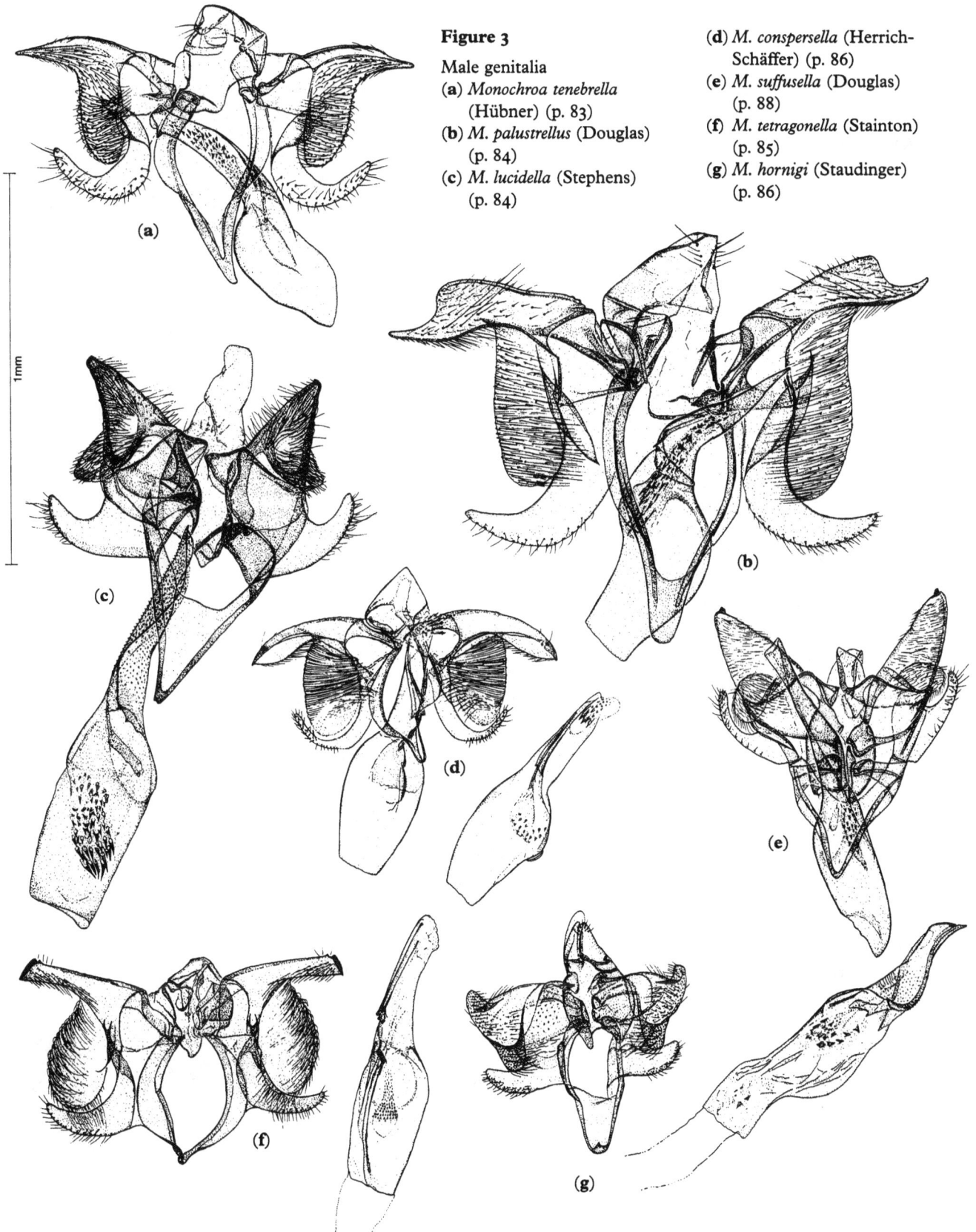

Figure 4

Male genitalia
(**a**) *Monochroa niphognatha* (Gozmány) (p. 88)
(**b**) *M. arundinetella* (Stainton) (p. 91)
(**c**) *M. lutulentella* (Zeller) (p. 89)
(**d**) *M. divisella* (Douglas) (p. 93)
(**e**) *M. elongella* (Heinemann) (p. 90)
(**f**) (i) *M. moyses* Uffen (p. 92), from slide of holotype;
    (ii) *M. moyses* Uffen (p. 92), reconstruction of genitalia.

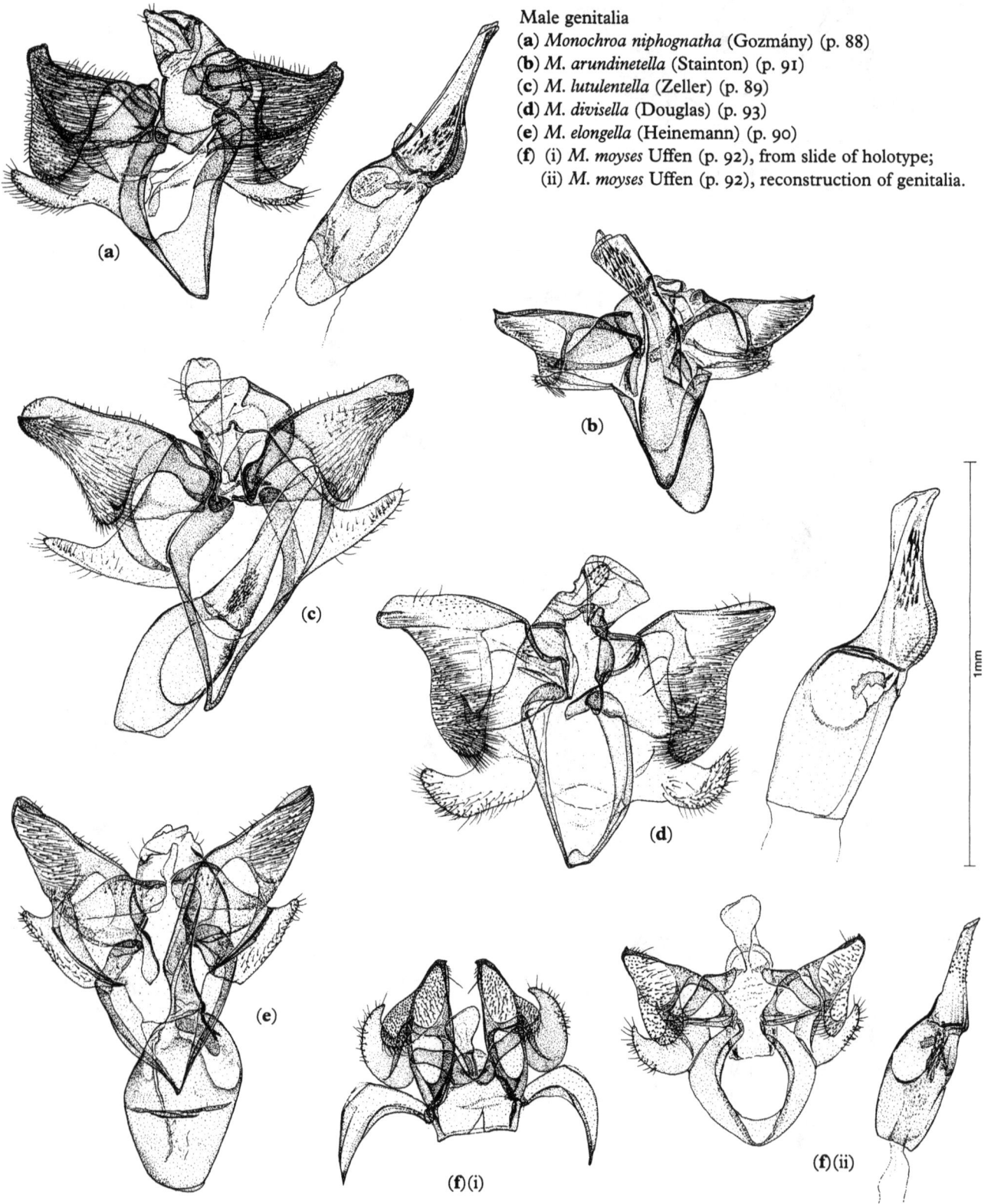

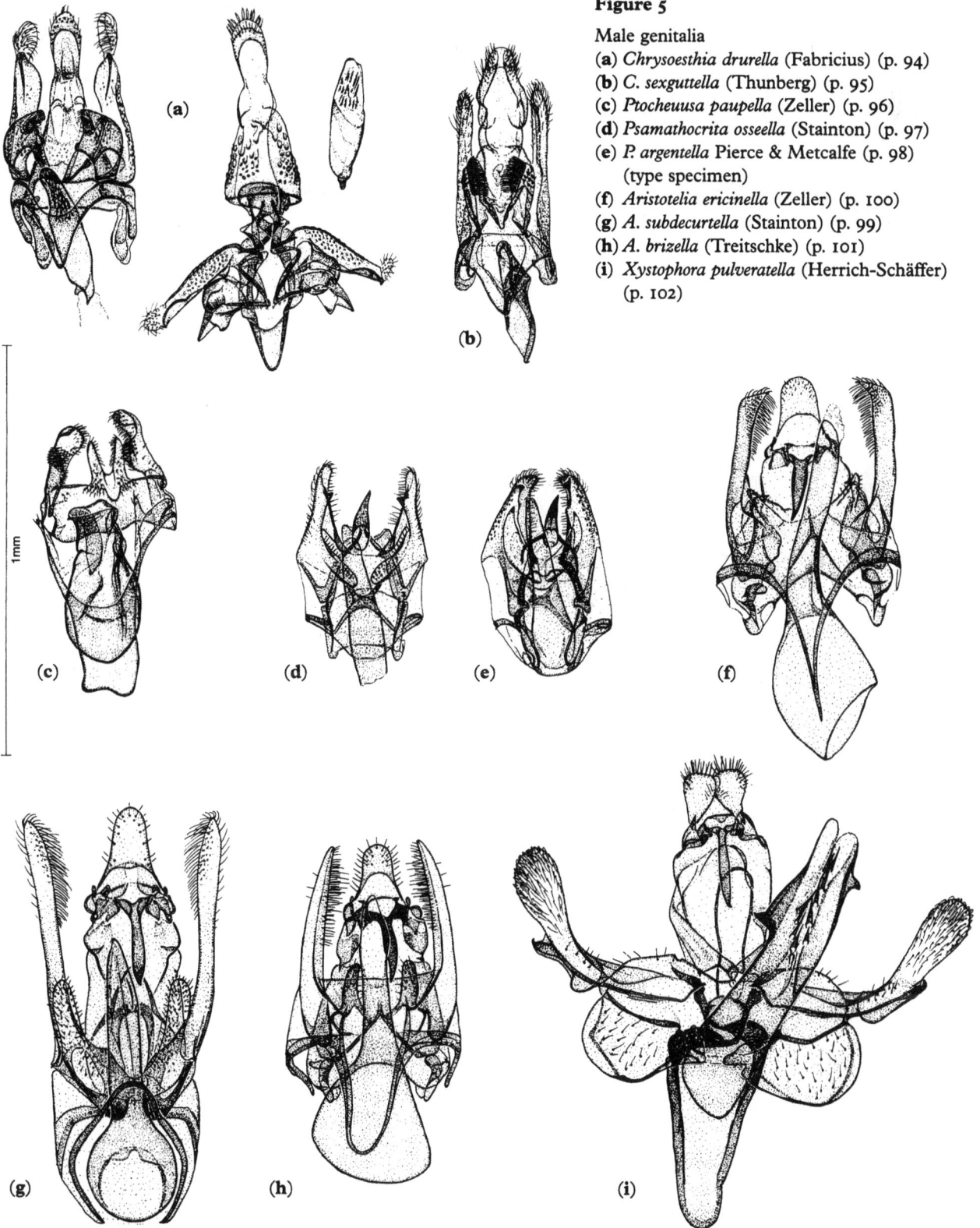

Figure 5

Male genitalia
(a) *Chrysoesthia drurella* (Fabricius) (p. 94)
(b) *C. sexguttella* (Thunberg) (p. 95)
(c) *Ptocheuusa paupella* (Zeller) (p. 96)
(d) *Psamathocrita osseella* (Stainton) (p. 97)
(e) *P. argentella* Pierce & Metcalfe (p. 98)
    (type specimen)
(f) *Aristotelia ericinella* (Zeller) (p. 100)
(g) *A. subdecurtella* (Stainton) (p. 99)
(h) *A. brizella* (Treitschke) (p. 101)
(i) *Xystophora pulveratella* (Herrich-Schäffer)
    (p. 102)

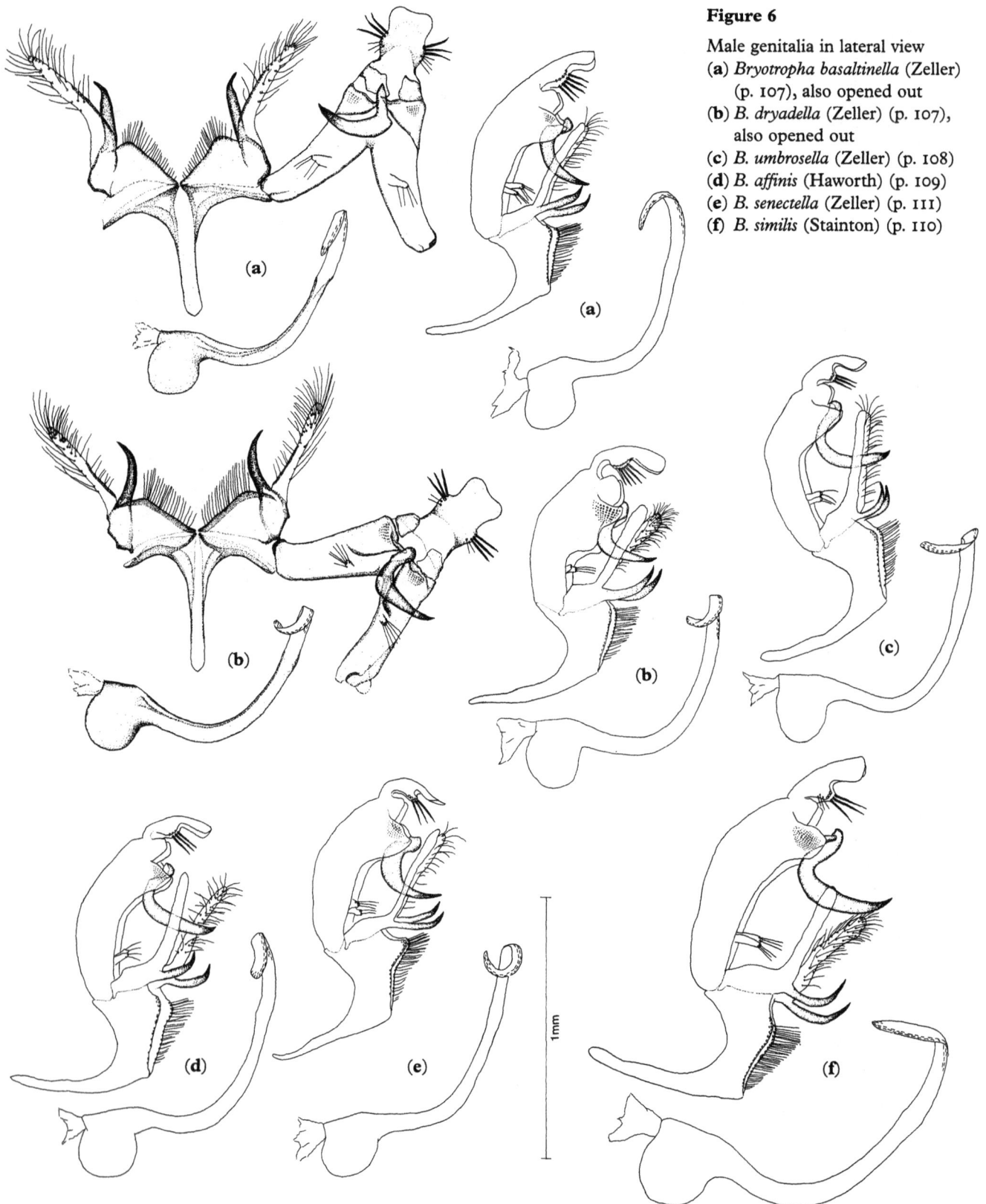

Figure 6

Male genitalia in lateral view
(a) *Bryotropha basaltinella* (Zeller) (p. 107), also opened out
(b) *B. dryadella* (Zeller) (p. 107), also opened out
(c) *B. umbrosella* (Zeller) (p. 108)
(d) *B. affinis* (Haworth) (p. 109)
(e) *B. senectella* (Zeller) (p. 111)
(f) *B. similis* (Stainton) (p. 110)

Figure 7

Male genitalia in lateral view
(a) *Bryotropha boreella* (Douglas) (p. 112) (larger scale)
(b) *B. galbanella* (Zeller) (p. 113) (larger scale)
(c) *B. politella* (Stainton) (p. 116)
(d) *B. desertella* (Douglas) (p. 114)
(e) *B. terrella* ([Denis & Schiffermüller]) (p. 115)
(f) *B. domestica* (Haworth) (p. 117), also in ventral view

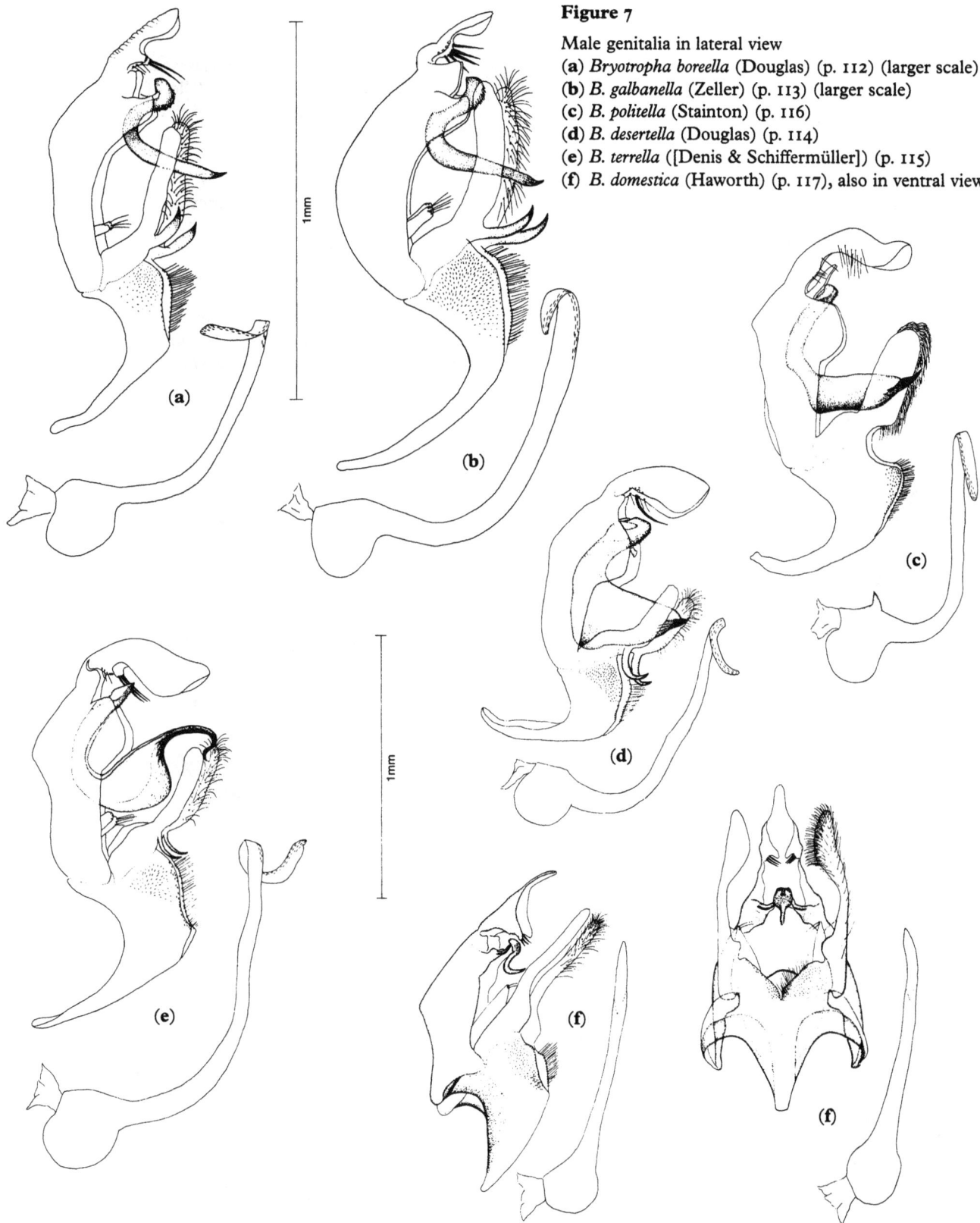

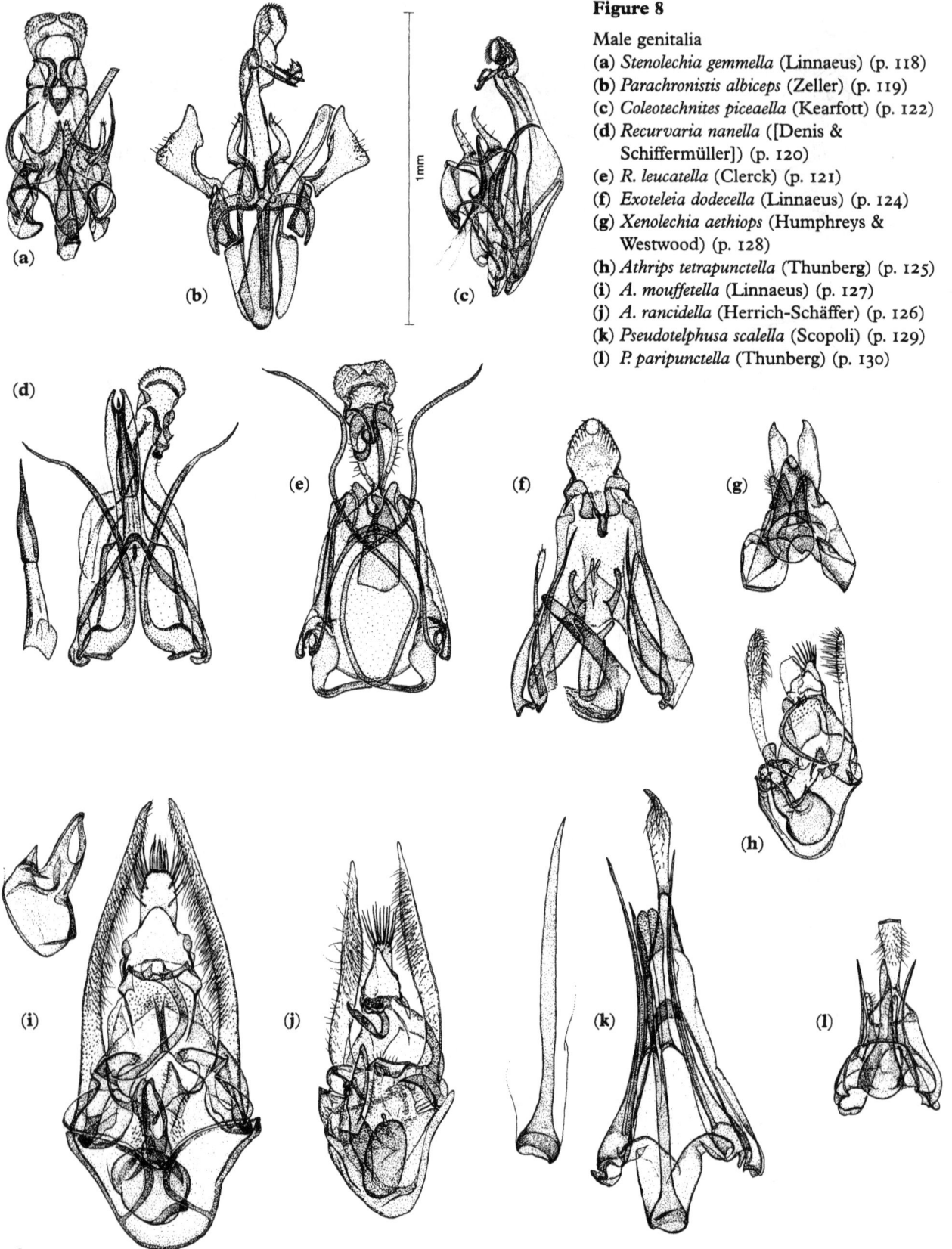

Figure 8

Male genitalia
(**a**) *Stenolechia gemmella* (Linnaeus) (p. 118)
(**b**) *Parachronistis albiceps* (Zeller) (p. 119)
(**c**) *Coleotechnites piceaella* (Kearfott) (p. 122)
(**d**) *Recurvaria nanella* ([Denis & Schiffermüller]) (p. 120)
(**e**) *R. leucatella* (Clerck) (p. 121)
(**f**) *Exoteleia dodecella* (Linnaeus) (p. 124)
(**g**) *Xenolechia aethiops* (Humphreys & Westwood) (p. 128)
(**h**) *Athrips tetrapunctella* (Thunberg) (p. 125)
(**i**) *A. mouffetella* (Linnaeus) (p. 127)
(**j**) *A. rancidella* (Herrich-Schäffer) (p. 126)
(**k**) *Pseudotelphusa scalella* (Scopoli) (p. 129)
(**l**) *P. paripunctella* (Thunberg) (p. 130)

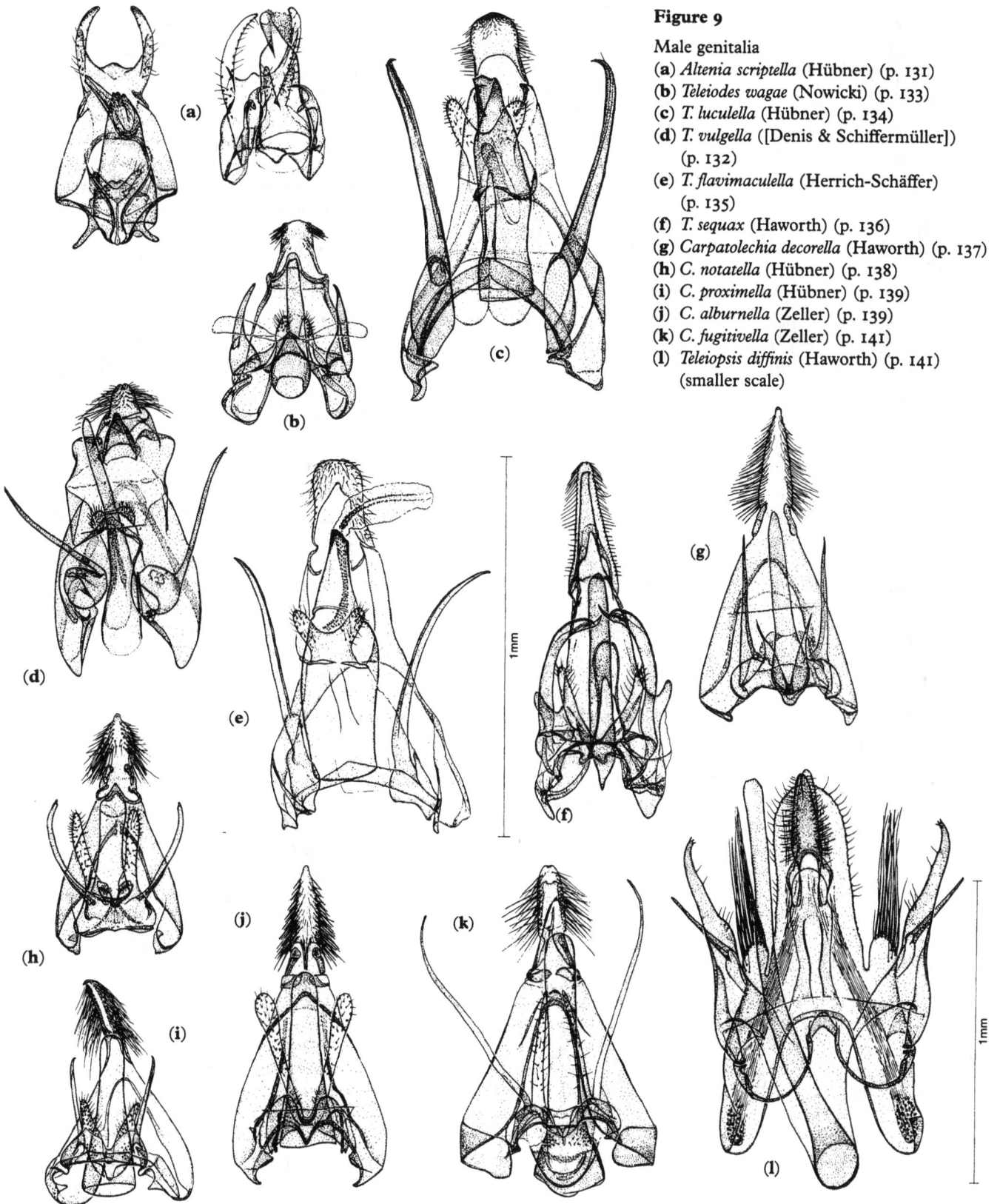

Figure 9

Male genitalia
(a) *Altenia scriptella* (Hübner) (p. 131)
(b) *Teleiodes wagae* (Nowicki) (p. 133)
(c) *T. luculella* (Hübner) (p. 134)
(d) *T. vulgella* ([Denis & Schiffermüller]) (p. 132)
(e) *T. flavimaculella* (Herrich-Schäffer) (p. 135)
(f) *T. sequax* (Haworth) (p. 136)
(g) *Carpatolechia decorella* (Haworth) (p. 137)
(h) *C. notatella* (Hübner) (p. 138)
(i) *C. proximella* (Hübner) (p. 139)
(j) *C. alburnella* (Zeller) (p. 139)
(k) *C. fugitivella* (Zeller) (p. 141)
(l) *Teleiopsis diffinis* (Haworth) (p. 141) (smaller scale)

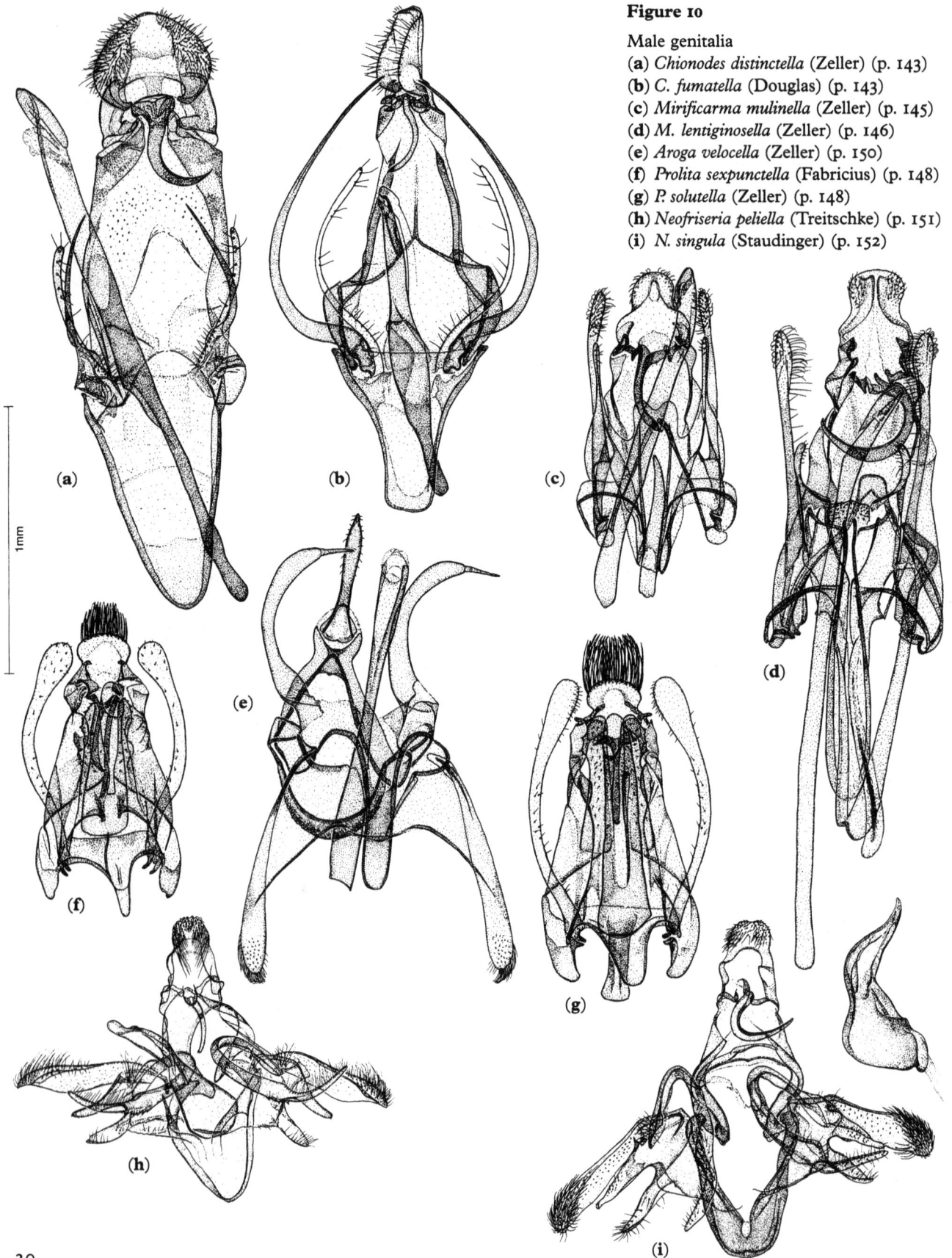

Figure 10

Male genitalia
(**a**) *Chionodes distinctella* (Zeller) (p. 143)
(**b**) *C. fumatella* (Douglas) (p. 143)
(**c**) *Mirificarma mulinella* (Zeller) (p. 145)
(**d**) *M. lentiginosella* (Zeller) (p. 146)
(**e**) *Aroga velocella* (Zeller) (p. 150)
(**f**) *Prolita sexpunctella* (Fabricius) (p. 148)
(**g**) *P. solutella* (Zeller) (p. 148)
(**h**) *Neofriseria peliella* (Treitschke) (p. 151)
(**i**) *N. singula* (Staudinger) (p. 152)

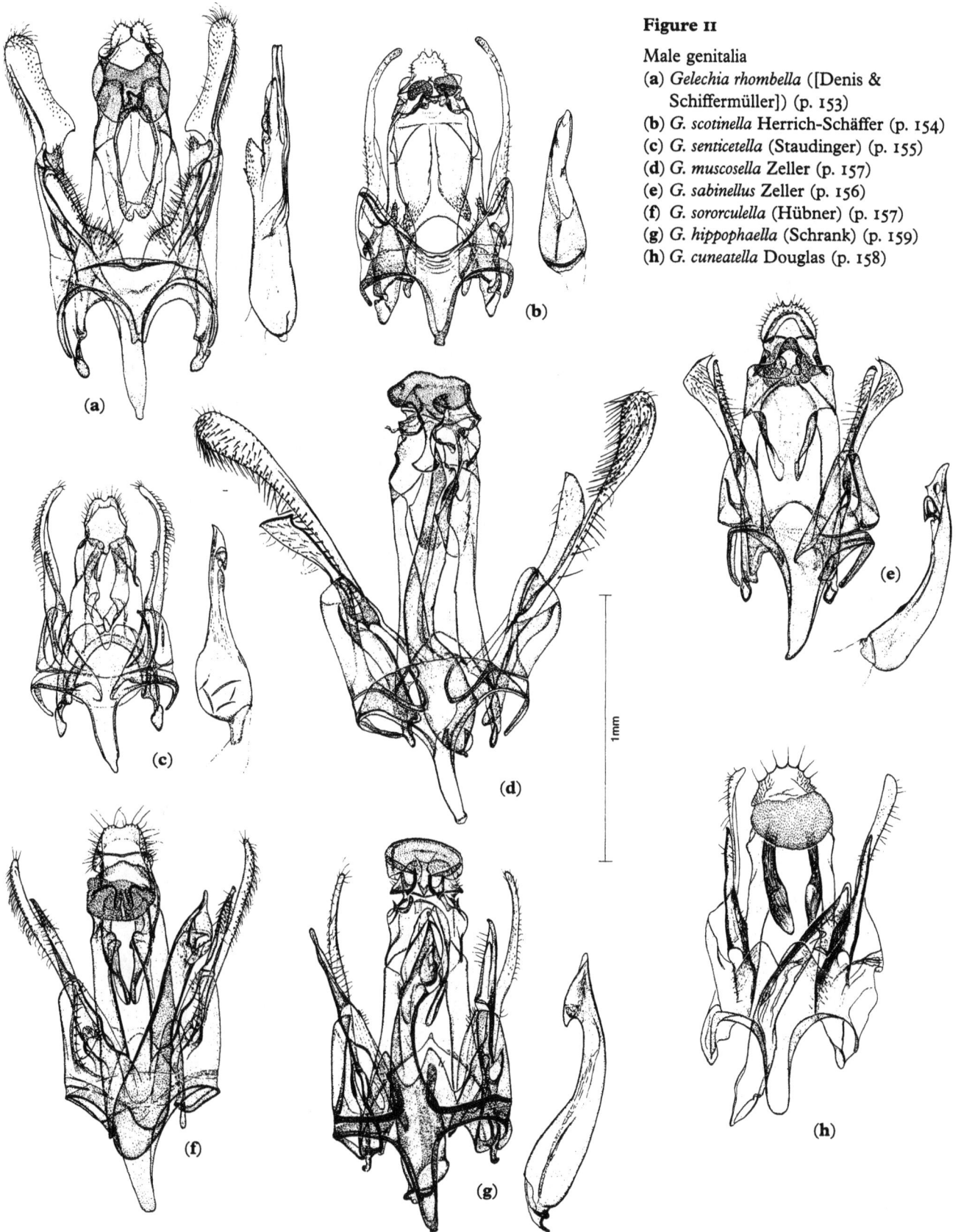

Figure 11

Male genitalia
(a) *Gelechia rhombella* ([Denis &
   Schiffermüller]) (p. 153)
(b) *G. scotinella* Herrich-Schäffer (p. 154)
(c) *G. senticetella* (Staudinger) (p. 155)
(d) *G. muscosella* Zeller (p. 157)
(e) *G. sabinellus* Zeller (p. 156)
(f) *G. sororculella* (Hübner) (p. 157)
(g) *G. hippophaella* (Schrank) (p. 159)
(h) *G. cuneatella* Douglas (p. 158)

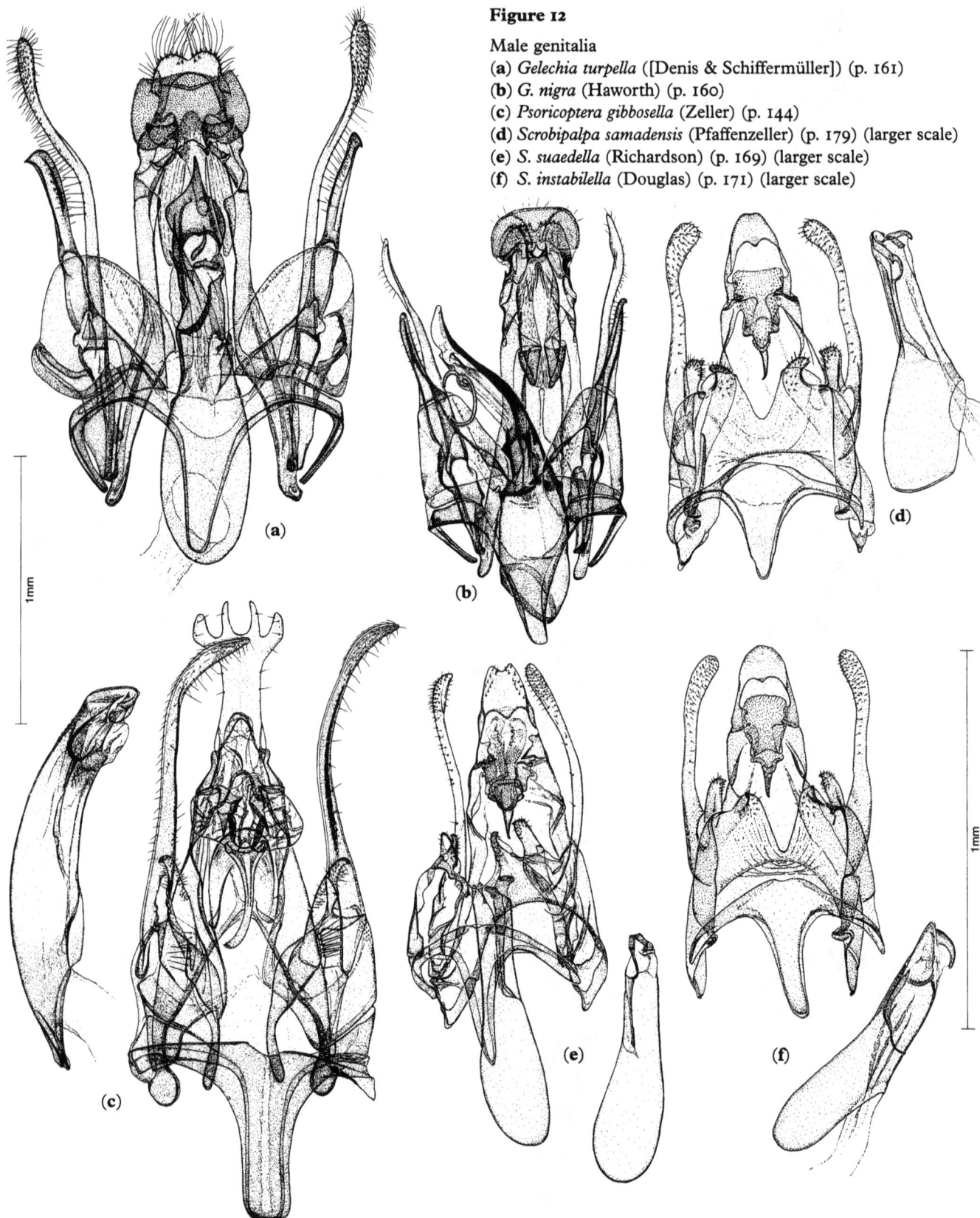

Figure 12

Male genitalia
(a) *Gelechia turpella* ([Denis & Schiffermüller]) (p. 161)
(b) *G. nigra* (Haworth) (p. 160)
(c) *Psoricoptera gibbosella* (Zeller) (p. 144)
(d) *Scrobipalpa samadensis* (Pfaffenzeller) (p. 179) (larger scale)
(e) *S. suaedella* (Richardson) (p. 169) (larger scale)
(f) *S. instabilella* (Douglas) (p. 171) (larger scale)

Figure 13

Male genitalia
(a) *Scrobipalpa salinella* (Zeller) (p. 171)
(b) *S. ocellatella* (Boyd) (p. 177)
(c) *S. pauperella* (Heinemann) (p. 167)
(d) *S. stangei* (E. Hering) (p. 180)
(e) *S. nitentella* (Fuchs) (p. 173)
(f) *S. obsoletella* (Fischer von Röslerstamm) (p. 175)
(g) *S. clintoni* Povolný (p. 181)
(h) *S. atriplicella* (Fischer von Röslerstamm) (p. 176)

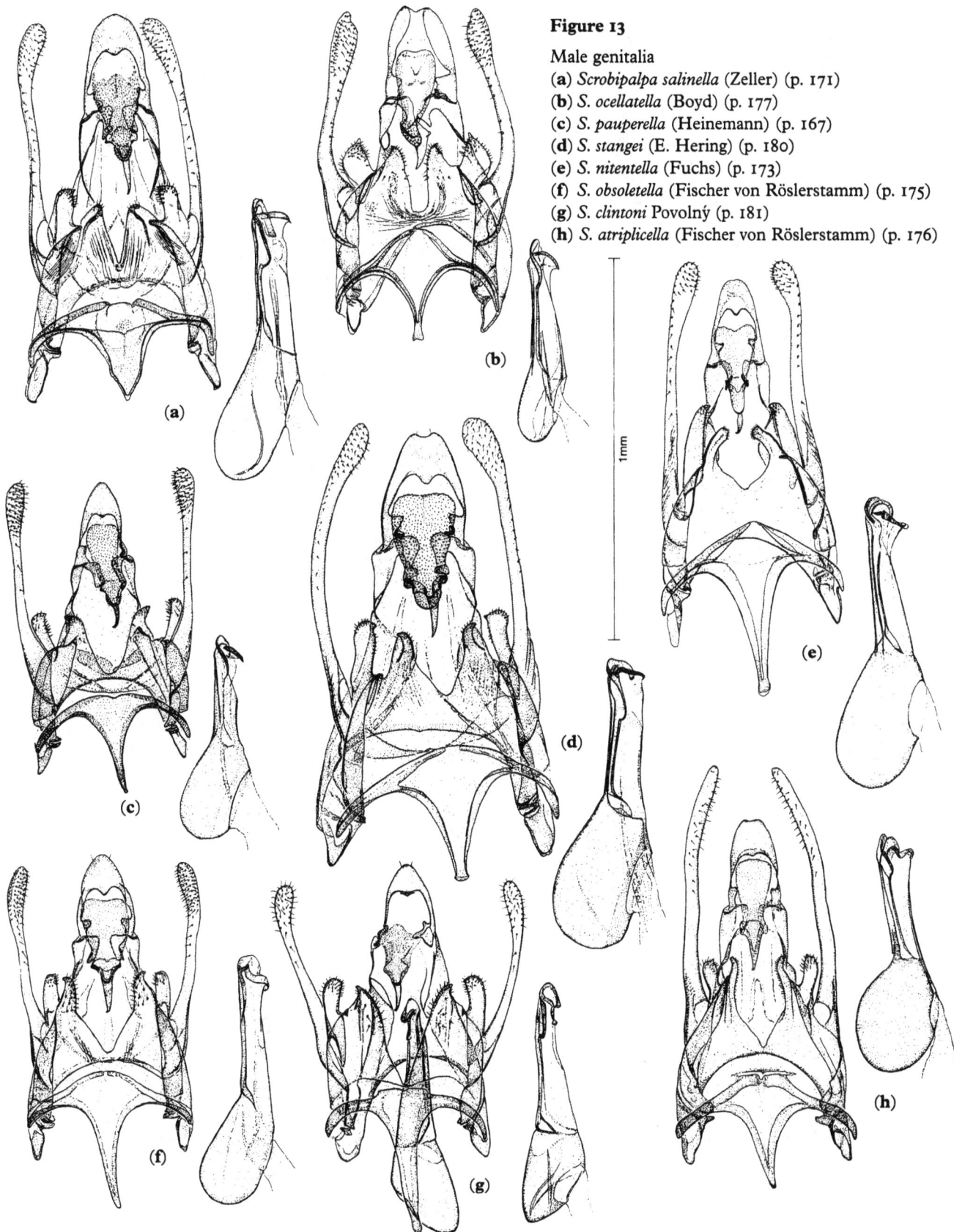

Figure 14

Male genitalia
(**a**) *Scrobipalpa costella* (Humphreys & Westwood) (p. 182)
(**b**) *S. artemisiella* (Treitschke) (p. 179)
(**c**) *S. murinella* (Duponchel) (p. 168)
(**d**) *S. acuminatella* (Sircom) (p. 166)
(**e**) *Phthorimaea operculella* (Zeller) (p. 188) (smaller scale)
(**f**) *Gnorimoschema streliciella* (Herrich-Schäffer) (p. 187) (smaller scale)
(**g**) *Scrobipalpula diffluella* (Frey) (p. 184)
(**h**) *S. diffluella* (Frey) (p. 184), variant
(**i**) *S. tussilaginis* (Stainton) (p. 186)

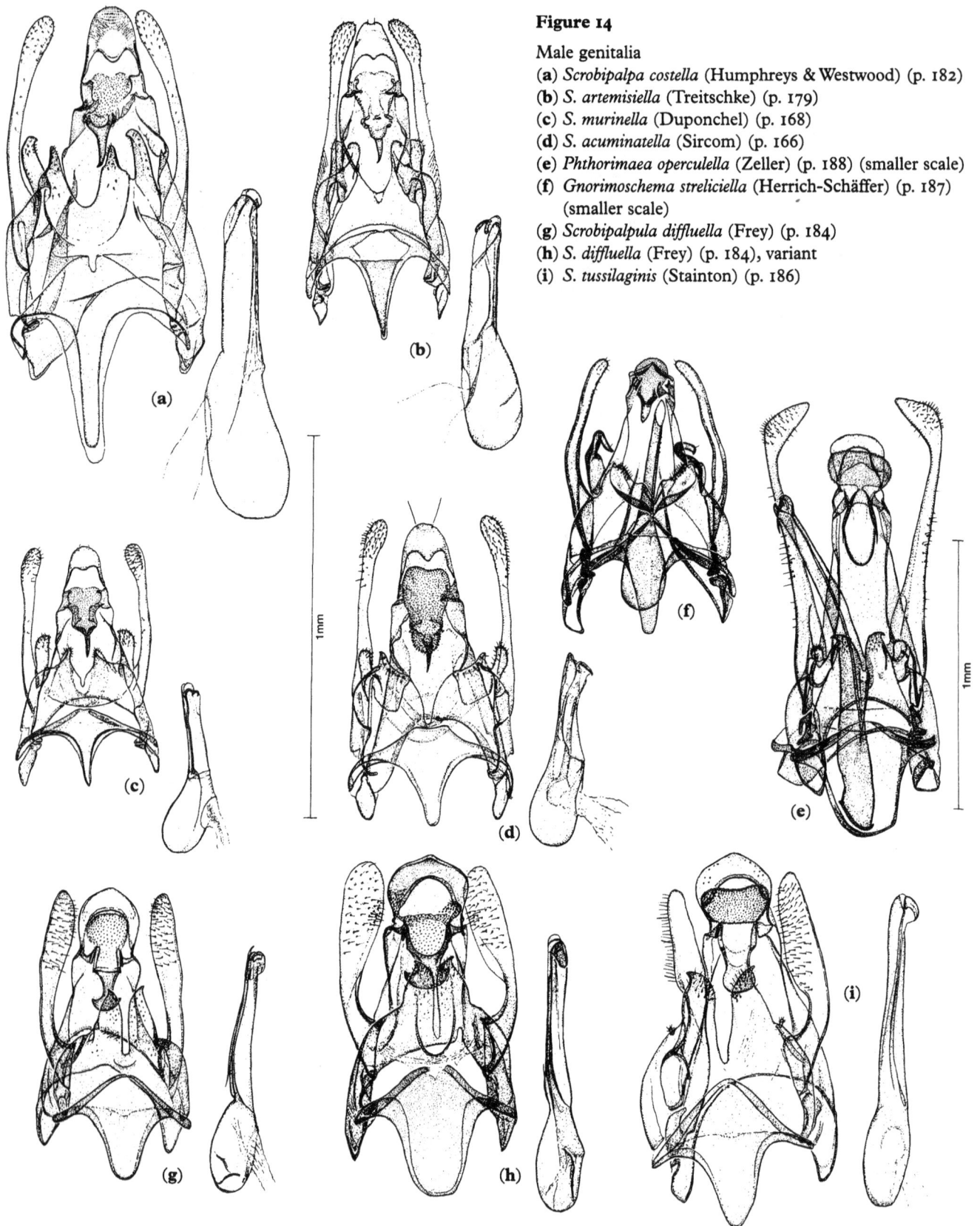

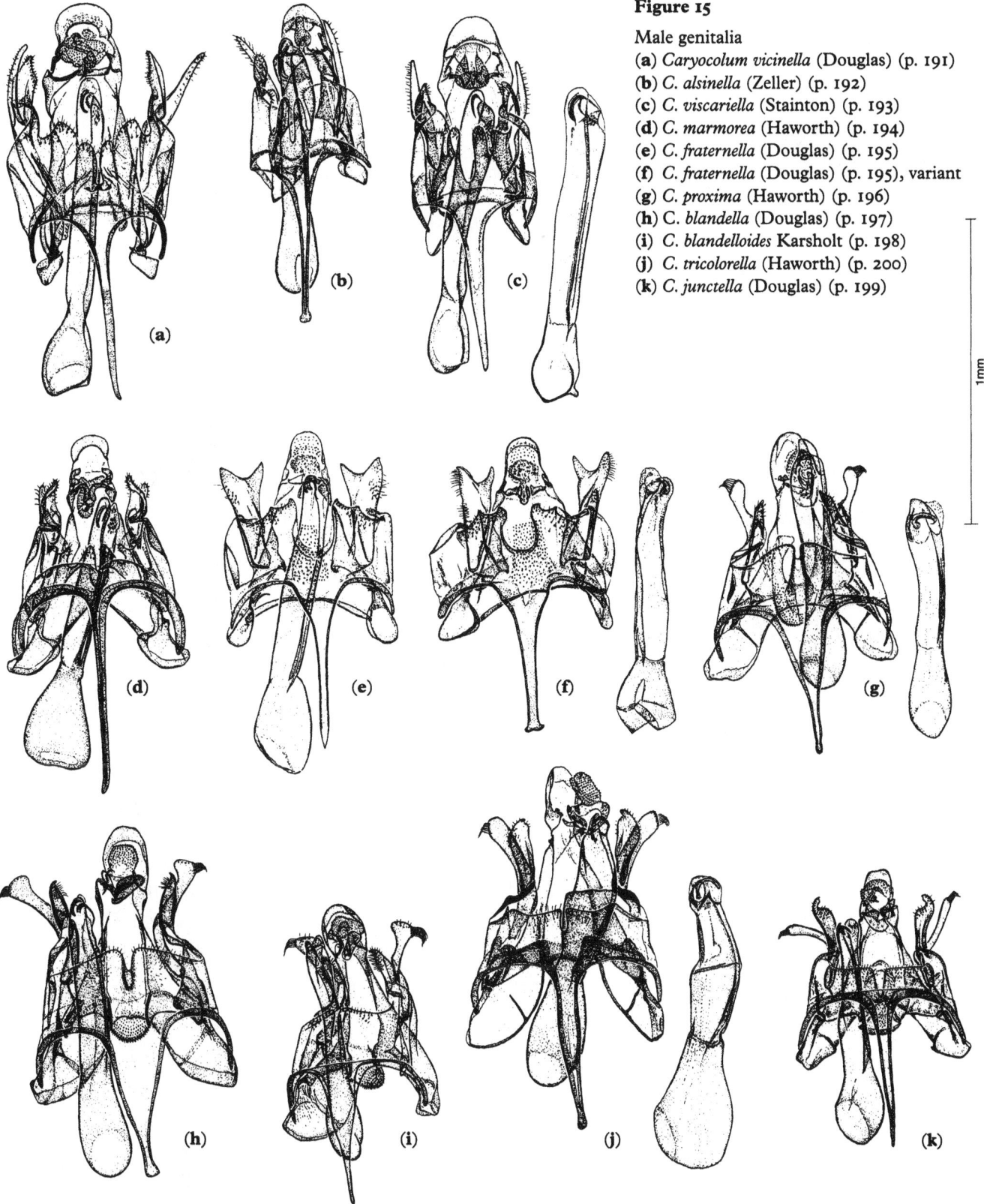

Figure 15

Male genitalia
(**a**) *Caryocolum vicinella* (Douglas) (p. 191)
(**b**) *C. alsinella* (Zeller) (p. 192)
(**c**) *C. viscariella* (Stainton) (p. 193)
(**d**) *C. marmorea* (Haworth) (p. 194)
(**e**) *C. fraternella* (Douglas) (p. 195)
(**f**) *C. fraternella* (Douglas) (p. 195), variant
(**g**) *C. proxima* (Haworth) (p. 196)
(**h**) C. *blandella* (Douglas) (p. 197)
(**i**) *C. blandelloides* Karsholt (p. 198)
(**j**) *C. tricolorella* (Haworth) (p. 200)
(**k**) *C. junctella* (Douglas) (p. 199)

1mm

35

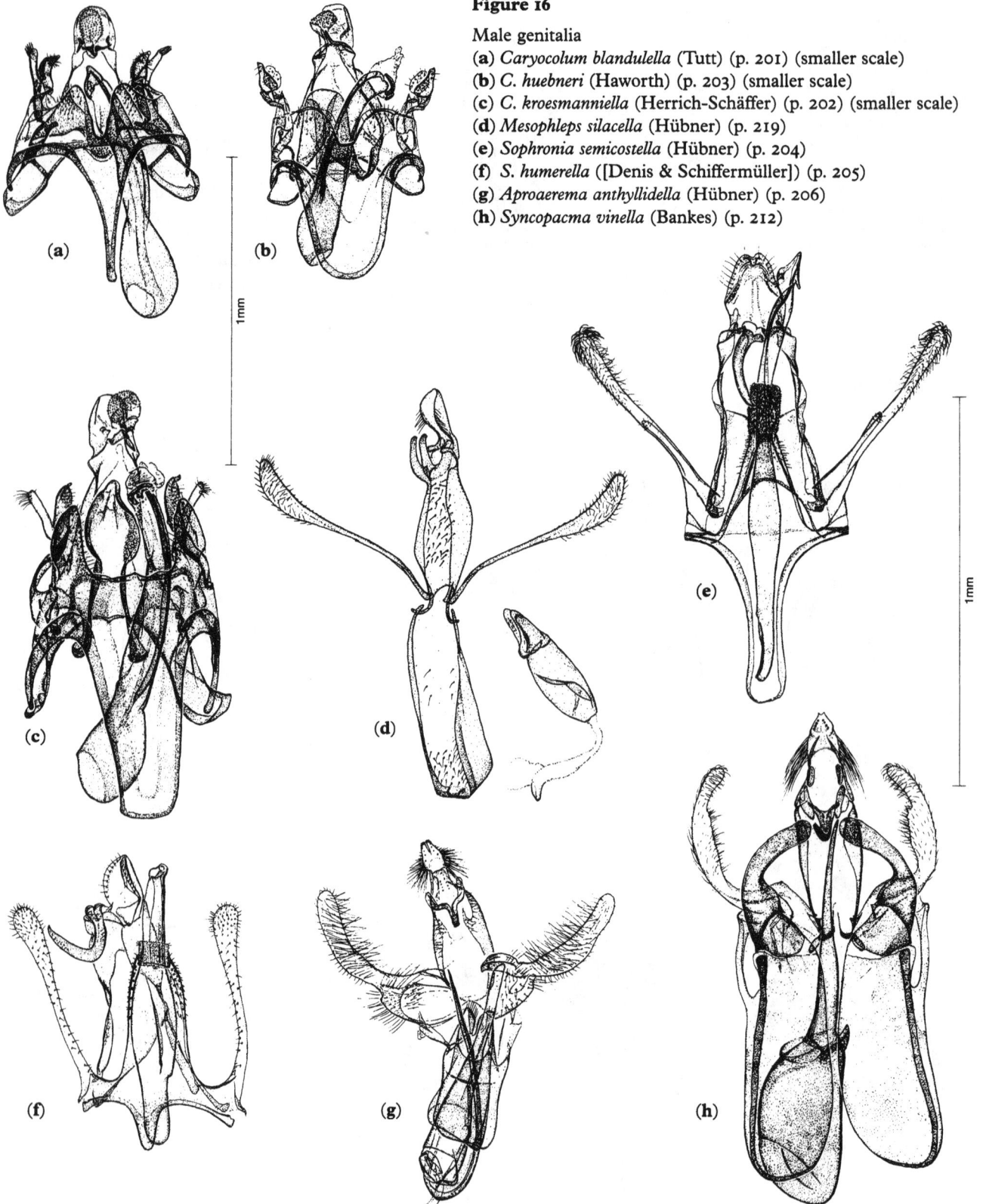

Figure 16

Male genitalia
(a) *Caryocolum blandulella* (Tutt) (p. 201) (smaller scale)
(b) *C. huebneri* (Haworth) (p. 203) (smaller scale)
(c) *C. kroesmanniella* (Herrich-Schäffer) (p. 202) (smaller scale)
(d) *Mesophleps silacella* (Hübner) (p. 219)
(e) *Sophronia semicostella* (Hübner) (p. 204)
(f) *S. humerella* ([Denis & Schiffermüller]) (p. 205)
(g) *Aproaerema anthyllidella* (Hübner) (p. 206)
(h) *Syncopacma vinella* (Bankes) (p. 212)

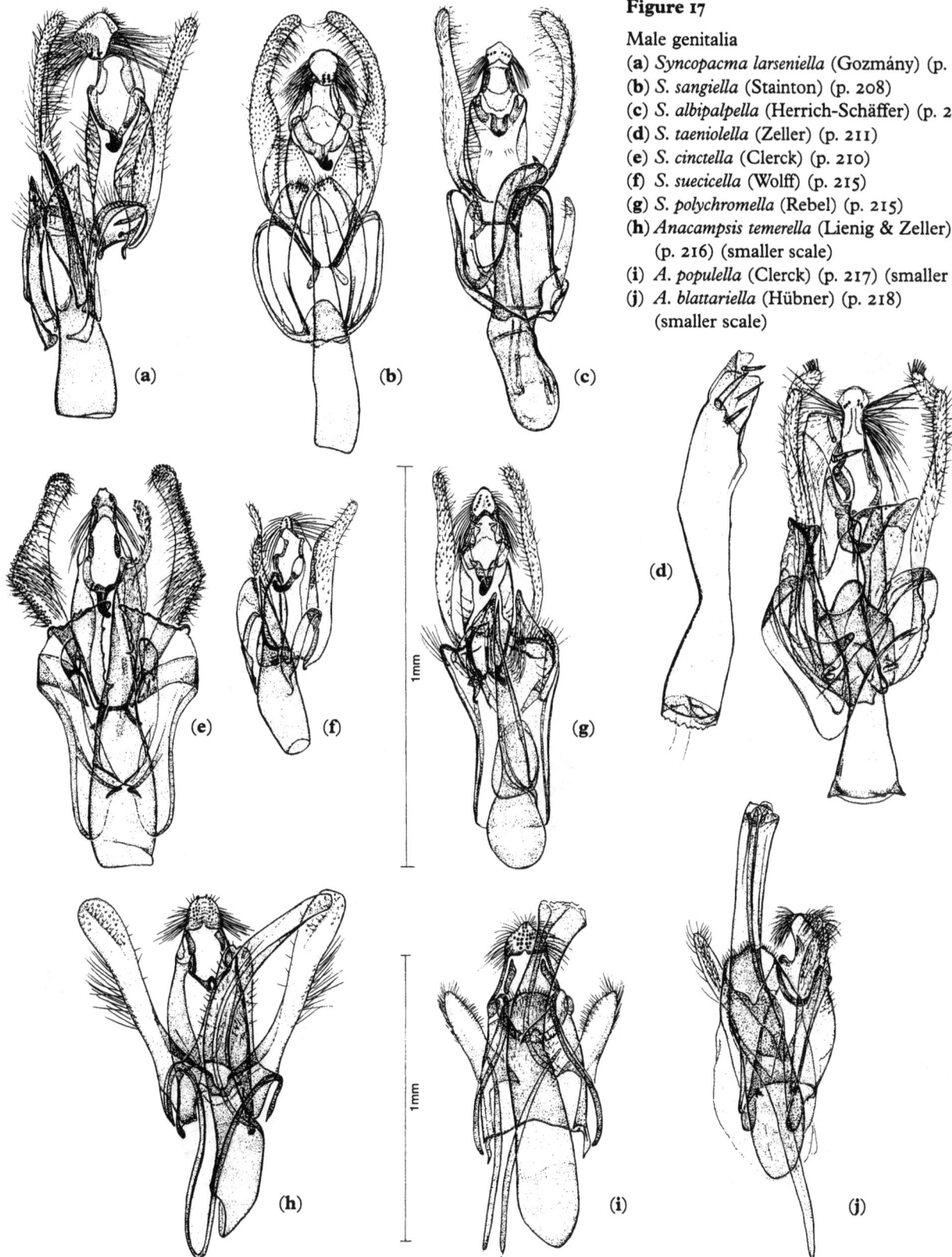

Figure 17

Male genitalia

(a) *Syncopacma larseniella* (Gozmány) (p. 209)
(b) *S. sangiella* (Stainton) (p. 208)
(c) *S. albipalpella* (Herrich-Schäffer) (p. 213)
(d) *S. taeniolella* (Zeller) (p. 211)
(e) *S. cinctella* (Clerck) (p. 210)
(f) *S. suecicella* (Wolff) (p. 215)
(g) *S. polychromella* (Rebel) (p. 215)
(h) *Anacampsis temerella* (Lienig & Zeller) (p. 216) (smaller scale)
(i) *A. populella* (Clerck) (p. 217) (smaller scale)
(j) *A. blattariella* (Hübner) (p. 218) (smaller scale)

Figure 18

Male genitalia
(**a**) *Neofaculta ericetella* (Geyer) (p. 220)
(**b**) *Nothris verbascella* ([Denis &
    Schiffermüller]) (p. 221)
(**c**) *N. congressariella* (Bruand) (p. 223)
(**d**) *Hypatima rhomboidella* (Linnaeus) (p. 226)
(**e**) *Anarsia spartiella* (Schrank) (p. 224)
(**f**) *A. lineatella* Zeller (p. 225)

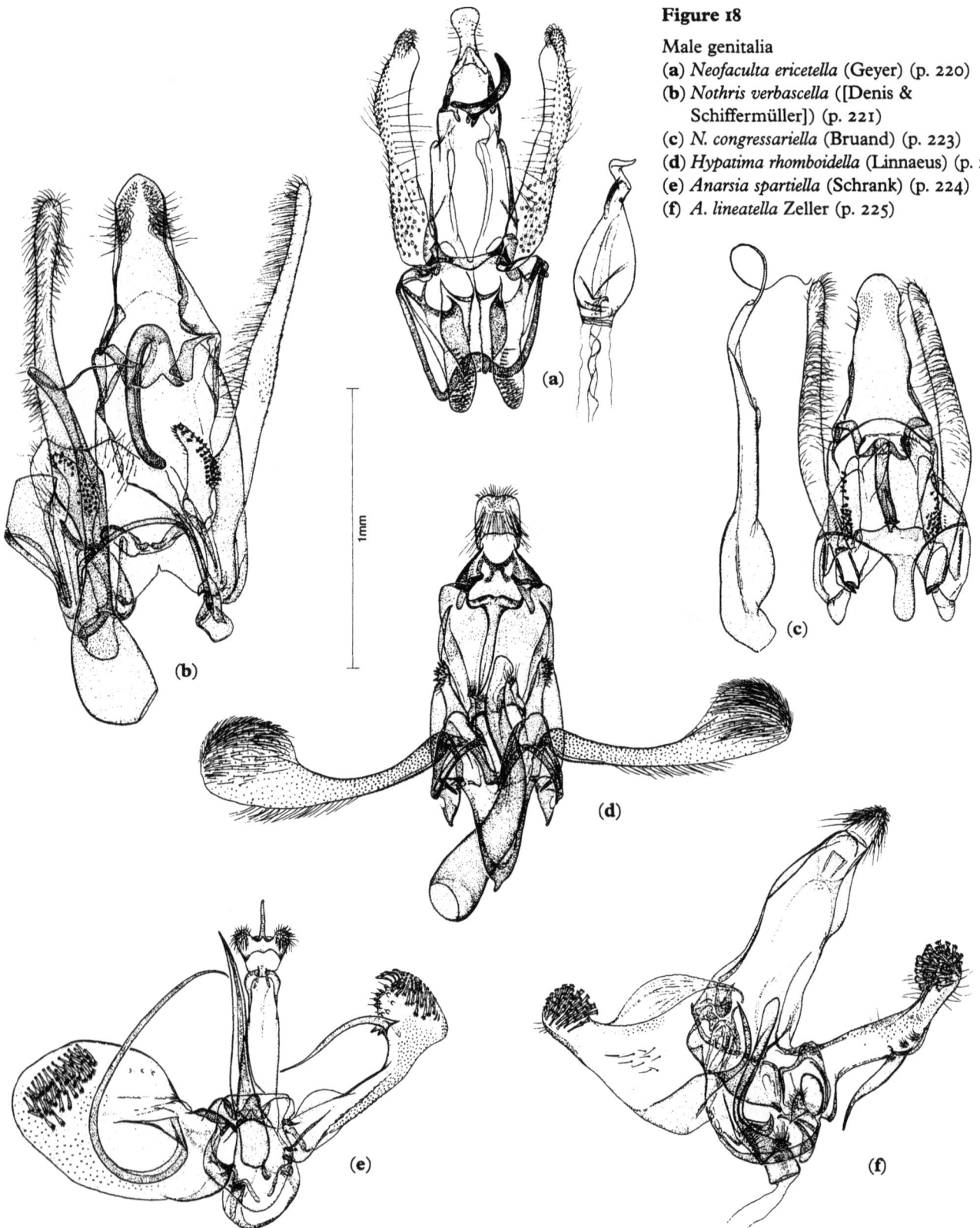

1mm

(a)

(b)

(c)

(d)

(e)

(f)

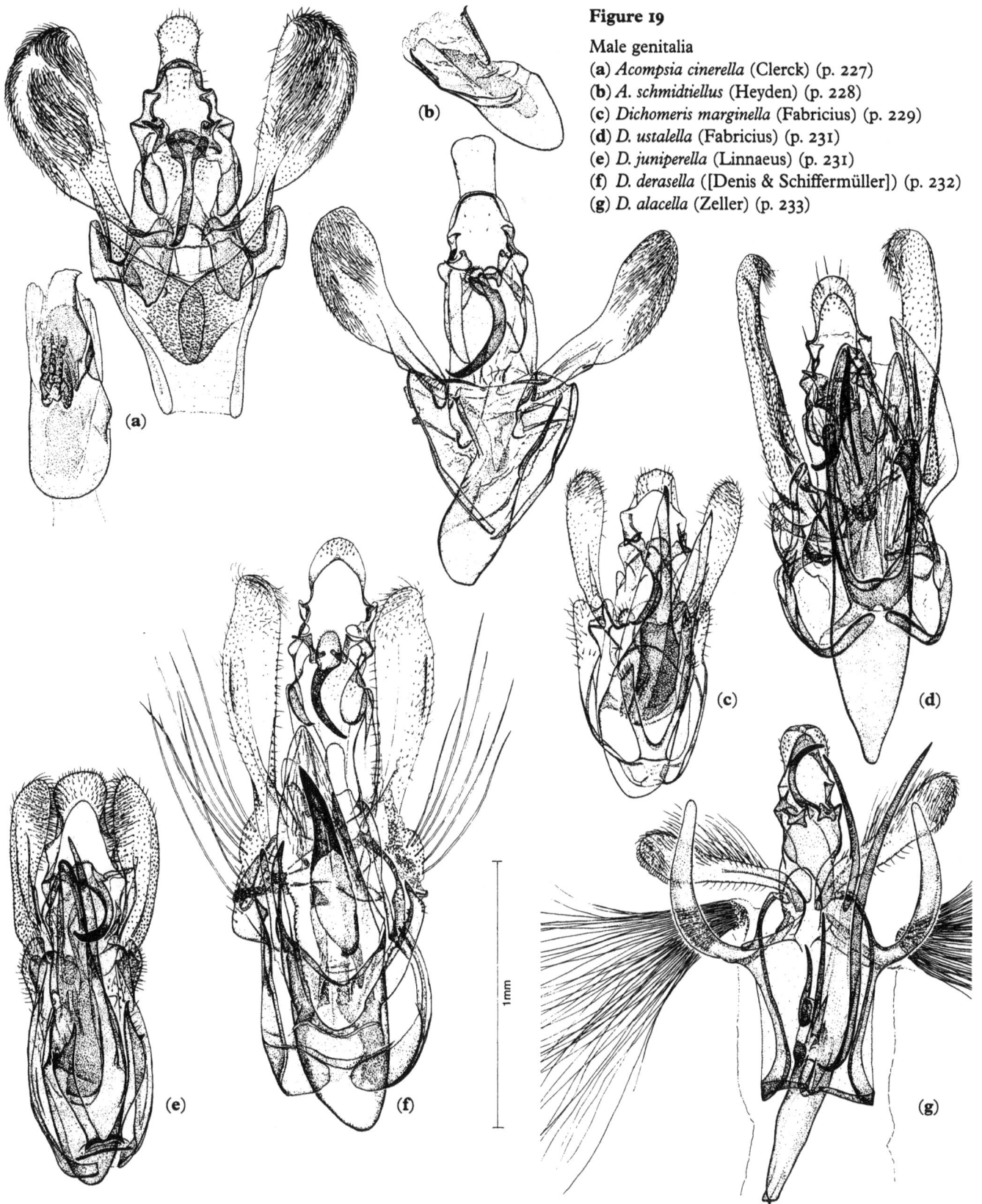

Figure 19

Male genitalia
(a) *Acompsia cinerella* (Clerck) (p. 227)
(b) *A. schmidtiellus* (Heyden) (p. 228)
(c) *Dichomeris marginella* (Fabricius) (p. 229)
(d) *D. ustalella* (Fabricius) (p. 231)
(e) *D. juniperella* (Linnaeus) (p. 231)
(f) *D. derasella* ([Denis & Schiffermüller]) (p. 232)
(g) *D. alacella* (Zeller) (p. 233)

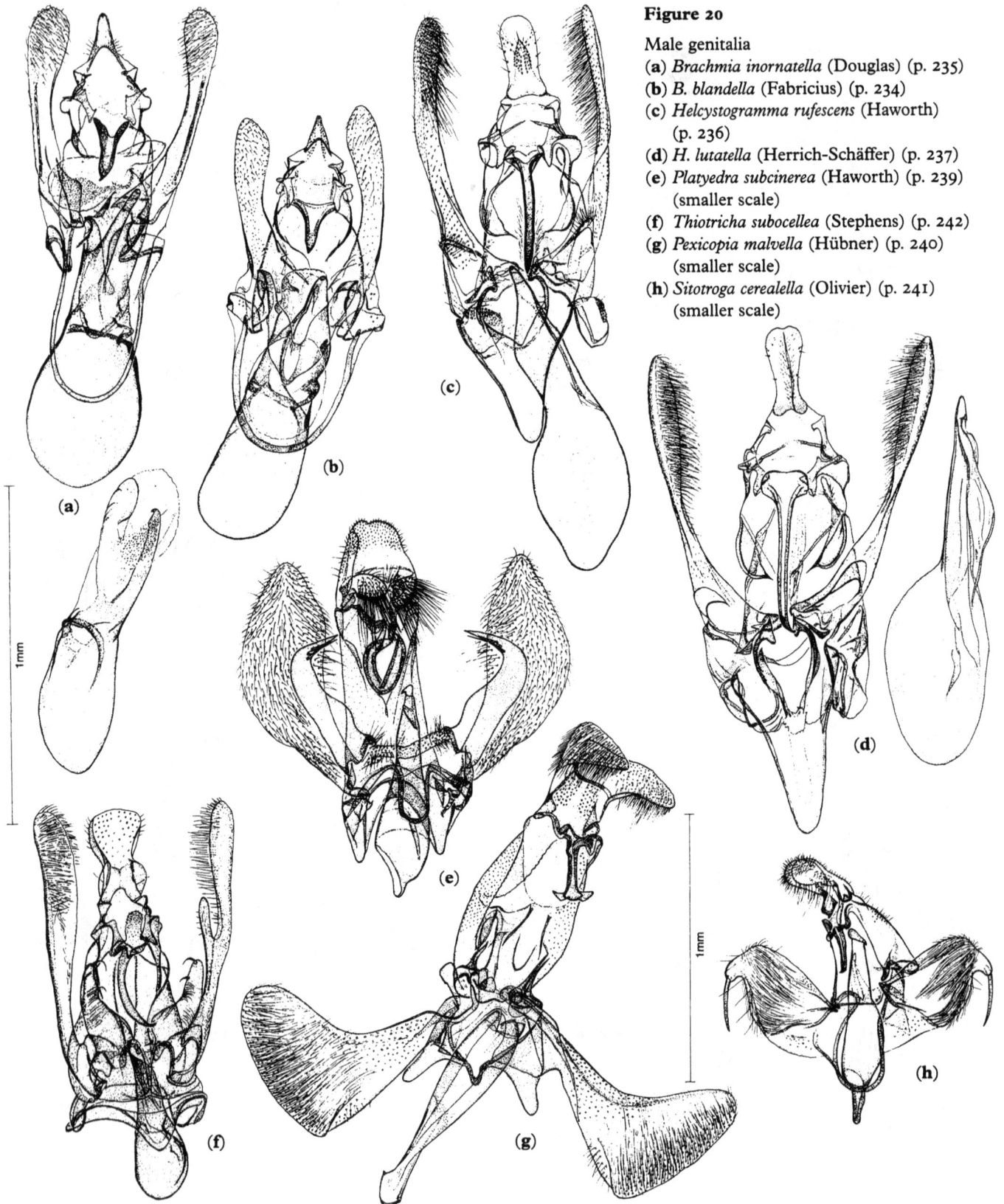

Figure 20

Male genitalia
(**a**) *Brachmia inornatella* (Douglas) (p. 235)
(**b**) *B. blandella* (Fabricius) (p. 234)
(**c**) *Helcystogramma rufescens* (Haworth)
    (p. 236)
(**d**) *H. lutatella* (Herrich-Schäffer) (p. 237)
(**e**) *Platyedra subcinerea* (Haworth) (p. 239)
    (smaller scale)
(**f**) *Thiotricha subocellea* (Stephens) (p. 242)
(**g**) *Pexicopia malvella* (Hübner) (p. 240)
    (smaller scale)
(**h**) *Sitotroga cerealella* (Olivier) (p. 241)
    (smaller scale)

Figure 21

Female genitalia
(**a**) *Metzneria lappella* (Linnaeus) (p. 68)
(**b**) *M. aestivella* (Zeller) (p. 69)
(**c**) *M. metzneriella* (Stainton) (p. 70)
(**d**) *M. littorella* (Douglas) (p. 68)
(**e**) *M. neuropterella* (Zeller) (p. 71)
(**f**) *M. aprilella* (Herrich-Schäffer) (p. 72)
(**g**) *Isophrictis striatella* ([Denis & Schiffermüller])
   (p. 73)
(**h**) *Apodia bifractella* (Duponchel) (p. 74)

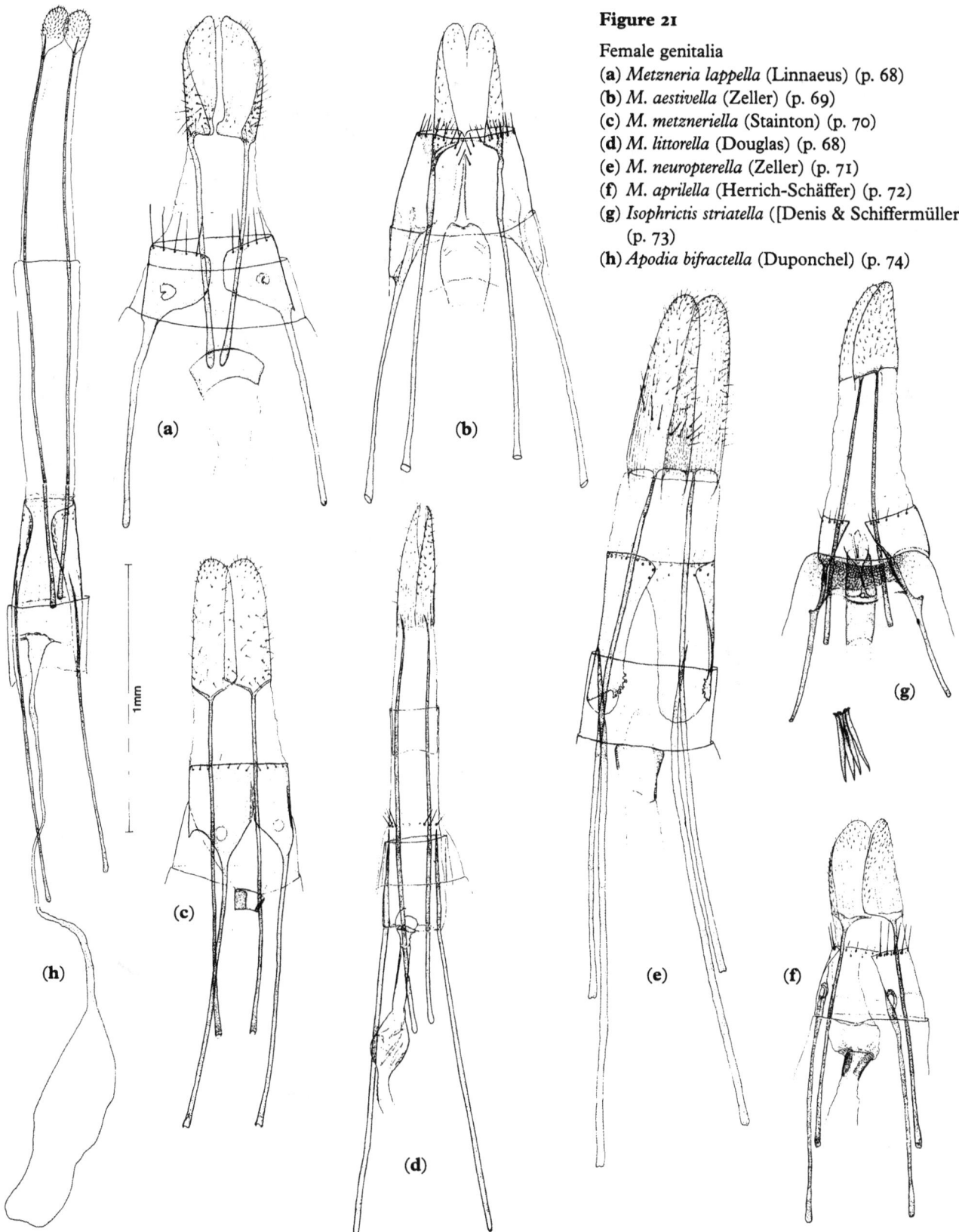

(**a**)

(**b**)

(**c**)

(**d**)

(**e**)

(**f**)

(**g**)

(**h**)

1mm

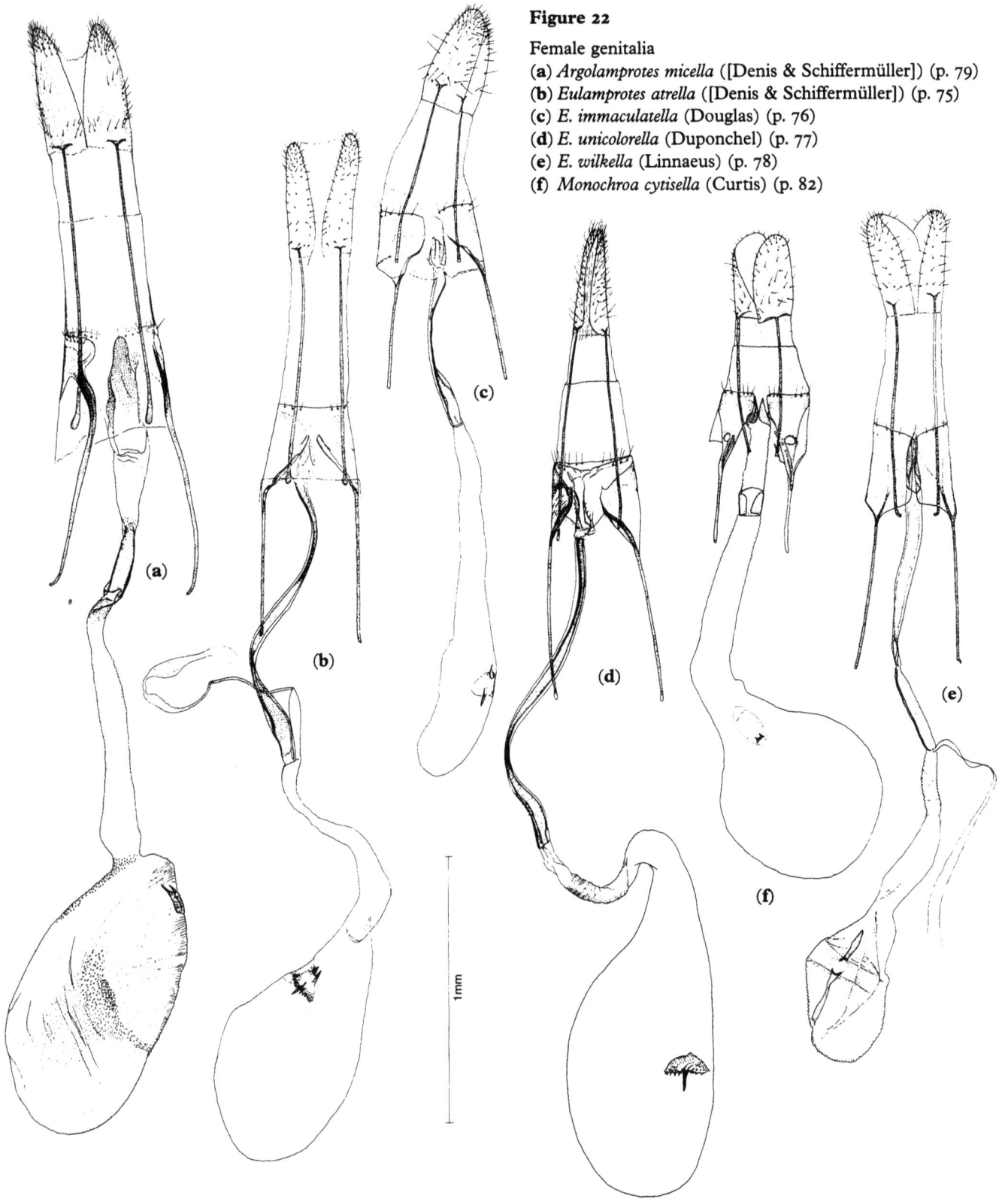

Figure 22

Female genitalia
(a) *Argolamprotes micella* ([Denis & Schiffermüller]) (p. 79)
(b) *Eulamprotes atrella* ([Denis & Schiffermüller]) (p. 75)
(c) *E. immaculatella* (Douglas) (p. 76)
(d) *E. unicolorella* (Duponchel) (p. 77)
(e) *E. wilkella* (Linnaeus) (p. 78)
(f) *Monochroa cytisella* (Curtis) (p. 82)

Figure 23

Female genitalia
(**a**) *Monochroa lucidella* (Stephens)
    (p. 84)
(**b**) *M. tenebrella* (Hübner) (p. 83)
(**c**) *M. palustrellus* (Douglas) (p. 84)
(**d**) *M. conspersella* (Herrich-Schäffer)
    (p. 86)
(**e**) *M. hornigi* (Staudinger) (p. 86)
(**f**) *M. tetragonella* (Stainton) (p. 85)

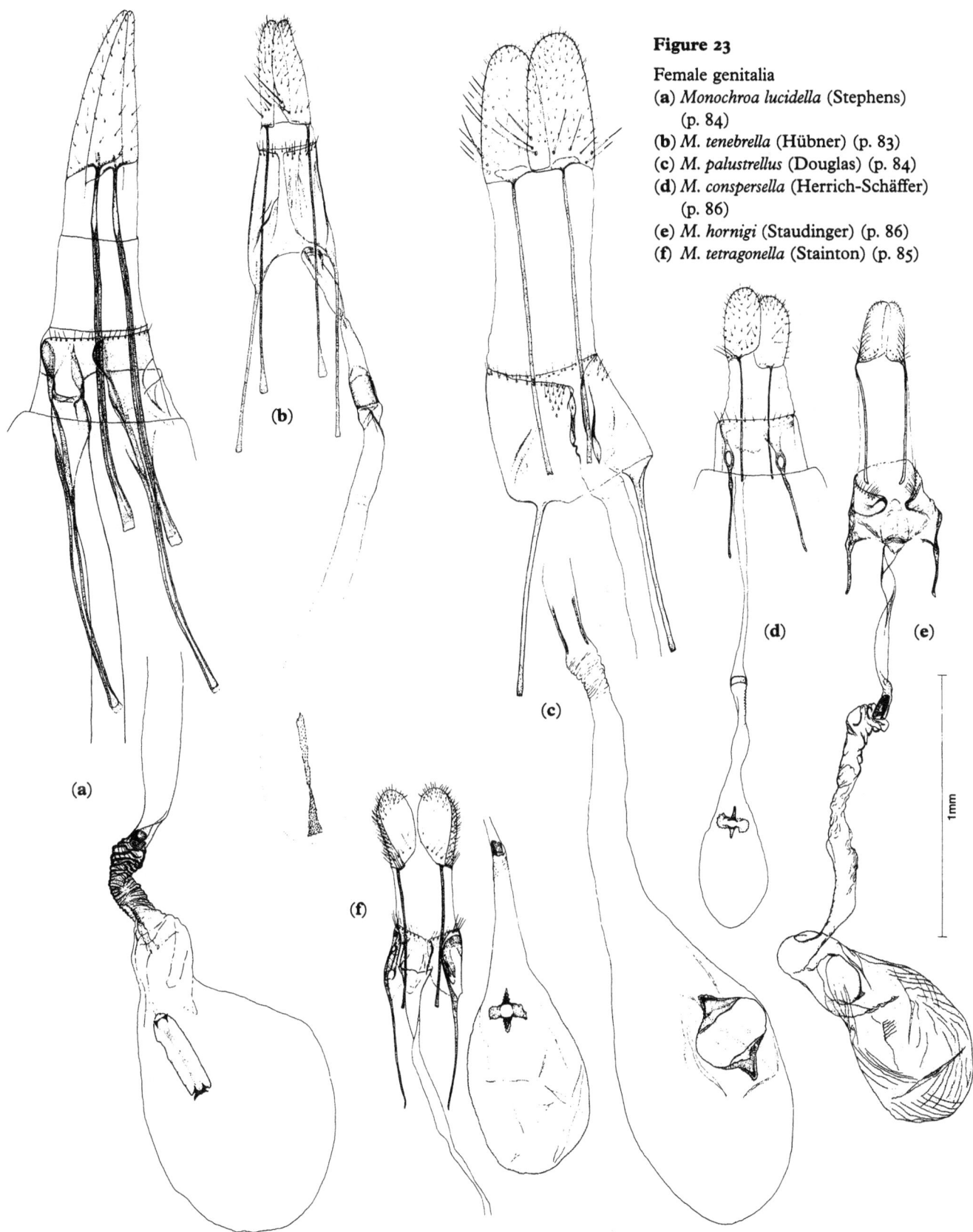

Figure 24

Female genitalia
(a) *Monochroa arundinetella* (Stainton) (p. 91)
(b) *M. moyses* Uffen (p. 92)
(c) *M. lutulentella* (Zeller) (p. 89)
(d) *M. elongella* (Heinemann) (p. 90)
(e) *M. niphognatha* (Gozmány) (p. 88)
(f) *M. suffusella* (Douglas) (p. 88)

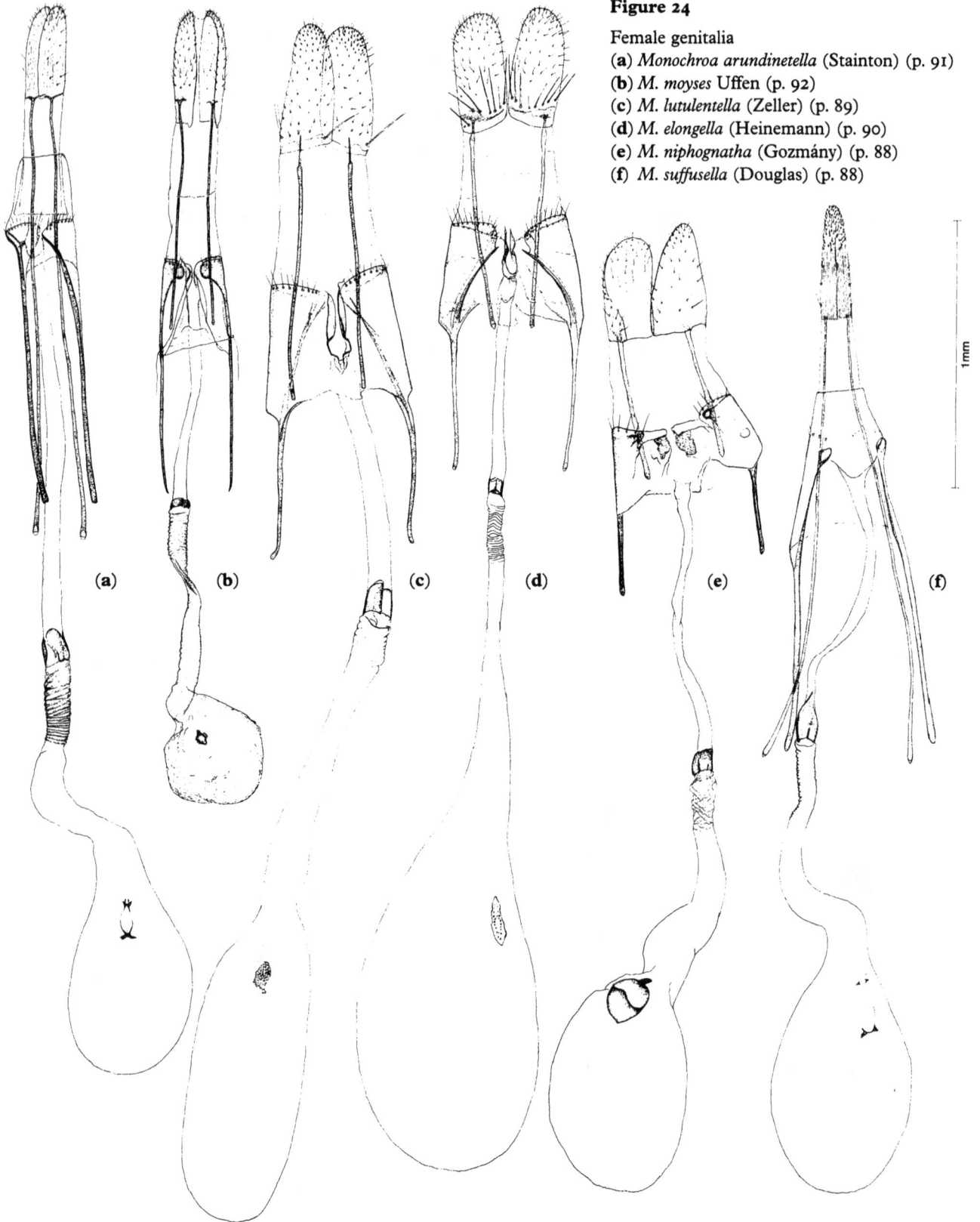

1mm

(a)   (b)   (c)   (d)   (e)   (f)

Figure 25

Female genitalia

(a) *Monochroa divisella* (Douglas) (p. 93)
(b) *Chrysoesthia drurella* (Fabricius) (p. 94)
(c) *C. sexguttella* (Thunberg) (p. 95)
(d) *Ptocheuusa paupella* (Zeller) (p. 96)
(e) *Aristotelia ericinella* (Zeller) (p. 100)
(f) *A. subdecurtella* (Stainton) (p. 99)
(g) *A. brizella* (Treitschke) (p. 101)
(h) *Psamathocrita argentella* Pierce & Metcalfe, holotype (in lateral view) (p. 98)
(i) *P. osseella* (Stainton) (in lateral view) (p. 97)

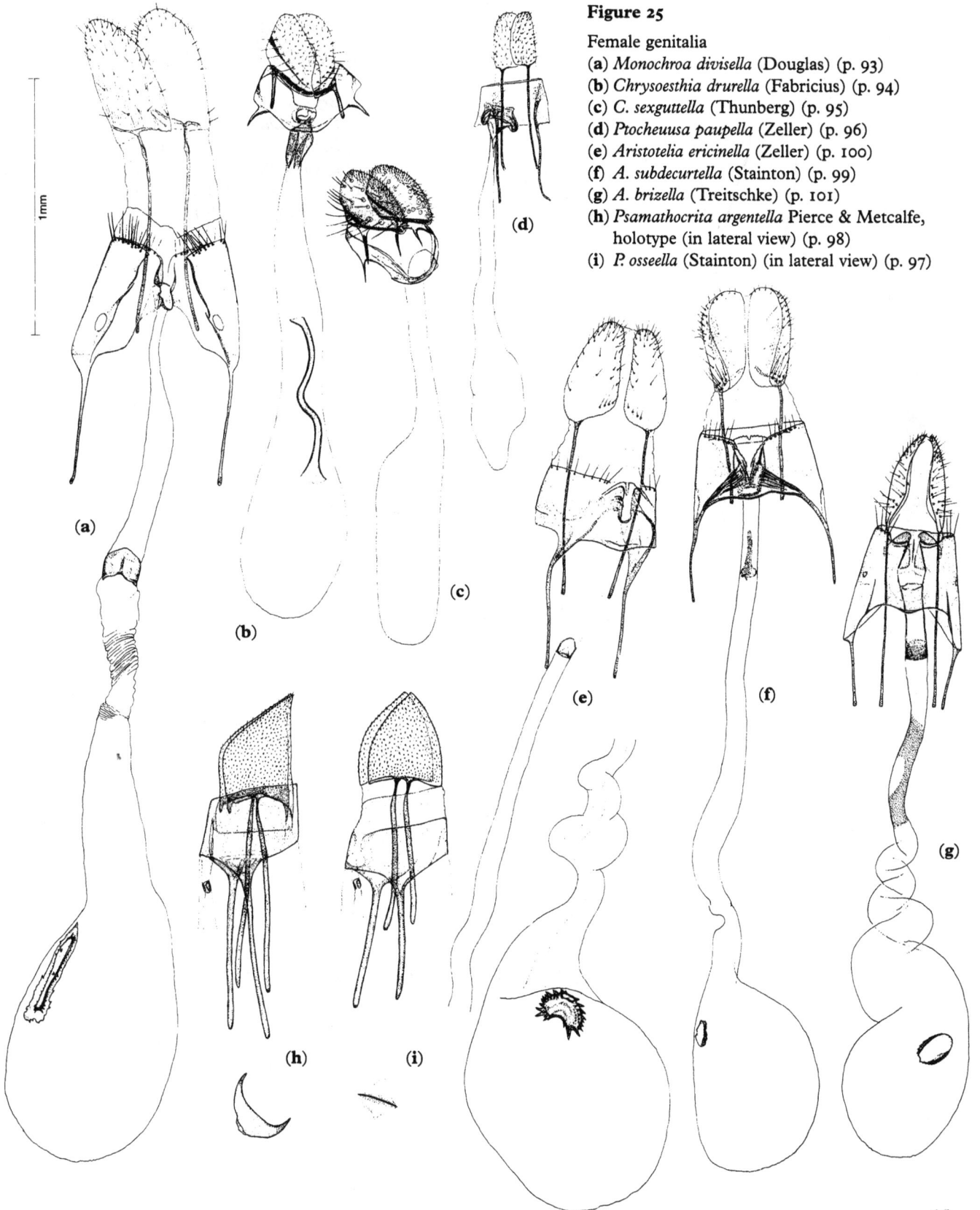

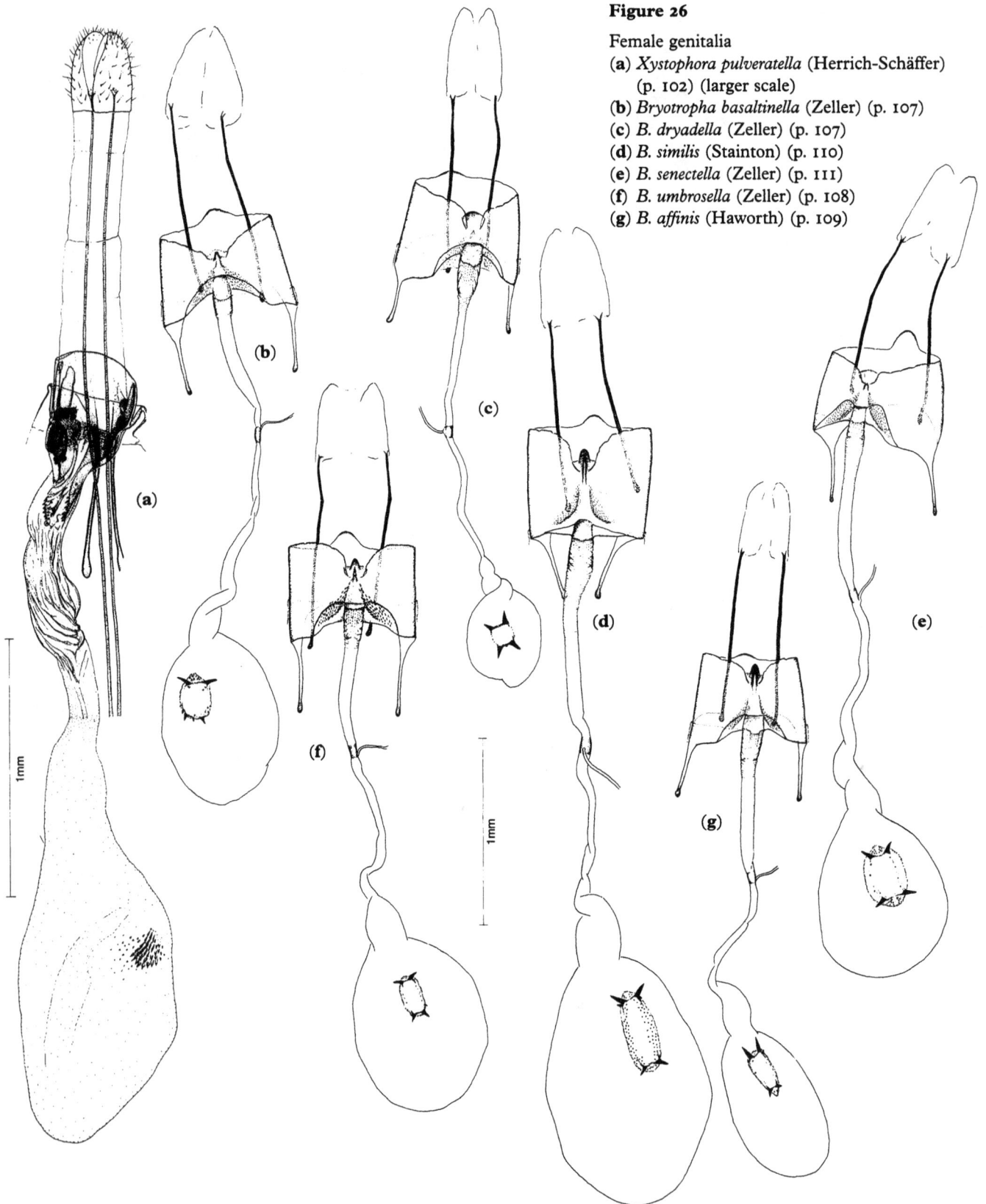

Figure 26

Female genitalia
(a) *Xystophora pulveratella* (Herrich-Schäffer)
(p. 102) (larger scale)
(b) *Bryotropha basaltinella* (Zeller) (p. 107)
(c) *B. dryadella* (Zeller) (p. 107)
(d) *B. similis* (Stainton) (p. 110)
(e) *B. senectella* (Zeller) (p. 111)
(f) *B. umbrosella* (Zeller) (p. 108)
(g) *B. affinis* (Haworth) (p. 109)

Figure 27

Female genitalia
(**a**) *Bryotropha desertella* (Douglas) (p. 114)
(**b**) *B. terrella* ([Denis & Schiffermüller]) (p. 115)
(**c**) *B. galbanella* (Zeller) (p. 113)
(**d**) *B. politella* (Stainton) (p. 116)
(**e**) *B. boreella* (Douglas) (p. 112)
(**f**) *B. domestica* (Haworth) (p. 117)

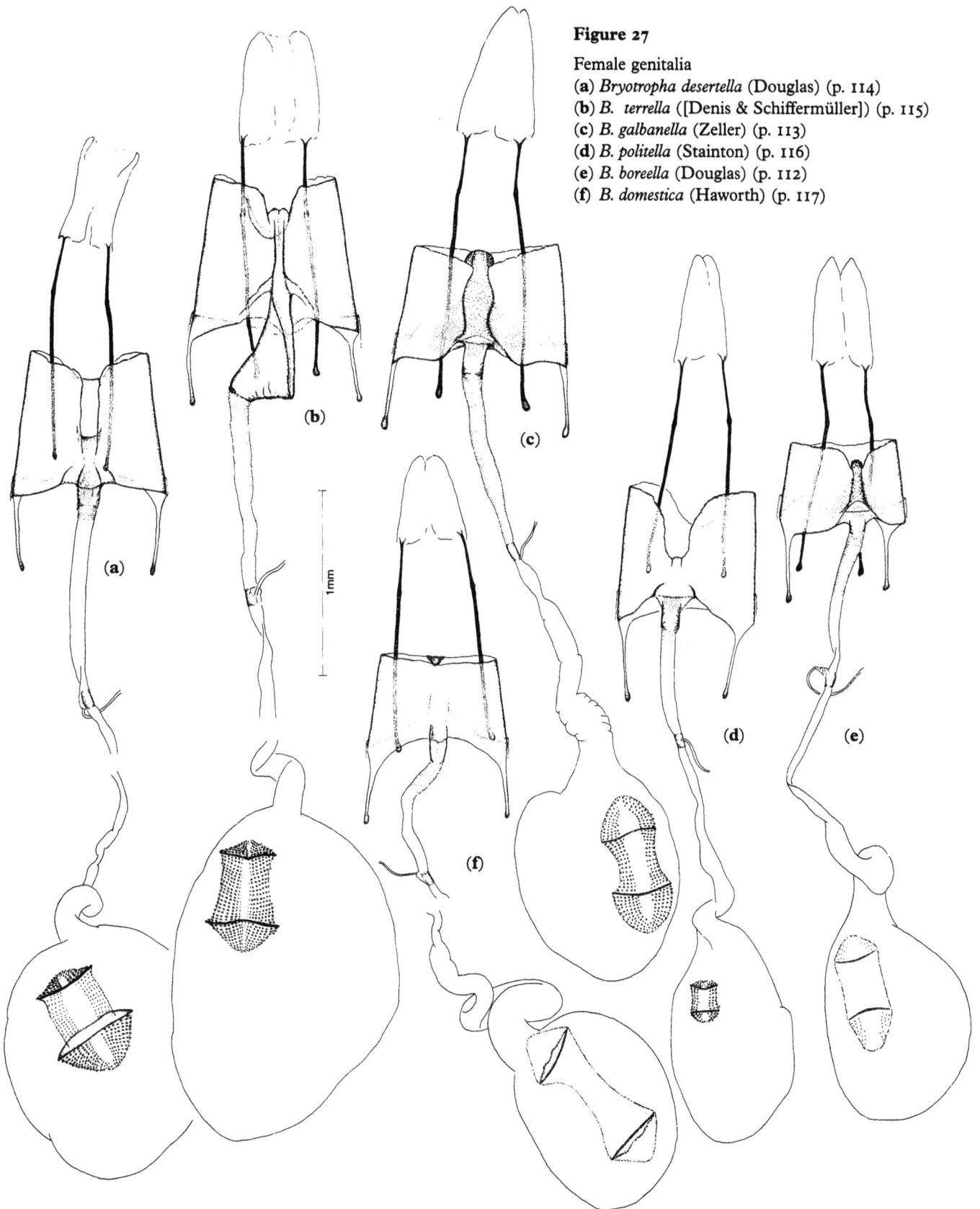

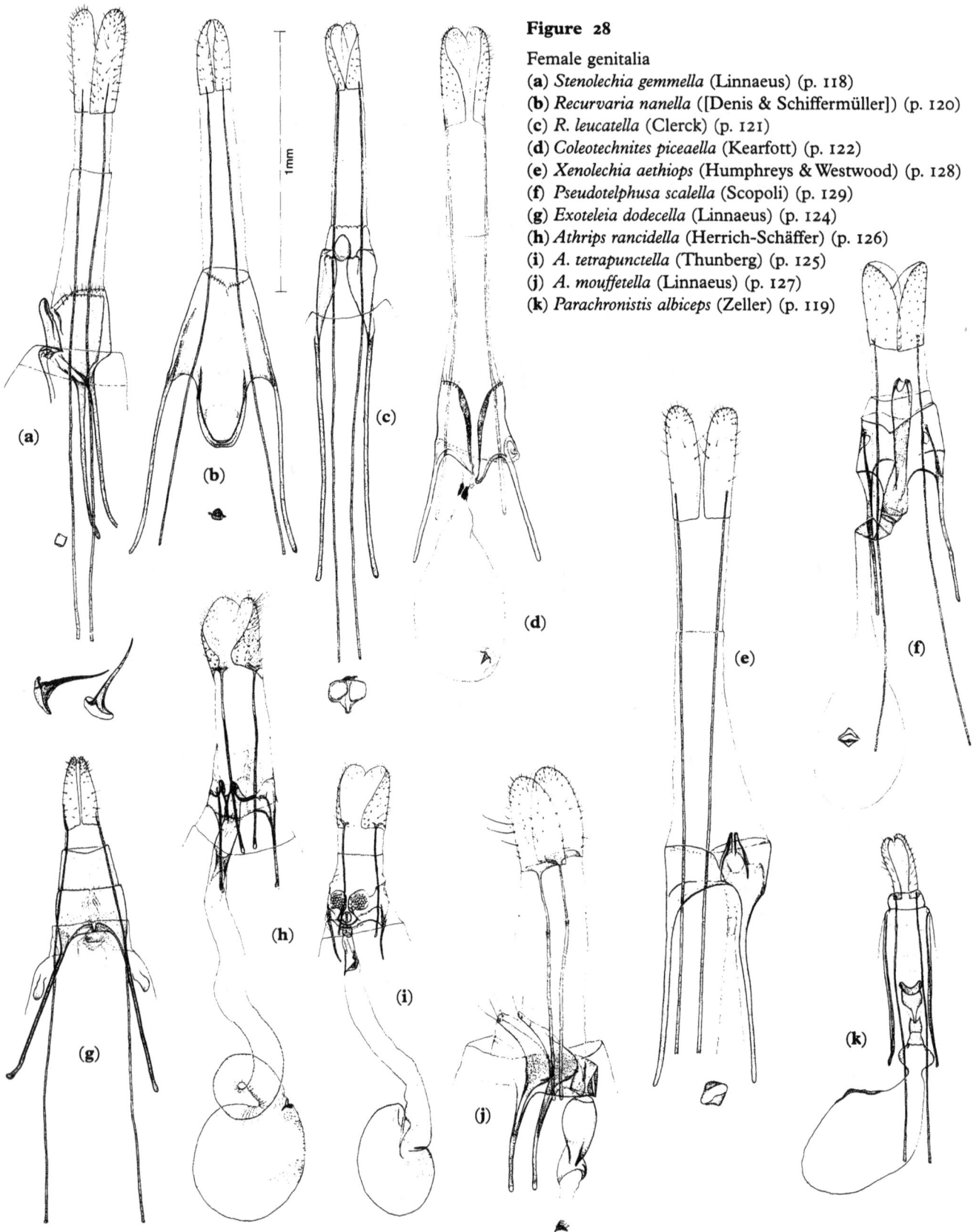

Figure 28

Female genitalia
(**a**) *Stenolechia gemmella* (Linnaeus) (p. 118)
(**b**) *Recurvaria nanella* ([Denis & Schiffermüller]) (p. 120)
(**c**) *R. leucatella* (Clerck) (p. 121)
(**d**) *Coleotechnites piceaella* (Kearfott) (p. 122)
(**e**) *Xenolechia aethiops* (Humphreys & Westwood) (p. 128)
(**f**) *Pseudotelphusa scalella* (Scopoli) (p. 129)
(**g**) *Exoteleia dodecella* (Linnaeus) (p. 124)
(**h**) *Athrips rancidella* (Herrich-Schäffer) (p. 126)
(**i**) *A. tetrapunctella* (Thunberg) (p. 125)
(**j**) *A. mouffetella* (Linnaeus) (p. 127)
(**k**) *Parachronistis albiceps* (Zeller) (p. 119)

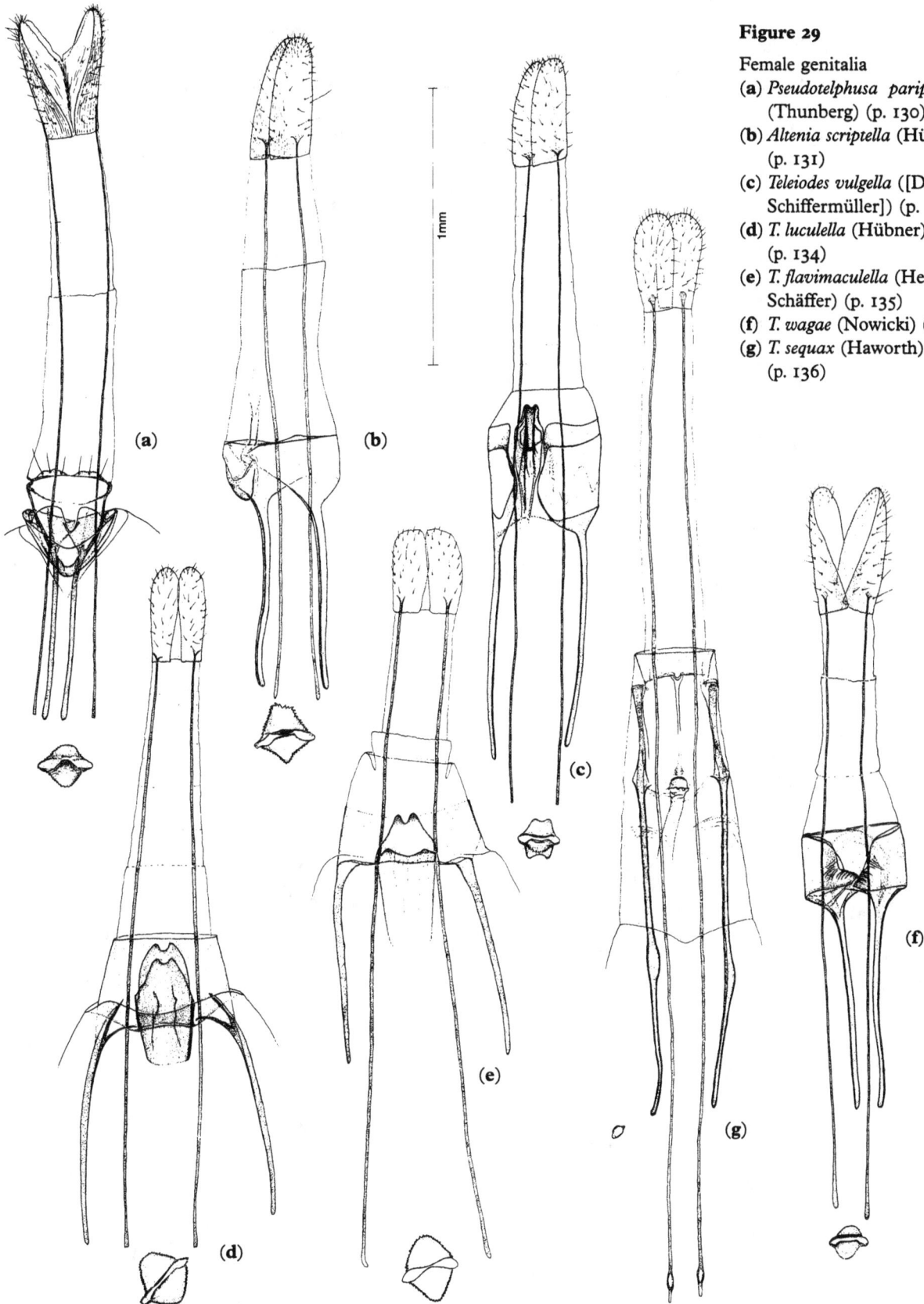

Figure 29

Female genitalia

(a) *Pseudotelphusa paripunctella* (Thunberg) (p. 130)

(b) *Altenia scriptella* (Hübner) (p. 131)

(c) *Teleiodes vulgella* ([Denis & Schiffermüller]) (p. 132)

(d) *T. luculella* (Hübner) (p. 134)

(e) *T. flavimaculella* (Herrich-Schäffer) (p. 135)

(f) *T. wagae* (Nowicki) (p. 133)

(g) *T. sequax* (Haworth) (p. 136)

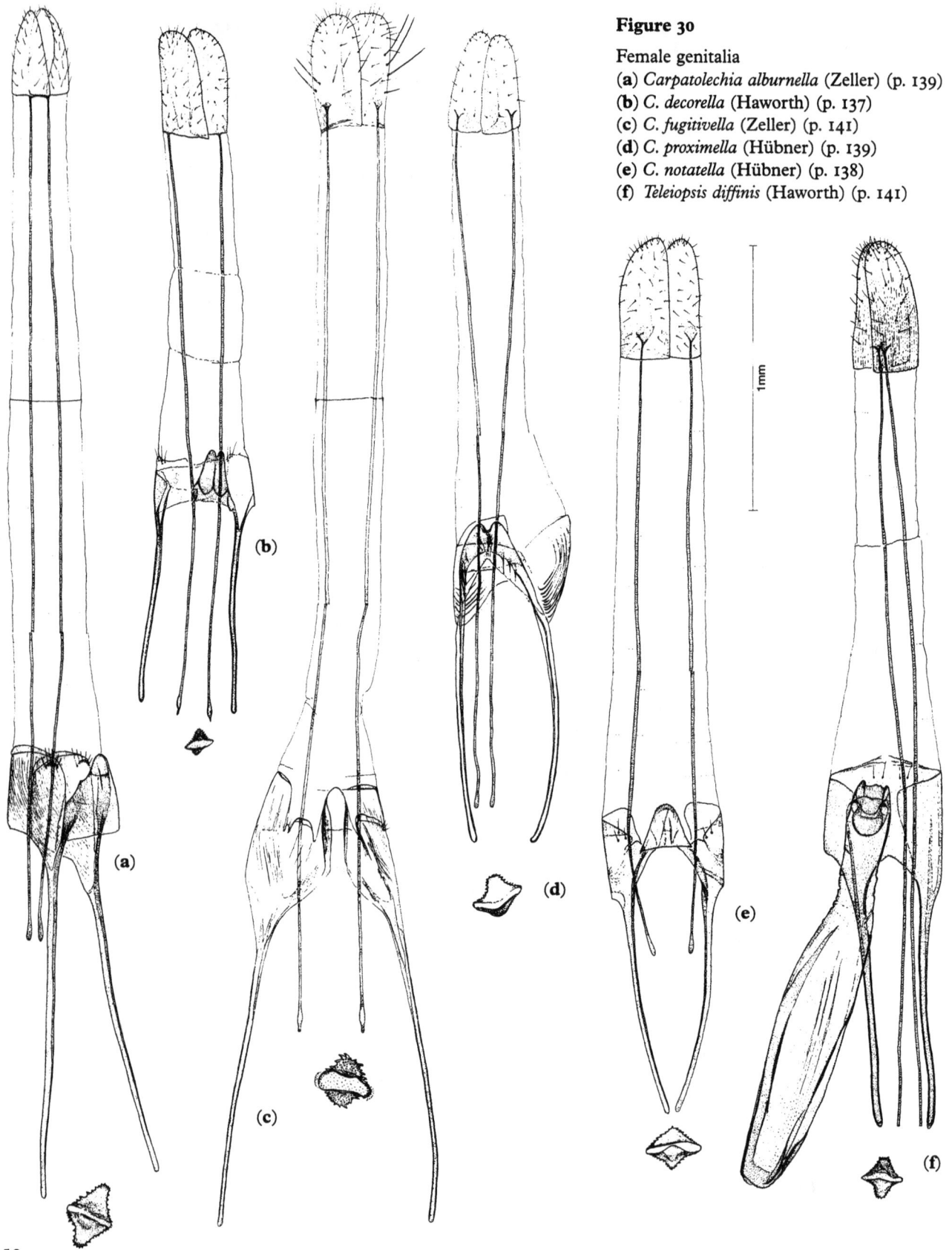

Figure 30

Female genitalia
(a) *Carpatolechia alburnella* (Zeller) (p. 139)
(b) *C. decorella* (Haworth) (p. 137)
(c) *C. fugitivella* (Zeller) (p. 141)
(d) *C. proximella* (Hübner) (p. 139)
(e) *C. notatella* (Hübner) (p. 138)
(f) *Teleiopsis diffinis* (Haworth) (p. 141)

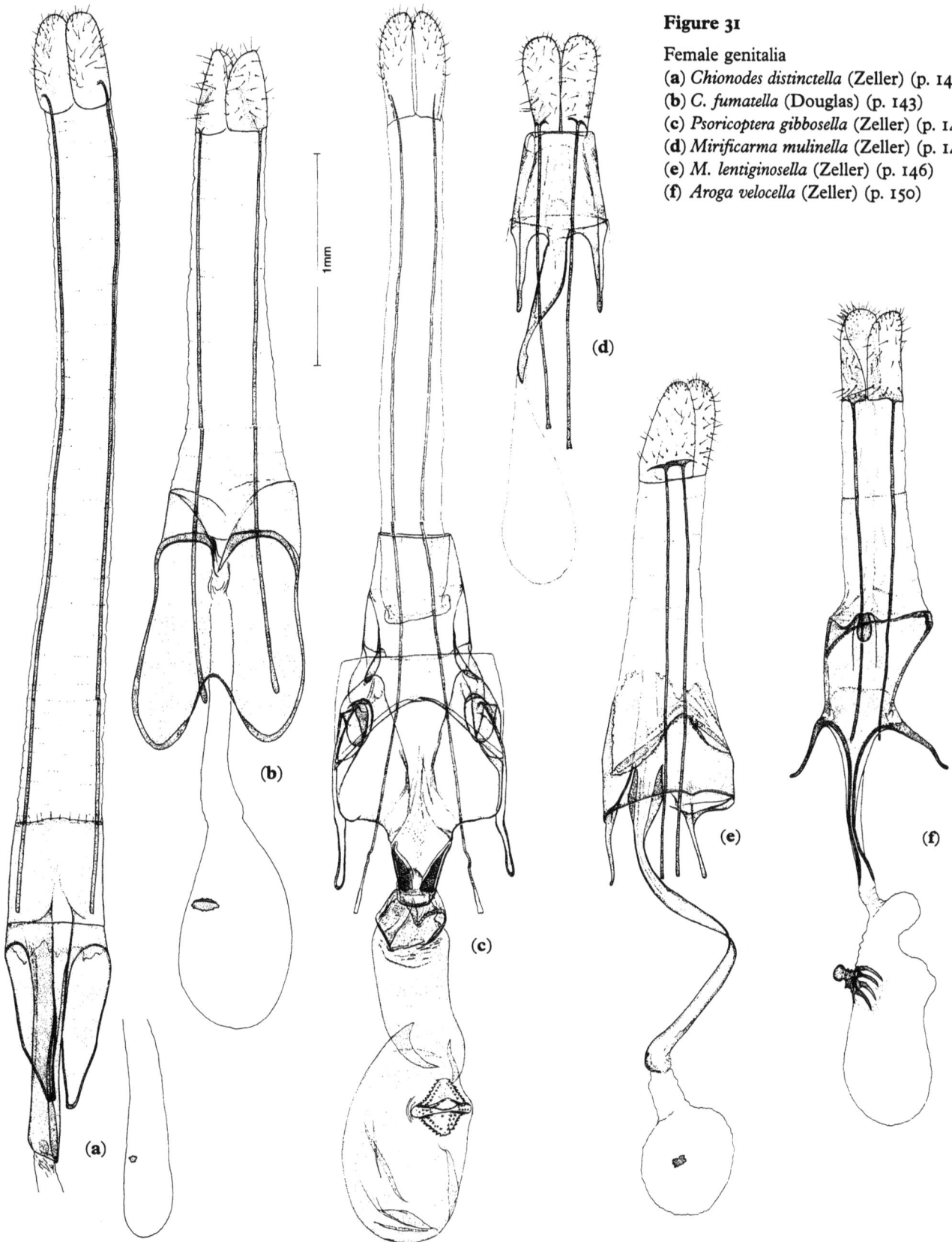

Figure 31

Female genitalia
(a) *Chionodes distinctella* (Zeller) (p. 143)
(b) *C. fumatella* (Douglas) (p. 143)
(c) *Psoricoptera gibbosella* (Zeller) (p. 144)
(d) *Mirificarma mulinella* (Zeller) (p. 145)
(e) *M. lentiginosella* (Zeller) (p. 146)
(f) *Aroga velocella* (Zeller) (p. 150)

Figure 32

Female genitalia
(a) *Prolita sexpunctella* (Fabricius) (p. 148)
(b) *P. solutella* (Zeller) (p. 148)
(c) *Neofriseria peliella* (Treitschke) (p. 151)
(d) *N. singula* (Staudinger) (p. 152)
(e) *Gelechia senticetella* (Staudinger) (p. 155)
(f) *G. scotinella* Herrich-Schäffer (p. 154)

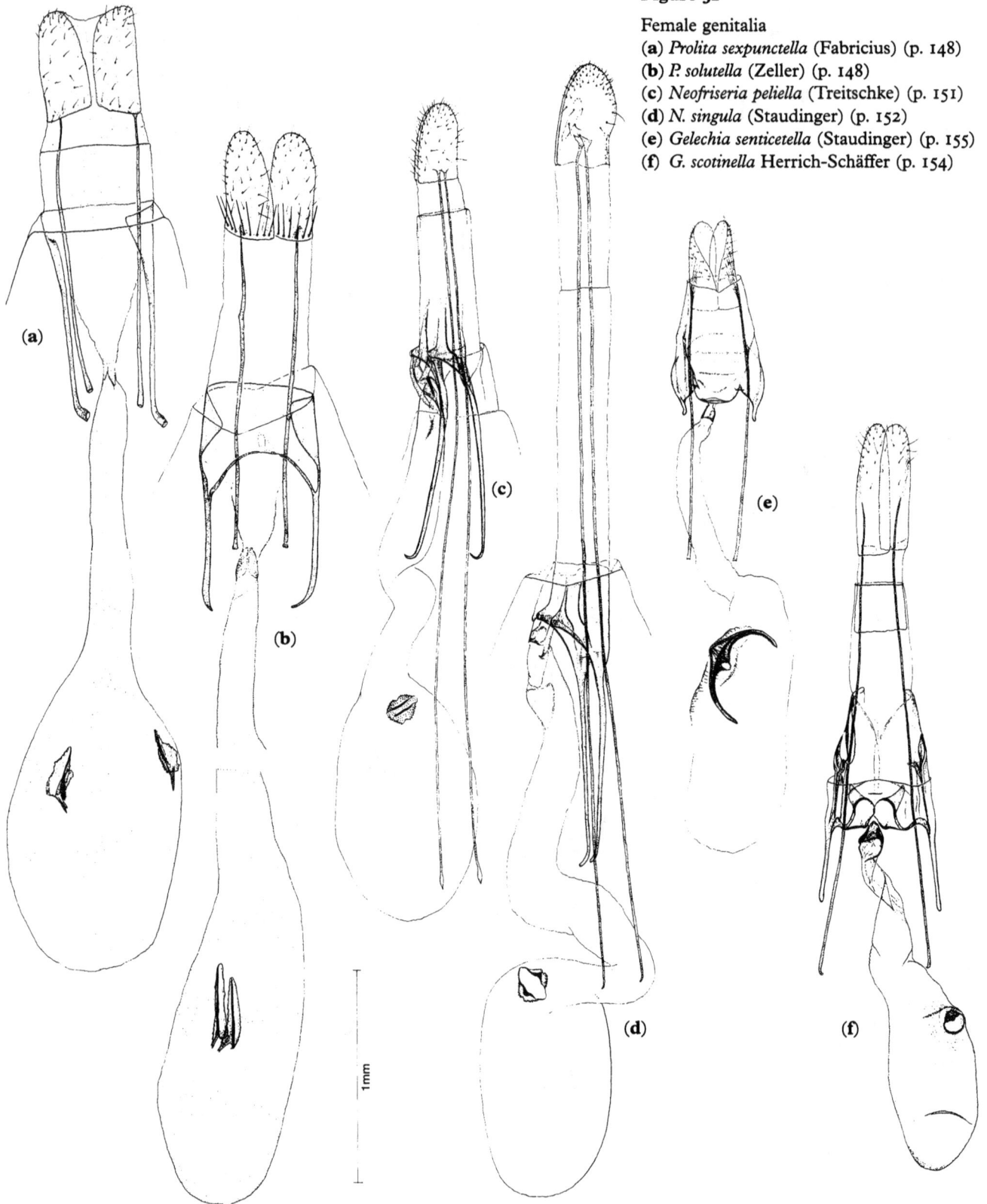

(a)

(b)

(c)

(d)

(e)

(f)

1mm

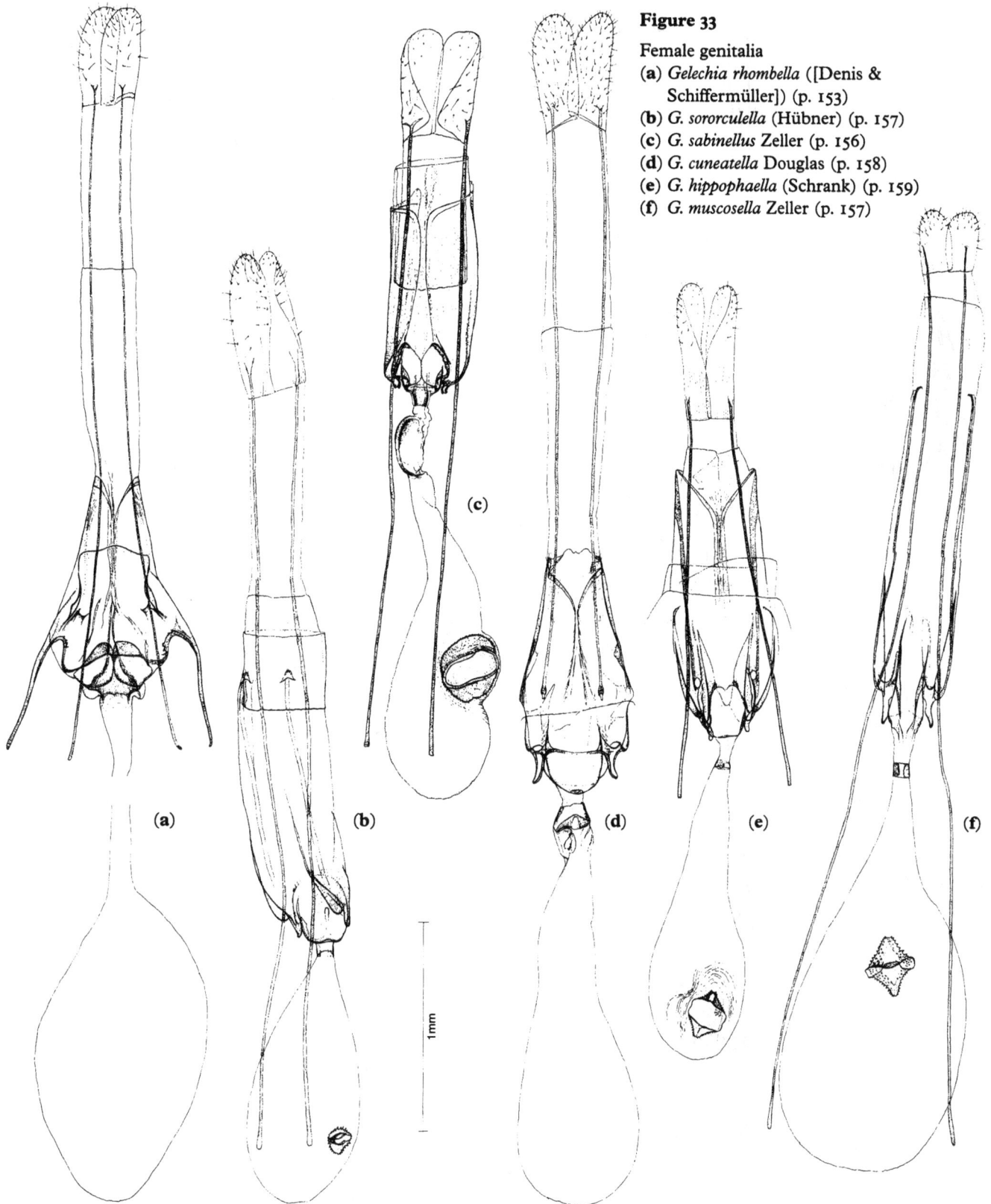

Figure 33

Female genitalia

(a) *Gelechia rhombella* ([Denis & Schiffermüller]) (p. 153)
(b) *G. sororculella* (Hübner) (p. 157)
(c) *G. sabinellus* Zeller (p. 156)
(d) *G. cuneatella* Douglas (p. 158)
(e) *G. hippophaella* (Schrank) (p. 159)
(f) *G. muscosella* Zeller (p. 157)

(a)

(b)

(c)

(d)

(e)

(f)

1 mm

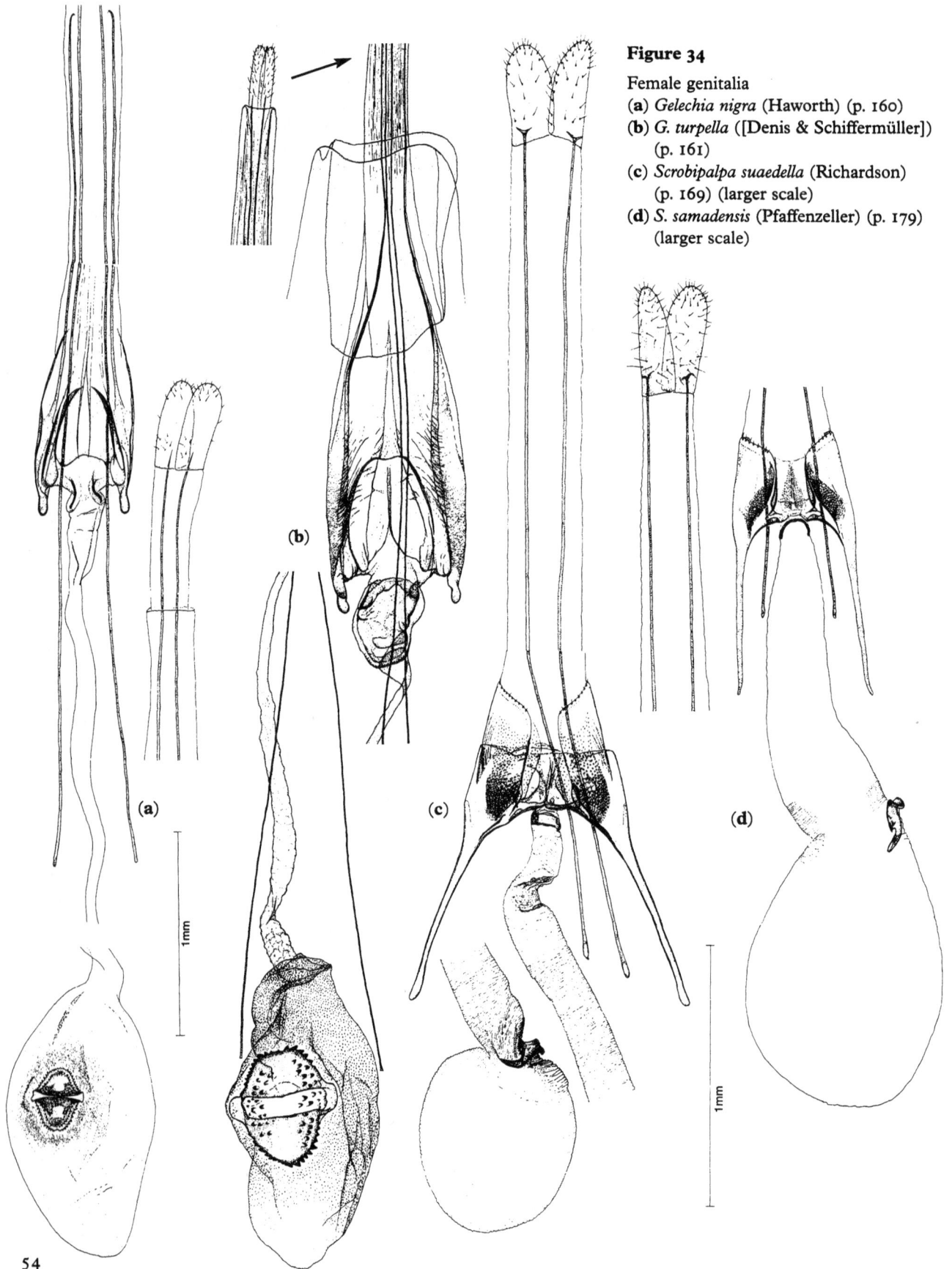

Figure 34

Female genitalia
(a) *Gelechia nigra* (Haworth) (p. 160)
(b) *G. turpella* ([Denis & Schiffermüller])
    (p. 161)
(c) *Scrobipalpa suaedella* (Richardson)
    (p. 169) (larger scale)
(d) *S. samadensis* (Pfaffenzeller) (p. 179)
    (larger scale)

(a)

(b)

(c)

(d)

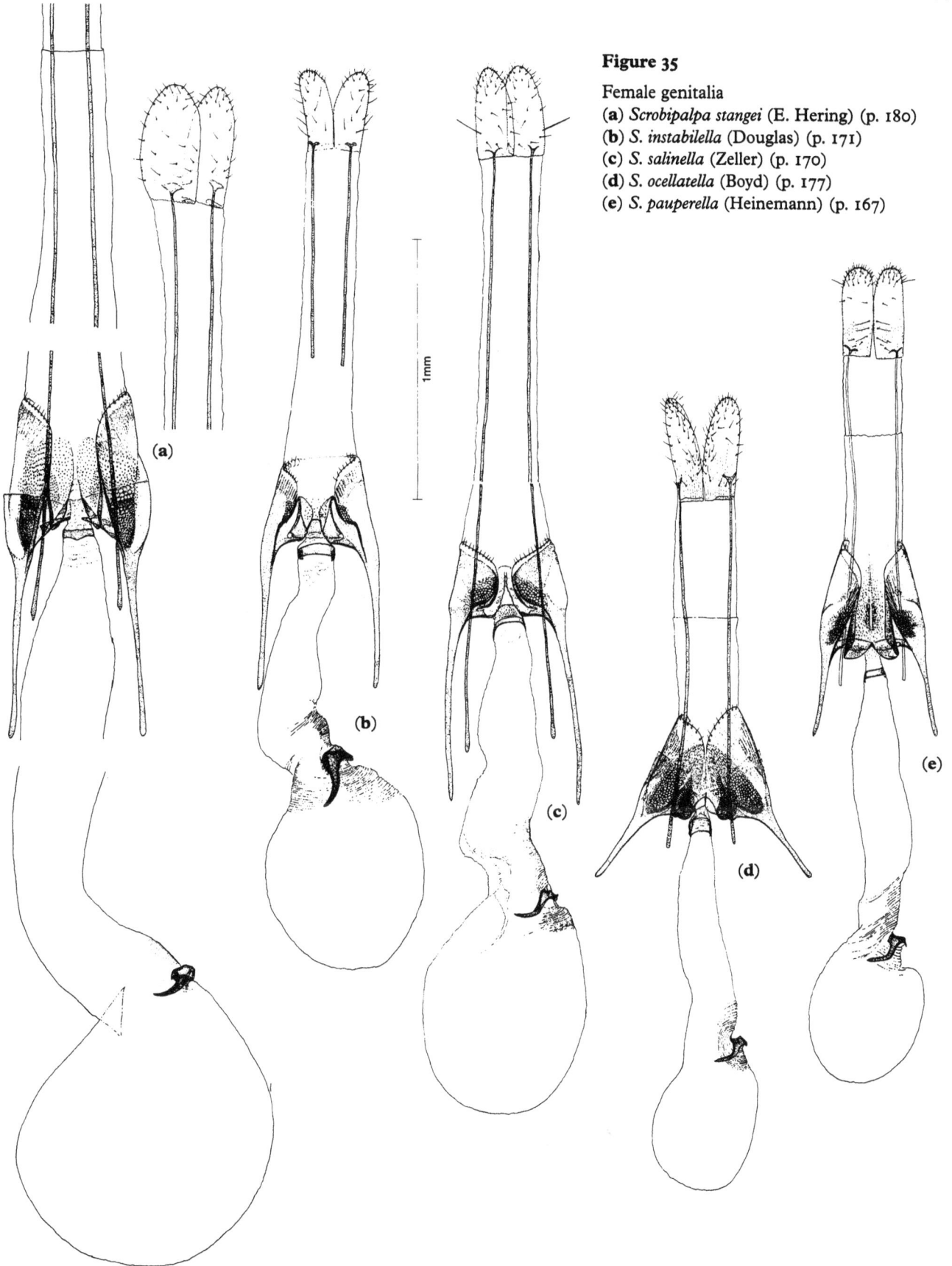

Figure 35

Female genitalia
(**a**) *Scrobipalpa stangei* (E. Hering) (p. 180)
(**b**) *S. instabilella* (Douglas) (p. 171)
(**c**) *S. salinella* (Zeller) (p. 170)
(**d**) *S. ocellatella* (Boyd) (p. 177)
(**e**) *S. pauperella* (Heinemann) (p. 167)

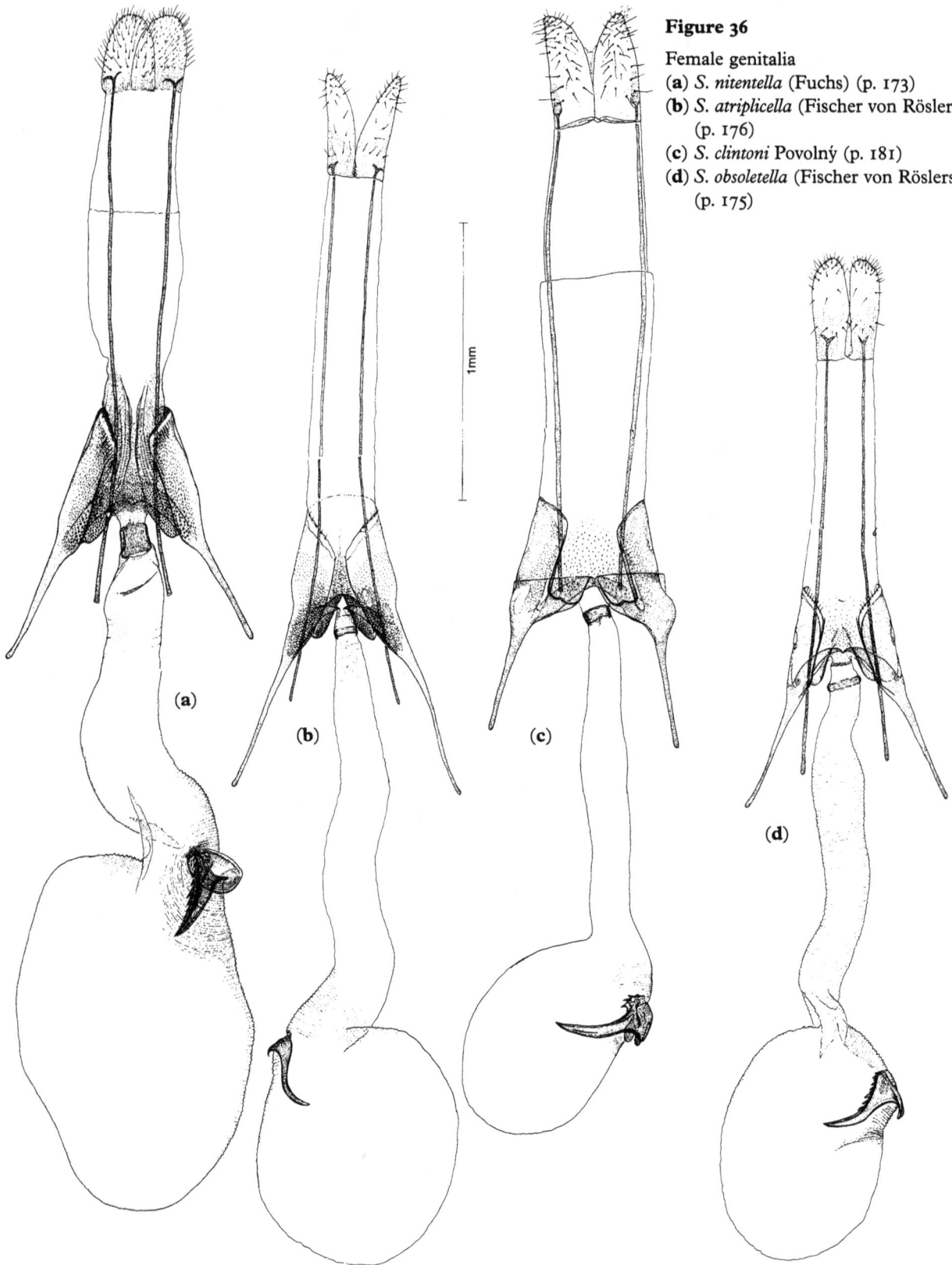

Figure 36

Female genitalia
(**a**) *S. nitentella* (Fuchs) (p. 173)
(**b**) *S. atriplicella* (Fischer von Röslerstamm)
(p. 176)
(**c**) *S. clintoni* Povolný (p. 181)
(**d**) *S. obsoletella* (Fischer von Röslerstamm)
(p. 175)

Figure 37

Female genitalia
(a) *Scrobipalpa costella* (Humphreys & Westwood) (p. 182)
(b) *S. murinella* (Duponchel) (p. 168)
(c) *S. artemisiella* (Treitschke) (p. 179)
(d) *S. acuminatella* (Sircom) (p. 166)
(e) *Scrobipalpula tussilaginis* (Stainton) (p. 186)

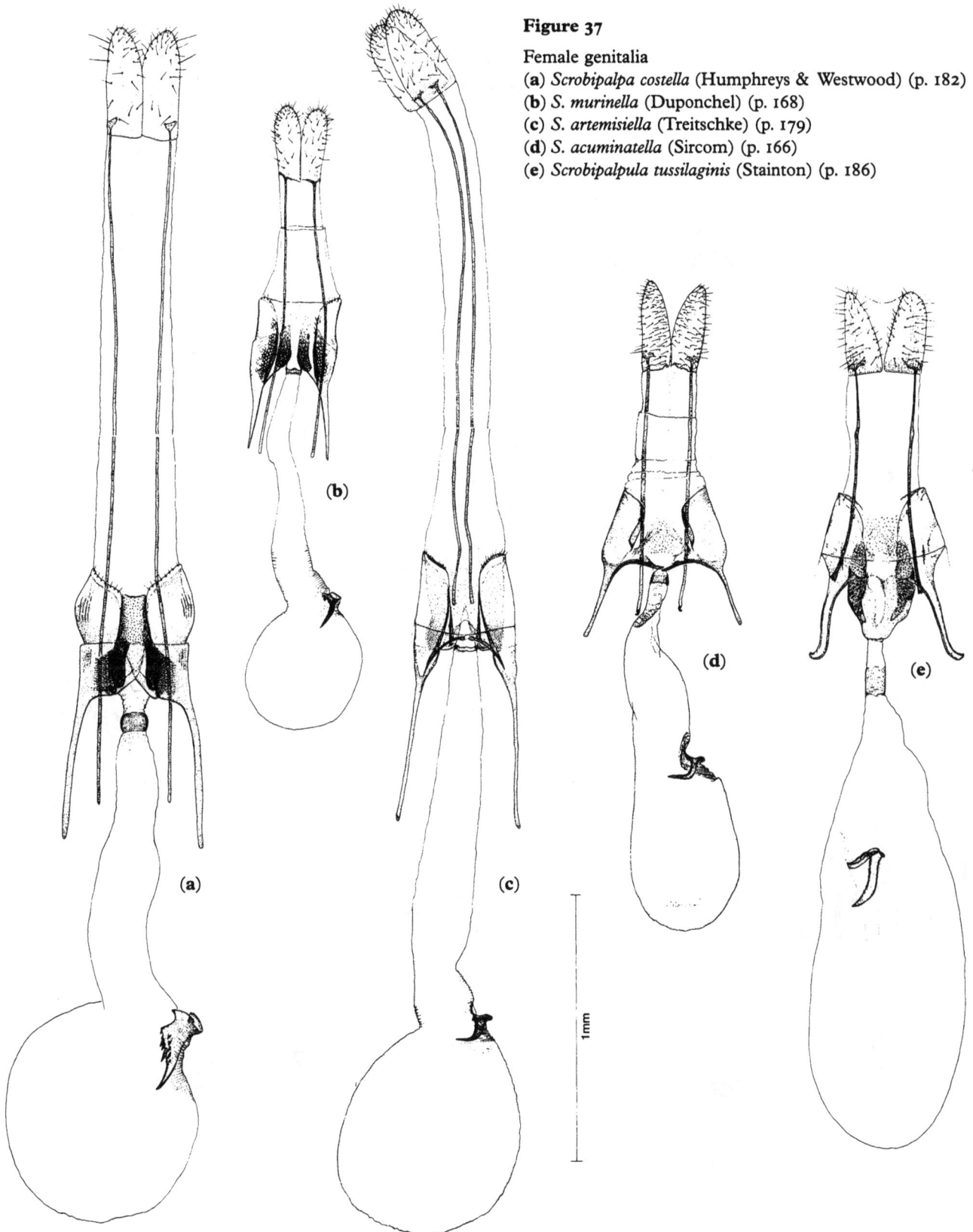

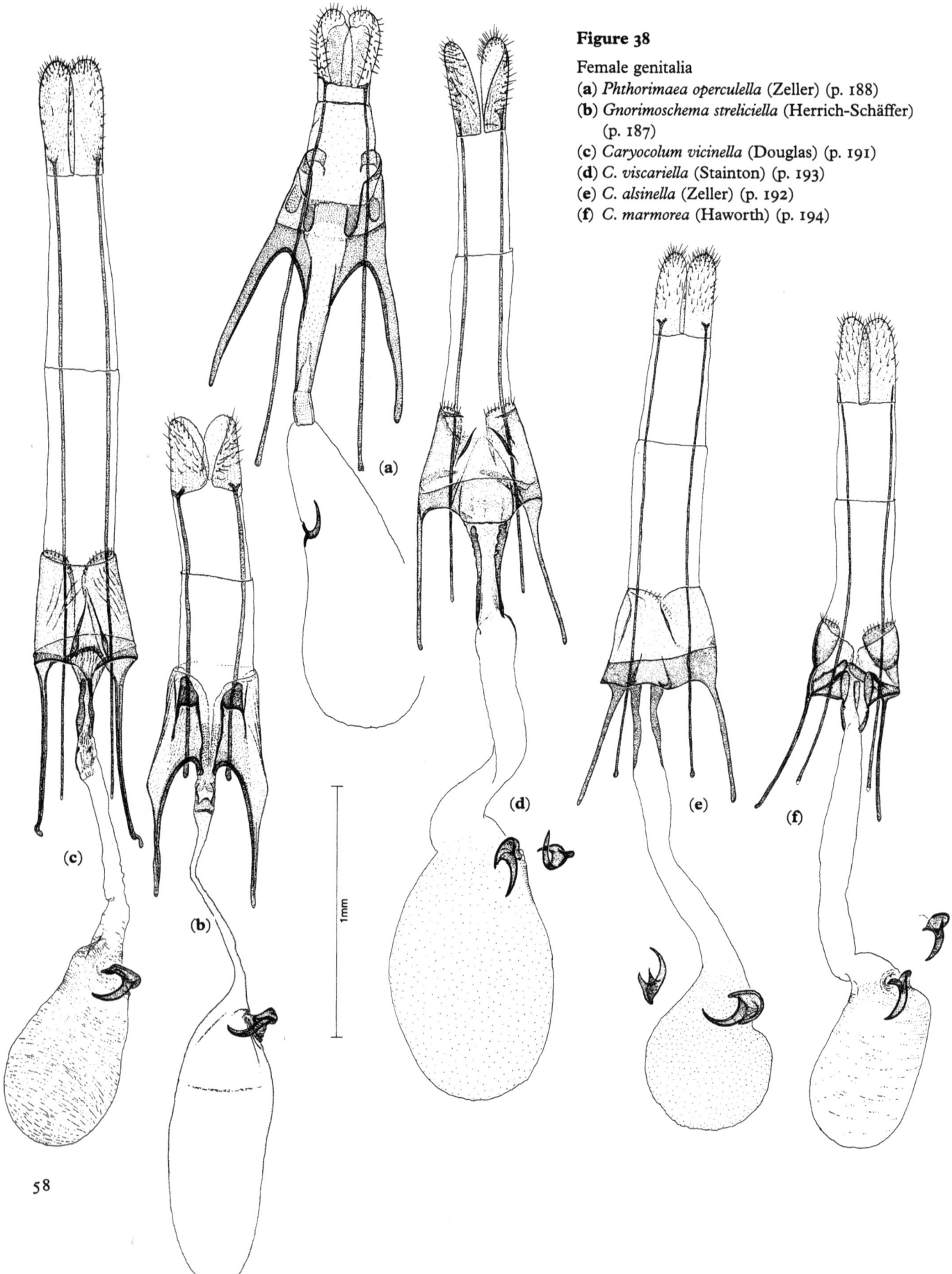

Figure 38

Female genitalia
(**a**) *Phthorimaea operculella* (Zeller) (p. 188)
(**b**) *Gnorimoschema streliciella* (Herrich-Schäffer)
    (p. 187)
(**c**) *Caryocolum vicinella* (Douglas) (p. 191)
(**d**) *C. viscariella* (Stainton) (p. 193)
(**e**) *C. alsinella* (Zeller) (p. 192)
(**f**) *C. marmorea* (Haworth) (p. 194)

Figure 39

Female genitalia
(**a**) *Caryocolum fraternella* (Douglas) (p. 195)
(**b**) *C. fraternella* (Douglas) (p. 195), variant
(**c**) *C. proxima* (Haworth) (p. 196)
(**d**) *C. blandelloides* Karsholt (p. 198)
(**e**) *C. blandella* (Douglas) (p. 197)
(**f**) *C. tricolorella* (Haworth) (p. 200)

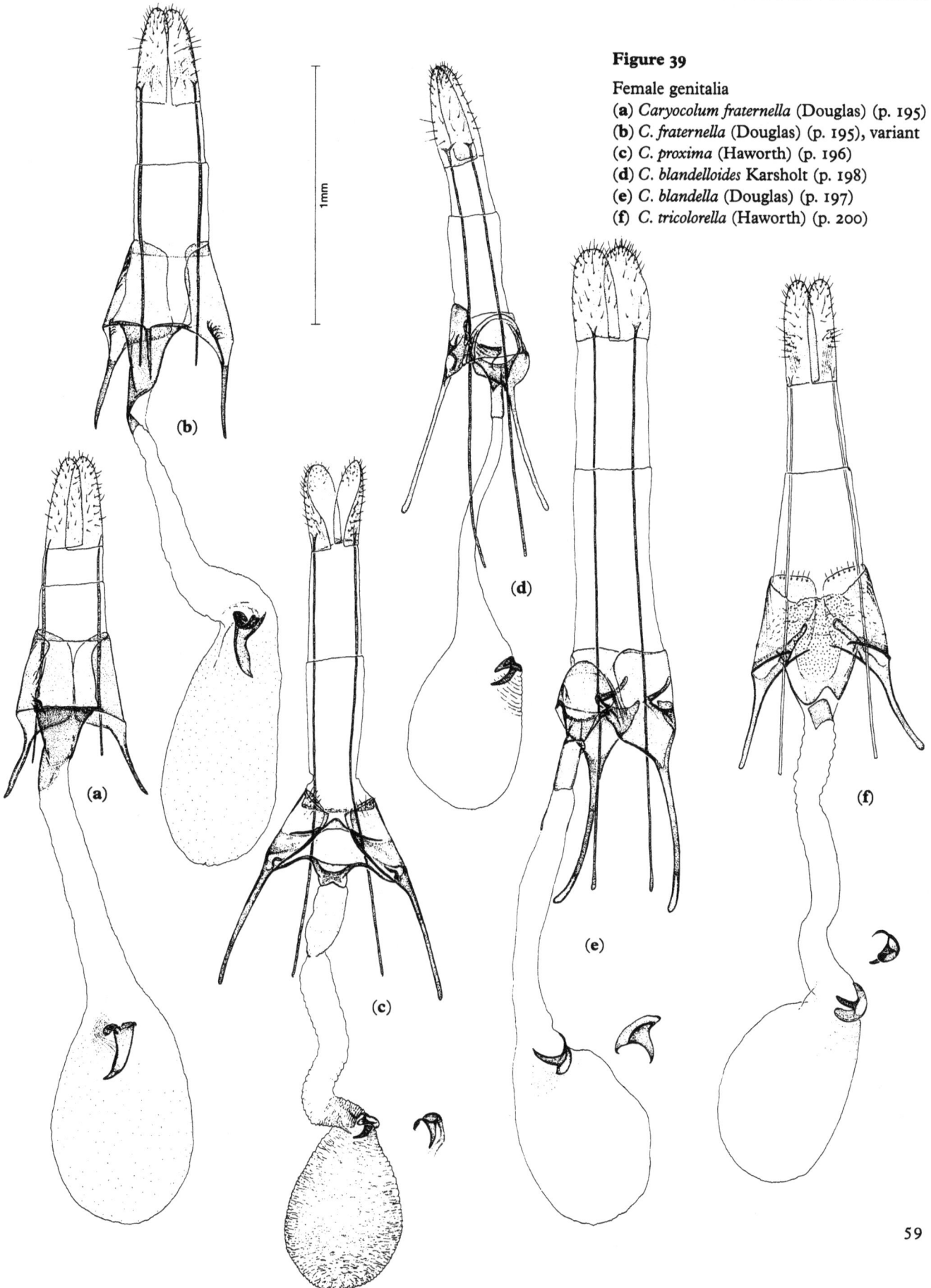

Figure 40

Female genitalia
(**a**) *Caryocolum kroesmanniella*
  (Herrich-Schäffer) (p. 202)
(**b**) *C. huebneri* (Haworth) (p. 203)
(**c**) *C. blandulella* (Tutt) (p. 201)
(**d**) *C. junctella* (Douglas) (p. 199)
(**e**) *Aproaerema anthyllidella* (Hübner) (p. 206)
(**f**) *Sophronia humerella* ([Denis & Schiffermüller])
  (p. 205) (smaller scale)
(**g**) *S. semicostella* (Hübner) (p. 204)
  (smaller scale)

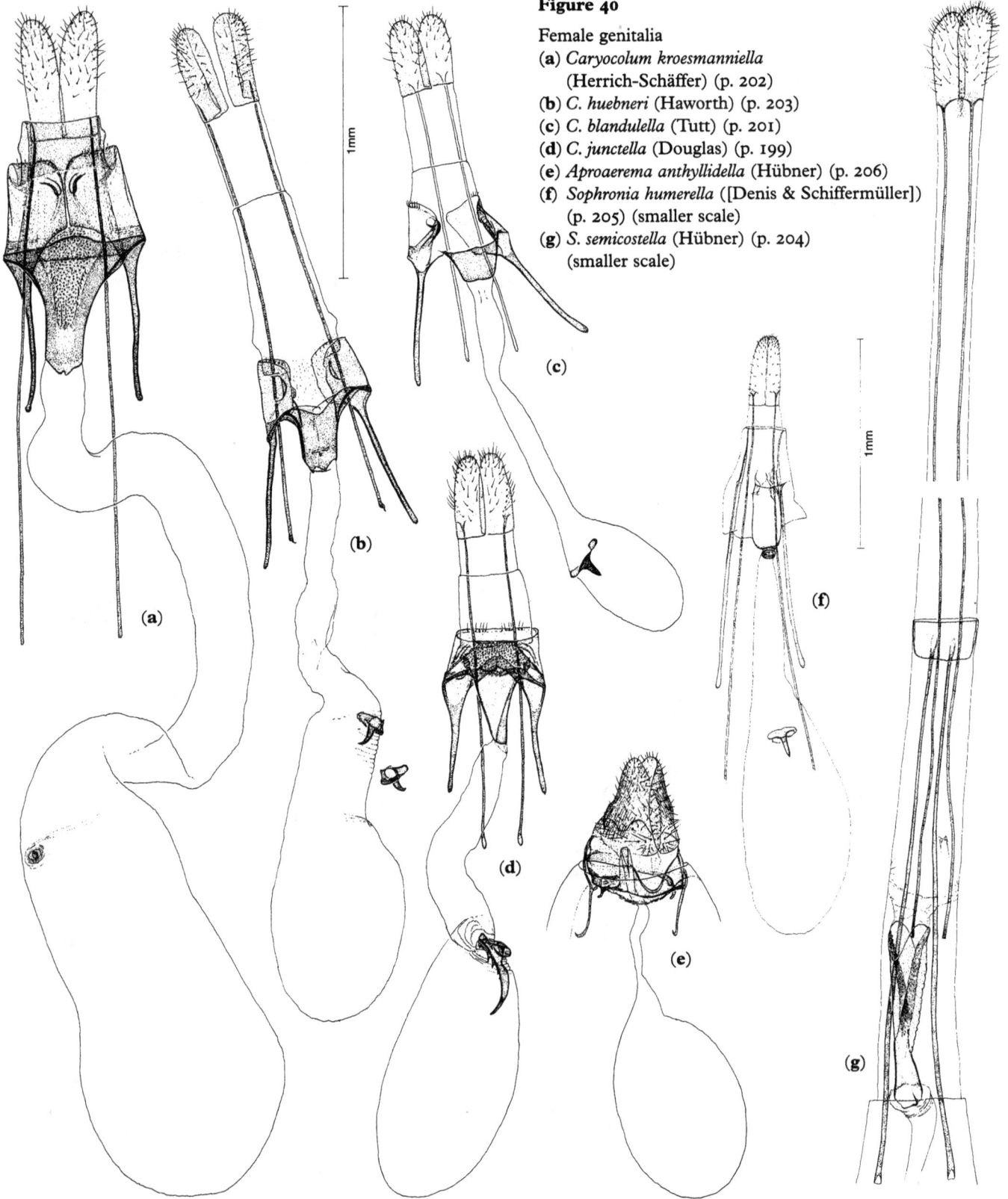

Figure 41

Female genitalia
(**a**) *Syncopacma sangiella* (Stainton) (p. 208)
(**b**) *S. larseniella* (Gozmány) (p. 209)
(**c**) *S. cinctella* (Clerck) (p. 210)
(**d**) *S. albipalpella* (Herrich-Schäffer) (p. 213)
(**e**) *S. polychromella* (Rebel) (p. 215)
(**f**) *S. vinella* (Bankes) (p. 212)
(**g**) *S. taeniolella* (Zeller) (p. 211)
(**h**) *S. suecicella* (Wolff) (p. 215)

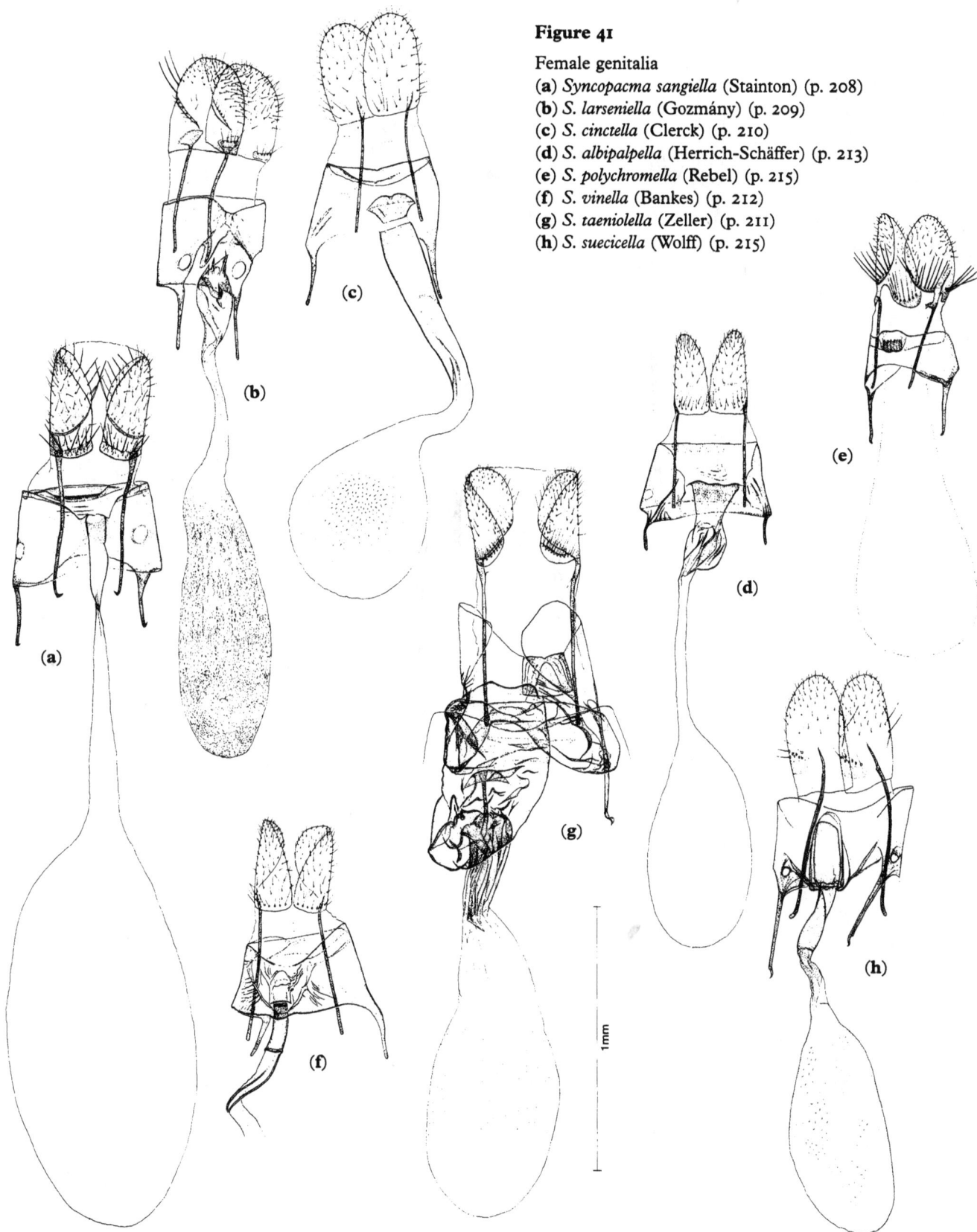

Figure 42

Female genitalia
(a) *Anacampsis temerella* (Lienig & Zeller) (p. 216)
(b) *A. populella* (Clerck) (p. 217)
(c) *A. blattariella* (Hübner) (p. 218)
(d) *Mesophleps silacella* (Hübner) (p. 219)
(e) *Neofaculta ericetella* (Geyer) (p. 220)
(f) *Anarsia spartiella* (Schrank) (p. 224) (larger scale)
(g) *A. lineatella* Zeller (p. 225)

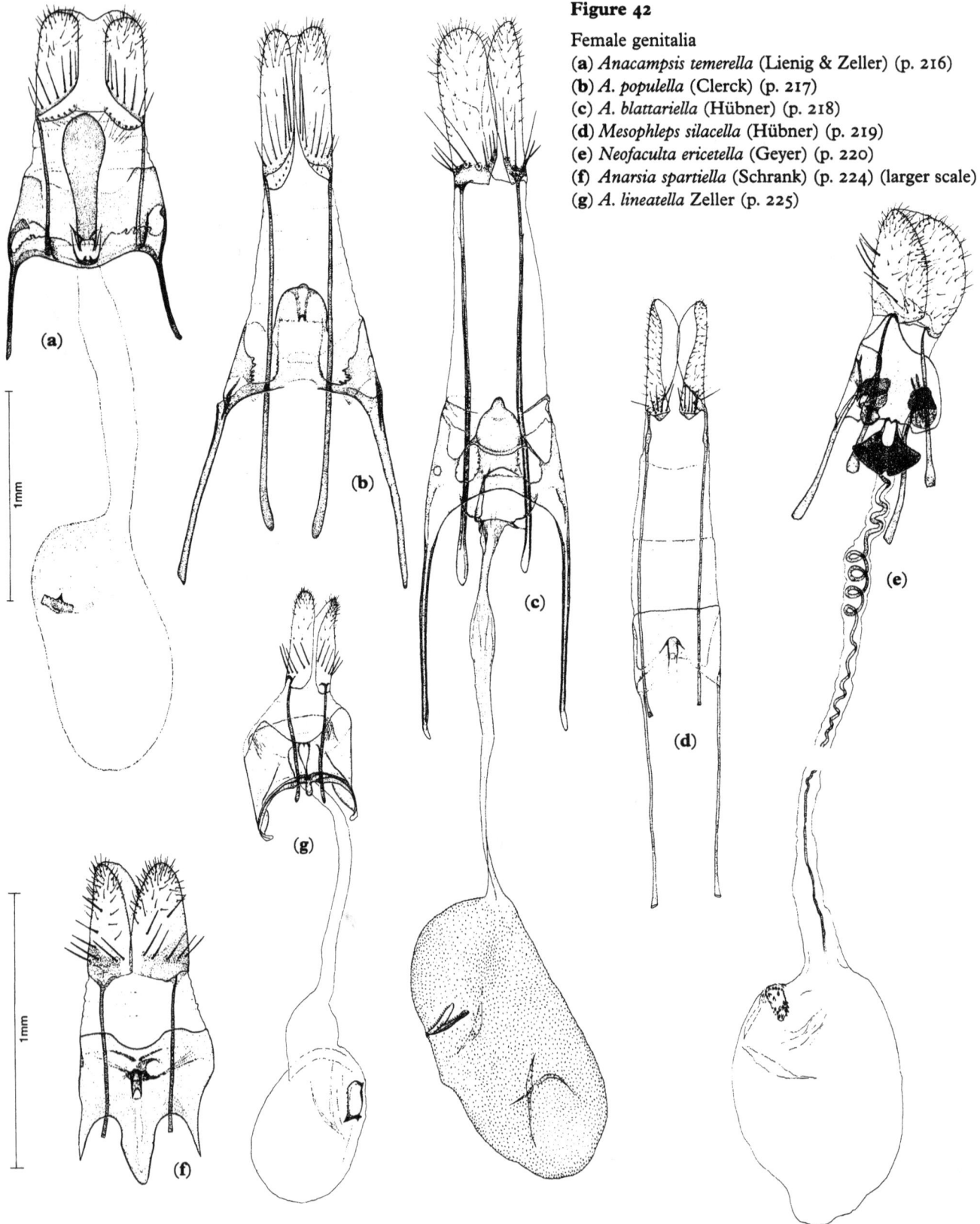

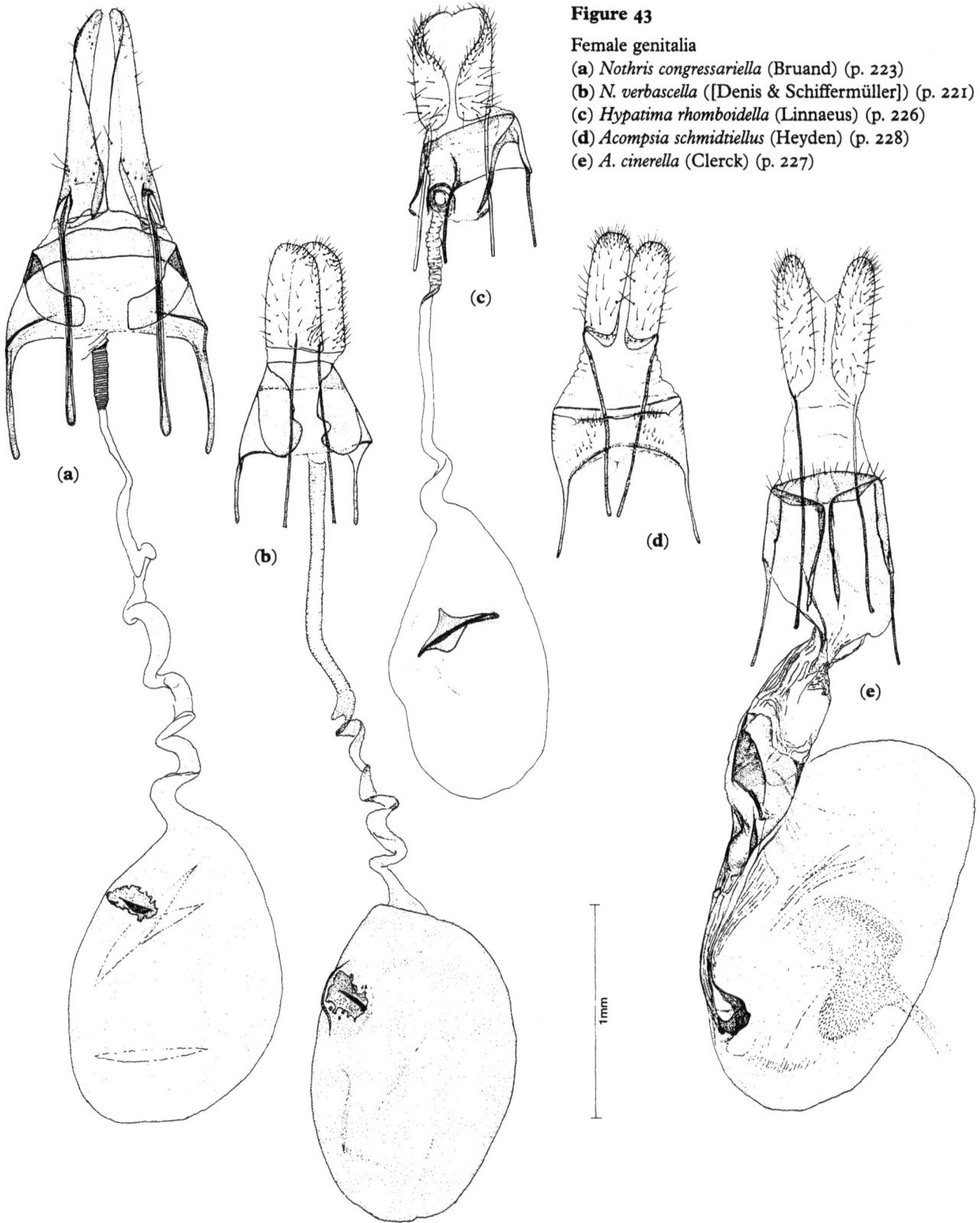

Figure 43

Female genitalia
(a) *Nothris congressariella* (Bruand) (p. 223)
(b) *N. verbascella* ([Denis & Schiffermüller]) (p. 221)
(c) *Hypatima rhomboidella* (Linnaeus) (p. 226)
(d) *Acompsia schmidtiellus* (Heyden) (p. 228)
(e) *A. cinerella* (Clerck) (p. 227)

Figure 44

Female genitalia
(a) *Dichomeris alacella* (Zeller) (p. 233)
(b) *D. juniperella* (Linnaeus) (p. 231)
(c) *D. ustalella* (Fabricius) (p. 231)
(d) *D. derasella* ([Denis & Schiffermüller]) (p. 232)
(e) *D. marginella* (Fabricius) (p. 229)

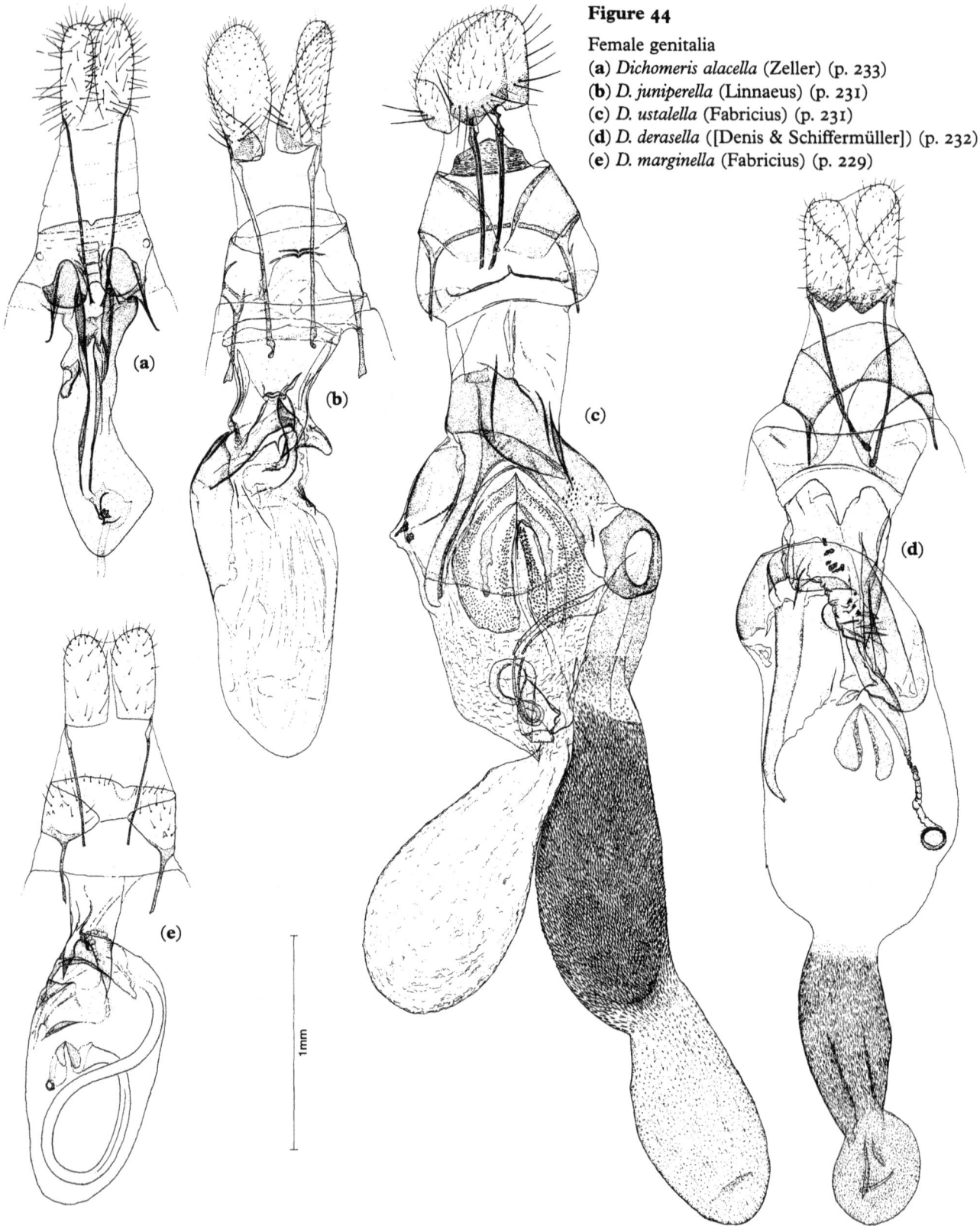

1mm

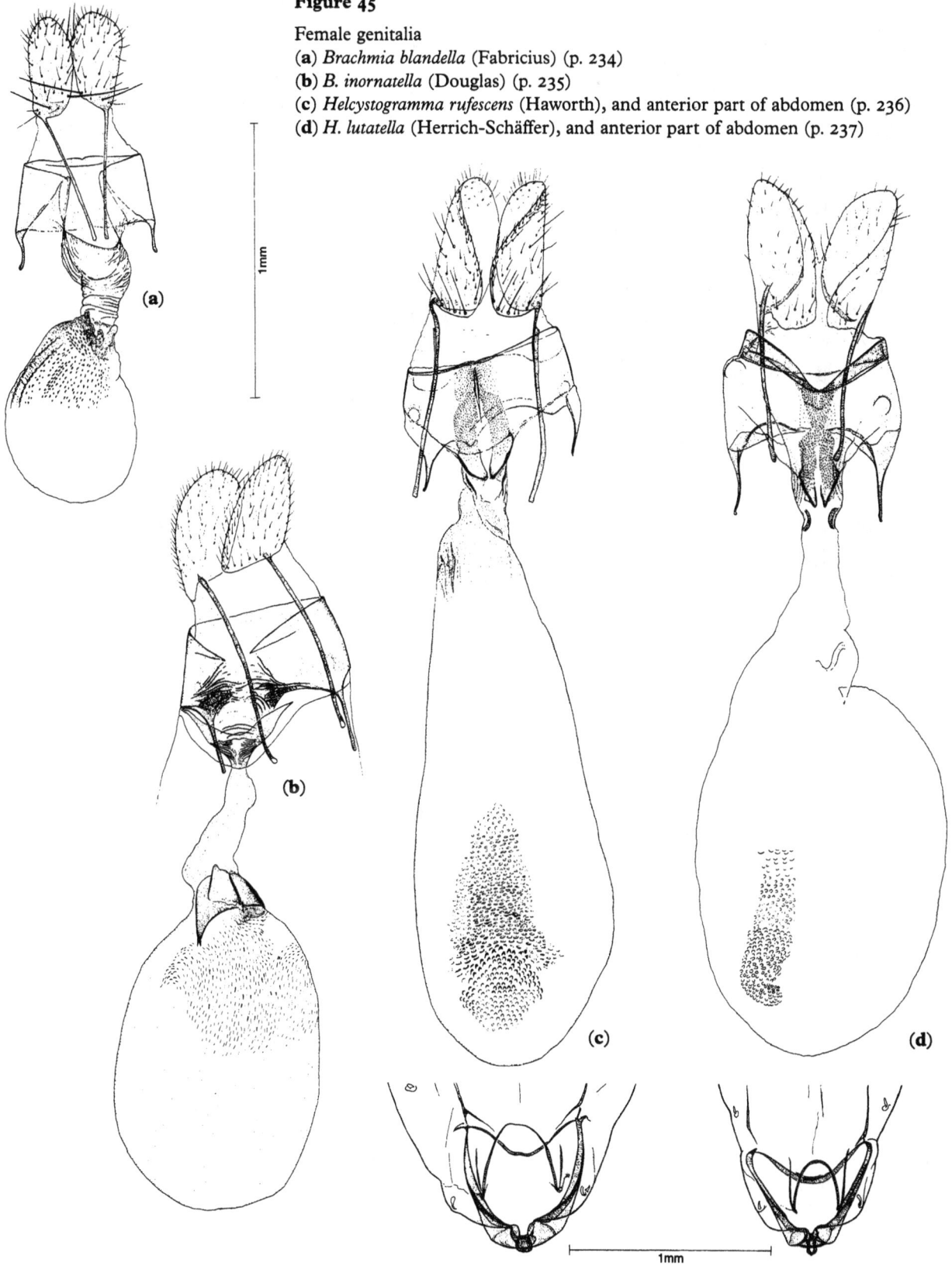

Figure 45

Female genitalia
(**a**) *Brachmia blandella* (Fabricius) (p. 234)
(**b**) *B. inornatella* (Douglas) (p. 235)
(**c**) *Helcystogramma rufescens* (Haworth), and anterior part of abdomen (p. 236)
(**d**) *H. lutatella* (Herrich-Schäffer), and anterior part of abdomen (p. 237)

65

Figure 46

Female genitalia
(**a**) *Pexicopia malvella* (Hübner) (p. 240)
(**b**) *Platyedra subcinerea* (Haworth) (p. 239)
(**c**) *Sitotroga cerealella* (Olivier) (p. 241)
(**d**) *Thiotricha subocellea* (Stephens) (p. 242)

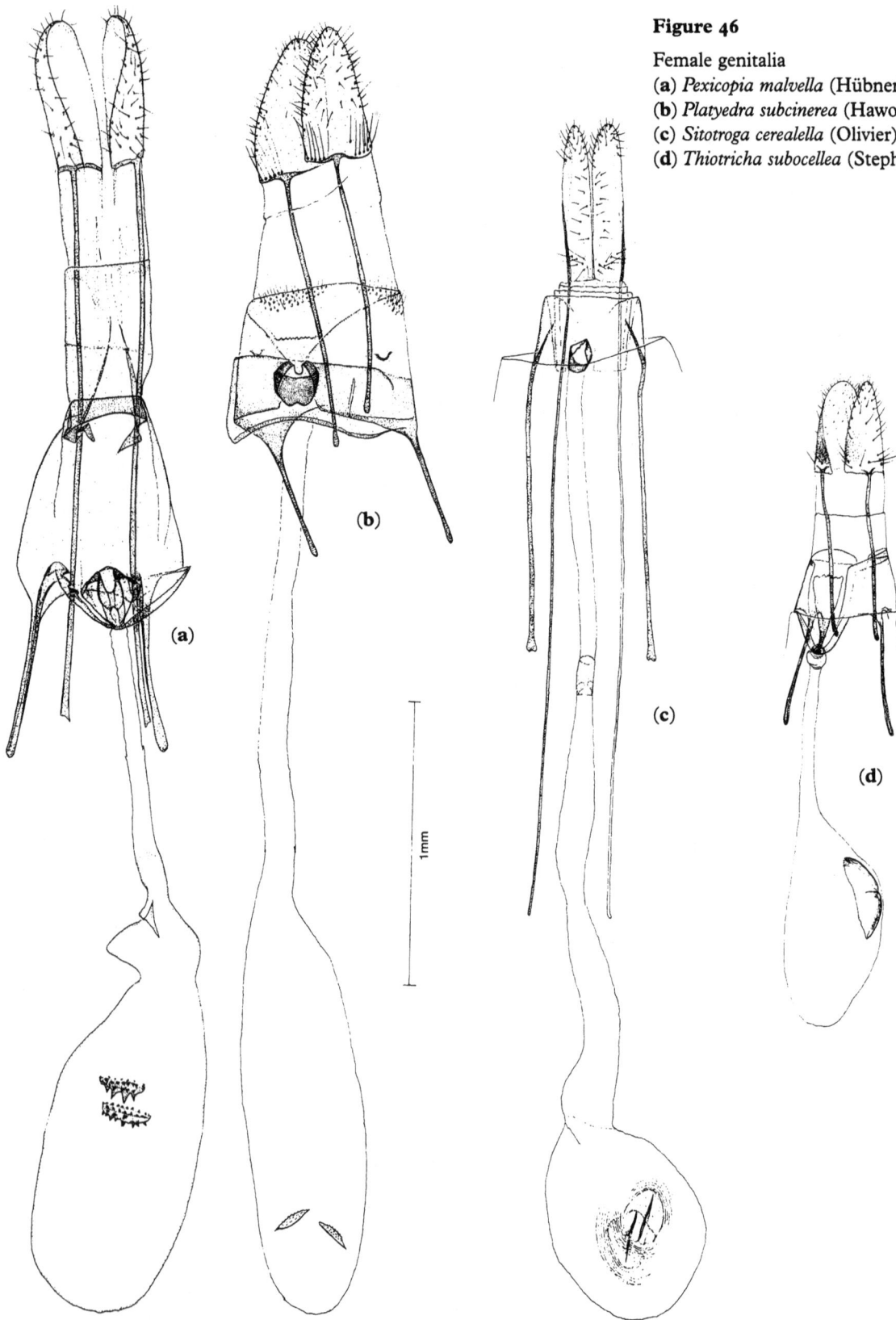

(a)

(b)

(c)

(d)

1mm

Anomologinae

METZNERIA Zeller

Metzneria Zeller, 1839, *Isis, Leipzig* **1839**: 197 (as sub-genus of *Gelechia* Hübner, [1825]).

Cleodora Stephens, 1834, *Ill.Br.Ent.* (Haust.) **4**: 220, *nec* Péron & Lesueur, 1810.

Parasia Duponchel, [1846], *Cat.Méthod.Lépid.Eur.*: 350.

A genus of about 40 species distributed over the western Palaearctic region; one, *M. lappella* (Linnaeus), has been accidentally introduced into North America. There are 22 species in Europe, of which six occur in Britain and two also in Ireland.

Imago. Antenna of male weakly serrate in apical half; labial palpus long, segment 3 half length of segment 2, segment 2 densely, segment 3 usually more sparsely, clad in appressed scales. Forewing lanceolate with acute apex; fully veined, veins R_5 (7) and R_4 (8) out of M_1 (6), R_5 to costa (figure 47). Hindwing trapezoidal,

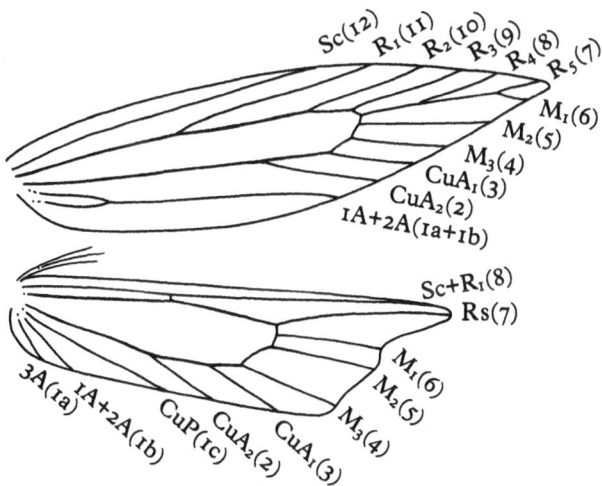

Figure 47 *Metzneria neuropterella* (Zeller), wing venation

dorsum straight, tornus bluntly angled, termen slightly sinuate, apex strongly produced; fully veined; cilia one and one-half to twice width of wing. Male genitalia with uncus reduced, gnathos absent; valva beaked, sacculus flap-like; aedeagus rather stout, with cornuti. Female genitalia with ovipositor extensile, ostium ill-defined, ductus and corpus bursae membranous, without signum.

The adults are nocturnal and come to light; otherwise they are rarely seen.

Ovum. Undescribed. Probably laid on the flower-head in which the larva will feed, since the latter is ill-equipped for walking.

Larva. Limaciform; head small; prothoracic plate present, sometimes pigmented; anal plate small, also sometimes pigmented; thoracic legs minute, abdominal prolegs vestigial or absent.

All but one of the British species feed on the seeds of Asteraceae (Compositae), the exception being *M. littorella* (Douglas), *q.v.* The whole larval stage of the Asteraceae-feeders is spent in a single seedhead, the larva feeding from late summer to early autumn and then overwintering in a cocoon or spun chamber in the feeding place until about April.

Pupa. In the cocoon or chamber in the feeding place.

The species of this genus are best obtained by collecting seedheads in late winter or early spring and keeping them out of doors until emergence is due.

Key to species (imagines) of the genus *Metzneria*

1	Forewing ground colour greyish white .... *littorella* (p. 68)	
–	Forewing ground colour ochreous, reddish or yellowish **2**	
2(1)	Labial palpus segment 3 thickened with appressed scales **3**	
–	Labial palpus segment 3 not so thickened .... *aestivella* (p. 69)	
3(2)	Forewing with first and second discal and plical stigmata present **4**	
–	Forewing without stigmata **5**	
4(3)	Forewing ochreous, mottled .... *lappella* (p. 68)	
–	Forewing yellowish, irrorate fuscous .... *metzneriella* (p. 70)	
5(3)	Forewing suffused orange-vermilion ....*aprilella* (p. 72)	
–	Forewing without orange-vermilion suffusion .... *neuropterella* (p. 71)	

METZNERIA LITTORELLA (Douglas)

Gelechia littorella Douglas, 1850, *Trans.ent.Soc.Lond.* (N.S.) **1**: 67.

Gelechia quinquepunctella Herrich-Schäffer, 1854, *Syst. Bearb.Schmett.Eur.* **5**: 172.

Type locality: England; Isle of Wight.

Description of imago (Pl.1, fig.1)

Wingspan 9–14mm. Head pale grey; antenna two-thirds length of forewing, grey, banded fuscous; labial palpus with segment 3 shorter than segment 2, segment 2 with appressed grey scales, segment 3 fuscous. Thorax concolorous with head. Forewing pale grey, irrorate fuscous; a rather obscure pale yellow longitudinal streak on fold and another, sometimes obsolete, in disc; plical and first and second discal stigmata black, first discal beyond plical; diffuse and often obscure blackish subcostal spot near base, and another more distinct spot on fold halfway between base and plical stigma; apical area with heavier fuscous irroration; cilia concolorous, with ill-defined darker ciliary line and obscure dark bars formed by extension of apical irroration. Hindwing concolorous with forewing; cilia pale grey with whitish sheen, one and one-half times width of wing. Foreleg pale fuscous, tibia pale grey; mid- and hindlegs pale grey. Abdomen pale grey; female in set specimens with yellow ovipositor protruding. Genitalia, see figures 1f,21d.

Life history

Ovum. Undescribed. Laid on buck's-horn plantain (*Plantago coronopus*), probably on the flower-spike.

Larva. Head dark brown. Body short and stout, sordid white; prothoracic plate dark brown with wide median sulcus; anal plate pale grey; thoracic legs and abdominal prolegs absent.

The larva chews an elongate recess in the stem, in which it rests head-upwards when not feeding. It eats the ripening seeds from within, without causing visible sign of its presence. September–March, overwintering fully fed in a brown silken cocoon spun on the surface of the stem covered by the seeds.

Pupa. Undescribed. April–May.

Imago. Probably flies at night, but is easily disturbed from its foodplant during the day. May–June.

Distribution (Map 1)

Occurs on sparsely vegetated, open ground close to the sea. Known in Britain only from the south coast of the Isle of Wight, where it has been found quite commonly for the last 150 years, apparently without ever having extended its range. Mainly Mediterranean, occurring

Metzneria littorella

in Spain, France, Italy and the western Mediterranean islands; south-western Russia; North Africa.

METZNERIA LAPPELLA (Linnaeus)

Phalaena (Tinea) lappella Linnaeus, 1758, *Syst.Nat.* (Edn 10) **1**: 537.

Recurvaria silacea sensu Haworth, 1828, *Lepid.Br.*: 555, *nec silacella* Hübner, 1796.

Cleodora silacella sensu Stephens, 1834, *Ill.Br.Ent.* (Haust.) **4**: 220, *nec* Hübner, 1796.

Type locality: not stated.

Description of imago (Pl.1, fig.2)

Wingspan 16–20mm. Head glossy creamy white; antenna with scape long, ochreous-fuscous, flagellum pale ochreous, annulate fuscous; labial palpus with segment 3 shorter than segment 2, segment 2 with long appressed scales, ochreous-brown on outer side, whitish on inner side, segment 3 with short, closely appressed scales, ochreous-brown. Thorax brown; tegulae creamy white. Forewing ochreous, variably mottled brown, especially in basal half and subcostally; veins greyish fuscous; stigmata black, first discal often obsolete; cilia pale ochreous with two darker ciliary

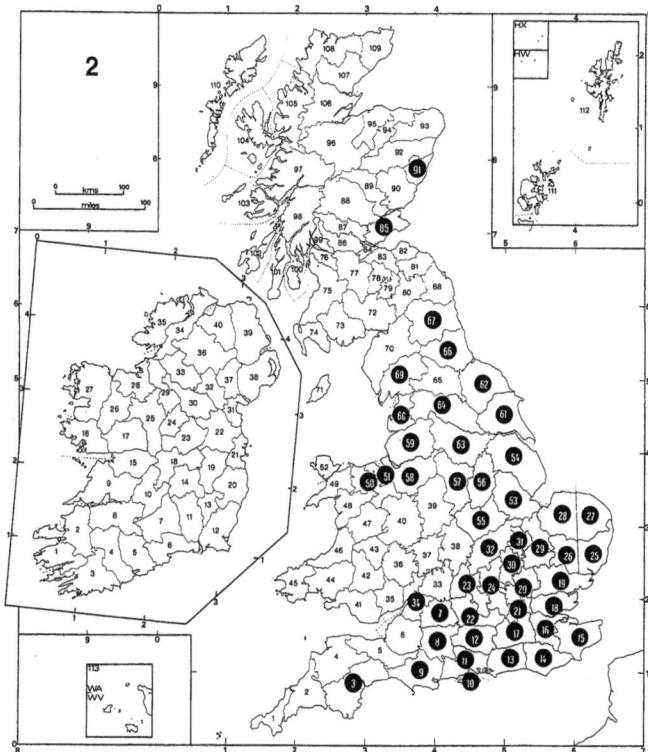

Metzneria lappella

lines. Hindwing brownish grey, cell and area between veins Cu₁ (3) and Cu₂ (2) paler; cilia whitish fuscous. Foreleg brown, tarsus whitish ochreous; mid- and hindlegs whitish ochreous each with darker streak on outer side. Abdomen brown, ventral surface of posterior segments whitish ochreous. Genitalia, see figures 1a,21a.

Life history

Ovum. Undescribed. Laid on greater burdock (*Arctium lappa*) or lesser burdock (*A. minus*), probably on a developing flower-head.

Larva. Head pale brown, sutures, mouth-parts and eye-spots dark brown. Body white, limaciform; prothoracic plate translucent yellowish white; anal plate and thoracic legs concolorous with body; thoracic legs and anal prolegs vestigial, abdominal prolegs obsolete.

The larva feeds internally on the seeds, without visible evidence of its presence, in September and October and then overwinters until April in a cocoon formed, in the example studied, within the seedhead in two hollowed-out seeds spun together.

Pupa. In the overwintering cocoon. May.

Imago. Univoltine; June–July. Nocturnal, occasionally coming to light.

Distribution (Map 2)

Occurs on waste ground, in open woodland and on downland. Widespread and fairly common in England northwards to Northumberland, but less frequent in the west; North Wales; very local in Scotland and restricted to the eastern counties; not recorded from Ireland. Europe; western Asia; North America, where it is an accidental introduction.

METZNERIA AESTIVELLA (Zeller)

Gelechia (Metzneria) aestivella Zeller, 1839, *Isis, Leipzig* **1839**: 202.
Gelechia carlinella Stainton, 1851, *Suppl.Cat.Br.Tineidae & Pterophoridae*: 5.

Type locality: Germany; Glogau (now Poland; Glogów).

Description of imago (Pl.1, fig.3)

Wingspan 13–16mm. Head ochreous-white; antenna whitish ochreous, annulate greyish fuscous; labial palpus segment 2 reddish ochreous with dense appressed scales, segment 3 whitish ochreous with few ochreous appressed scales. Thorax ochreous-brown. Forewing yellowish ochreous, suffused reddish brown; stigmata absent; variable irroration of whitish scales, especially on costa and in subcostal region; veins edged grey; ground colour without irroration forming streak on fold and streak along dorsum angled at tornus to form fascia not reaching costa; cilia pale ochreous-brown with darker ciliary line. Hindwing greyish brown; cilia concolorous, but with paler reflections. Foreleg ochreous-brown, tarsus whitish ochreous; mid- and hindlegs whitish ochreous. Abdomen ochreous-brown, paler between segments; anal tuft of male whitish ochreous. Genitalia, see figures 1c,21b.

Life history

Ovum. Undescribed. Laid on carline thistle (*Carlina vulgaris*), probably on a developing flower-head.

Larva. Head mottled pale and dark brown. Body white with pale yellowish tinge; prothoracic plate whitish with sparse irregular brown specks and mottling; anal plate concolorous with body, with scattered dark brown speckles; thoracic legs vestigial, whitish, mottled pale brown.

The larva feeds on the seeds, sometimes spinning parts of the pappus together and giving the head a rather untidy appearance, but often without external evidence of its presence. It feeds in September and October and then overwinters in a loosely spun, silk-lined chamber beneath the seeds.

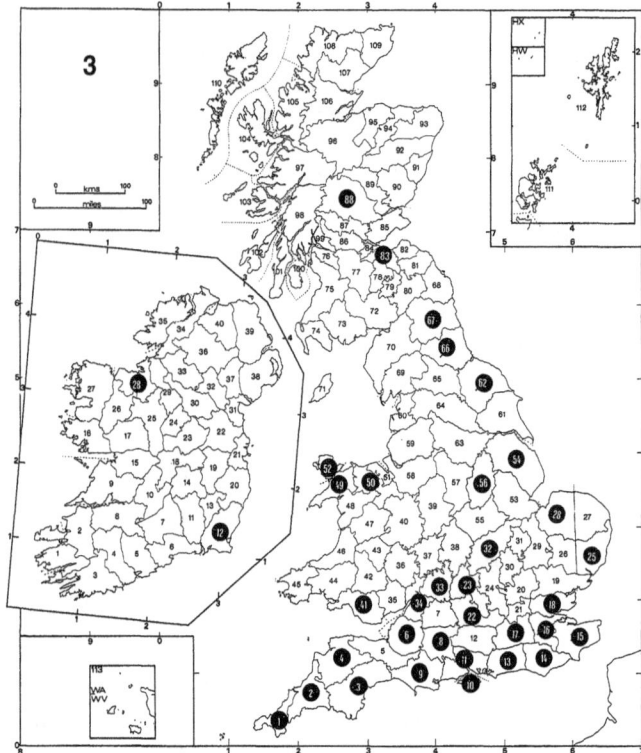

Metzneria aestivella

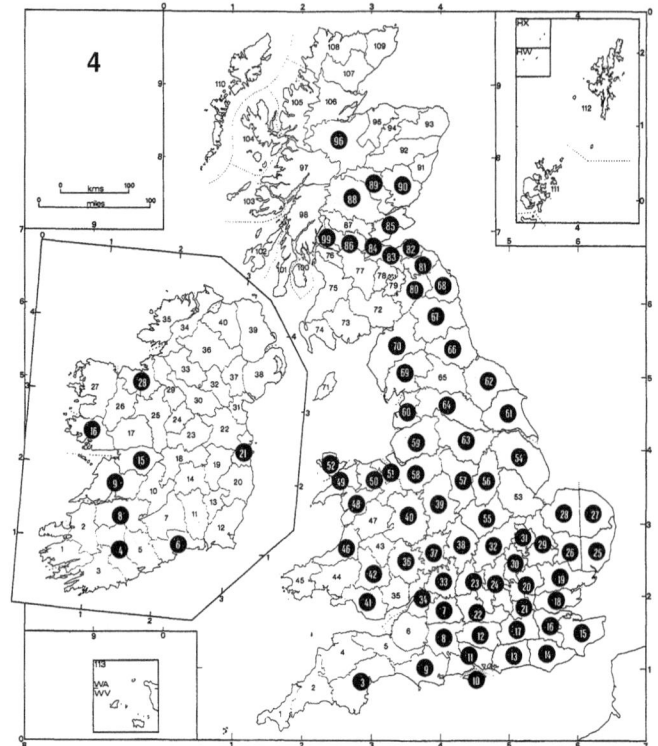

Metzneria metzneriella

Pupa. Undescribed. In the overwintering chamber. May.

Imago. Univoltine; June–July.

Distribution (Map 3)

Occurs on open grassland, especially on chalk and limestone. Locally common in southern England, extending more sparsely in eastern counties to Northumberland; North Wales; single records from Midlothian and Perthshire in Scotland; Co. Wexford, Ireland (Bond, 1999). There is also an unconfirmed record from Co. Sligo (Beirne, 1941). Europe, but absent from Norway; North Africa.

METZNERIA METZNERIELLA (Stainton)

Gelechia metzneriella Stainton, 1851, *Suppl.Cat.Br. Tineidae & Pterophoridae*: 5.

?*Cleodora ochroleucella* Stephens, 1834, *Ill.Br.Ent.* (Haust.) **4**: 221.

Gelechia paucipunctella sensu Douglas, 1850, *Trans.ent. Soc.Lond.* (N.S.) **1**: 14, *nec* Zeller, 1839, *Isis, Leipzig* **1839**: 202.

Type locality: [England].

Description of imago (Pl.1, fig.4)

Wingspan 14–19mm. Head pale ochreous; antenna pale ochreous, annulate fuscous; labial palpus segment 3 half length of segment 2, segment 2 densely, segment 3 more sparsely, thickened with appressed scales, segment 2 dark ochreous above, fuscous below, segment 3 fuscous. Thorax and tegulae greyish ochreous. Forewing yellowish ochreous, disc and subcostal region reddish ochreous, irrorate fuscous and white; stigmata black, first discal well beyond plical; veins variably suffused grey, irrorate fuscous and white; apical area mainly fuscous; terminal line fuscous, mixed white; cilia pale yellowish ochreous with darker ciliary line. Hindwing narrower than forewing, grey, sometimes paler in disc; costal and terminal cilia concolorous with wing, dorsal cilia grading to whitish ochreous at anal angle. Foreleg fuscous, tarsus paler; mid- and hindlegs with outer sides fuscous, banded ochreous, inner sides ochreous. Abdomen fuscous. Genitalia, see figures 1b,21c.

Life history

Ovum. Undescribed. Laid on common knapweed (*Centaurea nigra*) or saw-wort (*Serratula tinctoria*), probably on a flower-head.

Larva. Head dark brown, sutures and posterior margin darker. Body white with slight yellowish tinge; prothoracic plate with central area pigmented blackish, tapering posteriorly and with broad median sulcus not reaching anterior margin; anal plate brown; thoracic legs vestigial, pale brown; abdominal prolegs obsolete.

The larva feeds on the seeds in the lower part of the capsule from September to November, without external evidence of its presence. It then overwinters until April in a light brown silken cocoon spun at the base of the capsule.

Pupa. In the overwintering cocoon. April–May.

Imago. Univoltine; June–August. Flies at night and comes to light.

Distribution (Map 4)

Occurs more or less commonly on grassland, waste ground, downland and roadside verges throughout most of Britain as far north as Inverness-shire; in Ireland known only from Cos Dublin and Waterford and the west, but probably under-recorded. Throughout Europe, but apparently more sparingly in the east.

METZNERIA NEUROPTERELLA (Zeller)

Gelechia (Metzneria) neuropterella Zeller, 1839, *Isis, Leipzig* **1839**: 202.

Type locality: Hungary.

Description of imago (Pl.1, fig.5)

Wingspan 14–24mm. Head greyish brown; antenna ochreous, ringed fuscous, annulation more distinct beneath; labial palpus dark brown, inner surface of segment 2 paler, segment 3 thickened with appressed scales. Thorax greyish brown with three indistinct, darker brown longitudinal stripes; tegulae brown. Forewing ochreous; veins broadly edged dark greybrown, this colour more widely diffused at base of costa, in mid-wing and at apex; oval patch of ochreous ground colour at distal end of cell containing obscure fuscous stigma; cilia ochreous with darker ciliary line. Hindwing, including cilia, fuscous. Foreleg dark brown, tarsus greyish brown; mid- and hindlegs greyish brown. Abdomen dark brown; ovipositor projects in set female specimens. Genitalia, see figures 1e,21e.

Similar species. M. *aprilella* (Herrich-Schäffer), which is generally smaller, has the forewing suffused with orange-vermilion, a colour lacking in *M. neuropterella*, and lacks the pale area containing a stigma at the distal end of the cell.

Life history

Ovum. Undescribed. Laid on dwarf thistle (*Cirsium*

Metzneria neuropterella

acaule) or, less frequently, common knapweed (*Centaurea nigra*), probably on a flower-head. Greater knapweed (*C. scabiosa*) has also been cited as a foodplant (Ford, 1949b; Bradford *in* Emmet, [1979]), but probably in error.

Larva. Head mahogany-brown, sutures and mouthparts dark brown. Body white; prothoracic plate translucent whitish, marbled pale brown anteriorly and medially; anal plate concolorous with body; thoracic legs whitish, ringed pale brown.

Larvae collected on 31 August in seedheads of dwarf thistle were of all sizes, some being fully fed. They were invariably in heads still attached to the plants, not in the many that were lying loose. Most occurred in heads where the bracts were still greenish brown and the pappus still white, but there was no outward sign of infestation. The larvae overwintered in a chamber constructed from eaten-out seeds at the base of the seedhead (J. R. Langmaid, pers.comm.).

Pupa. In the chamber in which the larva overwintered. June.

Imago. Univoltine; June–August. Flies at night and comes to light.

Distribution (Map 5)

Occurs sparingly on short-turfed downland in south-eastern England and Dorset. Records from other counties are probably based on misidentification of *M. aprilella*, which was not recognized as British until 1981. Throughout Europe except for the Mediterranean islands; China.

METZNERIA APRILELLA (Herrich-Schäffer)

Parasia aprilella Herrich-Schäffer, 1854, *Syst. Bearb. Schmett. Eur.* **5**: 207, pl.118, fig.963.
Metzneria sanguinolentella de Joannis, 1910, *Bull. Soc. ent. Fr.* **1910**: 295.
Type locality: Turkey; Brussa [Bursa].

Description of imago (Pl.1, fig.6)
Wingspan 15–18mm. Head yellowish ochreous; antenna ochreous, annulate fuscous; labial palpus brown, inner surface of segment 2 paler, segment 3 thickened with appressed scales. Thorax yellowish ochreous with three longitudinal brown lines; tegulae brown. Forewing ochreous, suffused, sometimes totally, with orange-vermilion; veins grey-brown; stigmata absent; yellow streaks on fold near base, on costa at one-quarter, on dorsal edge of cell and at tornus; cilia greyish ochreous with darker ciliary line. Hindwing dark grey; cilia concolorous, with paler reflections. Foreleg brown, tarsus greyish ochreous; mid- and hindlegs greyish ochreous. Abdomen ochreous-brown. Genitalia, see figures 1d,21f.

Similar species. M. neuropterella (Zeller), *q.v.*

Life history
Ovum. Undescribed. Laid on greater knapweed (*Centaurea scabiosa*), probably on a flower-head.
Larva. Head pale honey-coloured medially, irregularly marbled mid-brown, shading to dark brown anteriorly, laterally and posteriorly; sutures, eye-spots and mouthparts very dark brown. Body white with slight yellowish tinge; prothoracic plate translucent yellowish white, faintly mottled pale brown on each side of mid-line; spiracles with fine brown perimeters; anal plate and thoracic legs concolorous with body.

The larva feeds on the seeds in September and October, causing the seedhead to remain closed and retain its seeds. It then overwinters until April in a silk-lined chamber in the woody part of the pedicel immediately beneath the seeds (J. R. Langmaid, pers. comm.).
Pupa. In the overwintering chamber. April.
Imago. Univoltine; May–August. Flies at night and comes to light.

Distribution (Map 6)
Occurs in grassland and on waste ground, especially

Metzneria aprilella

on calcareous soil. It was confused with *M. neuropterella* until a British specimen was detected by Sattler (1981) and it is likely that many early records of that species refer to *M. aprilella*, which has been present in Britain at least since the beginning of the twentieth century (Corley, 1988). It is local in England south of a line from the Severn estuary to The Wash and is also found, not uncommonly, in south Yorkshire (Sutton & Beaumont, 1989); not recorded from the West Country, Wales, Scotland or Ireland. Throughout most of Europe, but absent from Norway, Portugal and most of the Mediterranean islands; western Asia.

ISOPHRICTIS Meyrick

Isophrictis Meyrick, 1917, *Entomologist's mon. Mag.* **53**: 113.

A Holarctic genus with 18 species occurring in North America, 12 in Europe, mainly in the Mediterranean region, and one in Britain.

Imago. Antenna serrate, more strongly in male; labial palpus with segments 2 and 3 of equal length, segment 2 with long projecting scales beneath. Forewing elongate, apex acute. Hindwing narrow, trapezoidal, apex strongly produced. Venation as in *Metzneria* Zeller. Male genitalia with uncus rounded, gnathos absent, valva and sacculus similar to *Metzneria*; saccus divided; aedeagus bulbed, with cornuti. Female genitalia with signum present in corpus bursae.

Larva. In seedheads or stems of Asteraceae (Compositae).

ISOPHRICTIS STRIATELLA ([Denis & Schiffermüller])

Tinea striatella [Denis & Schiffermüller], 1775, *Schmett. Wien.*: 135.

Tinea tanacetella Schrank, 1802, *Fauna Boica* **2**(2): 122.

Type locality: [Austria]: Vienna district.

Description of imago (Pl.1, fig.7)

Wingspan 11–13mm. Head white, neck-tufts tinged ochreous; antenna brown with obscure paler annulation; labial palpus with segment 3 as long as segment 2, ochreous-white, segment 2 with long, loose tuft beneath and short tuft above, segment 3 slender, almost straight, ascending. Thorax and tegulae ochreous-white. Forewing lanceolate but appearing obtusely rounded owing to cilia; dark brown; stigmata black, plical and first discal elongate but obscure and often obsolete, first discal well beyond plical, second discal circular, distinct; obscure, darker-edged ochreous streak in disc, encircling first discal stigma but terminating before second; similar, more distinct, streak on fold; area between this streak and dorsum ochreous; outwardly oblique white streak from costa at four-fifths to middle of termen, extending into cilia; two or three white specks on costa between streak and apex; a slightly inwardly oblique black streak from apex into cilia in upper half of wing, bordered proximally by golden-ochreous patch; cilia grey with three dark ciliary lines and outward-directed whitish streaks in tornal area. Hindwing trapezoidal, elongate, termen sinuate below apex, apex strongly produced; brownish grey; cilia paler. Foreleg brown; mid- and hindlegs ochreous, hindleg with tarsus banded brown. Abdomen brown. Genitalia, see figures 2f,21g.

Life history

Ovum. Laid on tansy (*Tanacetum vulgare*) or sneezewort (*Achillea ptarmica*), probably on the pedicel *c*.5mm below the receptacle.

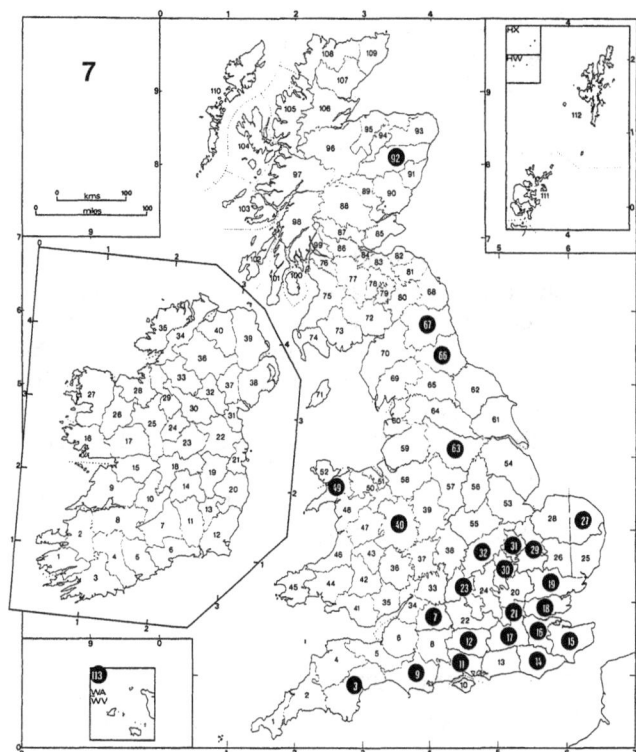

Isophrictis striatella

Larva. Head brown, sometimes marked or marbled blackish, sutures, eye-spots and mouth-parts dark brown. Body very pale pinkish yellow-brown or pinkish buff, posterior two-thirds with indistinct darker dorsal line edged by darker mottling; prothoracic plate with median sulcus, pinkish pale brown mottled dark brown or blackish posteriorly; anal plate brown or blackish; thoracic legs concolorous with body; anal prolegs well developed (J. R. Langmaid, pers.comm.). Prior to pupation the larva turns light red (Wakely, 1957).

Minute frass in the pedicel indicates that the larva mines upwards into the receptacle, from where it enters successive flower-heads through a small hole chewed in the side. When fully fed in September or October, many enter the upper part of the stem to overwinter. Several may be found in a single stem, close to each other but in separate chambers. There is only a single entry hole at a leaf-axil, which is conspicuous because of the whitish, sawdust-like chewed pith exuding from the hole. Others, however, overwinter in semicircular chambers in the seedheads above the receptacle, which they leave in late May or early June for pupation (J. R. Langmaid, pers.comm.; Wakely, *loc.cit.*).

Pupa. In the upper part of a stem where the larva over-wintered, or in a flimsy cocoon spun amongst debris on the ground.

Imago. Univoltine; July–August. Flies at night and comes to light; may also be swept from its foodplant by day.

Distribution (Map 7)

Occurs on waste ground, grass verges and in gardens on tansy, and in damp grassland and fens on sneeze-wort. Widespread but local south-east of a line from Devon to The Wash; Shropshire (Riley, 1991); south-west Yorkshire (R. I. Heppenstall, pers.comm.); Co. Durham and Northumberland (Dunn & Parrack, 1992); Bangor, North Wales (H. N. Michaelis, pers.comm.); Aberdeenshire (E. C. Pelham-Clinton, pers.comm.); the Channel Islands. Throughout Europe; Asia Minor; North America.

APODIA Heinemann

Apodia Heineman, 1870, *Schmett.Dtl.Schweiz* (2)**2**(1): 286.

A Palaearctic genus represented in Britain and currently considered to contain two species in Europe, but it is possible that they are conspecific (Karsholt & Razowski, 1996).

Imago. Forewing elongate, veins R_5 (7) and R_4 (8) stalked with M_1 (6). Hindwing narrower than forewing, trapezoidal with apex strongly produced. For other characters see *Apodia bifractella* (Duponchel). Male genitalia similar to *Isophrictis* Meyrick, but aedeagus without cornuti. Female genitalia without signum in corpus bursae.

Larva. Almost apodal.

APODIA BIFRACTELLA (Duponchel)

Lita bifractella Duponchel, [1843], in Godart & Duponchel, *Hist.nat.Lépid.Fr.* (Suppl.) **4**(1842): 292.
?*Apodia martinii* Petry, 1911, *Dt.ent.Z.Iris* **25**: 101.
Type locality: [France].

Description of imago (Pl.1, fig.8)

Wingspan 9–12mm. Head orange-yellow, frons sometimes whitish; antenna half length of forewing, dark fuscous, obscurely annulate paler; labial palpus orange-yellow, segment 3 shorter than segment 2, seg-ment 2 thickened with short scales beneath. Thorax and tegulae fuscous. Forewing fuscous, irrorate whitish grey; sometimes a few ferruginous-orange scales in dorsal area near base; inwardly oblique ferruginous-orange streak at tornus and outwardly oblique streak beyond on costa, often connected by scales of similar colour forming a Z-shaped fascia, these markings, however, sometimes obscure or obsolete; cilia greyish fuscous with two darker ciliary lines; underside with apical area blackish fuscous. Hindwing fuscous; cilia grey. Foreleg fuscous, tarsus ringed ochreous; mid- and hindlegs ochreous, hindleg with tarsus obscurely banded darker. Abdomen fuscous. Genitalia, see figures 2a,21h.

Life history

Ovum. Laid on common fleabane (*Pulicaria dysenterica*), ploughman's spikenard (*Inula conyzae*) or sea aster (*Aster tripolium*), probably on a flower-head.

Larva. Limaciform, apodal except for vestigial anal prolegs. Head white, sutures, eye-spots and mouth-parts brown. Body white, faintly tinged yellowish; prothoracic plate concolorous.

The larva feeds on the seeds within the head in October and November, spinning them together into a narrow tube. It then overwinters in the feeding place. There is no external evidence of its presence.

Pupa. In a tough cocoon of silk mixed with frass and chewed seed fragments, firmly attached to the receptacle, remnants of seeds and pappus remaining attached to the seedhead (J. R. Langmaid, pers.comm.). April–June.

Imago. Univoltine; July–August. It becomes active from late afternoon onwards and may sometimes be seen resting on the flowers of its foodplant.

Distribution (Map 8)

Depending on the foodplant, it may occur in damp meadows and ditches, on open downland or on salt-marshes. Widespread and fairly common in southern England north to Worcestershire and Norfolk; south Yorkshire (H. E. Beaumont, pers.comm.); Cheshire (I. F. Smith, pers.comm.); Lancashire (K. McCabe & S. M. Palmer, pers.comm.); South Wales (R. G. Warren, pers.comm.); North Wales (H. N. Michaelis, 1979 and pers.comm.). Not recorded from Scotland or Ireland. Throughout Europe, but records from northern and eastern Europe should be attributed to *A. martinii* Petry if that proves a distinct species (Karsholt & Razowski, 1996); Asia Minor; North Africa.

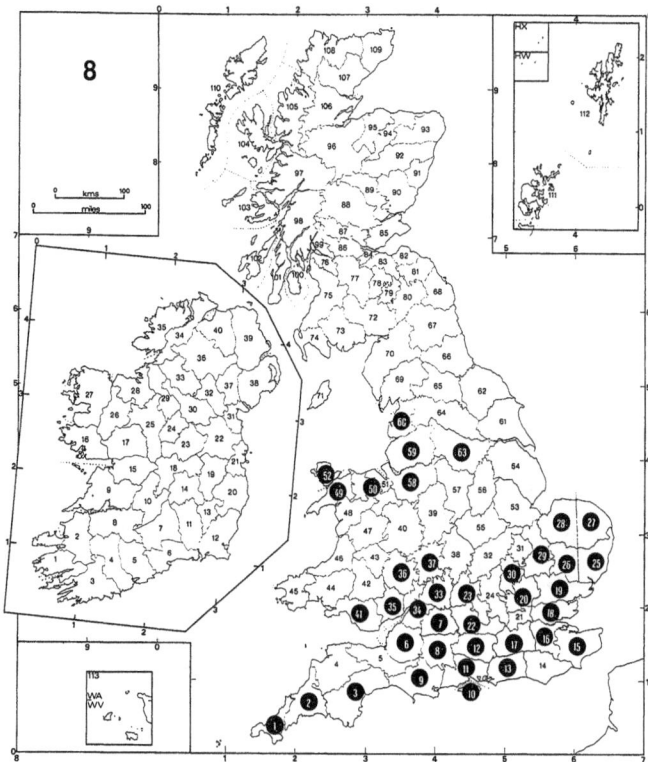

Apodia bifractella

EULAMPROTES Bradley

Eulamprotes Bradley, 1971, *Entomologist's Gaz.* **22**: 27.

Lamprotes Heinemann, 1870, *Schmett.Dtl.Schweiz* (2)**2**(1): 309, *nec* R. L., 1817.

Argyritis Heinemann, 1870, *Schmett.Dtl.Schweiz* (2)**2**(1): 283, *nec* Hübner, [1821].

A Palaearctic genus containing about a dozen species, four of which occur in Great Britain and Ireland.

Imago. Head without ocelli, except *E. wilkella* (Linnaeus) which has distinct ocelli; scape of antenna without pecten, except *E. wilkella* which has a single scale; labial palpus long and slender, segments 2 and 3 of approximately equal length, segment 2 slightly thickened with appressed scales. Forewing with veins R_4 (8) and R_5 (7) stalked, M_3 (4), CuA_1 (3) and CuA_2 (2) separate, M_1 (6) obsolete in *E. wilkella*. Male genitalia with small valva and cylindrical tegumen. Female genitalia with subtriangular signum in corpus bursae.

Larva. The known larvae mine stems of St John's-worts (*Hypericum* spp.) or feed on leaves of common mouse-ear (*Cerastium fontanum*).

Key to species (imagines) of the genus *Eulamprotes*

1   Forewing with one or more silver fasciae .... *wilkella* (p. 78)

–   Forewing without silver fasciae 2

2(1)   Forewing unicolorous bronzy fuscous; labial palpus segments 1 and 2 unicolorous purplish, segment 3 dark fuscous .... *unicolorella* (p. 77)

–   Forewing and labial palpus otherwise 3

3(2)   Labial palpus ochreous .... *atrella* (p. 75)

–   Labial palpus segments 2 and 3 greyish or yellowish white above and at sides, lower edge dark fuscous .... *immaculatella* (p. 76)

EULAMPROTES ATRELLA ([Denis & Schiffermüller])

Tinea atrella [Denis & Schiffermüller], 1775, *Schmett. Wien.*: 140.

?*Anacampsis umbriferella* Herrich-Schäffer, 1854, *Syst. Bearb.Schmett.Eur.* **5**: 195; 1853, *Ibid.*, pl.70, fig.524 [non-binominal].

Type locality: [Austria]; Vienna district.

Description of imago (Pl.1, fig.9)

Wingspan 11–13mm. Head bronzy fuscous, frons paler shining grey; antenna dark fuscous, distal three-quarters serrate; labial palpus ochreous, lower edge of segment 3 narrowly fuscous. Thorax and tegulae dark bronzy fuscous. Forewing dark fuscous to blackish fuscous, often with purplish sheen in apical quarter; small subtriangular spot at tornus and similar one just beyond on costa at two-thirds, both merging into cilia, three minute dashes between costal spot and apex, sometimes obsolete, and a few scattered scales between costal and tornal spots, all ochreous-yellow; cilia dark fuscous to dark grey. Hindwing fuscous to dark grey; cilia concolorous. Foreleg shining dark fuscous, tarsal segments banded yellowish white distally; mid- and hindlegs shining greyish fuscous. Abdomen dark fuscous. Genitalia, see figures 2c,22b.

Similar species. E. immaculatella (Douglas), *q.v.*

Life history

Ovum. Undescribed. Oviposition site unknown, but presumably on the foodplant, perforate St John's-wort

(*Hypericum perforatum*) or hairy St John's-wort (*H. hirsutum*).

Larva. Head very light yellowish brown, translucent. Prothoracic plate light yellowish brown, also translucent, six blackish dots anteriorly, one each side laterally, and a very narrow finely bisected blackish line posteriorly; body translucent, showing greenish body contents; purplish brown dorsal and subdorsal lines, sometimes indistinct, subdorsal lines fading posteriorly; pinacula large, blackish; anal plate greyish; thoracic legs and prolegs translucent.

On 22 March 1997, a few young stems of hairy St John's-wort, 5.0–7.5cm high, were found with the terminal two or three whorls of leaves slightly drooping (R. J. Heckford, pers.obs.). Two plants each had a very small hole edged reddish brown in one of the outer terminal leaves by which the larva presumably entered, the terminal leaves of the other plants having already withered too much for this to be seen. The larva later fed within the smaller terminal leaves enclosed by the outer, and deposited brownish frass within. It then mined the stem from the tip, causing the terminal leaves to wither. Shortly prior to pupation it vacated the stem and spun one edge of a leaf of the foodplant downwards to join the other edge and fed on the under surface from within. The larva then cut part of the folded leaf away to form a case which dropped to the ground; unlike a coleophorid, the larva does not feed once the case is formed. Stainton (1869b) stated that the case was made from a portion of the mined stem and that it was not unlike the then new-fashioned spectacle-cases, which were rather limp and open at both ends. Meyrick ([1928]) and others follow Stainton in stating that the case is made from the stem, but personal observations of the present author have not confirmed this. This is one of only two British gelechiids which are known to make a case and one of very few British species of Lepidoptera known to make a case solely for pupation. Larvae were first found by W. R. Jeffrey of Saffron Walden, Essex, in 1866 (Stainton, 1867b). March–May.

Pupa. In a reddish brown suboval case *c*.14mm long cut from a leaf of the foodplant. The pupa is not extruded on emergence of the moth. Stainton (1869b) records that de Grey (later Lord Walsingham) swept a brown cocoon in June, possibly 1867, amongst grass where there was much *Hypericum*, species not mentioned, and stated that it may have been attached to either this or the long grass. June.

Imago. Univoltine; July–August. Flies from early evening and comes to light.

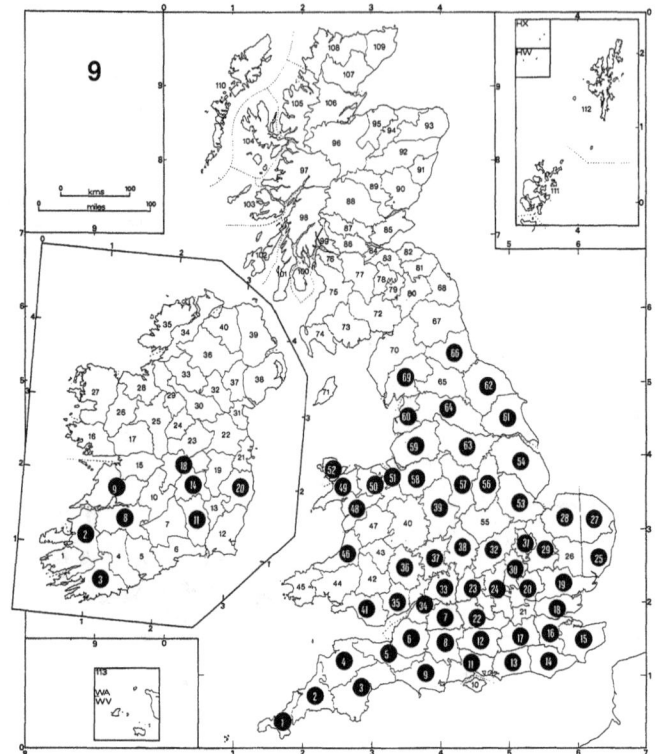

Eulamprotes atrella

Distribution (Map 9)

Frequents rough ground, dry pastures, clearings in woods and downland where the foodplant grows; local but widespread in England and Wales, absent from Scotland, all supposed Scottish specimens having proved to be *E. immaculatella*; local in Ireland. Western Europe to the former U.S.S.R., but has not been recorded from Corsica, Sardinia, Sicily, Malta or the former Yugoslavia.

EULAMPROTES IMMACULATELLA (Douglas)

Gelechia immaculatella Douglas, 1850, *Trans. ent. Soc. Lond.* (N.S.) **1**: 67.

Eulamprotes phaeella Heckford & Langmaid, 1988, *Entomologist's Gaz.* **39**: 1.

Type locality: England; West Wickham Wood, Kent.

Description of imago (Pl.1, fig.10)

Wingspan 8–13mm. Head bronzy fuscous, frons paler; antenna dark fuscous, distal three-quarters serrate; labial palpus segment 1 dark fuscous, segment 2 greyish white above and at sides, lower edge dark fuscous, segment 3 yellowish white, lower edge lined dark fuscous. Thorax and tegulae bronzy fuscous to blackish fuscous. Forewing dark fuscous to blackish fuscous, sometimes with darker stigmata visible; small yellowish white costal spot at two-thirds, most of which is in cilia; sometimes a few yellowish white scales at tornus and a few scattered yellowish white scales in apical area, but in the type specimen the yellowish white costal spot and scales at tornus are obsolete; cilia from costal spot to apex dark fuscous, paler from apex to tornus. Hindwing light fuscous; cilia concolorous, paler towards anal angle. Foreleg shining dark fuscous, tarsal segments banded yellowish white distally; mid- and hindlegs shining greyish fuscous. Abdomen bronzy fuscous. Genitalia, see figures 2d,22c.

Similar species. E. atrella ([Denis & Schiffermüller]) differs in having an ochreous labial palpus and flies only in July and August. *Aproaerema anthyllidella* (Hübner) has the scape of the antenna lined buffish fuscous anteriorly, the forewing has a ciliary line and sometimes yellowish ochreous scales on the fold, all absent in *E. immaculatella*. Further, in male *A. anthyllidella* the genitalia protrude from the abdomen; in *E. immaculatella* they are completely withdrawn and the eighth abdominal segment forms a blunt triangle.

Life history

Early stages. Unknown. Its foodplant has not yet been identified. Although Heckford & Langmaid (1988) suggested that it might be slender St John's-wort (*Hypericum pulchrum*), it is absent from some of the localities where the moth has since been found and in several no species of *Hypericum* was noticed.

Imago. Has been found in every month from May to September. Either univoltine with prolonged emergence, or bivoltine. Flies from early evening and comes to light.

Distribution (Map 10)

Although described as new to science from a specimen taken at Kynance Cove, Cornwall, in 1983 and given the name *E. phaeella* (Heckford & Langmaid, *loc.cit.*), it has since been discovered that another specimen had been taken over a century earlier by J. W. Douglas on 11 August 1849 at West Wickham Wood, Kent. The name Douglas gave it was subsequently treated as a junior synonym of *E. unicolorella* (Duponchel), *q.v.* (Bankes, 1899a). However, with the aid of remarkable new digital micrographic techniques, the genitalia slide prepared in 1951 from Douglas's specimen, now in the BMNH, has been critically re-examined. It has been shown beyond doubt that it is conspecific with *phaeella* and that *immaculatella* is therefore the senior synonym of *E. phaeella* Heckford & Langmaid and has priority (Heckford *et al.*, 1999).

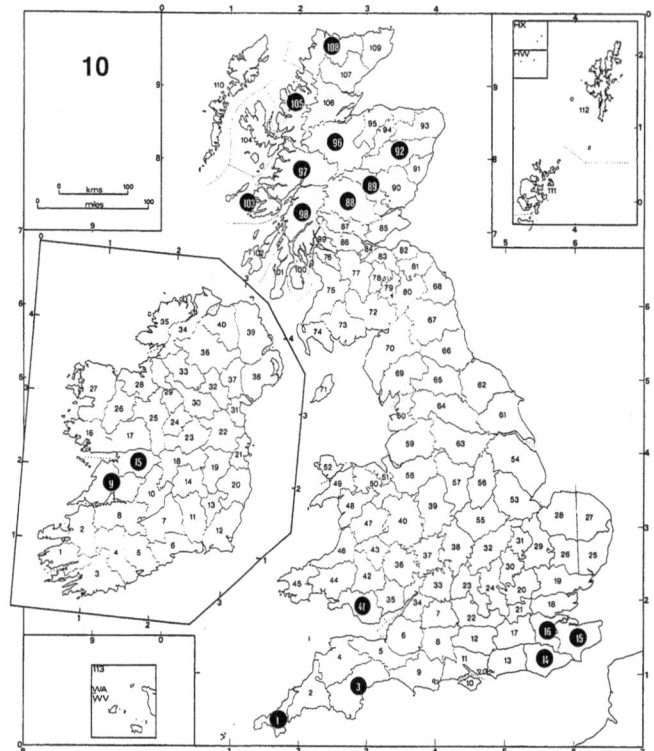

Eulamprotes immaculatella

The moth has been found in a wide range of habitats including dry coastal cliffs, damp low-lying fields, limestone pavement, areas with sandstone or slate as the underlying rocks, and herb-rich grassland in the Cairngorms, Scotland. Although the map shows a wide distribution, records so far suggest that it is extremely local. With the exception of the Burren, Ireland, it has not been recorded from more than two localities in any vice-county, and mostly only one. In continental Europe it has been recorded only from Denmark, Portugal and Majorca.

EULAMPROTES UNICOLORELLA (Duponchel)

Lita unicolorella Duponchel, [1843], *in* Godart & Duponchel, *Hist.nat.Lépid.Fr.* (Suppl.) **4**[1842]: 458, pl.85, fig.8.

Type locality: France.

Description of imago (Pl.1, fig.11)

Wingspan 10–13mm. Head shining purplish fuscous; antenna dark fuscous, distal three-quarters serrate; labial palpus segments 1 and 2 purplish, segment 3 dark fuscous, segments 2 and 3 of equal length. Thorax and tegulae shining purplish fuscous. Forewing bronzy

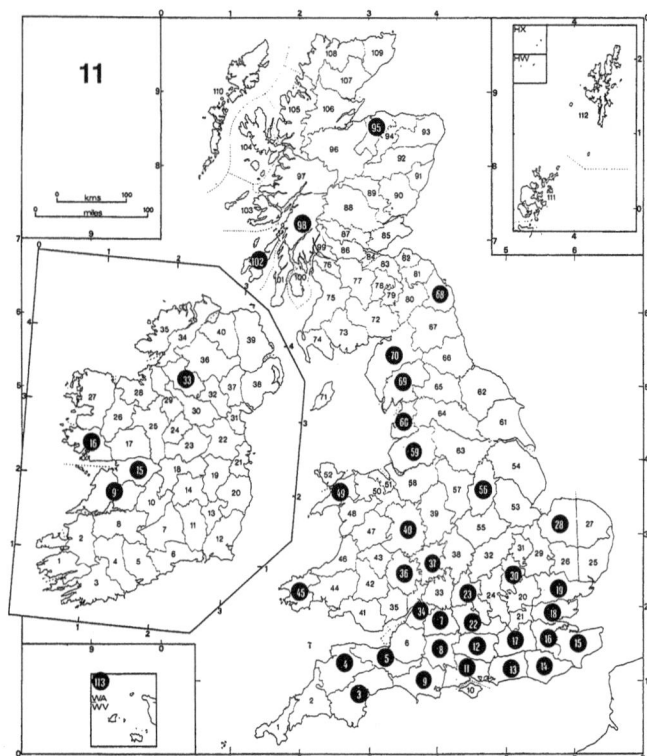

Eulamprotes unicolorella

fuscous, sometimes with purplish sheen on basal one-third especially towards costa; cilia concolorous. Hindwing light fuscous; cilia bronzy fuscous. Legs dark bronzy fuscous. Abdomen dark bronzy fuscous. Genitalia, see figures 2e,22d.

Similar species. Monochroa tenebrella (Hübner), *q.v.*, differs in having the labial palpus with segment 3 shorter than segment 2, and, in the female, the apical third of the antenna white.

Life history

Early stages. Apparently undescribed. Threlfall (1878) recorded that on 13 May 1877 he collected roots of sea plantain (*Plantago maritima*) on the banks of the river Wyre, Lancashire. From these on 30 June emerged a specimen of a little black gelechiid, unknown to him, which Stainton thought to be probably *immaculatella*. However, Bankes (1899a) made clear that he had seen the specimen and that it was not conspecific with the type specimen of *Eulamprotes immaculatella* (Douglas), although he did not say what he considered it to be! Adults have been found plentifully in a field with abundant perforate St John's-wort (*Hypericum perforatum*) (R. J. Heckford, pers.obs.) which may be the foodplant.

Imago. Univoltine. Late May–early July.

History and distribution (Map 11)

The species was introduced to the British list in 1899 by Bankes (*loc.cit.*) as *Aristotelia unicolorella* (Duponchel) on the basis of specimens, identified though not taken by him, from a few localities in England and Oban, Scotland. His paper also included a review of the status of the then unique specimen of *Gelechia immaculatella* Douglas (*q.v.*), taken in Kent in 1849. Bankes had mistakenly considered this to be 'without doubt an undersized male specimen [wingspan 10.5mm.] of *unicolorella*', adding that it agreed 'in every other respect with this species', but *immaculatella* is now treated as a valid species. Bankes also discussed the confusion between *E. unicolorella* and *Monochroa* (then *Aristotelia*) *tenebrella*.

Frequents dry fields and waste ground. Widely distributed in southern England and Wales but usually very local, extending to Moray but declining northwards, with few confirmed records from Scotland and Ireland; also from the Channel Islands (Guernsey). Western Europe to the former U.S.S.R.

EULAMPROTES WILKELLA (Linnaeus)

Phalaena (*Tinea*) *wilkella* Linnaeus, 1758, *Syst.Nat.* (Edn 10) **1**: 541.

Gelechia (*Brachmia*) *pictella* Zeller, 1839, *Isis, Leipzig* **1839**: 202.

Gelechia tarquiniella Stainton, 1862, *Entomologist's Annu.* **1862**: 112.

Type locality: Europe.

Description of imago (Pl.1, figs 12–14)

Wingspan 8–10mm. Head creamy white, line of dark fuscous scales around eye and posteriorly, frons creamy white; antenna with scape dark fuscous, flagellum dark fuscous to one-half, sometimes alternate segments creamy white to one-half, creamy white from one-half to apex; labial palpus segment 1 dark fuscous, segment 2 creamy white, basally dark fuscous, segment 3 creamy white, apically dark fuscous. Thorax and tegulae dark fuscous. Forewing dark fuscous, sometimes with a faint violet gloss; three silver fasciae, first outwardly oblique not reaching fold, second at right angles to costa reaching fold, third inwardly acute, complete; silver apical spot; cilia at third fascia and apex creamy white, otherwise dark fuscous, becoming lighter towards tornus. Hindwing light fuscous; cilia concolorous. Female differs from male in having forewing narrower distally, hindwing narrower overall.

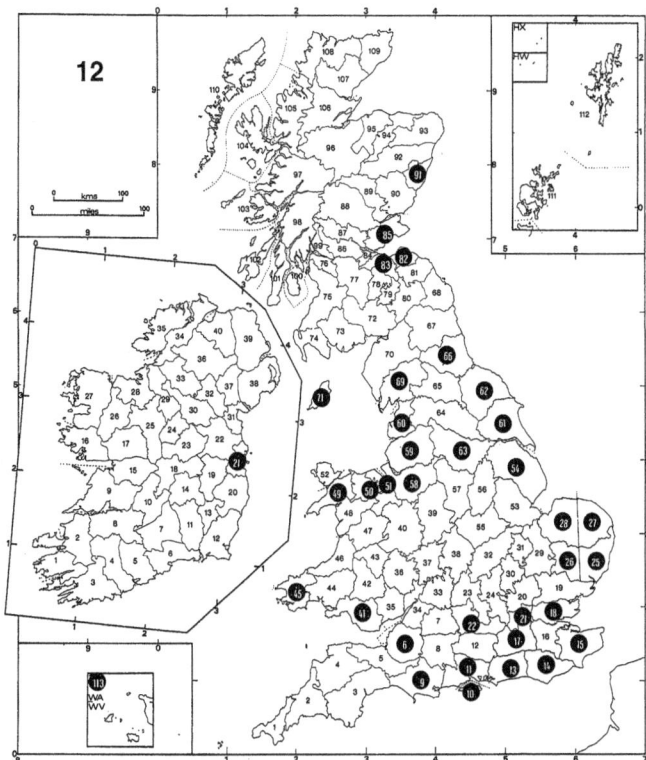

Eulamprotes wilkella

Fore- and midlegs with femora and tibiae greyish, hindleg femur grey, tibia dark fuscous with a median creamy white spot; tarsi of all legs dark fuscous, ringed creamy white. Abdomen fuscous; anal tuft ochreous. Genitalia, see figures 2b,22e.

Two other forms are known. One, f. *tarquiniella* Stainton (fig.13), has both the second and third fasciae obsolete, and the forewing of the female is acuminate; the other (unnamed) (fig.14) has only the third fascia obsolete. In both forms the silver apical spot is absent and the entire cilia dark fuscous. Both forms can occur together and with the nominate form, but both are very local and may occur in only a few localities (Bond, 1991).

Life history

Ovum. Undescribed. Oviposition site unknown, but presumably on the foodplant, common mouse-ear (*Cerastium fontanum*).

Larva. According to Stainton (1867a), head small, pale brown. Prothoracic plate brownish with a dark patch in centre; body pale rose; thoracic legs and prolegs pale yellow.

Stainton (*loc.cit.*) states that the larva makes webs along the stems, silken galleries amongst the roots and pale blotches in some of the leaves; when full-fed it forms a cocoon amongst the sand. May; July.

Pupa. Undescribed. May and June; July and August.

Imago. Bivoltine; June; August. Appears to be associated with thyme (*Thymus* spp.).

Distribution (Map 12)

Frequents shingle beaches and sandy areas, usually coastal but is also known inland. The habitat designation in Emmet (1991) of '9b,d' (mosses, peat-bogs; vegetation in water) is an obvious error for '10b,d' (sand-dunes; shingle beaches). Widespread but local in England, except the Midlands and the far south-west, and Wales; the Channel Islands; eastern Scotland, Cos Dublin and Cork in Ireland. Throughout most of Europe to the former U.S.S.R. and Asia Minor.

ARGOLAMPROTES Benander

Argolamprotes Benander, 1945, *Ent.Tidskr.* **66**: 126, 128 (key), 135.

A monotypic Palaearctic genus which is represented in England and Wales.

Imago. Head without ocelli; scape of antenna without pecten; labial palpus long and slender, segments 2 and 3 of approximately equal length, segment 2 slightly thickened with appressed scales. Forewing bronzy fuscous with a purple or violet iridescence with silvery blue fasciae and spots; with veins R_4 (8) and R_5 (7) stalked, M_3 (4), CuA_1 (3) and CuA_2 (2) separate. Male genitalia approaching *Monochroa* Heinemann. Female genitalia closer to *Eulamprotes* Bradley.

Larva. Feeds in shoots of raspberry (*Rubus idaeus*) and, in England, also on blackberry (*R. fruticosus* agg.).

ARGOLAMPROTES MICELLA ([Denis & Schiffermüller])

Tinea micella [Denis & Schiffermüller], 1775, *Schmett. Wien.*: 140.

Oecophora asterella Treitschke, 1833, *Schmett.Eur.* **9**(2): 172.

Type locality: [Austria];Vienna district.

Description of imago (Pl.1, fig.15)

Wingspan 10–14mm. Head bronzy purple with violet iridescence; antenna dark fuscous with slight violet iridescence; labial palpus ochreous-yellow becoming dark fuscous towards apex of segment 3. Thorax and tegulae bronzy purple with violet iridescence. Forewing bronzy fuscous with purple or violet iridescence; oblique fascia from costa at one-sixth sometimes to beyond fold and usually interrupted on fold; plical spot, three or four spots in disc, scattered subcostal spots, tornal spot, spot just beyond on costa and several terminal spots all silvery blue; cilia dark fuscous. Hindwing dark fuscous; cilia concolorous. Legs dark fuscous, tarsal segments banded yellowish white. Abdomen dark fuscous. Genitalia, see figures 2g,22a.

Life history

Ovum. Undescribed. Oviposition site unknown, but presumably on the foodplant, raspberry (*Rubus idaeus*) or blackberry (*R. fruticosus* agg.) in England, but apparently known only from raspberry in continental Europe.

Larva. Head and prothoracic plate black, finely bisected whitish. Body very pale, almost translucent, yellow-white, anterior half of each segment reddish, except thoracic segments 2 and 3, interrupted laterally; dorsal line interrupted on thoracic segments 2 and 3, interrupted subdorsal line and two lateral lines, converging posteriorly; all lines reddish, broken at intersegmental divisions and more or less obsolete around pinacula which are very small and black; anal plate black; thoracic legs blackish, a black semicircular line above each leg; prolegs concolorous with body, anal legs translucent, anteriorly blackish. Late April–mid-May. Emmet (1991) gives only June, but this was not based on British observations.

In the British Isles the larva remained undetected until 1997. Between 3 and 28 April 1997, larvae were found at five sites near Plymouth, Devon (Heckford, 1998). They were 3–4mm in length, each occurring only within the bud and the adjoining unexpanded leaflet. They all fed initially in flowering shoots 5–8cm long, although a few were considerably longer, up to a maximum of 30cm, with up to five tenanted shoots on a stem; all were in sunny positions on the edges of tracks. At first, the larva skeletonizes the leaflet, betraying its presence by a small amount of blackish brown frass at the tip; the leaflet initially appears whitish, owing to the hairs on the underside, but eventually turns brown. The larva then feeds on part of the bud itself and sometimes burrows into the shoot up to or just over its own length. In due course the affected part of the shoot withers and turns brown, and may then fall off. The rest of the shoot and leaves are not affected and noticeably they do not droop, so there is little outward sign of a larva whilst it is feeding, and almost none once it has finished.

Argolamprotes micella

Pupa. Light yellowish; in a slight silken cocoon. In captivity the larvae pupated in tissue. May–June.

Imago. Univoltine; July, but in captivity the first moth emerged on 30 April. Can be disturbed in late afternoon, flies in the evening and comes to light.

Distribution (Map 13)

Frequents hedgerows, open woodland and gardens and is not uncommon locally. Unknown in the British Isles until 1963 when Kennard (1965) took one specimen at light at Ide, near Exeter, Devon. The next was taken in 1971 at Saltash, Cornwall. It has since been found in other localities in Devon and Cornwall, and also Dorset, Somerset, Hampshire and Glamorgan. As it is a distinctive species it seems unlikely to be overlooked and so its current distribution in the British Isles suggests that it is slowly spreading eastwards. Widely distributed in Europe except the extreme north; Japan.

MONOCHROA Heinemann

Monochroa Heinemann, 1870, *Schmett. Dtl. Schweiz* (2)**2**(1): 308.

Paltodora Meyrick, 1894, *Entomologist's Mon. Mag.* **30**: 230.

A Holarctic genus with 11 species represented in North America and 26 in Europe, of which 14 occur in Britain and five in Ireland.

Imago. Head smooth-scaled; antenna about three-quarters length of forewing, in some species with apical one-third weakly serrate, flagellum in most species with pale annulations, these in some being irregularly or more widely disposed towards apex, conspicuously in *M. elongella* (Heinemann), less distinctly in *M. hornigi* (Staudinger), *M. suffusella* (Douglas), *M. niphognatha* (Gozmány), *M. lutulentella* (Zeller) and *M. arundinatella* (Stainton) (Svensson, 1992). Labial palpus with segment 3 as long as or slightly shorter than segment 2, segment 2 thickened with appressed scales; ocelli present (figure 48). Forewing elongate, apex sub-

Figure 48 *Monochroa lucidella* (Stephens), side view of head showing palpus

acute but appearing rounded owing to disposition of cilia; fully veined, vein R_5 (7) stalked with R_4 (8). Hindwing elongate trapezoidal except in *M. palustrellus* (Douglas); tornus bluntly angled at 100°–110°, termen slightly sinuate in most species, apex strongly and abruptly produced; vein M_1 (6) and Rs (7) remote at origin; cilia up to three times width of wing. Male genitalia with uncus reduced, gnathos absent, aedeagus bulbed, flagon-shaped, with cornuti. Female genitalia with signum present in corpus bursae.

Larva. Cylindrical with the full number of functional prolegs. Most feed internally, mining stems, roots or leaves; *M. conspersella* (Herrich-Schäffer) makes a spinning in a terminal shoot, but feeds mainly on the young stem. Most species give little or no evidence of their presence and in consequence the larvae are rarely observed.

Pupa. In a flimsy cocoon spun in the feeding place or externally amongst detritus. It appears that some species can use either strategy.

Key to species (imagines) of the genus *Monochroa*

1	Forewing with blackish longitudinal streaks; hindwing with tornus rounded and apex only moderately produced .... *palustrellus* (p. 84)	
–	Forewing otherwise; hindwing trapezoidal 2	
2(1)	Forewing without discal or plical stigmata 3	
–	At least one of the discal and plical stigmata present 5	
3(2)	Head whitish; forewing reddish ochreous, variably suffused fuscous .... *cytisella* (p. 82)	
–	Head otherwise; forewing not reddish ochreous 4	
4(3)	Forewing glossy dark purplish fuscous .... *tenebrella* (p. 83)	
–	Forewing dark fuscous, irrorate grey .... *conspersella* (part) (p. 86)	
5(2)	Forewing with only second discal stigma present ... 6	
–	Forewing with two or three stigmata present 10	
6(5)	Forewing whitish ochreous; a white spot on costa above second discal stigma .... *suffusella* (p. 88)	
–	Forewing otherwise 7	
7(6)	Larger species; wingspan 12–16mm 8	
–	Smaller species; wingspan 8–9mm ...... *moyses* (p. 92)	
8(7)	Abdomen with segments 1–4 ochreous-yellow above .... *lutulentella* (p. 89)	
–	Abdomen wholly fuscous above 9	
9(8)	Antenna with irregularly distributed silvery white rings in apical area .... *hornigi* (part) (p. 86)	
–	Antenna without such rings in apical area .... *lucidella* (p. 84)	
10(5)	Abdomen with dorsal surface of segments 2 and 3 ochreous-yellow .... *elongella* (p. 90)	
–	Abdomen otherwise 11	
11(10)	Forewing with costal two-fifths pale ochreous, dorsal three-fifths fuscous, colour change abrupt .... *divisella* (p. 93)	
–	Forewing otherwise 12	
12(11)	Head and labial palpus mostly whitish; larger species, wingspan 13–15mm .... *niphognatha* (p. 88)	
–	Head and labial palpus not whitish; smaller species, wingspan 9–12mm 13	
13(12)	Antenna with white annulations irregularly or more widely spaced in apical area 14	
–	Antenna with white annulations evenly spaced or absent in apical area 15	
14(13)	Forewing with ground colour dark fuscous .... *hornigi* (part) (p. 86)	

–   Forewing with ground colour ochreous-fuscous ....
.... *arundinetella* (p. 91)

15(13)   Forewing pale fuscous irrorate whitish ....
.... *tetragonella* (p. 85)

–   Forewing dark fuscous irrorate grey ....
.... *conspersella* (part) (p. 86)

MONOCHROA CYTISELLA (Curtis)

Cleodora cytisella Curtis, 1837, *Br.Ent.* **14**: 671.

Anacampsis fuscipennis Humphreys & Westwood, 1845, *Br.Moths Transform.* **2**: 192.

Gelechia walkeriella Douglas, 1850, *Trans.ent.Soc.Lond.* (N.S.)**1**: 21.

Aristotelia clinosema Meyrick, 1935, *Entomologist* **68**: 285.

Type localities: Ireland; Glengarriff, Co. Cork, and England; Isle of Wight and the London district.

Description of imago (Pl.1, figs 16,17)

Wingspan 10–12mm. Head creamy white, sometimes with yellowy tinge; antenna black, weakly serrate in distal one-third; labial palpus with segments 2 and 3 of equal length, ochreous-white, segment 2 loosely long-scaled beneath, segment 3 ascending, slightly curved, slender. Thorax, including tegulae, ochreous. Forewing with area between dorsum and fold ochreous, between fold and radial vein reddish ochreous and between radial vein and costa fuscous, fuscous sometimes extending over central area or over whole wing; outwardly oblique whitish streak at three-quarters, not extending to tornus; often indistinct, suffused whitish spot on dorsum opposite; series of white subterminal dots round apex, sometimes extending towards tornus; cilia brownish grey with three fuscous ciliary lines. Hindwing slightly narrower than forewing, trapezoidal, apex strongly produced; brownish fuscous; cilia paler, slightly longer than breadth of wing. Foreleg fuscous, tarsal segments paler distally; mid- and hindlegs greyish ochreous, hindleg with tarsus obscurely banded darker. Abdomen fuscous, paler ventrally. Genitalia, see figures 2h,22f.

Life history

Ovum. The foodplant is bracken (*Pteridium aquilinum*). The oviposition site is unknown, nor is it known whether the winter is passed as an egg or as a young larva. Bracken dies in the autumn and larvae have not been observed until the new growth appears in the spring.

Larva. Head fuscous. Body dark red; prothoracic plate blackish brown; spiracles white.

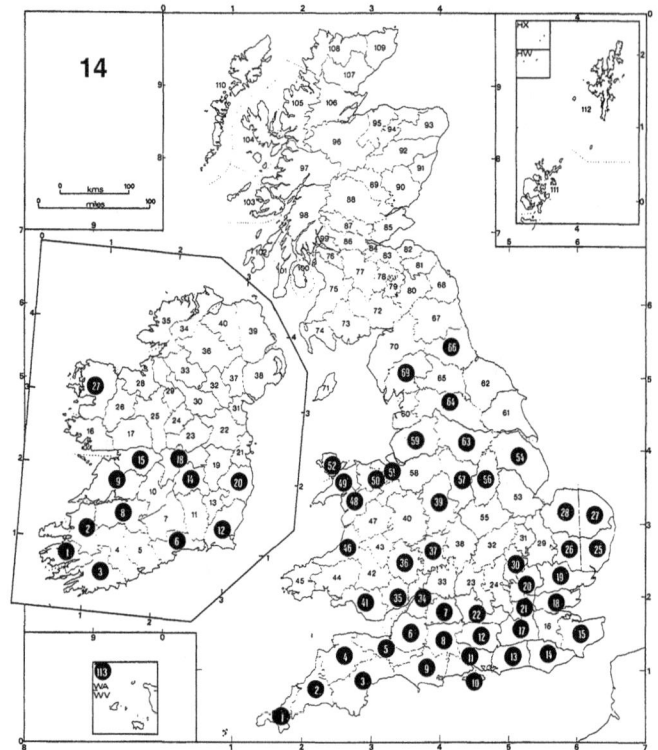

Monochroa cytisella

The larva feeds in the upper part of a stem, often in a slight swelling, or in a side shoot, causing it to droop. Rearing is difficult, because the stems quickly rot if enclosed in a container or dry out if left exposed. Success has been achieved by wrapping the bottom ends of the stems in a bunch of sphagnum moss enclosed in a nylon stocking, hanging it out of doors and watering the sphagnum daily (A. M. Emmet, pers.obs.). May–June.

Pupa. Amongst detritus on the ground; June–July.

Imago. Univoltine; July. Occasionally comes to light.

Distribution (Map 14)

Occurs in woods, on heaths and hillsides, mainly on acid soils. Widespread and fairly common in England and Wales as far north as Westmorland and Co. Durham; southern Ireland; the Channel Islands; not recorded from Scotland. Throughout Europe; eastern Siberia; Assam.

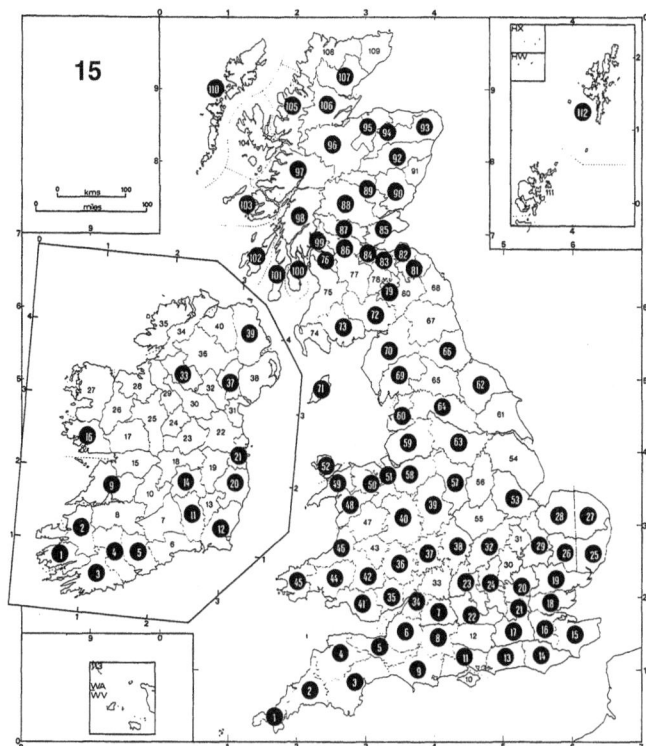

15

Monochroa tenebrella

MONOCHROA TENEBRELLA (Hübner)

Tinea tenebrella Hübner, [1817], *Samml.eur.Schmett.* **8**: pl.65, fig.434.
Tinea fuscocuprella Haworth, 1828, *Lepid.Br.*: 569.
Glyphipterix subcuprella Stephens, 1834, *Ill.Br.Ent.* (Haust.) **4**: 273.
Gelechia tenebrosella Zeller, 1839, *Isis, Leipzig* **1839**: 201.
Type locality: [Europe].

Description of imago (Pl.1, fig.18)

Wingspan 10–12mm. Head glossy black; antenna in male black, in female with apical one-third white; labial palpus black, segment 3 shorter than segment 2, segment 2 moderately thickened with appressed scales, segment 3 slender, curved, ascending. Thorax and tegulae blackish fuscous. Forewing unicolorous, glossy dark purplish fuscous; cilia concolorous at apex, paler towards tornus. Hindwing slightly narrower than forewing, grey; cilia one and one-half times width of wing, grey, paler towards anal angle. Foreleg purplish fuscous, tarsus paler; mid- and hindlegs fuscous. Abdomen, including anal tuft of male, fuscous; ovipos-itor of female whitish, protruding in set specimens. Genitalia, see figures 3a,23b.

Similar species. Eulamprotes unicolorella (Duponchel) (*q.v.*), in which segments 2 and 3 of the labial palpus are of equal length, and the antenna of the female is wholly dark fuscous.

Life history

Ovum. Laid on sheep's sorrel (*Rumex acetosella*).

Larva. Head and prothoracic plate blackish brown. Body reddish; pinacula brown (Meyrick [1928]).

The larva feeds in the rootstock from September onwards, overwintering in the feeding place when fully grown.

Pupa. In the larval feeding place. May.

Imago. Univoltine; June–July. The moth flies in sunshine amongst its foodplant, keeping close to the ground. It may be obtained by sweeping.

Distribution (Map 15)

Occurs in scrubland, poor grassland and on heaths. Widespread and locally common throughout the British Isles from the Channel Islands to Shetland, though more scarce in northern Scotland. Throughout Europe.

MONOCHROA LUCIDELLA (Stephens)

Cleodora lucidella Stephens, 1834, *Ill.Br.Ent.* (Haust.) **4**: 221.

Type locality: England; Brockenhurst, [Hampshire].

Description of imago (Pl.1, figs 19,20)

Wingspan 12–14mm. Head fuscous; antenna fuscous; labial palpus with segment 3 shorter than segment 2, ochreous to fuscous, segment 2 moderately thickened with appressed scales, segment 3 slender, almost straight, ascending. Thorax ochreous to fuscous. Forewing variable; ground colour ranging from pale ochreous-brown to fuscous with paler ochreous blotches, more prominent in lighter-coloured specimens, the most strongly expressed being an elongate blotch in disc distal of middle but proximal of black second discal stigma, the only stigma present; other pale blotches may occur along costa and dorsum, especially on costa at three-quarters, but these are obscure or absent in dark specimens; cilia greyish fuscous to ochreous, paler than wing, three pale streaks in costal cilia and weakly expressed darker ciliary line in terminal cilia. Hindwing equal in width to forewing, elongate, trapezoidal, tornal angle rather more obtuse than in most *Monochroa* spp., apex strongly produced; grey-

83

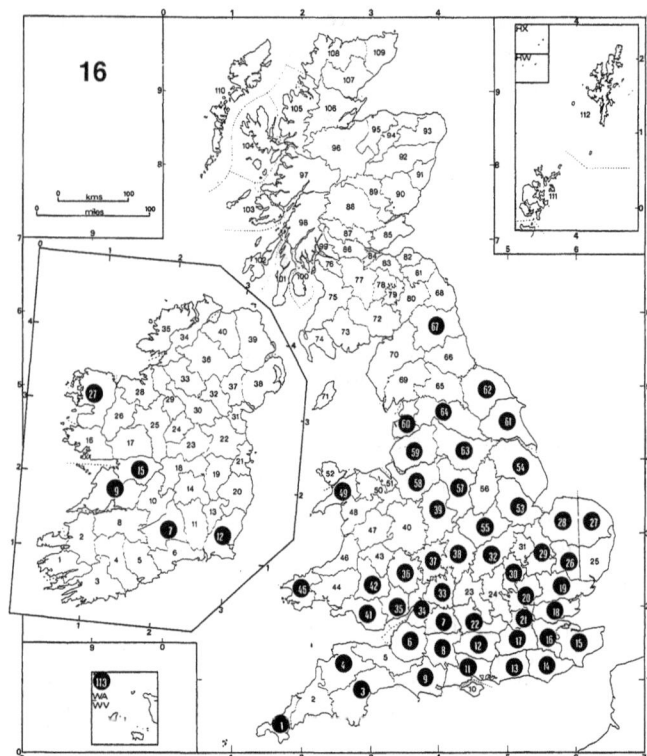

Monochroa lucidella

ish fuscous, slightly darker in specimens with fuscous forewings; cilia one and one-half times width of wing, glossy pale grey. Foreleg fuscous; mid- and hindlegs ochreous, tarsus of hindleg pale-banded. Abdomen fuscous, paler ventrally. Genitalia, see figures 3c,23a.

Life history

Ovum. Undescribed. Laid on common spike-rush (*Eleocharis palustris*); recorded in continental Europe also on bulrush (reedmace) (*Typha* spp.).

Larva. Head black. Prothoracic plate brownish grey; body pale greenish grey with ochreous tinge; pinacula black, minute. (Meyrick, [1928]).

The larva feeds in the stem without external evidence of its presence. May–June.

Pupa. Undescribed. Probably in the feeding place.

Imago. Univoltine; June–July. The moth flies in late afternoon and may sometimes be swept in numbers from the foodplant. It also flies by night and comes to light.

Distribution (Map 16)

Occurs in freshwater situations such as fenland, marshes, water-meadows and shallow ponds. Widespread and locally common in England and Wales as far north as Northumberland; southern Ireland; the Channel Islands. Not recorded from Scotland. Northern and central Europe; Sicily; eastern Siberia.

MONOCHROA PALUSTRELLUS (Douglas)

Ypsolophus? palustrellus Douglas, 1850, *Proc.ent.Soc. Lond.* (N.S.) **2**: 14.

Type locality: England; Yaxley, [Suffolk].

Description of imago (Pl.1, fig.21)

Wingspan 15–19mm. Head whitish ochreous; antenna with scape fuscous, flagellum proximally fuscous, distally paler, banded fuscous; labial palpus with segments 2 and 3 of equal length, segment 2 strongly thickened with appressed scales forming apical tuft, whitish ochreous above, fuscous beneath, segment 3 angled upwards, slightly curved, slender, outer side fuscous, inner side whitish ochreous with black apex. Thorax whitish ochreous with longitudinal black line. Forewing yellowish ochreous, veins marked white; interneural spaces streaked blackish fuscous, especially in subapical area; plical stigma represented by elongate black streak, first discal well beyond, elongate and sometimes broken into two spots, second discal round, all stigmata strongly white-edged; subdorsal area sometimes wholly suffused blackish fuscous; cilia whitish, costal cilia with one, terminal cilia with two darker ciliary lines. Hindwing differing from those of other British *Monochroa* to such an extent that Douglas, when he named the species, did not recognize it as a gelechiid; slightly broader than forewing, dorsal margin very slightly arched, tornus gently curved, termen oblique, apex shortly and bluntly produced; light grey, subdorsal area sometimes tinged ochreous; cilia equal in length to width of wing, whitish ochreous with darker ciliary line at base. Foreleg fuscous, annulate whitish ochreous; mid- and hindlegs whitish ochreous. Abdomen fuscous, paler ventrally; anal tuft of male whitish ochreous. Genitalia, see figures 3b,23c.

Life history

Ovum. Undescribed. Laid on curled dock (*Rumex crispus*), possibly sometimes on other *Rumex* spp.

Larva. Head deep orange. Prothoracic plate brownish ochreous to dark brown with blackish markings; body pinkish white with irregular dorsal, subdorsal and subspiracular dull crimson lines; pinacula brown.

The larva feeds internally in the rootstock, stem or leaf petioles of its foodplant from April to June. It is not known whether the winter is passed in the egg stage or as a small larva.

Pupa. In the rootstock; June–July.

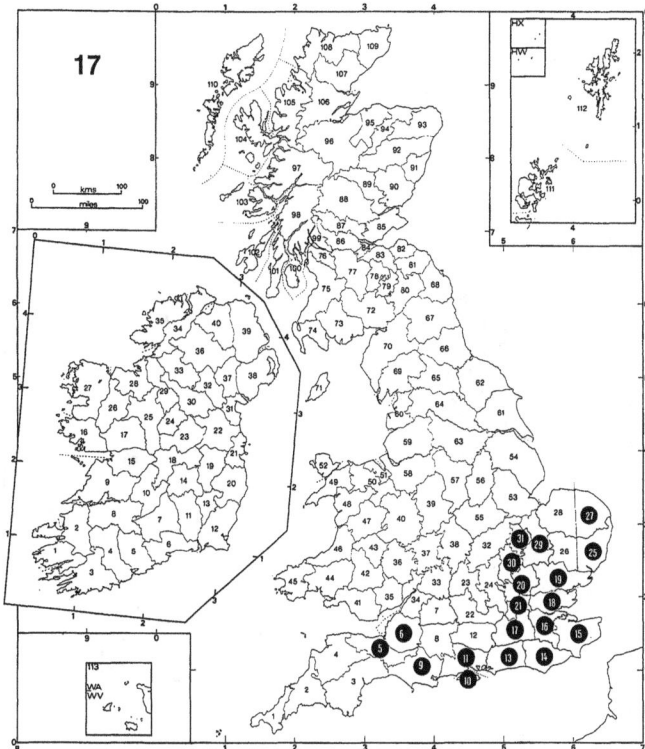

Monochroa palustrellus

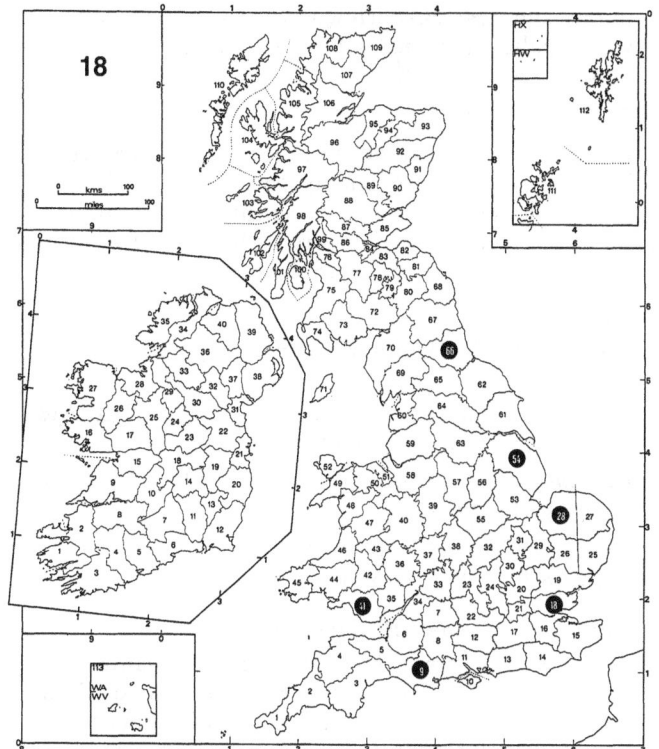

Monochroa tetragonella

Imago. Univoltine; late June to August. It flies at night and readily comes to light.

Distribution (Map 17)

Occurs on waste ground, dry pastures and sand-dunes. Widespread but local south-east of a line from Somerset to Norfolk. Western, central and northern Europe.

MONOCHROA TETRAGONELLA (Stainton)

Gelechia tetragonella Stainton, 1885, *Entomologist's mon. Mag.* 22: 99.

Type locality: England; Greatham Marsh, Co. Durham.

NOTE. Owing to misinformation, Stainton gave the type locality as Redcar, Yorkshire (Sutton & Beaumont, 1989).

Description of imago (Pl.1, fig.22)

Wingspan 9–11mm. Head brownish fuscous, frons mixed whitish; antenna dark fuscous; labial palpus with segments 2 and 3 of equal length, white, segment 2 mixed fuscous on outer side, segment 3 with apex dark fuscous. Thorax and tegulae brownish fuscous. Forewing brownish fuscous, sparsely irrorate whitish; short blackish elongate mark on fold at one-fifth basad

of plical stigma; plical and first discal stigmata elongate, blackish, first discal well beyond plical; second discal stigma a blackish dot; above this on costa indistinct whitish outwardly oblique dash, and similar one opposite this at tornus, the two almost meeting to form an acutely angled fascia; series of indistinct pale spots around apex and along termen; cilia ochreous-grey with indistinct interrupted darker ciliary line, fading toward tornus. Hindwing light fuscous; cilia pale greyish ochreous. Fore- and midlegs brownish fuscous, tarsi indistinctly ringed whitish at joints; hindleg pale greyish ochreous, outer tibial spurs dark fuscous. Abdomen brownish fuscous, paler ventrally. Genitalia, see figures 3f,23f.

Life history

Ovum. Laid on sea-milkwort (*Glaux maritima*).

Larva. Head ochreous-yellow. Prothoracic plate dark brown; body crimson-reddish; pinacula minute, black (Meyrick [1928]).

The larva mines the stem and the root, causing brown discoloration and often killing the plant. April–May.

Pupa. Undescribed.

Imago. Univoltine; June–July. Rests on salt-marsh

plants such as sea-wormwood with its head up and wings tightly wrapped round its abdomen, then flies between 20.00 and 21.00hrs BST.

Distribution (Map 18)

Frequents the drier borders of salt-marshes. It has been recorded from widely separated localities in Dorset, Essex, Norfolk, Lincolnshire, Co. Durham – the type locality, and Glamorgan; there are few recent records. Northern and north-western Europe.

MONOCHROA CONSPERSELLA (Herrich-Schäffer)

Gelechia conspersella (Herrich-Schäffer), 1854, *Syst. Bearb.Schmett.Eur.* **5**: 177, pl.78, fig.591.

Anacampsis quaestionella (Herrich-Schäffer), 1854, *Ibid.* **5**: 193, pl.77, fig.587.

Gelechia morosa Mühlig, 1864, *Stettin.ent.Ztg* **25**: 101.

Type locality: Austria; Vienna.

Description of imago (Pl.1, fig.23)

Wingspan 11–12mm. Head dark grey; antenna three-quarters length of forewing, blackish fuscous; labial palpus with segments 2 and 3 of equal length, segment 2 thickened with appressed fuscous scales, segment 3 slender, curved, grey. Thorax grey. Forewing fuscous, irrorate grey; stigmata, when visible, black, first discal spot well beyond plical; cilia fuscous, paler towards tornus. Hindwing narrower than forewing, fuscous; cilia concolorous at apex, grading to pale fuscous at anal angle. Foreleg fuscous, tarsus narrowly annulate white; mid- and hindlegs greyish fuscous, tarsus of hindleg tinged ochreous, annulate whitish, tibia annulate whitish at spurs. Abdomen fuscous. Genitalia, see figures 3d,23d.

In f. *quaestionella* Herrich-Schäffer the forewing has a whitish, outwardly oblique streak on the costa at three-quarters.

Life history

Ovum. Undescribed. Laid on yellow loosestrife (*Lysimachia vulgaris*).

Larva. Head pale yellowish, mouth-parts and ocelli black. Prothoracic plate whitish, central area with mixed black and brown pigmentation, tapered posteriorly with broad median sulcus; body green with purplish brown dorsal and lateral stripes and transverse markings, giving a reticulate appearance, more strongly expressed on posterior segments and almost obscuring ground colour; anal plate marbled whitish and blackish; thoracic legs dark brown.

In spring the larva makes a loose spinning in a terminal shoot. Feeding consists mainly in mining down the

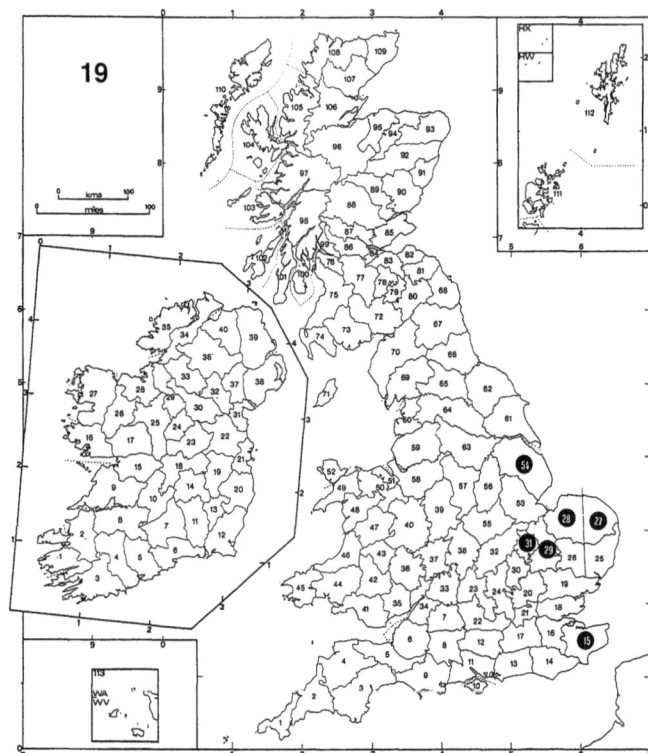

Monochroa conspersella

stem for a distance of up to 2cm. The larva probably first mines downwards in a narrow gallery, then turns and mines back in a broader one, since there is frass at both ends. Young leaves are eaten to a lesser extent. Larvae are fully fed between the middle and end of May. The overwintering stage is unknown.

Pupa. Probably in a cocoon spun amongst detritus; in captivity in tissue. June.

Imago. Univoltine; July–August. Comes to light.

Distribution (Map 19)

Occurs in fens and damp situations. Very local, but can be plentiful where it is found. Restricted to suitable sites in eastern counties from Kent to Lincolnshire. Northern and central Europe; Siberia.

MONOCHROA HORNIGI (Staudinger)

Doryphora hornigi Staudinger, 1883, *Stettin.ent.Ztg* **44**: 184.

Type locality: Austria; Vienna.

Description of imago (Pl.1, fig.24)

Wingspan 9–12mm. Head fuscous, frons beige; antenna dark fuscous, indistinctly annulate greyish white in proximal one-half, distally with irregularly distributed silvery white rings; labial palpus with segments 2 and 3 of equal length, segment 2 moderately thickened with appressed scales, fuscous, brindled whitish on inner side and with apex whitish, segment 3 fuscous with median whitish annulation. Thorax and tegulae dark fuscous. Forewing dark fuscous with obscure paler irroration; sometimes obscure darker outwardly oblique fascia from costa at one-half not reaching dorsum; plical and first discal stigmata elongate, blackish, rather indistinct and sometimes obsolete, first discal stigma well beyond plical; second discal stigma a black dot; series of indistinct greyish white spots along distal one-third of costa, around apex and along termen to tornus; cilia slightly paler than wing with darker basal ciliary line. Hindwing narrower than forewing, dark brownish fuscous; cilia brownish fuscous becoming paler on dorsum towards anal angle. Foreleg fuscous, tarsal joints annulate whitish; mid- and hindlegs greyish fuscous with obscure paler annulations. Abdomen dark fuscous dorsally, greyish white ventrally. Genitalia, see figures 3g,23e.

Life history

Based mainly on the description by Heckford & Langmaid (1999) who, in 1997, found six larvae by splitting open dead and prostrate stems of *Persicaria* in a boggy area beside a ditch at a locality where an adult had previously been taken at light.

Ovum. Laid on pale persicaria (*Persicaria* (=*Polygonum*) *lapathifolia*); or possibly other *Polygonum* spp. The dead persicaria from which Heckford & Langmaid reared adults could not be identified to species.

Larva. Head dark reddish brown. Prothoracic plate with median sulcus, slightly paler brown, speckled darker posteriorly; body light orange-brown; interrupted whitish subdorsal, sublateral and lateral lines from abdominal segment 1; pinacula dark brown, ringed whitish on thoracic segments 2 and 3; anal plate concolorous with prothoracic plate, with some darker speckling; thoracic legs translucent, a black semicircular line on body at base of each leg; prolegs concolorous with body.

The larva mines a stem in the autumn without showing any sign of its presence. By mid-November it spins a hibernaculum at a node at the lower end of its mine, in which it rests curled in a semicircle. The hibernaculum is semitransparent, covered with reddish brown material which is either frass or the chewed inner portion of the stem or a combination of both.

Pupa. Reddish brown; wings paler. According to Buhl

Monochroa hornigi

et al. (1996) pupation takes place in the hibernaculum, but this was not the case with the British material. Stems containing larvae had been overwintered in linen bags lying on the ground, but when these were brought in at the beginning of May, it was found that the material had rotted and four of the six larvae had escaped. One larva had left its hibernaculum and was walking about. When placed in a container with tissue, it promptly spun a cocoon. For a similar apparent discrepancy in pupation habits between British and Continental populations, see *M. suffusella* (Douglas) (p.89).

Imago. Univoltine; July–August, but in captivity adults emerged at the end of May. Comes sparingly to light.

Distribution (Map 20)

Known in Britain mainly from specimens taken in light-traps; the foodplant grows on damp bare ground and in ditches. First recorded in England from Buckingham Palace garden in 1963 (Bradley & Mere, 1964). Since then it has been taken, mostly as single specimens, in ten vice-counties south of a line from Wiltshire to Bedfordshire and in south Yorkshire. The indications are that it is widespread and possibly fairly common but elusive. Northern and central Europe.

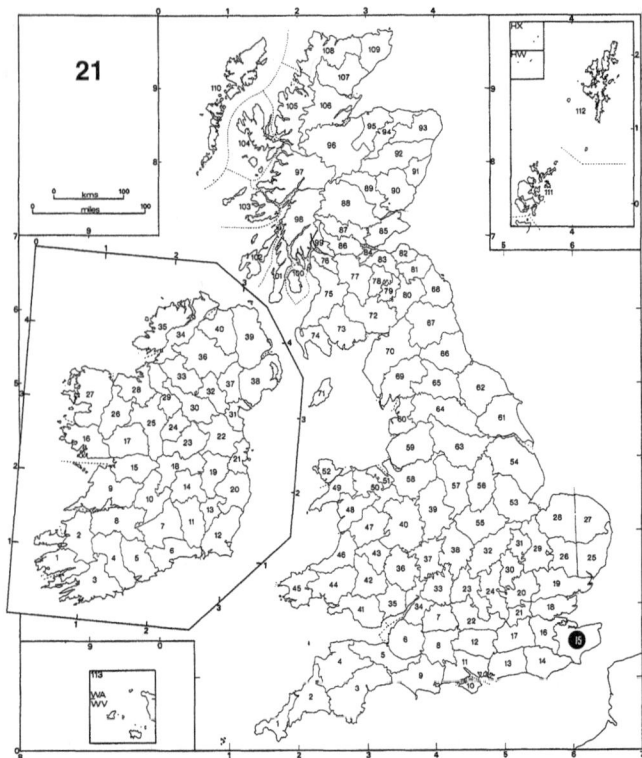

Monochroa niphognatha

MONOCHROA NIPHOGNATHA (Gozmány)

Aristotelia (Xystophora) niphognatha Gozmány, 1953, *in* Székessy, *Bátorliget élövilaga*: 390, 485, fig.38.

Type locality: Hungary.

Description of imago (Pl.1, fig.25)

Wingspan 13–15mm. Head whitish, mixed pale brown on vertex; antenna brown, basal two-thirds annulate dark fuscous, apical one-third with four to six whitish rings regularly spaced every fourth segment; labial palpus with segments 2 and 3 of equal length, segment 2 mixed whitish and dark fuscous, paler on inner side, segment 3 whitish beneath, above and at sides whitish in basal half and fuscous in apical half. Thorax pale brown; tegulae paler. Forewing brown, paler towards base and costa, scales with whitish bases giving a granular appearance; subcostal brownish fuscous dots at one-sixth and one-third; between and dorsad of these, ill-defined elongate brown mark above fold, sometimes obsolete; plical stigma elongate, brownish fuscous; first discal stigma absent; second discal stigma a black dot; cilia concolorous with wing, paler towards tornus, with two darker ciliary lines, the outer rather indistinct, fading toward tornus. Hindwing light grey; cilia light grey-

ish buff. Fore- and midlegs fuscous, ringed paler at joints; hindleg buff. Abdomen fuscous dorsally, whitish buff ventrally. Genitalia, see figures 4a,24e.

Life history

The early stages are unknown in Britain. In Denmark the larva has been found in September feeding in the stem of amphibious bistort (*Persicaria* (=*Polygonum*) *amphibia*), a small hole indicating its presence (Buhl *et al.*, 1989).

Imago. Univoltine; June–July. Comes to light.

Distribution (Map 21)

Known in Britain only from a single marshy locality in Kent where it was first taken in June 1984 by J. M. Chalmers-Hunt (1985) and has occurred regularly since. Very local in north-western and central Europe; Latvia.

MONOCHROA SUFFUSELLA (Douglas)

Gelechia suffusella Douglas, 1850, *Trans.ent.Soc.Lond.* (N.S.) **1**: 64.

Gelechia oblitella Doubleday, 1859, *Zoologist synonymic List Br.Lepid.*: 30.

Type locality: England; Whittlesea Mere, Norfolk.

Description of imago (Pl.1, figs 26,27)

Wingspan 10–12mm. Head whitish ochreous; antenna fuscous, annulate whitish ochreous, more distinctly in apical half; labial palpus segment 3 slightly shorter than segment 2, segment 2 slightly thickened with appressed scales, whitish ochreous mixed fuscous on outer side, segment 3 slender, whitish ochreous with apex fuscous. Thorax whitish ochreous. Forewing whitish ochreous; plical and first discal stigmata occasionally indicated by one or two black scales; second discal stigma well distad of plical, black, distinct; black spot on costa directly above second discal stigma; apical area slightly darker ochreous; cilia whitish ochreous with obscure darker ciliary line. Hindwing narrower than forewing, trapezoidal, tornus roundly angled almost at a right-angle, termen weakly sinuate, apex strongly produced; pale grey; cilia slightly longer than breadth of wing, concolorous. Foreleg ochreous, tarsus paler; mid- and hindlegs pale ochreous. Abdomen ochreous. Genitalia, see figures 3e,24f.

Specimens from Wicken Fen, Cambridgeshire, where the recorded foodplant does not occur, differ as follows.

Wingspan greater (12–14mm); forewing deeper ochreous; hindwing with tornal angle slightly more obtuse, giving more oblique termen and with apex more abruptly produced. Wakely (1936) stated that moths reared from Burnham-on-Crouch, Essex, (see below) were much darker than normal specimens. The adults fly later in the year. No genitalic differences have been observed.

Life history

Based on information received from P. H. Sterling (*in litt.*)

Ovum. Laid on common cotton-grass (*Eriophorum angustifolium*). Foodplant at Wicken Fen unknown. Two specimens in the Wakely collection (Cambridge), determined after dissection by F. N. Pierce, were reared on 3 July 1935 from dead stems collected on a salt-marsh at Burnham-on-Crouch, Essex. Wakely (*loc.cit.*) thought the stems might have been those of triangular club-rush (*Schoenoplectus* (=*Scirpus*) *triqueter*), but this is not recorded in the area. Sea club-rush (*Bolboschoenus* (=*Scirpus*) *maritimus*) is plentiful, but the stems may have been a pupation site rather than the foodplant.

Larva. Head translucent dark brown with mandibles visible beneath. Prothoracic plate small, subtriangular, confined to dorsal area, pale brown, darker anteriorly; body dull yellow with internal organs appearing whitish; anal plate subtriangular, pale brown, darker laterally; thoracic legs translucent yellowish brown; prolegs dull yellowish, crotchets brown. According to Buhl *et al.* (1992), in early instars the larva is almost transparent.

The larva mines the stem and the lowest part of the leaves in the autumn and overwinters in the mine (Buhl *et al.*, *loc.cit.*). In spring it at first mines the reddish portion at the base of an overwintering leaf, making an obscure, pale red, short, sinuous gallery in which the larva itself is invisible. When it progresses into the green part of the leaf, the gallery becomes pale green and widens, filling all or most of the width of the leaf. The larva and its greenish frass are still only barely visible and only when viewed from above. The length of the green mine varies from 20–65mm; the larva appears not to change leaves. When full-fed, it leaves through an exit-hole on the upperside of the leaf.

Pupa. In a cocoon spun amongst detritus. 'In the mine' (Buhl *et al.*, *loc.cit.*); *cf.* the varying pupation habits of *M. hornigi* (Staudinger) (p.87) and *M. moyses* Uffen (p.92).

Imago. Univoltine; June; at Wicken Fen, July–mid-August. The adults may be disturbed by day or swept from the foodplant. Michaelis (1951) observed the moths flying freely on the Cheshire mosses in late evening sunshine from about 21.00 hrs BST onwards, generally amongst cotton-grass. Later they come to light.

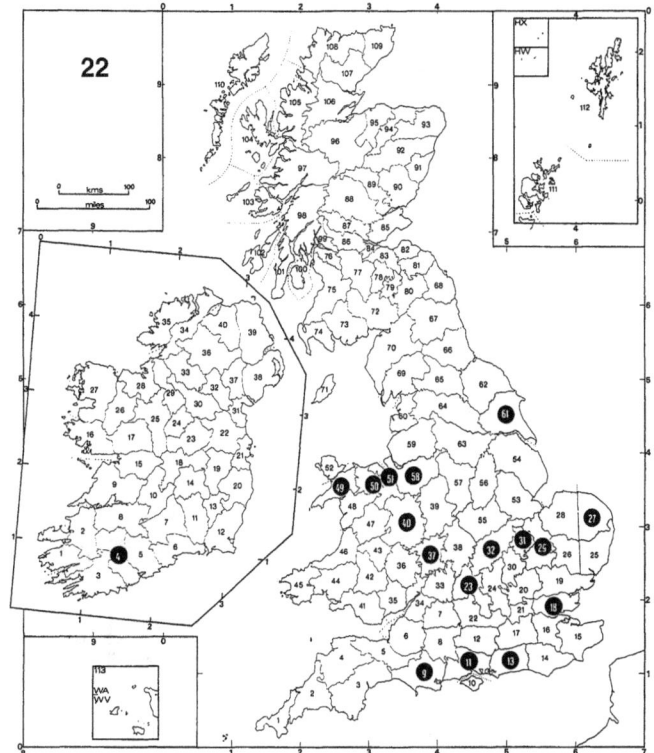

Monochroa suffusella

Distribution (Map 22)

Occurs on bogs, fens, swamps and salt-marshes; the race on cotton-grass is found mainly on acid soils. Local in England as far north as south-east Yorkshire; North Wales; Douglas River, Co. Cork, Ireland (Bond, 1996). Northern and central Europe.

MONOCHROA LUTULENTELLA (Zeller)

Gelechia (*Brachmia*) *lutulentella* Zeller, 1839, *Isis, Leipzig* **1839**: 201.

Type locality: Germany; Oberweisen.

Description of imago (Pl.1, figs 28,29)

Wingspan 14–16mm. Head ochreous-brown to fuscous, slightly glossy; antenna brown to fuscous, annulate whitish, irregularly in apical area; labial palpus segment 3 slightly shorter than segment 2, segment 2 thickened with brown to fuscous scales, paler above, segment 3 curved, slender. Thorax brown to fuscous. Forewing dichromatic, either light ochreous-brown or dark fuscous-brown, both colour forms occurring in the same populations; only second discal stigma present, black, round, obscure in dark forms; sometimes obscure paler streaks or blotches mainly in distal half

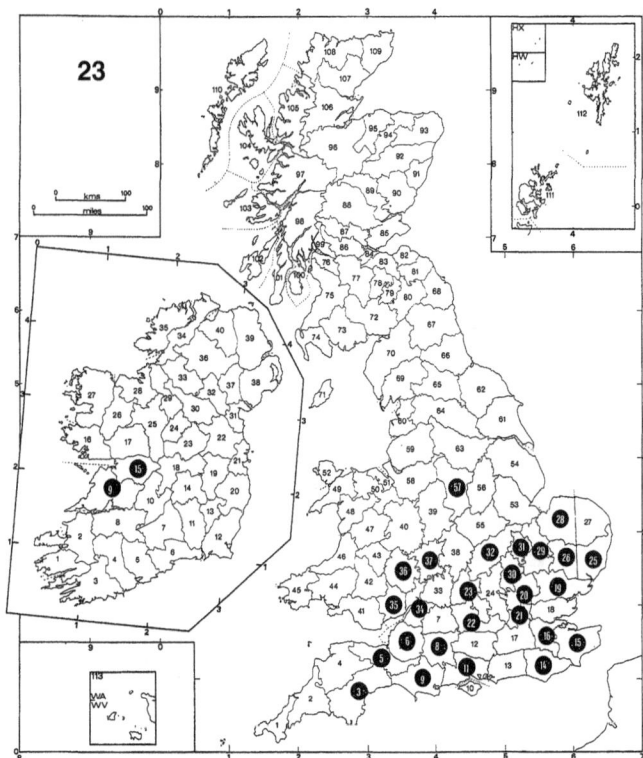

Monochroa lutulentella

of wing; dark spot at apex in lighter forms; cilia concolorous, with darker ciliary line in terminal cilia and sometimes whitish flecks in costal cilia. Hindwing as broad as forewing, trapezoidal, tornus obtusely angled, apex strongly produced; light grey, cilia concolorous, on dorsum with whitish basal line. Foreleg concolorous with forewing, tarsus pale-ringed; mid- and hindlegs pale greyish ochreous. Abdomen with segments 1–4 ochreous-yellow above, other segments fuscous; ventral surface paler; anal tuft of male ochreous. Genitalia, see figures 4c,24c.

Life history

Unknown in Britain. On the Continent the larva is reported to feed in the rootstock of meadowsweet (*Filipendula ulmaria*).

Imago. Univoltine; late June–early August. Comes readily to light.

Distribution (Map 23)

Occurs principally in fens and damp localities, but also in dry fields (Meyrick, [1928]) and has even been taken at light in suburban gardens. Widespread but local south and east of a line from Devon to Herefordshire and The Wash; Derbyshire; South Wales (Mon-

mouthshire); western Ireland (the Burren, Co. Clare and the adjacent part of Co. Galway). In some localities such as Wicken Fen, Cambridgeshire, it is common. Northern and central Europe; Asia Minor.

MONOCHROA ELONGELLA (Heinemann)

Doryphora elongella Heinemann, 1870, *Schmett.Dtl. Schweiz* (2)**2**(I): 307.
Aristotelia servella sensu auctt., *nec* Zeller, 1839.
Aristotelia micrometra Meyrick, 1935, *Entomologist* **68**: 121.
Type locality: Germany; Brunswick.

Description of imago (Pl.1, fig.30)

Wingspan 12–15mm. Head dark ashy grey with slight purplish tinge, frons in male paler grey, in female whitish ochreous; antenna of male dark grey, of female whitish, obscurely ringed fuscous, apical one-third in both sexes darker with three narrow rings and apex whitish; labial palpus blackish grey, apex of segments 2 and 3 whitish, in female with inner surface whitish. Thorax dark ashy grey. Forewing fuscous, finely irrorate whitish; stigmata darker, obscure, first discal much beyond plical, both elongate, second discal rather more distinct; indistinct darker elongate marks in disc towards base, beneath costa before middle and before first discal stigma; small whitish spot on costa at two-thirds and similar smaller spot slightly proximal on dorsum; four minute transverse streaks of whitish irroration between costal spot and apex; four minute white dots at base of cilia on termen; cilia brownish grey with two obscure darker ciliary lines. Hindwing slightly narrower than forewing, trapezoidal, apex produced; grey; cilia light brownish grey; in male small erect hair-pencil on dorsum at base. Foreleg dark fuscous, apex of tibia and tarsus sharply ringed white; mid- and hindlegs pale fuscous. Abdomen grey; dorsal surface of segments 2 and 3 suffused ochreous-yellow. Genitalia, see figures 4e,24d.

Distinguished from related species by the white rings on the antenna and the ochreous-yellow base of the abdomen.

Life history

Unknown in Britain; in continental Europe, the larva is reported to feed on silverweed (*Potentilla anserina*).

Imago. Univoltine; June–July. Flies by night and comes to light.

Distribution (Map 24)

Mostly coastal in Britain, associated primarily with sand-dunes (Parsons, 1995[1996]), but in 2001 it was found in some numbers on Salisbury Plain, Wiltshire

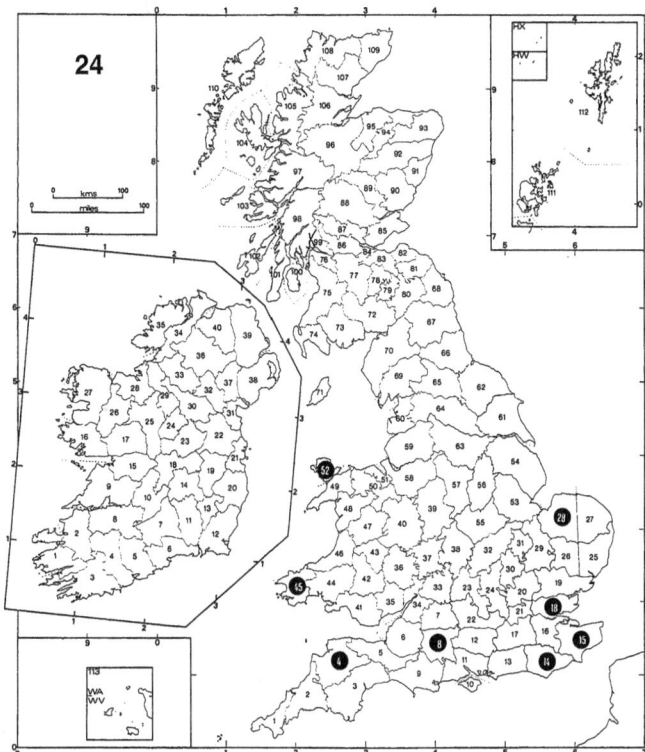

Monochroa elongella

Monochroa arundinetella

(M. S. Parsons, pers.comm.) In the past it has been recorded singly or in small numbers from north Devon, south-eastern England, Norfolk, Pembrokeshire and Anglesey; the scattered nature of these captures suggests that it has been overlooked or under-recorded. Northern and central Europe as far east as Russia, occurring also inland in continental Europe.

MONOCHROA ARUNDINETELLA (Stainton)

Gelechia arundinetella Stainton, [1857], *Entomologist's Annu.* **1858**: 91.

Type locality: England; Hackney Marshes, Essex (now Greater London).

Description of imago. (Pl.1, fig.31)

Wingspan 9–10mm. Head fuscous, frons whitish grey; antenna fuscous, obscurely annulate whitish; labial palpus whitish, segments 2 and 3 fuscous at apex. Thorax fuscous. Forewing elongate, apex acute; ochreous-fuscous; subbasal dark costal streak and another slightly distad on fold; elongate blackish streak in disc at two-thirds; immediately above this, whitish spot on costa preceded and followed by darker scales; sometimes similar spot on dorsum opposite, preceded by

darker suffusion; cilia pale grey, costal cilia with four pale spots, terminal cilia with slightly darker ciliary line. Hindwing slightly narrower than forewing, trapezoidal, tornus roundly angled, apex sharply produced; grey; cilia pale grey. Fore- and midlegs with femora and tibiae fuscous, tarsi dark fuscous annulate greyish ochreous; hindleg greyish ochreous, paler inwardly, tarsus with paler annulation. Abdomen fuscous. Genitalia, see figures 4b,24a.

Life history

Ovum. Laid on a leaf of greater pond-sedge (*Carex riparia*) or, less often, lesser pond-sedge (*C. acutiformis*). Possibly overwinters in this stage.

Larva. Head blackish brown. Body slender, dull whitish; thoracic segment 1 with black lateral spot.

The larva mines the leaves in a long, slender whitish or pale brown gallery, upwards or downwards, changing direction and also changing leaves. When fully fed, it spins a white cocoon within the leaf just above water-level. March–May. A species difficult to rear because mined leaves picked in March are hard to keep in fresh condition, and if left until April the leaves begin to wither, rendering the mines inconspicuous (Stainton, 1867a).

Pupa. In the cocoon described above. May–June.

Imago. Univoltine; June–July. Flies at dusk and into the night and occasionally comes to light.

Distribution (Map 25)

Occurs on river banks and in swamps or fens in England south of a line from the Severn estuary to The Wash. Scarce, but possibly under-recorded. North-western Europe, extending eastwards to Hungary.

MONOCHROA MOYSES Uffen

Monochroa moyses Uffen, 1991, *Br.J.Ent.nat.Hist.* **4**: 1.
Type locality: England; East Mersea, Essex.

Description of imago (Pl.1, fig.32)

Wingspan 8–9mm. Head grey; antenna grey with faint paler annulation; labial palpus with segments 2 and 3 of equal length, segment 2 thickened with grey appressed scales, segment 3 slender, grey, apex fuscous. Thorax grey. Forewing grey, bases of scales paler; only second discal stigma present, black; strongly oblique whitish streak from costa at three-quarters, soon curving towards apex; costa narrowly fuscous distal to streak; cilia grey with darker ciliary line in costal cilia and round apex where it terminates. Hindwing narrower than forewing, termen slightly oblique, grey; costal and apical cilia slightly darker, dorsal cilia fading to whitish grey at anal angle; cilia one and one-half times width of wing. Legs grey, spurs darker. Abdomen fuscous, ventrally with pale grey longitudinal stripe. Genitalia, see figures 4f,24b.

Similar species. M. *hornigi* (Staudinger), which has the forewing brownish fuscous and obscure paler annulations on the tibia of the hindleg which are not present in *M. moyses*.

Life history

Ovum. Laid on sea club-rush (*Bolboschoenus* (=*Scirpus*) *maritimus*). The oviposition site is not known.

Larva. c. 6mm long, very slender. Head pale brown, sutures, mouth-parts and eye-spots darker brown. Body pinkish white; prothoracic plate concolorous and translucent, the musculature posterior to the head showing through as a brownish spot; paired oval brown gonads showing through at abdominal segment 5; anal plate, pinacula and thoracic legs concolorous with body (J. R. Langmaid, pers.comm.).

The larva mines the leaves, generally upwards. A film of silk is spun over the entrance hole and the early part of the mine is also lined with silk; the larva, which is highly active, leaves the mine to excrete and the silk provides a secure foothold. It sometimes changes

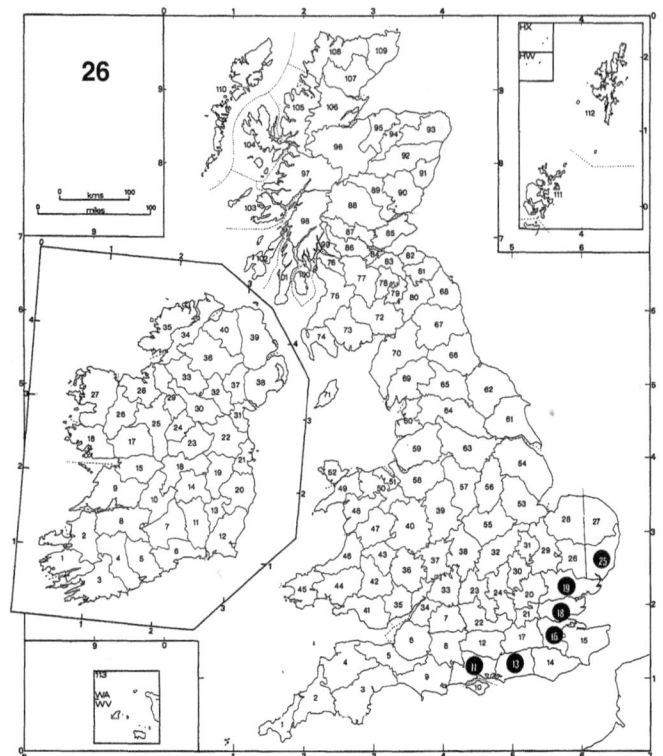

Monochroa moyses

mines, generally after ecdysis, the new mine usually being in the same leaf. Mid-August to mid- or late September.

Pupa. Undescribed. Pupation habits vary. Langmaid (1992) observed an over-wintered larva in early May walking around in its container prior to pupating in tissue. He found small holes in the sea club-rush stems leading to chambers containing a little frass, and concluded that this was where hibernation had taken place. However, in 1996 Langmaid (1997) found mines still tenanted in mid-October at another locality, the larvae being in slight cocoons spun in any part of the mine. The larvae overwintered and in spring pupated in their mines, where he found the pupal exuviae after emergence had taken place. He concludes 'It appears that *M. moyses* has different habits in different localities, but constant for each individual population in any one year. Whether the habit varies from year to year will necessitate annual monitoring of different colonies'. It appears that *M. suffusella* (Douglas) (p.000) has similar behavioural flexibility, since Sterling (*in litt.*) records that his larvae from Dorset vacated their mines prior to overwintering, whereas Buhl *et al.* (1992) state that larvae reared in Denmark overwintered and pupated in their mines. See also *M. hornigi* (Staudinger) (p.86).

Imago. Univoltine; June–July. Little is known about the moth's habits in the wild; one specimen was taken at light.

History and distribution (Map 26)

The earliest specimen was taken by E. S. Bradford on 3 July 1971 at Mucking, Essex, but remained unidentified. The larvae were known for a number of years before one was successfully overwintered and the adult reared (Uffen, 1991). The specimen, however, became too damaged for description. The first moths in good condition were reared by J. R. Langmaid in 1987. The species occurs where the foodplant grows in brackish ditches and borrow-dykes behind sea-walls. Locally common in coastal areas of the southern and eastern counties from Hampshire to Suffolk. The Netherlands.

MONOCHROA DIVISELLA (Douglas)

Gelechia divisella Douglas, 1850, *Trans. ent. Soc. Lond.* (N.S.) **I**: 60.

Type locality: England; Whittlesea Mere, Norfolk.

Description of imago (Pl.1, fig.33)

Wingspan 15–16mm. Head smooth-scaled, ochreous; antenna with apical one-third serrate in male, pale fuscous ringed whitish ochreous, scape with anterior margin blackish; labial palpus with segment 3 shorter than segment 2, segment 2 with appressed scales, ochreous above, fuscous with whitish apex below, segment 3 ochreous. Thorax and tegulae ochreous. Forewing bicoloured; costal two-fifths pale ochreous lightly irrorate fuscous, dorsal three-fifths ochreous-fuscous, paler subdorsally in basal one-quarter and darker in apical area, this colour change being abrupt along axis extending from middle of base to apex; minute elongate black subcostal spot at one-fifth; plical stigma elongate, second discal round, both black, first discal absent; cilia fuscous at apex, grading to whitish ochreous at tornus, with obscure, diffuse darker ciliary line. Hindwing trapezoidal, tornus rounded, apex moderately produced; pale grey, veins darker; cilia concolorous. Foreleg dark fuscous, tarsal joints obscurely ringed ochreous; midleg and hindleg with femora fuscous, tibiae and tarsi ochreous. Abdomen pale fuscous; anal tuft of male ochreous. Genitalia, see figures 4d,25a.

Life history

Unknown in Britain. It is stated that the larva feeds on iris (*Iris* spp.) in continental Europe.

Imago. Univoltine; June–July. Comes to light.

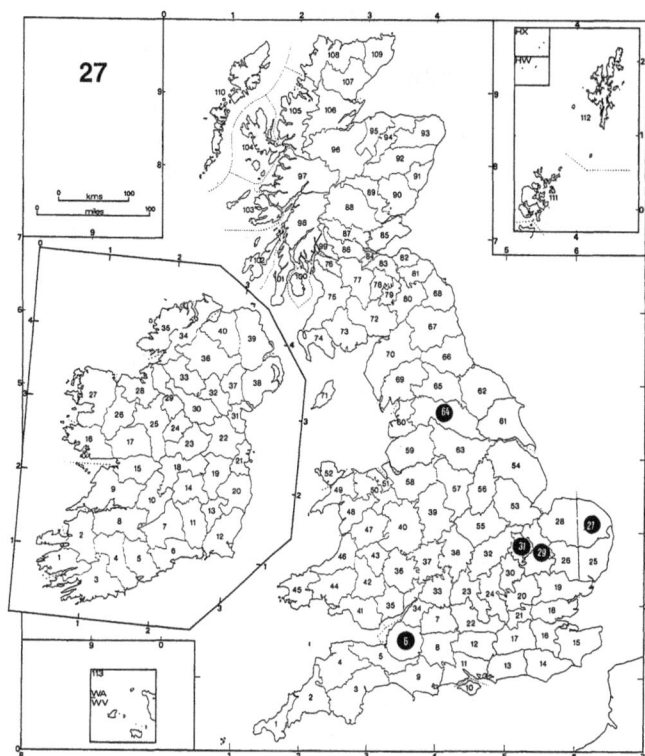

Monochroa divisella

Distribution (Map 27)

Occurs in fenland and similar damp localities. It has been recorded regularly from the Norfolk Broads, the type locality; also intermittently from the fens of Huntingdonshire and Cambridgeshire; Shapwick Fen, Somerset, two in 1936 (Turner, 1955), but not since; Yorkshire (Meyrick, [1928]), but the source of this record was not traced by Sutton & Beaumont (1989); however, the late D. Hall-Smith (pers.comm.), formerly in charge of entomology at the Leicester Museum where the collection of the Yorkshire entomologist A. Smith is housed, reported it from mid-west Yorkshire (VC64). Western and central Europe; Latvia.

CHRYSOESTHIA Hübner

Chrysoesthia Hübner [1825], *Verz. bekannt. Schmett.*: 422.

Microsetia Stephens, 1829, *Nom. Br. Insects*: 49.

A Holarctic genus represented by four species in North America and six in Europe, of which two are found in Britain and one in Ireland. One of these also occurs in South Africa.

Imago. The British representatives are bivoltine. They are small moths with orange or yellow markings on the forewing. The hindwing is narrower than the forewing, relatively short and trapezoidal. Male genitalia with gnathos absent; sacculus spined; aedeagus with or without cornuti. Female genitalia with ovipositor a pair of spined pads; corpus bursae without signum.

Larva. The British representatives are leaf-miners, feeding on Chenopodiaceae.

Key to species (imagines) of the genus *Chrysoesthia*

–	Forewing with ground colour reddish orange .... *drurella* (p. 94)
–	Forewing with ground colour dark purplish fuscous .... *sexguttella* (p. 95)

CHRYSOESTHIA DRURELLA (Fabricius)

Tinea drurella Fabricius, 1775, *Syst.Ent.*: 666.

Tinea hermannella sensu auctt., *nec* Fabricius, 1781.

Phalaena schaefferella sensu Donovan, 1796, *Nat.Hist. Br.Insects* **5**: 99, pl.175, *nec* Linnaeus, 1758.

Tinea zinckeella Hübner, [1813], *Samml.eur.Schmett.* **8**: pl.59, figs 401, 402.

Type locality: Denmark; Copenhagen.

Description of imago (Pl.2, fig.1)

Wingspan 8–9mm. Head metallic leaden grey; antenna black, distally dark grey; labial palpus short, porrect, segments 2 and 3 of approximately equal length, ochreous, segment 2 with appressed scales beneath, segment 3 slender, slightly curved. Thorax and tegulae leaden metallic. Forewing reddish orange; black-edged silvery markings arranged as follows: subbasal spots on costa and dorsum; costal and dorsal spots at one-quarter, often joined to form outwardly oblique fascia; streak on costa at one-half; inwardly oblique streak from costa in subapical area; interrupted streak on termen from below apex to tornus; costa black between second and third silvery spots; black streak above cell and black spot below cell, both sometimes carrying a few silvery scales; apex and apical cilia black, cilia grading to dark grey at tornus. Hindwing narrower than forewing, trapezoidal, short, with almost vertical termen and very long apical extension; dark grey; cilia one and one-half times width of wing, dark grey, grading to whitish grey at anal angle. Foreleg black, tarsus

grey; mid- and hindlegs grey. Abdomen blackish, silvery white ventrally. Genitalia, see figures 5a,25b.

Life history

Ovum. Laid on various species of orache (*Atriplex* spp.) or goosefoot (*Chenopodium* spp.), probably near the midrib on the upperside of the leaf, where the mine begins, but there is no trace of the chorion which is probably eaten by the young larva.

Larva. Head pale brown. Body unicolorous greenish white or yellowish white in early instars. In the final instar it is spotted with crimson and has prominent black pinacula.

The larva mines in a contorted gallery which often turns back on itself in a gut-like configuration, amalga-

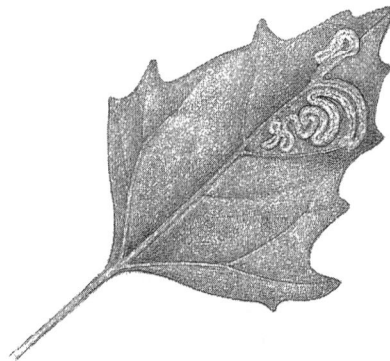

Figure 49 *Chrysoesthia drurella* (Fabricius), larval mine on *Atriplex* sp. (natural size)

mating with the earlier working to form a blotch (figure 49). Plentiful frass is retained in the mine; it is at first greenish, later blackish. Small leaves may be completely mined out and the larva can change to a fresh leaf. July–August; September, overwintering until May in a cocoon which it leaves to pupate.

Pupa. July larvae pupate without a cocoon. Overwintering larvae vacate their cocoons and pupate amongst detritus.

Imago. Bivoltine; late May–early July; mid-August–September. It is seldom observed in the adult stage, but may be obtained by sweeping the foodplant.

Distribution (Map 28)

Occurs on waste ground, on arable land and where the foodplants grow amongst root-crops. Locally common in England as far north as Yorkshire; North Wales; rare in central Scotland; the Channel Islands. Not recorded from Ireland. Throughout Europe; North America.

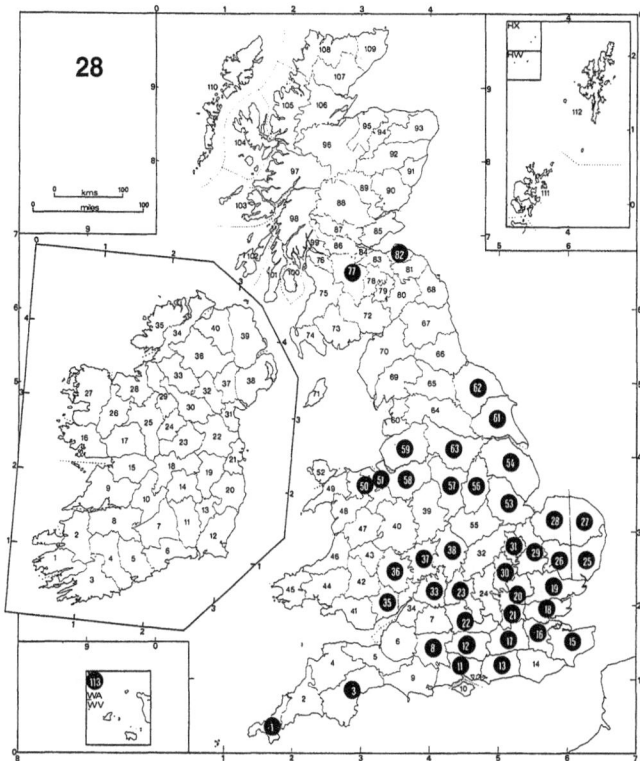

Chrysoesthia drurella

CHRYSOESTHIA SEXGUTTELLA (Thunberg)

Tinea sexguttella Thunberg, 1794, *Diss.ent.sistens Insecta suecica* **7**: 88, fig.6.

Tinea stipella sensu Hübner, 1796, *Samml.eur.Schmett.* **8**: 57, *nec* Linnaeus, 1758.

Tinea knockella sensu Haworth, 1828, *Lepid.Br.*: 568, *nec* Fabricius, 1794.

Tinea miscella sensu Haworth, 1828, *Ibid.*: 580, *nec* [Denis & Schiffermüller], 1775.

Microsetia aurofasciella Stephens, 1834, *Ill.Br.Ent.* (Haust.) **4**: 270.

Lita naeviferella Duponchel, [1843] 1842, *in* Godart & Duponchel, *Hist.nat.Lépid.Fr.* (Suppl.) **4**: 455.

Type locality: Sweden.

Description of imago (Pl.2, figs 2,3)

Wingspan 8–10mm. Head submetallic leaden grey; antenna black, annulate whitish; labial palpus rather short, segments 2 and 3 of equal length, segment 2 only slightly thickened with appressed scales, dark fuscous, segment 3 with base and apex whitish. Thorax and tegulae blackish grey. Forewing dark purplish fuscous weakly irrorate white; basal area, central fascia and tornal and apical patches obscurely darker; sometimes yel-

low streak on fold at one-quarter; subdorsal yellow spot at one-half and similar smaller spot in disc beyond cell; whitish ochreous subapical spot, sometimes extending as irregular fascia to termen and similar but smaller subtornal spot placed more proximally; cilia grey mixed fuscous, forming three ciliary lines, the inner two very irregular, tips beyond third line whitish. Hindwing narrower than forewing, trapezoidal, rather short with tornus almost right-angled and apical extension very long; greyish fuscous; cilia rather longer than width of wing, concolorous with wing on costa and at apex, grading to paler grey at anal angle. Foreleg blackish, tarsal joints banded whitish; mid- and hindlegs grey with white tarsal annulations, spurs white. Abdomen fuscous, banded white ventrally. Genitalia, see figures 5b,25c.

There is considerable variation in the extent of the yellow markings of the forewing. In extreme examples there is a yellow streak along the dorsum from near the base to the subtornal whitish spot.

Life history

Ovum. Laid on the underside of one of various species of orache (*Atriplex* spp.) or goosefoot (*Chenopodium* spp.), plants growing in sheltered or shaded places being preferred.

Larva. Head pale brown. Prothoracic plate blackish brown; body yellowish white; dorsal line brown; lateral line broken into a series of orange-red spots; thoracic legs brown.

The larva mines the leaves of its foodplant. The mine starts as a slender gallery, leading abruptly to a large whitish blotch, often occupying most of the leaf (figure 50). If a small leaf is insufficient, the larva changes to

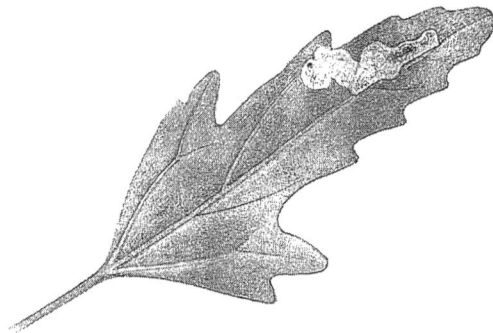

Figure 50 *Chrysoesthia sexguttella* (Thunberg), larval mine on *Atriplex* sp. (natural size)

another. Most of the frass is expelled from the mine; that which remains is generally stacked at the edges. June; September–October.

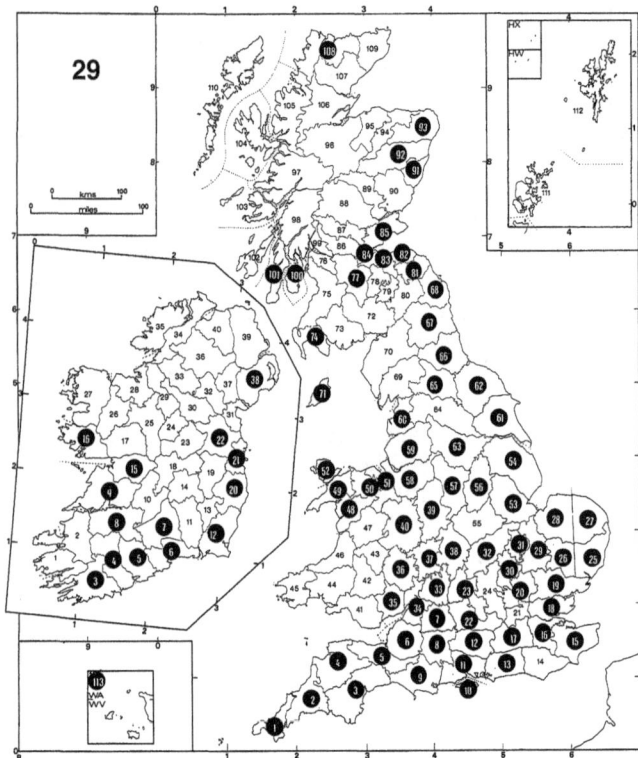

Chrysoesthia sexguttella

Pupa. In a cocoon spun amongst detritus on the ground. April–May; July.

Imago. Bivoltine; May–June; August. The adult is rarely observed but may be obtained by sweeping the foodplant.

Distribution (Map 29)

Occurs on waste ground and arable land, more especially where the foodplants grow in sheltered situations. Locally common throughout the British Isles as far north as Sutherland, but not recorded from the Scottish islands; the Isle of Man; the Channel Islands. Throughout Europe; eastern Siberia; North America; South Africa.

PTOCHEUUSA Heinemann

Ptocheuusa Heinemann, 1870, *Schmett. Dtl. Schweiz* (2)2(1): 288.

A Palaearctic genus represented by eight species in Europe, one of which occurs in the British Isles.

Imago. Labial palpus rather short, recurved, segment 3 shorter than segment 2, segment 2 slightly thickened with appressed scales. Forewing elongate; veins R_5 (7)

and R_4 (8) stalked with M_1 (6). Hindwing narrower than forewing, trapezoidal, apex strongly produced. Male genitalia with uncus bilobed, gnathos absent, sacculus spined, saccus broad, aedeagus bulbed. Female genitalia without signum in corpus bursae.

Larva. Feeds on seeds, internally in the seedhead.

PTOCHEUUSA PAUPELLA (Zeller)

Gelechia paupella Zeller, 1847, *Isis, Leipzig* **1847**: 858.
Aphelosetia inulella Curtis, 1850, *Ann. Mag. nat. Hist.* (2)**5**: 117.
Ptocheuusa inopella sensu auctt., *nec* Zeller, 1839.
Type locality: Sicily; Syracuse.

Description of imago (Pl.2, fig.4)

Wingspan 10–12mm, adults of the August generation being larger (Stainton, 1867a). Head white; antenna with scape white, flagellum greyish white, obscurely annulate fuscous; labial palpus rather short, subascending, segment 3 shorter than segment 2, segment 2 slightly thickened with appressed scales, white. Thorax whitish. Forewing ochreous-yellow irrorate fuscous; variable white markings consisting of costal and subcostal streaks, inwardly oblique fasciae before and beyond middle and subapical patch; some or nearly all of these markings may be absent and they may even be differently expressed on the left and right forewings of the same individual; elongate black plical stigma sometimes present; often black dots on termen at ends of veins; cilia grey, irregularly spotted fuscous. Hindwing slightly narrower than forewing, trapezoidal, tornus angled, apex strongly produced; pale grey, costal and dorsal areas slightly darker; cilia one and one-half times breadth of wing, grey. Foreleg greyish ochreous, tarsus greyish white; mid- and hindlegs greyish white, tibia of hindleg with darker streak on outer side. Abdomen fuscous; anal tuft of male yellowish white. Genitalia, see figures 5c,25d.

Life history

Ovum. Laid on common fleabane (*Pulicaria dysenterica*) or golden samphire (*Inula crithmoides*); sometimes on common knapweed (*Centaurea nigra*) or mint (*Mentha* spp.) (Bradford *in* Emmet [1979]); probably on the seedhead.

Larva. Limaciform, 3.0mm long by 1.5mm broad; thoracic legs absent; abdominal prolegs minute; intersegmental divisions deeply incised; three longitudinal furrows along length of body latero-ventrally, giving appearance of two rows of 'pads' on each side of venter where they cross intersegmental divisions. Head yel-

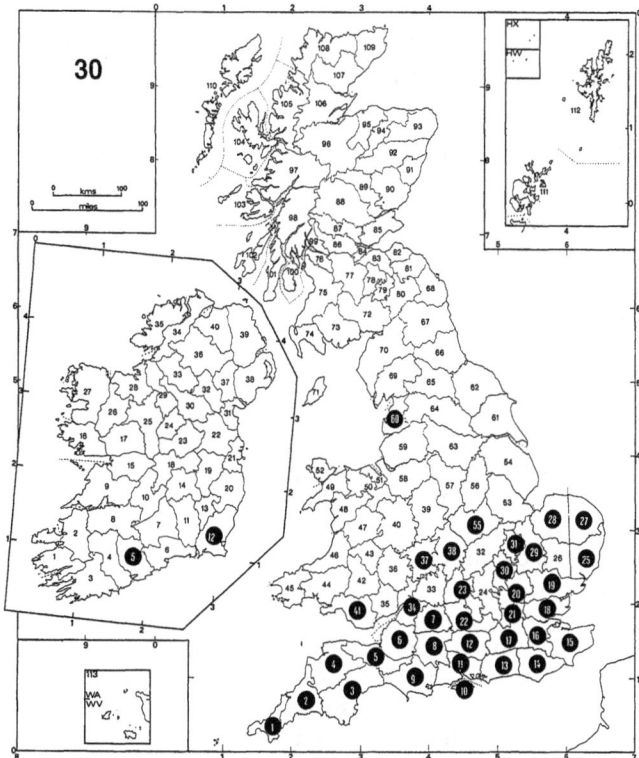

Ptocheuusa paupella

lowish white, mouth-parts dark brown, eyespots black. Body dorsally mid-brown, segmental divisions very dark brown or even black; laterally and ventrally pale yellowish brown; anal plate concolorous with venter (J. R. Langmaid, pers.comm.). This description, confirmed by the present author, was made in February from larvae that had overwintered. However, Stainton (*loc.cit.*), followed by Meyrick ([1928]) and Sokoloff & Bradford (1993), describes the larva as 'pale amber; the head dark brown, and two darker spots on the back of the second segment'. His figure, however, shows the head and prothoracic plate as pale brown. It is possible that the larva darkens prior to pupation and this accounts for the discrepancy in the descriptions.

The larva lives in the seedhead, eating the seeds. When it feeds on common fleabane, a patch of raised and discoloured florets indicates its presence. Late June–early August; September–October, then overwintering in the seedhead fully fed until May.

Pupa. In a tough cocoon of silk mixed with seeds, frass and chewed material, firmly attached to the upper side of the receptacle of the flowerhead and, in some cases, piercing it so that the cocoon is partly below the receptacle (J. R. Langmaid, pers.comm.). May–June; August–early September.

Imago. Bivoltine; June; late August–September. The adult comes to light, sometimes in places where its main foodplants are not present.

Distribution (Map 30)

Occurs on damp grassland, ditches, woodland rides and at the edge of salt-marshes. Locally common in southern England as far north as Leicestershire; Lancashire; South Wales; southern Ireland. Western, central and southern Europe; Asia Minor; North Africa.

PSAMATHOCRITA Meyrick

Psamathocrita Meyrick, 1925, *in* Wytsman, *Genera Insect.* **184**: 15 (key), 40.

A very small Palaearctic genus represented by three European species of which two occur in Britain and one is endemic to England.

Imago. Small to very small gelechiids with labial palpus short, stout and porrect; antenna with pecten; forewing without scale tufts, sharply lanceolate; hindwing less broad than forewing, trapezoidal with apex produced into a sharp point; abdominal tergites with modified scales which are concolorous with rest of abdomen, modified scales on three basal tergites in female and five basal tergites in male. Valva of male genitalia large, angular and not very mobile; signum of female variable.

Larva. Early stages poorly known.

Key to species (imagines) of the genus *Psamathocrita*

–   Forewing white lightly streaked yellowish .... *argentella* (p. 98)
–   Forewing pale brown .... *osseella* (p. 97)

PSAMATHOCRITA OSSEELLA (Stainton)

Gelechia osseella Stainton, [1860], *Entomologist's Annu.* **1861**: 87.

Type locality: England; Yedmandale and Forge Valley, near Scarborough, Yorkshire.

Description of imago (Pl.2, fig.5)

Wingspan 11mm. Head, antenna and labial palpus reddish buff becoming greyish white in worn individuals, pecten present. Thorax and tegulae reddish buff. Forewing reddish buff with pale greyish white scale bases becoming more apparent with wear; cilia concolorous but paler about tornus. Hindwing reddish fusc-

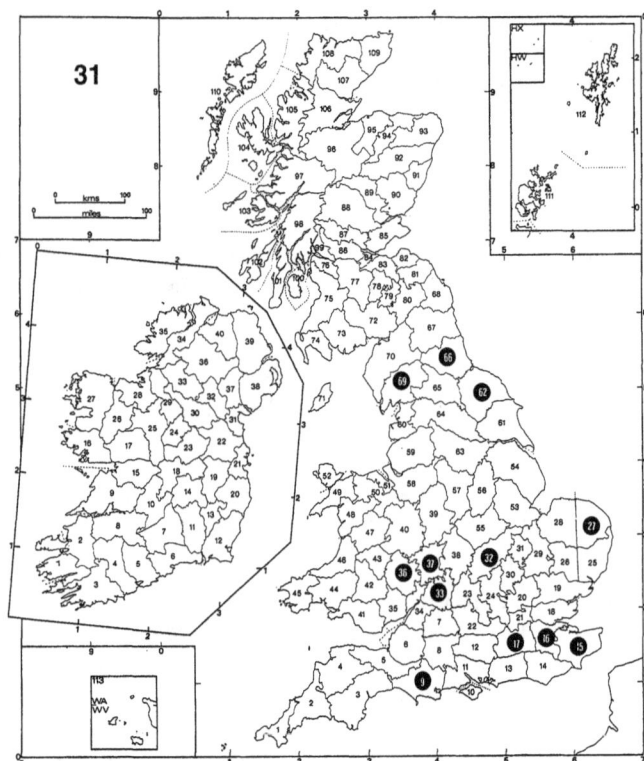

Psamathocrita osseella

ous; cilia concolorous. All legs reddish buff, hind femur noticeably whitish below. Abdomen reddish fuscous above, reddish buff below. Worn specimens will require examination of the genitalia. Genitalia, see figures 5d,25i.

Life history
Ovum. Laid during June, possibly on grass (Ffennel, 1974) or marjoram (*Origanum vulgare*) (Bradford, [1979]).
Larva. Unknown.
Pupa. Unknown.
Imago. Flies in bright sunshine in June and early July (Sutton & Beaumont, 1989).

Distribution (Map 31)
Recorded from widely separated localities throughout England as far north as Cumbria and Durham. The moth does not appear to have been recorded in Britain since 1926 when it was taken at Cranham, east Gloucestershire (Fletcher & Clutterbuck, 1939). It is probably still resident but overlooked due to its similarity to a coleophorid. Not yet reported from Wales, Scotland or Ireland. The moth is ascribed RDB1 (Endangered) conservation status by Parsons

(1995[1996]) in view of the absence of recent records. On the Continent known from Sweden, Estonia, Germany, France, Italy; Algeria.

PSAMATHOCRITA ARGENTELLA Pierce & Metcalfe
Psamathocrita argentella Pierce & Metcalfe, 1942, *Entomologist* **75** : 255.
Type locality: England; Cracknore Hard, near Southhampton.

Description of imago (Pl.2, fig.6)
Wingspan 10–11mm. Head smooth-scaled, varying from almost pure white to a pinkish buff; antenna, with pecten, white sometimes vaguely barred with buff; labial palpus white above, buff to pale fuscous below. Thorax white with a midline stripe and anterior third of tegulae buff. Forewing very *Coleophora*-like in shape, white with basal part of costa narrowly fuscous, with patches of pale ochreous suffusion and streaks of pinkish buff, apical quarter also pinkish buff when fresh; cilia whitish. Hindwing fuscous, apex forming an extended point; cilia whitish-fuscous; underside of both fore- and hindwing uniform fuscous; cilia concolorous. All legs fuscous above, white below. Abdomen dark fuscous dorsally, whitish below. Worn specimens will require examination of the genitalia. Genitalia, see figures 5e,25h. Easily mistaken for a coleophorid except for shape of the hindwing.

Life history
Based on observations by J. R. Langmaid and I. R. Thirlwell in 2001.
Ovum. Laid deep inside a floret of sea couch (*Elytrigia atherica*). Ovipositing must be a hazardous act, as several moths were found dead or dying with their ovipositors firmly trapped in the spikelet by the stiff scale-like glumes and lemmas. Sea couch was originally suggested as the foodplant by Ffennell (1974) who observed many moths running up the grass-stems and resting with wings partially extended.
Larva. Limaciform; full-fed 4.0mm long by 1.5mm broad at its broadest point. Head dark greyish brown. Body rather bright pinkish orange; prothorax without sclerotized plate but with a pair of paramedian pale brown longitudinal marks dorsally; anal plate greyish brown; thoracic legs absent; anal prolegs modified into a pair of greyish brown sclerotized setaceous pads, other prolegs absent.
    The larva feeds on the developing flower and seed, extruding whitish frass which can sometimes be seen adhering to the spikelet. It is not known whether the

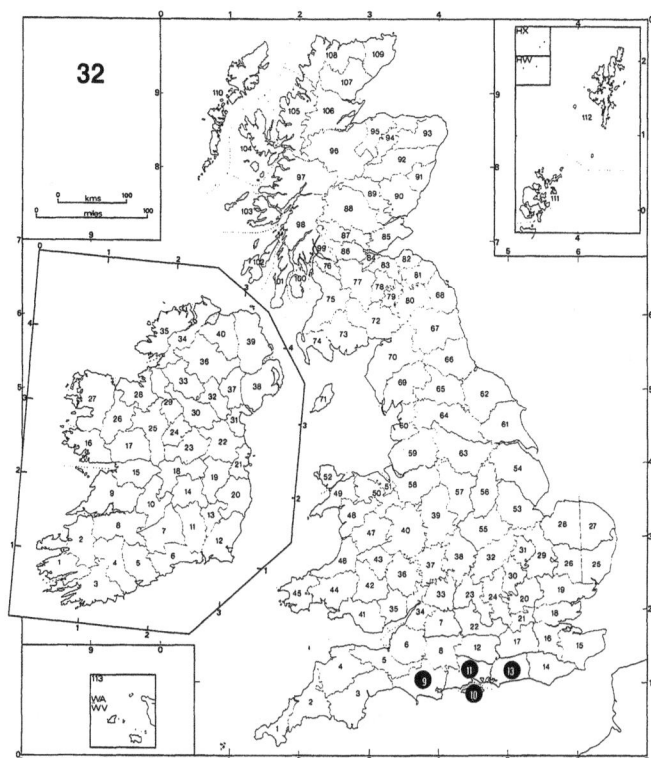

Psamathocrita argentella

larva leaves the feeding-place to pupate, or remains inside when the spikelet drops to the ground in the late summer. Larvae were found throughout July and August.

Pupa. Undescribed. Presumably on the ground, but whether in the fallen spikelet or amongst other detritus is not known. Neither is it known when pupation takes place.

Imago. Univoltine, occurring mid-June to early July. On the wing at dawn and dusk, but main flight-time 04.45–05.45 GMT (Ffennell, *loc.cit.*) It has also been recorded flying freely after rain at about 16.00 GMT (Bradford & Sokoloff, 1988).

History and Distribution (Map 32)
First taken by Fassnidge in late June 1937 on a saltmarsh at Cracknore Hard, on the west side of Southampton Water, but not recognized as new until 1942 (Pierce & Metcalfe, 1942). The original site was destroyed in 1971 but the species was rediscovered on a salt-marsh near St Helen's, Isle of Wight, in June, 1973 (Ffennell, *loc.cit.*). Now also known from south Hampshire, west Sussex (Parsons, 1995[1996]) and Dorset. Endemic to the British Isles. In view of its very restricted distribution it has been ascribed RDB2 (Vulnerable) conservation status.

ARISTOTELIA Hübner

Aristotelia Hübner, [1825], *Verz. bekannt. Schmett.*: 424.

Ergatis Heinemann, 1870, *Schmett. Dtl. Schweiz* (2)**2**(1): 295, *nec* Blackwall, 1841.

Eucatoptus Walsingham, 1897, *Proc. zool. Soc. Lond.* **1897**: 69.

A large genus, even as presently constituted, of worldwide distribution. Some 14 species occur in Europe but the ranges of only three of these extend into Britain.

Imago. Small to medium-sized gelechiids with labial palpus long, slender, segment 2 thickened with scales ventrally; antenna with pecten reduced to a single scale which is easily dislodged. Forewing elongate, lanceolate and without scale-tufts. Hindwing same breadth as forewing, trapezoidal with apex extended into a point. Abdominal tergites concolorous. Valva of male genitalia with elongate digitate costa, a much reduced digitate sacculus and with little mobility, aedeagus with large bulbous base; signum of female genitalia a sclerotized disc with serrate edges.

Larva. Early stages on various low plants.

Key to species (imagines) of the genus *Aristotelia*

1	Forewing buff; partially suffused reddish brown; markings inconspicuous .... *brizella* (p. 101)
–	Forewing dark fuscous to chocolate with conspicuous white markings 2
2(1)	Forewing markings consisting of four oblique transverse fasciae .... *ericinella* (p. 100)
–	Forewing markings consisting of a longitudinal band along dorsum............................ *subdecurtella* (p. 99)

ARISTOTELIA SUBDECURTELLA (Stainton)

Gelechia subdecurtella Stainton, [1858], *Entomologist's Annu.* **1859**: 152.

Type locality: England; Cambridge Fens.

Description of imago (Pl.2, fig.7)
Wingspan 14–15mm. Head and frons with flattened pale fuscous scales, tipped darker; antenna dark fuscous, ringed pale orange-brown; labial palpus dark fuscous with segments 2 and 3 ringed white at both ends and in the middle. Thorax and tegulae dark fuscous-

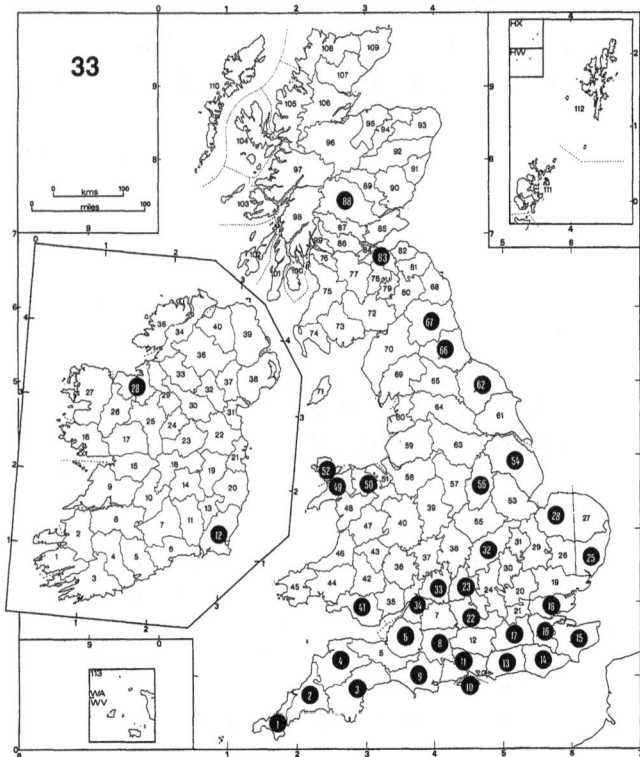

Aristotelia subdecurtella

brown with paler scale-bases showing. Forewing dark chocolate-brown sprinkled with white; white band along dorsum stretching one-third across wing except just before and at termen where it reaches halfway across and almost makes contact with square white spot on costa at three-quarters; these white markings overlaid by patches of yellowish orange which sometimes almost obscure white ground colour; other variable patches of yellowish orange overlaying the chocolate-brown areas; row of white dots around apex just below cilia; cilia dark fuscous-brown along costa to apex, then ochreous with two dark fuscous-brown ciliary lines. Hindwing pale brownish grey; cilia ochreous-grey. Legs dark fuscous, ringed white except hind tibial tufts which are ochreous. Abdomen dark grey. Genitalia, see figures 5g,25f.

Life history

Ovum. Laid on purple-loosestrife (*Lythrum salicaria*) (Stainton, 1862).

Larva. Undescribed, feeding in spun shoots during June.

Pupa. Length 9–10mm. Dark brown, rather slender and tapering; dorsum of each abdominal segment with two transverse rows of short blunt spines, in the cranial row the spines larger and widely spaced while in the posterior row less than half the size and more numerous; caudal tip of pupa tapered to a point with a small triangular flange projecting on each side just before apex, and with several ochreous hooked hairs (from specimen in BMNH). June–July.

Imago. Univoltine; late June and July (Bradford & Sokoloff, 1988).

Distribution (Map 33)

Originally described and reared from the Cambridgeshire fens. Meyrick (1895) included Norfolk in its distribution. A very rare and local fenland species, which is considered to be extinct by Parsons (1995[1996]) in view of the absence of records since the 19th century. However, it may still be present on some fenland sites. Throughout most of Europe and Scandinavia but not known from Norway or the Iberian Peninsula.

ARISTOTELIA ERICINELLA (Zeller)

Tinea micella sensu Hübner, 1796, *Samml. eur. Schmett.* 8: 31, fig.210, *nec* [Denis & Schiffermüller], 1775.
Gelechia (Brachmia) ericinella Zeller, 1839, *Isis, Leipzig* **1839**: 202.
Type locality: not stated.

Description of imago (Pl.2, fig.9)

Wingspan 12–13mm. Head pale fuscous; antenna dark fuscous, ringed paler; labial palpus dark fuscous, edged white below, both segments with white band at base, middle and apex. Thorax dark fuscous with ferruginous sheen and with broad pale fuscous longitudinal band either side of midline; tegulae pale fuscous. Forewing dark fuscous, shining ferruginous or velvet black according to light, with silvery white markings; four straight almost parallel-sided oblique transverse fasciae, two in basal half almost parallel meeting costa basally at an obtuse angle, two in apical half almost parallel meeting costa at an acute angle, middle two meeting to form a V-shape with apex on dorsum; other markings a spot at base of dorsum, another at extreme apex and other small irregular spots in apical area often involving cilia; cilia with three bands, basal one dark fuscous, central band pale fuscous darkening towards outer edge and distal band buff. Hindwing fuscous; cilia concolorous. Underside of both fore- and hindwing uniform fuscous. All legs fuscous with apical part of each segment whitish. Abdomen fuscous above, darker beneath and edged paler. Genitalia, see figures 5f,25e.

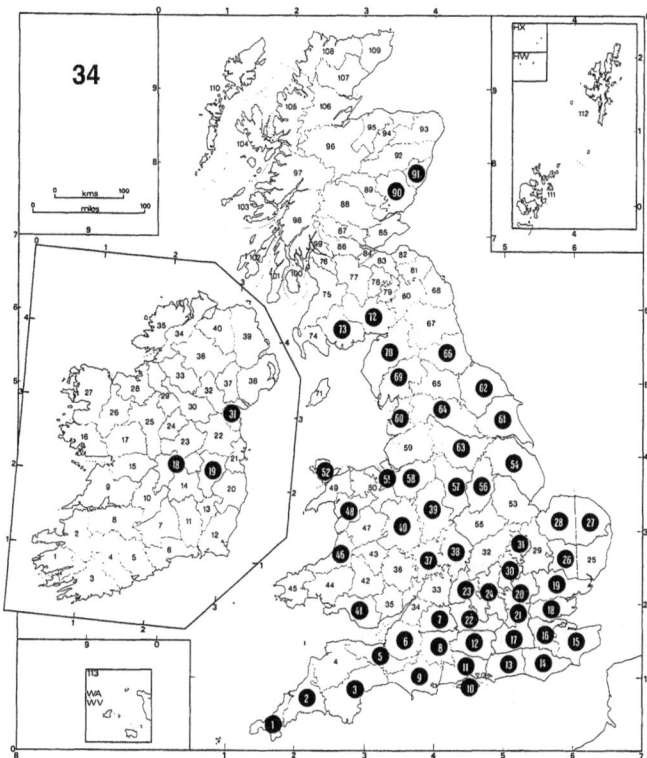

Aristotelia ericinella

Life history

Ovum. Laid on heather (*Calluna vulgaris*) in July or August, probably overwintering and hatching in the spring.

Larva. Length in final instar about 10mm. Head pale brown. Prothoracic plate brown anteriorly, becoming darker posteriorly. Body ochreous-brown, rosy-tinged, with dark brown, rather broad, dorsolateral longitudinal band on each side, edged above and below with pale yellowish, and with pale yellowish interrupted line along middle of each band; pale yellowish slender spiracular line; prothorax yellowish white laterally, both meso- and metathorax with transverse dorsal yellowish white slender band anteriorly and lateral yellow-white patches posteriorly. Legs black-brown; prolegs greyish (Stainton, 1867a).

The larva can be found from mid-June to early July when it makes a slender silk tube amongst the twigs of heather. It retreats into this tube when not feeding on the heather leaves. It is easily disturbed, falling to the ground if the plant is jarred. On climbing back up the plant it constructs a new retreat tube, thus deserted empty tubes may be quite numerous. Larvae are best collected by sweeping heather (Stainton, *loc.cit.*).

Pupa. Yellow-brown, covered with fine short hairs; hairs longest at caudal tip but not modified into a specialized cremaster. In a loose white cocoon in a spun shoot of the foodplant; July.

Imago. Univoltine; throughout July and into early August. On the wing during the afternoon and evening. Best obtained by sweeping, but does occasionally come to light.

Distribution (Map 34)

Widespread on heaths in England and Wales, just extending into the extreme south-west of Scotland. Elsewhere in Scotland there are isolated colonies near Kirriemuir, Angus (Bland, 1990), and at St Cyrus, Kincardineshire (M. R. Young, pers.comm.). Central and eastern Ireland (Bond, 1984). Widespread throughout Europe.

ARISTOTELIA BRIZELLA (Treitschke)

Oecophora brizella Treitschke, 1833, *Schmett.Eur.* **9**(2): 173.

Type locality: Germany; Dresden.

Description of imago (Pl.2, fig.8)

Wingspan 9–10mm. Head buff, scales often edged fuscous; antenna dark fuscous, ringed paler; labial palpus dark fuscous, regularly banded whitish. Thorax and tegulae buff, the latter more fuscous anteriorly. Forewing buff tending to dark fuscous, especially on basal half of costa, suffused reddish brown to a greater or lesser extent especially on costal half and apical third; narrow irregular transverse lines of silvery grey at intervals; three black spots (all or any of which may be absent) at two-thirds along fold from base, at apical end of fold and in centre of wing just beyond this; cilia buff with two bands, basal band tipped with dark fuscous, the other tipped fuscous. Hindwing fuscous; cilia slightly paler. Underside of fore- and hindwings fuscous-brown. Legs fuscous to dark fuscous, distal end of segments paler. Abdomen dark fuscous. Genitalia, see figures 5h,25g.

Life history

Ovum. Laid on a flower of thrift (*Armeria maritima*) or, less frequently, on common sea-lavender (*Limonium vulgare*).

Larva. Final instar length 6–8 mm. Head brown. Prothoracic plate dark brown to black, divided centrally; body olive-brown to yellowish brown with greenish line along dorsum; pinacula whitish.

The first generation larva makes a tubular silk tunnel in a seedhead from which it feeds on the unripe seeds

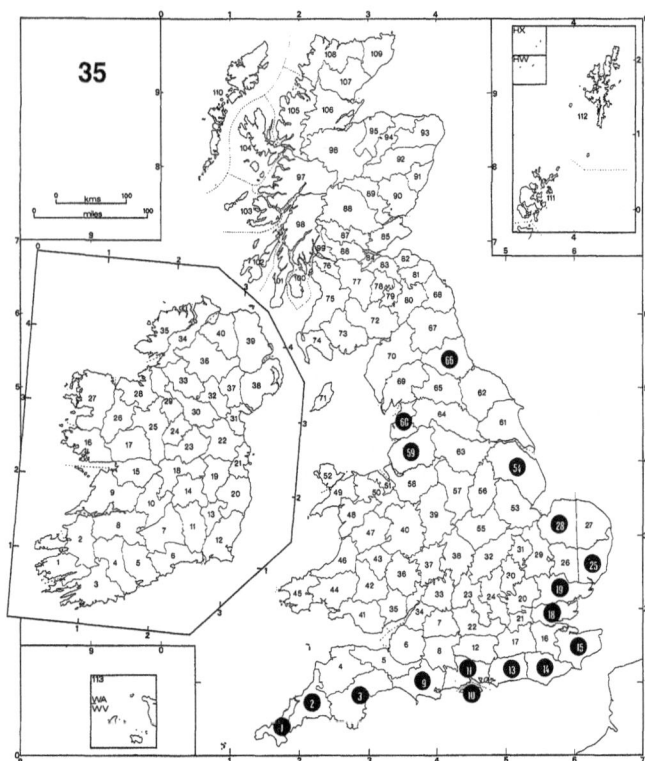

Aristotelia brizella

in June and early July. Several larvae may occupy a single flower-head. The second generation larva occurs in a similar tunnel in a dead dry seedhead in September and October, overwintering as a pupa.

Pupa. Undescribed. Both generations pupate in a slight cocoon in the larval workings in the seedhead (Stainton, 1859, 1867a). October–May; July.

Imago. Bivoltine; mid-May–early June and late July–late August. The moth has been seen flying at sunrise and in the evening. It has also been taken at light (Parsons, 1995[1996]).

Distribution (Map 35)

Widespread but local on the south and east coasts of England from Cornwall to Durham and recorded on the west coast only from Lancashire, by L. T. Ford (Russell & Turner, 1941) and H. N. Michaelis (pers. comm.). In view of its restricted distribution and absence of records north of west Norfolk in the last 30 years it has been ascribed Notable B conservation status by Parsons (*loc.cit.*). Not recorded from Scotland or Ireland. Throughout Scandinavia and central and western Europe, both on the coast and inland.

XYSTOPHORA Wocke

Doryphora Heinemann, 1870, *Schmett. Dtl. Schweiz* (2)**2**(1): 298, *nec* Illiger, 1807.

Xystophora Wocke, [1876]1877, *in* Heinemann, *Schmett. Dtl. Schweiz* (2)**2**(2) Tabelle der Gattungen: 6.

Doryphorella Cockerell, 1888, *Entomologist* **21**: 163.

A very small Palaearctic genus containing two European species. Only one of these occurs in Britain.

Imago. Small gelechiids with labial palpus long, slender, segment 2 thickened with scales; antenna without pecten. Forewing without scale-tufts, lanceolate, narrower in female. Hindwing less broad than forewing, trapezoidal, tornus rounded, apex slightly produced. Abdomen with all tergites concolorous. Valva of male genitalia complex, aedeagus large with thorn-like projections and multiple cornuti; signum of female genitalia a small spinose plate, ostial and antral area heavily sclerotized.

Larva. In spun leaves of Fabaceae.

XYSTOPHORA PULVERATELLA (Herrich-Schäffer)

Anacampsis pulveratella Herrich-Schäffer, 1854, *Syst. Bearb. Schmett. Eur.* **5**: 199.

Gelechia intaminatella Stainton, 1854, *Ent. wkly Intell.* **7**: 140.

Type locality: Austria; Rodaun, near Vienna.

Description of imago (Pl.2, fig.10)

Wingspan 10–11mm. Sexes similar though wings of female slightly narrower. Head mixed light fuscous and buff, although fuscous sometimes almost absent; antenna dark fuscous, ringed buff; labial palpus buff, tinged reddish towards base of segment 2 when fresh and more whitish towards apex of segment 3. Thorax and tegulae mixed light fuscous and buff. Forewing without markings, mixed light fuscous and buff due to buff-tipped fuscous scales, often tending to be more fuscous in apical third; cilia ochreous-fuscous; underside dark fuscous. Hindwing greyish fuscous becoming more buff at extreme apex; cilia ochreous-fuscous. Fore- and midlegs dark fuscous, apical edge of tarsi edged whitish; hindleg buff, tarsus dark fuscous, edged whitish. Abdomen dark fuscous; male with buff genital tuft. Genitalia, see figures 5i,26a.

Life history

Ovum. Laid on bird's-foot-trefoil (*Lotus corniculatus*) or bitter-vetch (*Lathyrus linifolius*) during late May or

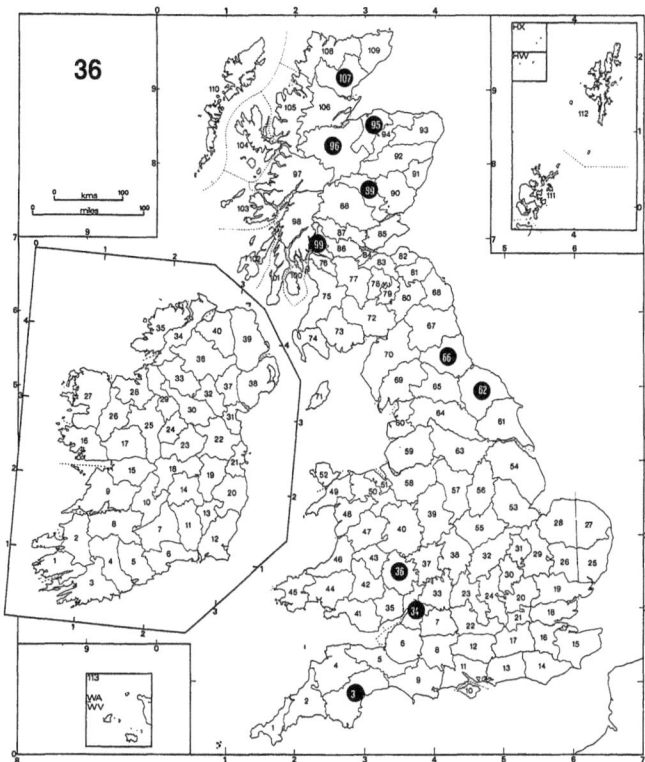

Xystophora pulveratella

June. Meyrick (1895) also gives crown vetch (*Coronilla varia*) and medick (*Medicago* spp.) as foodplants but these are unlikely in Britain. A European report of yarrow (*Achillea millefolium*) as a foodplant (Stainton, 1864) seems most improbable.

Larva. Final instar length 8–9mm. Head and prothoracic plate yellowish. Body yellowish green, with faint reddish tinge; pinacula small and black; earlier instar larvae are greener (Stainton, 1867b).

The larva makes a refuge of spun leaves of the foodplant from which it feeds on the surrounding foliage during August and September. After the final moult in September, it spins a dense white cocoon in a rolled leaf of its foodplant, in which it overwinters before pupation.

Pupa. Length 4.0–4.5mm. Yellowish brown, rather short and stubby; cremaster with dorsal pair of hooked bristles and two clumps each of four hooked bristles ventrally. Within a cocoon, late March–late April.

Imago. Univoltine; late May–June. Flies low over the foodplant at dusk (Bland *et al.*, 1987b).

Distribution (Map 36)

Widespread but very local in the Highlands of Scotland; in England there has been no record during the last 30 years. In view of the paucity of recent records,

Parsons (1995[1996]) ascribed this species RDB 1 (Indeterminate) conservation status. Not known from Wales or Ireland. Abroad, widespread in most of Europe, except the Iberian Peninsula.

BRYOTROPHA Heinemann

Bryotropha Heinemann, 1870, *Schmett.Dtl.Schweiz* (2)**2**(1): 233.
Mniophaga Pierce & Daltry, 1938, *Entomologist* **71**: 226.
Adelphotropha Gozmány, 1955, *Annls hist.-nat.Mus. natn.hung.* (S.N.) **6**: 310.

A genus with a Holarctic distribution. About 25 species are known from Europe, 12 of which occur in the British Isles.

Imago. The members of this genus are recognizable externally by a single erect scale on the first antennal segment (figure 51). This scale is the remnant of a pecten (a row of bristles) which is very significant and unusual in the family Gelechiidae, found in only a few genera such

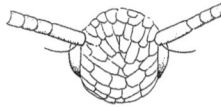

Figure 51 *Bryotropha* sp., head showing pecten hair on antenna

as *Platyedra* Meyrick (figure 62, p.239) and *Pexicopia* Common. A single pecten scale is also present in some *Monochroa* Heinemann, *Eulamprotes* Bradley, *Aristotelia* Hübner and others, but there it is deciduous whilst in *Bryotropha* it is stronger and persistent, though sometimes absent in *B. boreella* (Douglas) and *B. galbanella* (Zeller).

The labial palpus has a conspicuous furrowed brush on the underside of segment 2, segment 3 being as long as or longer than segment 2 (figure 52a). The exceptions are *B. galbanella* and *B. boreella*; here segment 2 does not have a furrowed brush while segment 3 is shorter than segment 2 (figure 52b). Forewing

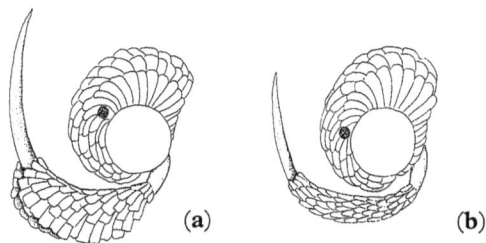

Figure 52 Head and labial palpus of *Bryotropha* spp.
(a) *B. terrella* ([Denis & Schiffermüller])
(b) *B. galbanella* (Zeller) and *B. boreella* (Douglas)

elongate, tapering from middle; veins R_5 (7) and R_4 (8) stalked, ending at costa. The wing-markings comprise a plical and two discal spots. The costal patch is opposite or slightly basad of the tornal spot, often the two fused to form a fascia. Hindwing as broad as forewing, subtrapezoidal, apex pointed. Veins M_1 (6) and R_3 (7) stalked, veins CuA_1 (3) and M_3 (4) connate or short-stalked.

The species within this genus are often difficult to separate because of individual variation and a tendency to form ecological forms. Genitalia, see figures 53,54.

Ovum. Undescribed for all British species. The paragraph is therefore omitted below.

Larva. Body with four pairs of abdominal prolegs. The caterpillar lives in a silken gallery which it leaves in order to feed. Most species feed on mosses but in at least one case, *B. terrella* ([Denis & Schiffermüller]), grasses are also a foodplant. The life history of all species within this genus is poorly known and the larvae of some have yet to be described.

Pupa. See under *B. domestica* (Haworth).

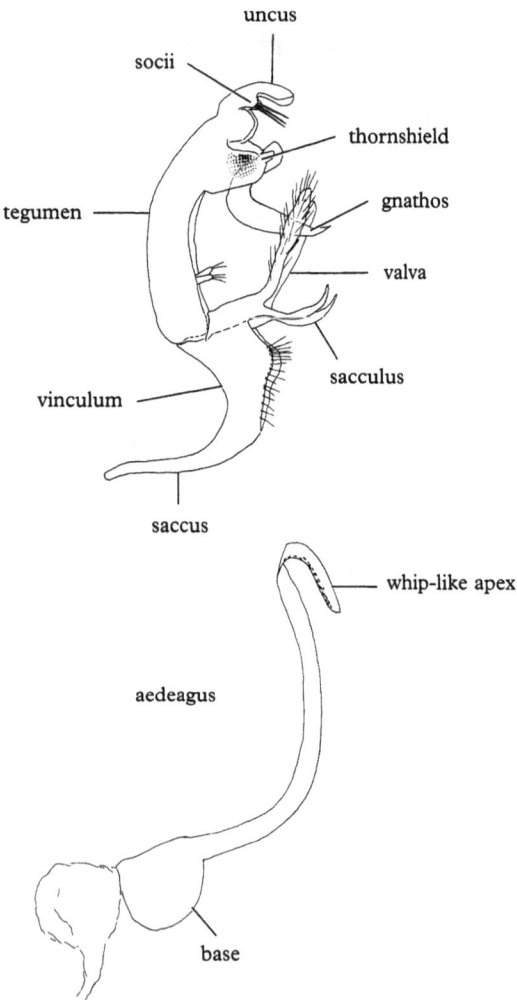

Figure 53 *Bryotropha similis* (Stainton), stylized drawing of male genitalia

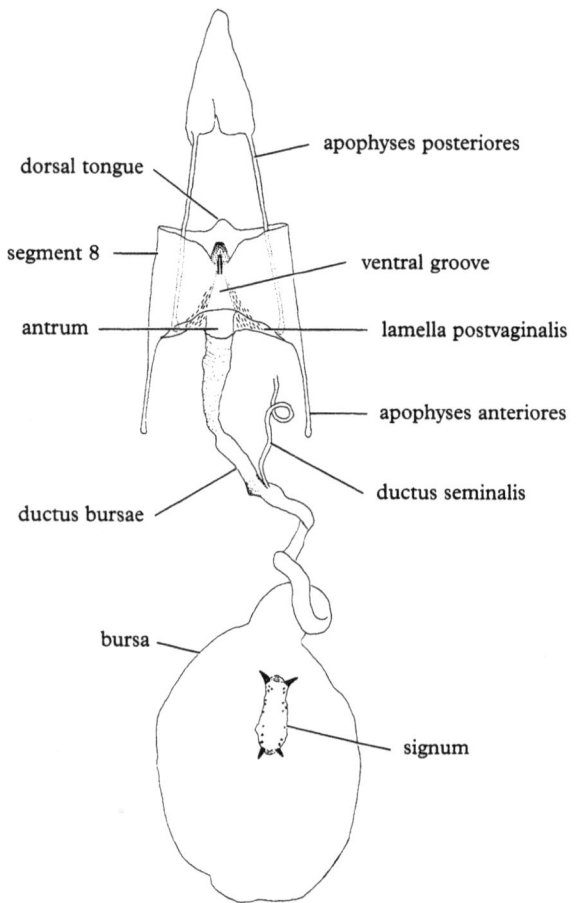

Figure 54 *Bryotropha similis* (Stainton), stylized drawing of female genitalia

Key to species (imagines) of the genus *Bryotropha* based on external features

1   Labial palpus with segment 3 shorter than segment 2; segment 2 without conspicuous furrowed brush 2

–   Labial palpus with segment 3 longer than segment 2; segment 2 with conspicuous furrowed brush 3

2(1)   Forewing with ground colour dark yellowish grey, fascia yellowish; hindwing with blunt apex; wingspan 15.0–17.5mm ....*galbanella* (p. 113)

–   Forewing with ground colour dark fuscous with purple gloss in female, fascia creamy white; hindwing with acute apex; wingspan 12–15mm .... *boreella* (p. 112)

3(1)   Small species, wingspan 9–13mm 4

–   Large species, wingspan 14–17mm 15

4(3)   Forewing with first discal stigma directly above plical 5

–   Forewing with first discal stigma distal to plical 7

5(4)   Forewing with ground colour ochreous-grey, markings very prominent; wingspan 12–13mm .... *domestica* (p. 117)

–   Forewing with ground colour dark greyish brown, markings moderately prominent; wingspan 11–12mm 6

6(5)   Forewing with fascia usually bent outwards .... *basaltinella* (p. 107)★

–   Forewing with fascia usually straight or bent inwards .... *dryadella* (p. 107)★

7(4)   Forewing with ground colour clear grey or whitish .... *umbrosella* (part) (p. 108)

–   Forewing with ground colour otherwise 8

8(7)   Forewing with ground colour blackish 9

–   Forewing with ground colour whitish brown, ochreous, light or dark brown 13

9(8)   Forewing with plical stigma bordered outwardly by cluster of pale scales 10

–   Forewing with plical stigma not bordered outwardly by pale scales 11

10(9)   Forewing with pale scales beyond plical stigma white; costal and tornal patches well developed, white, seldom fused; ground colour usually very dark without pale irroration ...... *umbrosella* (part) (p. 108)

–   Forewing with pale scales beyond plical stigma yellowish; costal and tornal patches yellow to ochreous and often fused to form slender fascia; whole wing with pale irroration .... *affinis* (part) (p. 109)

11(9)   Forewing with pale irroration ...*affinis* (part) (p. 109)

–   Forewing without pale irroration 12

12(11)   Forewing with costal and tornal patches well developed, white, contrasting with very dark ground colour .... *umbrosella* (part) (p. 108)

–   Forewing with costal and tornal patches absent or ill-defined, yellowish or whitish; if well developed, yellow .... *similis* (p. 110)

13 (8)   Forewing with ground colour ochreous, whitish brown or very light brown; markings prominent; small species, wingspan 9–11mm .... *affinis* (part) (p. 109)

–   Forewing with ground colour brown to dark brown, markings obscure; wingspan 9–14mm 14

14(13)   Fringe of forewing with single dark ciliary line; head with bright yellow cheeks (genae), contrasting with the darker vertex; small species, wingspan 9–12mm .. .. *senectella* (p. 111)

–   Fringe of forewing with two dark ciliary lines; head with vertex, frons and cheeks concolorous; wingspan 12–14mm .... *desertella* (part) (p. 114)

15(3)   Forewing with ground colour ochreous 16

–   Forewing with ground colour light or dark brown, greyish brown or ferruginous 17

16(15)   Forewing with ground colour greyish ochreous; apex rather pointed; fringe with one ciliary line; wingspan 13–15mm .... *politella* (♀) (p. 116)

–   Forewing with ground colour ochreous; apex rounded; fringe with two or more ciliary lines; wingspan 12–14mm .... *desertella* (part) (p. 114)

17(15)   Forewing with ground colour greyish brown; fascia absent or very faint; subapical area not darkened; fringe of hindwing without ciliary line; wingspan 12–16mm .... *politella* (♂) (p. 116)

–   Characters otherwise 18

18(17)   Forewing with ground colour light to dark brown, occasionally ferruginous; rather broad, without indication of median streak; wingspan 13–17mm .... *terrella* (p. 115)

–   Forewing with ground colour brown to dark brown; rather narrow, often with faint median streak; wingspan 12–14mm .... *desertella* (part) (p. 114)

Key to the male genitalia of the genus *Bryotropha* (see figures 6,7, pp.26,27)

(The terms used are shown on figure 53, p.104)

1   Aedeagus without 'whip-like' apex .... *domestica* (p. 117)

–   Aedeagus with 'whip-like' apex 2

★ Examination of the genitalia is essential for reliable identification

2(1) Sacculus absent .... *politella* (p. 116)

– Sacculus present 3

3(2) Base of gnathos and surface of vinculum covered with microtrichia 4

– Base of gnathos and surface of vinculum without microtrichia 7

4(3) Gnathos extremely large and broad 5

– Gnathos long and slender 6

5(4) Gnathos with S-shaped apex .... *terrella* (p. 115)

– Gnathos without S-shaped apex .... *desertella* (p. 114)

6 (4) Gnathos sinuous ....*galbanella* (p. 113)

– Gnathos straight ....*boreella* (p. 112)

7(3) Gnathos comparatively small and squat, length about twice that of its greatest width 8

– Gnathos slender, length at least three times that of its greatest width 10

8(7) Thornshield rectangular, gnathos with a 90° bend halfway .... *dryadella* (p. 107)

– Thornshield triangular, gnathos otherwise 9

9(8) Gnathos gradually narrowing to a sharp sclerotized point, base not protruding; thornshield with up to 40 spikes of different sizes .... *basaltinella* (p. 107)

– Gnathos angular, point weakly sclerotized, base ball-shaped, clearly protruding; thornshield with up to 80 spikes of different sizes .... *senectella* (p. 111)

10(7) Gnathos markedly thickened at or before middle, thence gradually narrowing to a sharp point; thornshield with up to 100 spikes of different sizes .... *similis* (p. 110)

– Gnathos otherwise; thornshield with up to 40 small spikes 11

11(10) Gnathos with gradual curve, thornshield with up to 20 small spikes .... *umbrosella* (p. 108)

– Gnathos with rather abrupt curve, thornshield with up to 40 small spikes .... *affinis* (p. 109)

Key to the female genitalia of the genus *Bryotropha* (see figures 26,27, pp. 46,47)

(The terms used are shown on figure 54, p. 104)

1 Signum consisting of a plate with two transverse ridges 2

– Signum consisting of a plate with strong spines on the corners 7

2(1) Segment 8 ventrally covered with microtrichia 3

– Segment 8 ventrally not covered with microtrichia ..
.... 4

3(2) Ventral groove of segment 8 gradually widening towards antrum, laterally with a moderate cover of tiny microtrichia .... *boreella* (p. 112)

– Ventral groove of segment 8 with sinuate margins, laterally with a strong cover of large needle-shaped microtrichia .... *galbanella* (p. 113)

4(2) Posterior margin of segment 8 deeply excavate ventrally 5

– Posterior margin of segment 8 not or shallowly excavate ventrally 6

5(4) Excavation on segment 8 with converging margins; signum markedly shorter than apophyses anteriores .... *politella* (p. 116)

– Excavation on segment 8 with parallel margins; signum markedly longer than apophyses anteriores .... *desertella* (p. 114)

6(4) Antrum large, triangular; posterior margin of segment 8 dorsally with shallow invagination with heavily sclerotized rim .... *terrella* (p. 115)

– Antrum short, tube-shaped; segment 8 dorsally with small but distinctive heavily sclerotized triangular scutulum .... *domestica* (p. 117)

7(1) Microtrichia on segment 8 wedge-shaped ....
.... *umbrosella* (p. 108)

– Microtrichia on segment 8 needle-shaped 8

8(7) Lamella postvaginalis with two lobes ....
.... *senectella* (p. 111)

– Lamella postvaginalis without lobes 9

9(8) Signum large, elongate, twice as long as wide and longer than apophyses anteriores .... *similis* (p. 110)

– Signum no longer than apophyses anteriores 10

10(9) Signum trapezoidal with two conspicuously strong spines at posterior end .... *affinis* (p. 109)

– Signum square, rounded or slightly oval 11

11(10) Signum with four very strong spines, segment 8 ventrally with semicircular excavation covered by indented frame, distal end of ventral groove not marked.. *dryadella* (p. 107)

– Signum with rather weak spines, ventral groove marked by sclerotized oval rim ....
.... *basaltinella* (p. 107)

BRYOTROPHA BASALTINELLA (Zeller)

Gelechia basaltinella Zeller, 1839, *Isis, Leipzig* **1839**: 198.
Type locality: Germany; Spitzberg.

Description of imago (Pl.2, fig.11)

Wingspan 10–12mm. Head with vertex greyish yellow to dark greyish yellow, frons and cheeks paler, usually ochreous; antenna ochreous, ringed fuscous; labial palpus with segment 3 longer than segment 2, segment 2 yellow to ochreous, suffused fuscous on outer side, segment 3 usually dark fuscous, sometimes suffused ochreous on inner side, occasionally ochreous all over. Thorax fuscous-brown. Forewing light to dark greyish brown, often mixed with large patches of yellow to ochreous scales; discal and plical stigmata blackish brown, usually clearly visible, first discal directly above plical and often fused, forming large blackish patch just before middle of wing; costal and tornal patches ochreous and usually well developed, often fused to form irregular fascia, its inner margin angled at about 120°; subapical area irrorate with very dark scales, occasionally darkening this area, near termen these scales being somewhat larger revealing their light-coloured bases; cilia ochreous to ferruginous with broad darker ciliary line and conspicuous pale yellow tips. Hindwing of equal width to forewing, pale fuscous, darker towards apex; cilia ochreous with broad dark ciliary line and pale yellowish tips in dark specimens, uniformly ochreous in light specimens. Legs variably mixed dark brown and ochreous. Abdomen pale greyish brown above, pale ochreous below; anal tuft of male pale ochreous. Genitalia, see figures 6a,26b.

Similar species. Only to be confused with *B. dryadella* (Zeller), *q.v.*

Life history

Larva. According to Spuler (1910), the head and prothoracic plate are blackish and the body creamy white with pale brown dorsal line and dark spots but Heckford & Sterling (2002) describe the body of a reared specimen as brown. March–May in mosses on walls and buildings.

Pupa. Undescribed.

Imago. Univoltine; late May–late August. A nocturnal species which comes to light.

Distribution (Map 37)

Occurs near buildings. Very local in southern England but sometimes common where it occurs; a few scattered records from northern England; the Channel Islands. The recorded distribution may not be accurate, owing to confusion with *B. dryadella*, and records

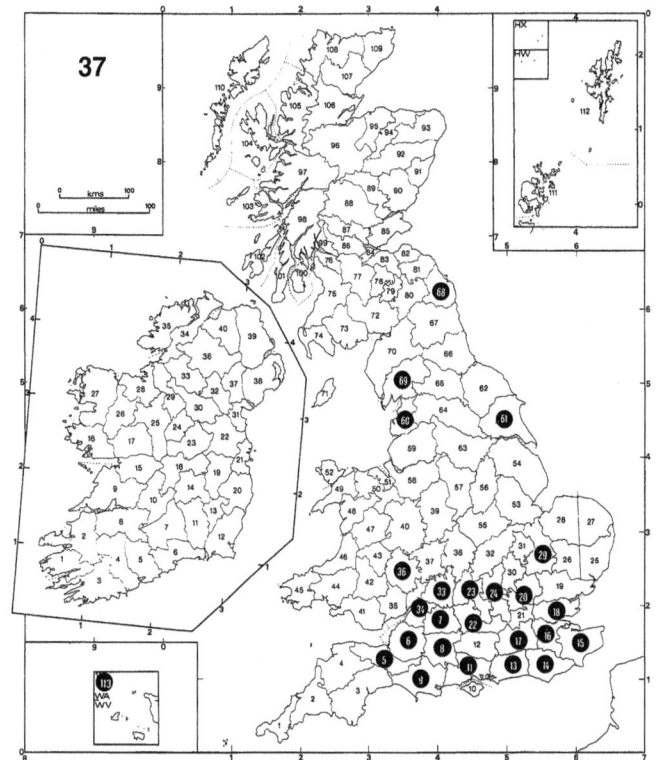

Bryotropha basaltinella

should be considered valid only if supported by dissection of the genitalia. Southern and central Europe, extending northwards to The Netherlands and Latvia; not recorded from Scandinavia.

BRYOTROPHA DRYADELLA (Zeller)

Gelechia dryadella Zeller, 1850, *Stettin.ent.Ztg.* **11**: 152.
Type localities: Italy; Poppi and Bibberia, Tuscany.

Description of imago (Pl.2, fig.12)

Wingspan 10–12mm. Head with frons dark ochreous, vertex ochreous-brown to dark brown; antenna pale ochreous, ringed fuscous; labial palpus with segment 3 longer than segment 2, creamy white to pale ochreous on inner side, weakly to heavily suffused fuscous on outer side. Thorax with posterior half ochreous-brown to fuscous-brown, anterior half ochreous to orange-brown. Forewing grey-brown to dark grey-brown; discal and plical stigmata blackish, moderately prominent, first discal above the often much smaller plical, never fused, often with ochreous-brown to brown scales basad and beyond plical and first discal stigmata; costal and tornal patches ochreous-brown to brown, well developed, often fused to form a straight

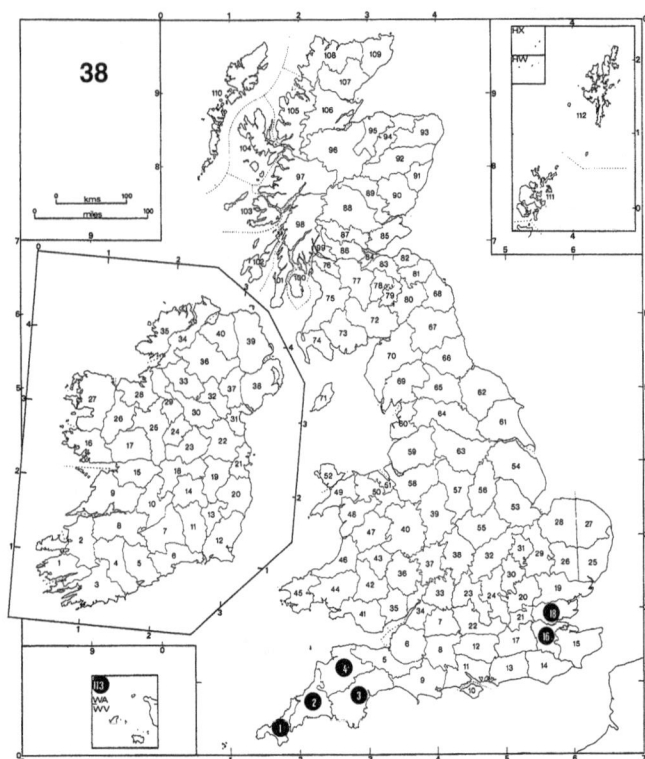

Bryotropha dryadella

or slightly inwardly bent fascia; subapical area irrorate with dark scales, darkening this area; cilia ferruginous to dark grey with dark ciliary line and pale yellow tips. Hindwing of equal width to forewing, pale fuscous, darker towards apex; cilia concolorous, with pale yellow tips. Legs variably mixed dark brown and ochreous. Abdomen pale to dark ochreous-grey above, mixed ochreous and pale to dark ochreous-grey below; anal tuft of male pale ochreous. Genitalia, see figures 6b,26c.

Similar species. B. dryadella strongly resembles *B. basaltinella* (Zeller). In *B. dryadella* the costal and tornal patches never point outwards, and when fused they form a straight or inwardly bent fascia. In *B. basaltinella* the costal and tornal patches never point inwards, and when fused they form an outwardly bent fascia. At least in western Europe, *B. dryadella* lacks bright ochreous scales which makes it a less vividly coloured species here than *B. basaltinella*. These observation were based on the examination of Continental specimens of *B. dryadella*. In the British specimens so far examined they do not appear to apply, and the species seems to be indistinguishable from *B. basaltinella* on external characters but readily separa-

ble in the female genitalia, though less clearly so in the male genitalia (R. J. Heckford, pers.comm.).

Life history

Accounts of the early stages (and also distribution) are in press (Heckford & Sterling, 2002; Heckford & Sattler, 2002).

Imago. Univoltine; late May–late August. Flies from early evening and at night when it comes to light.

Distribution (Map 38)

The first known record from the British Isles was made on Jersey, Channel Islands, in the late 1800s or early 1900s. Several specimens from England, taken between 1983 and 1999 and previously identified as *B. basaltinella*, have been dissected and identified by R. J. Heckford and K. Sattler. They are from Cornwall, Devon and Essex. Specimens from Blisland, Cornwall, were taken at rest on an old stone wall; other known habitats being coastal limestone cliffs, coastal serpentine, sand dunes and disused chalk quarries. Further specimens from Essex and Devon, and also Kent, were reared in 2001 (R. J. Heckford, pers.comm). France and the Mediterranean region.

BRYOTROPHA UMBROSELLA (Zeller)

Gelechia umbrosella Zeller, 1839, *Isis, Leipzig* **1839**: 201.
Gelechia mundella Douglas, 1850, *Trans.ent.Soc.Lond.* (N.S.) **1**: 64.
Gelechia portlandicella Richardson, 1890, *Entomologist's mon.Mag.* **26**: 29.
Bryotropha anacampsoidella Hering, 1924, *Notul.ent.* **4**: 80.
Type locality: Germany; Glogau (now Poland; Glogów).

Description of imago (Pl.2, figs 13,14)

Wingspan 9.5–11.5mm. Head with vertex very dark greyish brown, frons yellow to ochreous; antenna dark fuscous; labial palpus with segment 3 longer than segment 2, ochreous on inner side, heavily suffused fuscous on outer side. Thorax concolorous with forewing. Forewing very dark greyish brown, discal and plical stigmata blackish, but barely visible, first discal slightly beyond plical, plical stigma often bordered distally by a small number of clear white scales; a streak of white scales sometimes present between and beyond the two discal stigmata; costal and tornal patches clear white and usually well developed, rarely fused to form irregular fascia; subapical area irrorate with very dark scales; cilia dark grey. Hindwing of equal width to forewing, pale fuscous, darker towards apex; cilia concolorous. Legs variably mixed dark greyish brown, ochreous and creamy white. Abdomen dark brownish

Bryotropha umbrosella

grey above, light grey below; anal tuft of male ochreous to pale fuscous. Genitalia, see figures 6c,26f.

Specimens found in coastal areas are often not as dark as their inland counterparts owing to a weak or heavy irroration of greyish white scales; in cases of heavy irroration, specimens may appear greyish (f. *portlandicella* Richardson) or even whitish (f. *mundella* Douglas (fig.14)) instead of blackish. In these moths the blackish plical and discal stigmata become very distinct while the costal and tornal patches may no longer be visible.

Similar species. The well-developed and clear white costal and tornal patches contrasting with the very dark forewing distinguish *B. umbrosella* from *B. affinis* (Haworth) and *B. similis* (Stainton). Dark specimens in which the stigmata are not bordered by white scales and in which the costal and tornal patches are rather indistinct and whitish instead of clear white, are difficult to separate from certain forms of *B. similis*. When the costal patch is present but the tornal patch absent or very indistinct, the specimen is *B. similis*; in other cases it may be necessary to study the genitalia. Very light forms of *B. umbrosella* are greyish or greyish white, whereas light forms of *B. affinis* are ochreous to brownish.

Life history

Larva. Head and prothoracic plate dark brown. Body reddish brown. The full-grown caterpillar can be found from April to early June feeding on mosses growing on coastal sandhills; less often inland on walls (Spuler, 1910).

Pupa. Undescribed.

Imago. Univoltine; late May–mid-August. Becomes active near dusk and comes to light.

Distribution (Map 39)

Mainly coastal and often common in open dune areas, local and rare inland. Occurs on sandy coasts throughout most of the British Isles, but absent from western Scotland, the Inner and Outer Hebrides, Orkney and Shetland. Throughout most of Europe, including landlocked central European countries, but rare or absent from much of the Mediterranean region and southeastern Europe.

BRYOTROPHA AFFINIS (Haworth)

Recurvaria affinis Haworth, 1828, *Lepid.Br.*: 551.
Gelechia tegulella Herrich-Schäffer, 1855, *Syst.Bearb. Schmett.Eur.* **5**: 182.
Gelechia affinella Doubleday, 1859, *Zoologist synonymic List Br.Lepid.*: 30.

Type locality: Great Britain.

Description of imago (Pl.2, fig.15)

Wingspan 9.5–12.5mm. Head (figure 55) with vertex dark blackish grey, sometimes speckled due to light coloured bases of scales, frons pale yellow to ochreous,

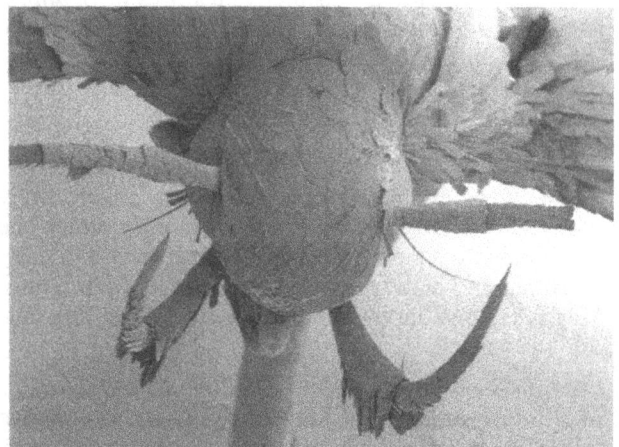

Figure 55 *Bryotropha affinis* (Haworth), scanning electromicrograph of head clearly showing pecten hair on antenna and palpi

cheeks ochreous, dorsally darkened; antenna ochreous, ringed fuscous; labial palpus with segment 3 longer than segment 2, ochreous on inner side, more or less heavily suffused fuscous on outer side, segment 3 usually darker than segment 2. Thorax concolorous with forewing. Forewing dark greyish brown, greyish tone sometimes prominent, bright yellowish bases of scales often exposed, giving forewing a speckled appearance; discal and plical stigmata blackish brown, first discal slightly beyond plical; beyond plical stigma nearly always a small patch of conspicuous yellowish or ochreous scales; similar scales sometimes present on inner side of plical stigma, and between and beyond discal stigmata ; costal and tornal patches yellow, sometimes ochreous, occasionally whitish, often fused to form irregular fascia, its inner margin angled at about 120°; subapical area irrorate with very dark scales, slightly darkening this part of wing; cilia brownish with dark ciliary line and pale yellow tips. Hindwing pale fuscous, darker towards apex; cilia concolorous. Legs variably mixed dark brown and ochreous. Abdomen greyish to greyish brown above, pale brown below; anal tuft of male whitish brown. Genitalia, see figures 6d,26g.

As in *B. umbrosella* (Zeller), specimens of *B. affinis* found in coastal areas are often not as dark as their inland counterparts owing to a weak to heavy irroration of yellowish scales; some specimens can be completely ochreous in which case the blackish plical and discal stigmata are very distinct, while the costal and tornal patches may no longer be visible.

Similar species. The speckled forewing, the yellowish scales beyond the plical stigma and the yellowish costal and tornal patches are distinctive features of *B. affinis*. Both *B. umbrosella* and *B. similis* (Stainton) lack the speckling on the forewing; *B. umbrosella* has clear white costal and tornal patches and *B. similis* has no conspicuously light scales beyond the plical stigma.

Life history

Larva. Head and prothoracic plate dark brown. Body brown. Feeds on mosses on walls, amongst which the larva makes a silken gallery in which it lives. On early mornings and cloudy days from the end of March until May full-grown larvae can often be found feeding outside the gallery.

Pupa. Light brown; normally in the larval gallery.

Imago. Univoltine; mid-May–mid-August. It is active from dusk onwards and frequently comes to light.

Distribution (Map 40)

Occurs in open country and urban areas, frequenting

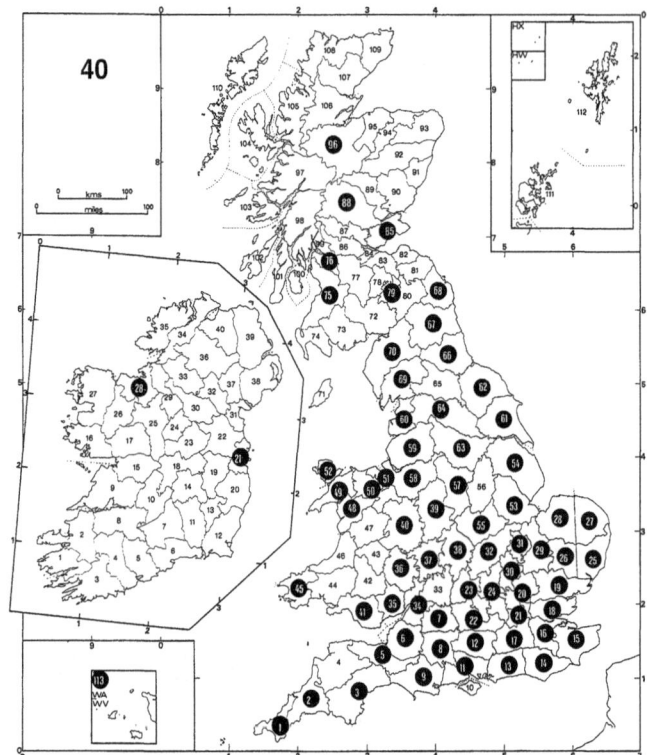

Bryotropha affinis

mossy walls and old thatch (Stainton, 1865b). Widespread and common in southern Britain, extending to Inverness-shire, but more local in the north; the Channel Islands; records from Ireland (Beirne, 1941) were either based on misidentifications or are unconfirmed (Bond, 1995). Throughout most of Europe, but more rare in the south-east; Syria.

BRYOTROPHA SIMILIS (Stainton)

Gelechia similis Stainton, 1854, *Ins.Br.Lepid*: 115.

Gelechia thuleella Zeller, 1857, *in* Staudinger, *Stettin.ent. Ztg* **18**: 276.

Gelechia similella Doubleday, 1859, *Zoologist synonymic List Br.Lepid.*: 30.

Gelechia confinis Stainton, 1871, *Entomologist's Annu.* **1871**: 98.

Gelechia stolidella Morris, 1872, *Br.Moths* **4**: 82.

Gelechia obscurecinerea Nolcken, 1872, *Fauna Livld.* **2**: 573.

Bryotropha fuliginosella Snellen, 1882, *Vlind.Nederland*: 645.

Type locality: England; London, Surrey.

Description of imago (Pl.2, fig.16)

Wingspan 10.5–13.5mm. Head with vertex glossy dark grey-brown, frons whitish yellow to yellow-grey, cheeks dark yellow-grey or as vertex; antenna fuscous; labial palpus with segment 3 longer than segment 2, segment 2 ochreous on inner side, weakly to heavily suffused fuscous on outer side, segment 3 ochreous-grey. Thorax dark fuscous. Forewing glossy greyish brown to blackish brown, bases of scales pale brown and almost transparent yet seldom visible; discal and plical stigmata blackish brown, often barely visible against the dark ground colour, first discal slightly beyond plical, sometimes a few almost transparent pale whitish scales beyond plical stigma; costal and tornal patches usually whitish and weakly expressed, sometimes absent in very dark specimens, but occasionally well developed and yellowish to bright yellow, sometimes fused to form irregular fascia, its inner margin angled at about 120°; subapical area often darkened by irroration of very dark scales; cilia variable, dark grey in very dark specimens, ochreous with darker ciliary line and pale yellow tips in paler specimens. Hindwing of equal width to forewing, fuscous, darker towards apex; cilia concolorous. Legs variably mixed dark greyish brown and ochreous. Abdomen dark fuscous above, pale fuscous below; anal tuft of male pale fuscous. Genitalia, see figures 6f,26d.

Similar species. The dark and glossy forewing and the absence of conspicuous light scales beyond the plical stigma distinguish *B. similis* from *B. affinis* (Haworth) and most forms of *B. umbrosella* (Zeller), *qq.v.* In very dark forms of *B. senectella* (Zeller) the head often has prominent ochreous cheeks contrasting with the greyish brown vertex; in *B. similis* the cheeks are concolorous with the vertex.

Life history

Larva. Head and prothoracic plate dark brown. Body brown. Feeds on mosses growing on walls (Meyrick, [1928]). Possible foodplants in Britain are *Hypnum cupressiforme*, *H. rutabulum*, *Tortula ruralis*, *T. intermedia*, *Grimmia pulvinata*, *Bryum capillare* and *B. caespititium* (Stainton, 1871). April–May.

Pupa. Undescribed.

Imago. Univoltine; early June–late August. A nocturnal species which comes to light in small numbers.

Distribution (Map 41)

Widespread but seldom common in open, sometimes sparsely vegetated, country and urban areas. Throughout Britain except for the north of Scotland, the Hebrides, Orkney and Shetland; very local in southern

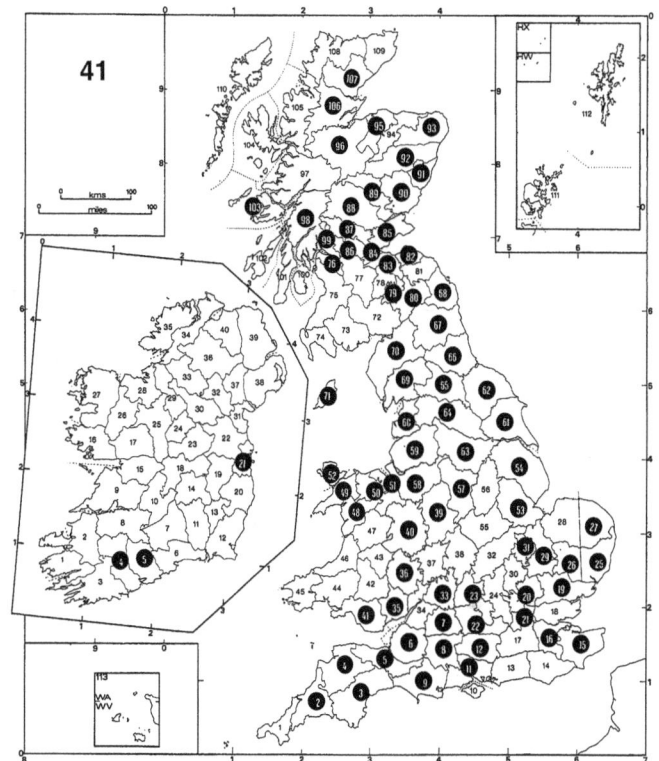

Bryotropha similis

and eastern Ireland; the Isle of Man. Throughout most of Europe except for the eastern Mediterranean region.

BRYOTROPHA SENECTELLA (Zeller)

Gelechia senectella Zeller, 1839, *Isis, Leipzig* **1839**: 199.
Bryotropha obscurella Heinemann, 1870, *Schmett.Dtl. Schweiz* (2)**2**(1): 240.

Type locality: Germany; Glogau (now Poland; Głogów).

Description of imago (Pl.2, fig.17)

Wingspan 9.0–12.5mm. Head with frons ochreous, cheeks yellow to brown, usually ochreous, vertex usually dark grey-brown, contrasting with cheeks, when viewed from above the vertex looking like a dark line extending over middle of head, flanked by ochreous cheeks; antenna dark brown; labial palpus with segment 3 longer than segment 2, ochreous, segment 3 suffused fuscous on outer side. Thorax brown to dark brown. Forewing brown to very dark brown; pale brownish, almost transparent bases of scales occasionally visible; discal and plical stigmata blackish, often rather obscure, especially second discal stigma, first discal slightly

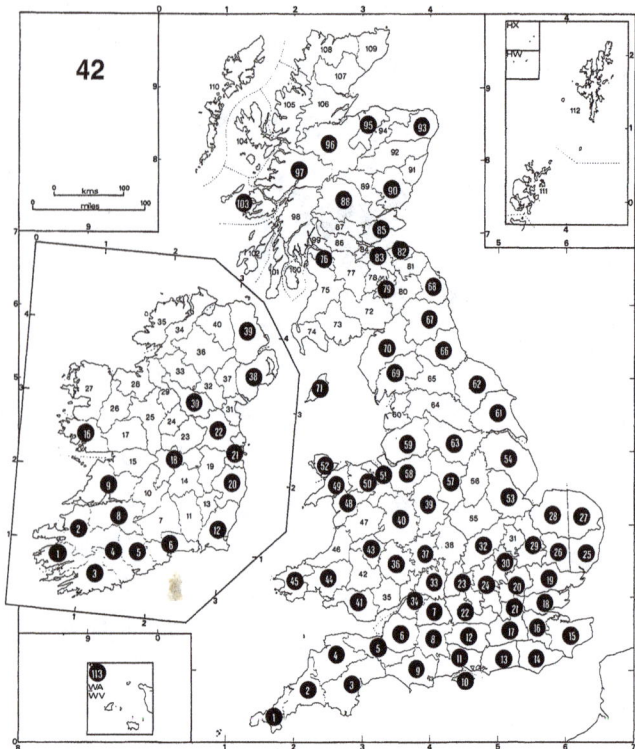

Bryotropha senectella

beyond plical, plical stigma usually bordered distally by small number of ochreous scales, ochreous scales sometimes present on inner side of plical stigma and between first and second discal stigmata; costal and tornal patches usually weakly expressed, sometimes absent, light brown, occasionally orange-brown, when fused to form fascia its inner margin being angled at less than 90°, fascia often accentuated by dark scales in subapical area, sometimes darkening this area considerably; cilia brown with slender dark ciliary line beyond middle and pale yellow tips. Hindwing of equal width to forewing, pale brownish, darker towards apex; cilia concolorous. Legs variably mixed brown and ochreous. Abdomen yellowish grey to greyish above, ochreous below; anal tuft of male ochreous. Genitalia, see figures 6e,26e.

Similar species. The brownish colour of the forewing and the conspicuous ochreous cheeks differentiate *B. senectella* from *B. umbrosella* (Zeller), *B. affinis* (Haworth) and *B. similis* (Stainton). Very small forms of *B. desertella* (Douglas) have a more rounded forewing with several dark ciliary lines, compared to only one dark ciliary line in *B. senectella*.

Life history
Larva. Head and prothoracic plate dark brown. Body

brown. April–May on mosses on walls (Spuler, 1910).
Pupa. Undescribed.

Imago. Univoltine; mid-June–early September. The moth becomes active towards dusk and comes to light.

Distribution (Map 42)
Widespread and common, both inland and on sandy coasts, throughout the British Isles except for the north of Scotland, the Hebrides, Orkney and Shetland. Throughout Europe.

BRYOTROPHA BOREELLA (Douglas)

Gelechia boreella Douglas, 1851, *Trans.ent.Soc.Lond.* (N.S.) 1: 105.
Type locality: Scotland; Holy Loch, Argyll.

Description of imago (Pl.2, figs 18,19)
Sexual dimorphism pronounced. *Male* (fig.18). Wingspan 13.5–15.0mm. Head dark grey; antenna dark fuscous, scape usually with pecten; labial palpus with segment 3 shorter than segment 2, segment 2 without furrowed ventral brush, cream-coloured on inner side, heavily suffused dark fuscous on outer side. Thorax dark grey. Forewing broad, dark brown-grey to dark fuscous; plical and discal stigmata blackish brown and rather obscure, first discal slightly beyond plical, second discal usually more pronounced; costal and tornal patches creamy white, often fused to form irregular fascia, occasionally ill-defined or even absent; termen lined with small groups of very dark scales, terminal area also often darkened by irroration of very dark scales; cilia concolorous, often with one or two ciliary lines. Hindwing of equal width to forewing, fuscous; cilia concolorous, with two diffuse ciliary lines. Legs variably mixed dark grey, greyish white and ochreous. Abdomen dark grey above and below; anal tuft lighter grey. *Female* (fig.19). Wingspan 12.0–13.5mm. Head dark fuscous; labial palpus wholly dark fuscous. Thorax dark fuscous. Forewing widening to about one-third, tapering beyond; dark fuscous with distinct purple gloss; plical and discal stigmata blackish but rarely visible against dark colour of wing; costal and tornal patches creamy white, usually well developed and fused to form broad fascia; subapical area irrorate with even darker scales; cilia concolorous. Hindwing widest near base, gradually tapering beyond; dark ochreous-grey; cilia concolorous, often with one or two ciliary lines. Legs variably mixed dark fuscous and creamy white. Genitalia, see figures 7a,27e.

Life history.
The early stages are undescribed.

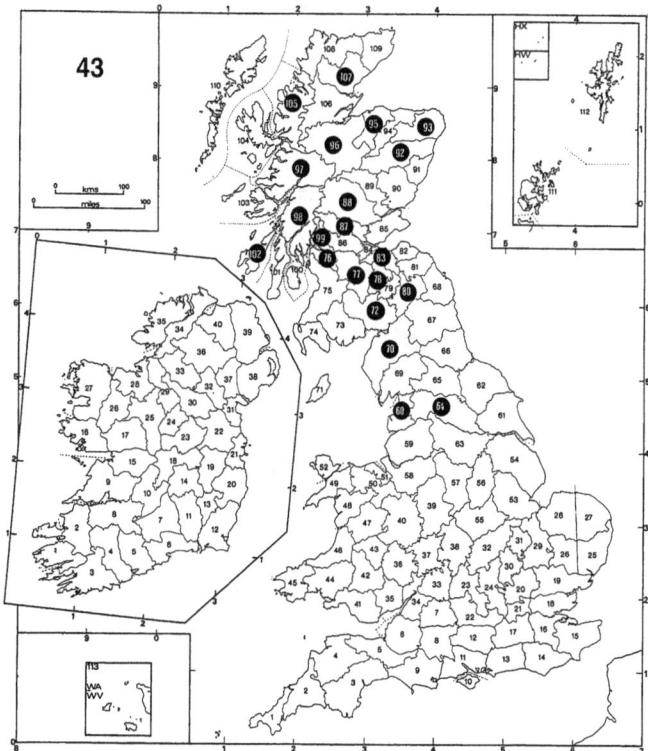

Bryotropha boreella

Bryotropha galbanella

Imago. Univoltine; June–July.

Distribution (Map 43)

A rare and local species found on heather in the north of England and Scotland, absent from the rest of England, Wales, Ireland, Inner and Outer Hebrides, Orkney and Shetland. Central Europe and northern Scandinavia; absent from western Europe.

BRYOTROPHA GALBANELLA (Zeller)

Tinea galbanella Zeller, 1839, *Isis, Leipzig* **1839**: 200.
Gelechia angustella Heinemann, 1870, *Schmett. Dtl. Schweiz* (2)**2**(1): 217.
Gelechia ilmatariella Hoffmann, 1893, *Stettin. ent. Ztg* **54**: 138.
Type locality: Germany (now Poland); Salzbrunn.

Description of imago (Pl.2, fig.20)

Wingspan 15.0–17.5mm. Head with vertex and cheeks yellowish grey, frons light to dark grey; antenna ochreous, ringed fuscous, scape usually with pecten at base; labial palpus ochreous on inner side, suffused greyish brown on outer side, segment 3 shorter than segment 2, segment 2 without ventral brush (see figure 52b, p.103). Thorax dark greyish yellow to dark greyish brown. Forewing dark greyish yellow, irrorate with a large number of yellowish or orange-yellow scales and a small number of very dark scales; discal and plical stigmata blackish, first discal slightly beyond plical, second discal very prominent; costal and tornal patches yellowish, fused to form fascia; terminal area marked by small groups of very dark scales, separated by yellowish scales; cilia dark grey. Hindwing of equal or slightly greater width than forewing, apex blunt; dark grey all over, without local darkening; cilia dark grey with yellow-orange base. Legs variably mixed greyish yellow and ochreous. Abdomen dark yellowish grey above, yellowish grey below; anal tuft of male ochreous. Genitalia, see figures 7b,27c.

Life history

Larva. Head and prothoracic plate dark fuscous-brown. Body brownish with dark spots. Feeds on mosses (Spuler, 1910). May–June.

Pupa. Undescribed.

Imago. Univoltine; late May–mid-August. Comes to light.

Distribution (Map 44)

Local and seldom common, usually found in forested areas. Recorded mainly from Scotland, but absent from the Hebrides, Orkney and Shetland. There is a single English record from Malham Tarn, mid-west Yorkshire (det. J. D. Bradley). Records from southern England are based on misidentifications. Central and northern Europe.

[BRYOTROPHA FIGULELLA (Staudinger)
Gelechia figulella Staudinger, 1859, *Stettin.ent.Ztg* **20**: 282.

Type locality: Spain; Chiclana.

A gelechiid, taken by C. T. Crutwell on a salt-marsh or waste ground near the sea at Aldeburgh, Suffolk, between 6 and 20 July 1892, was determined by Walsingham and Durrant as this southern European species (Barrett, 1893). The district is frequently visited by collectors, but no further specimen has been recorded and the moth is not migratory.

It is now generally accepted that the record was based on misidentification and that *B. figulella* does not occur in Britain.]

BRYOTROPHA DESERTELLA (Douglas)
Gelechia desertella Douglas, 1850, *Trans.ent.Soc.Lond.* (N.S.) **1**: 62.

Gelechia decrepidella Herrich-Schäffer, 1854, *Syst.Bearb. Schmett.Eur.* **5**: 177.

Gelechia decrepitella Heinemann, 1870, *Schmett.Dtl. Schweiz* (2)**2**(1): 236, misspelling.

Type locality: England; Weymouth, Dorset.

Description of imago (Pl.2, fig.21)
Wingspan 11–15mm. Head ochreous to dark brown, sometimes greyish, vertex, frons and cheeks concolorous with forewing; antenna ochreous, ringed fuscous to dark brown; labial palpus ochreous on inner side, weakly to heavily suffused fuscous on outer side, segment 3 longer than segment 2. Thorax concolorous with forewing. Forewing variable, ochreous, light brown, orange-brown, dark brown or greyish, sometimes with faint indication of dark median streak; discal and plical stigmata blackish, often rather diffuse; costal and tornal patches fused to form fascia, its inner angle at less than 90°, being often accentuated by dark scales both on inner and especially on outer margins, but when these dark scales are absent it is often impossible to discern a fascia as its colour is usually identical to that of forewing; subapical area irrorate with very dark scales, sometimes darkening this area, in light

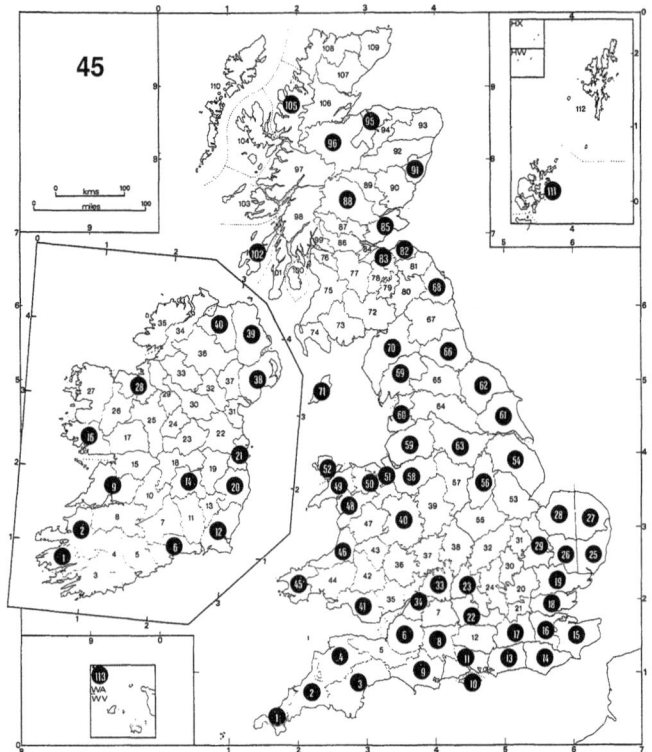

Bryotropha desertella

forms dark scales being restricted to termen; cilia concolorous, with several ciliary lines. Hindwing pale ochreous to pale fuscous, slightly darker towards apex; cilia concolorous, in light forms two or more ciliary lines visible. Legs variably mixed light to dark brown and ochreous. Abdomen yellowish grey to grey above, pale ochreous to pale grey below; anal tuft of male pale ochreous to pale grey. Genitalia, see figures 7d,27a.
Similar species. *B. terrella* ([Denis & Schiffermüller]) and *B. politella* (Stainton), *qq.v.*

Life history
Based on Sterling & Heckford (2001).
Ovum. Undescribed.

Larva. Length 6–9mm in final instar. Head shiny black, adfrontal sutures narrowly lined yellowish brown. Prothoracic plate black, divided by narrow white median line with brown, sometimes black-marked border; body dark reddish brown, dorsal line ochreous and divided by narrow reddish brown line, subdorsal, lateral and subspiracular lines indistinct and interrupted, grey; pinacula minute, black; anal plate pale yellowish brown, with a pair of blackish brown oval marks anterolaterally; thoracic legs black, ringed yellowish brown at joints, with shiny black elongate sclerite just anterior to, and another just posterior to, base of each leg; prolegs concolorous with body.

Larvae were found in whitish silken tubes, incorporating a few grains of sand, amongst and attached to moss. They were observed feeding on the mosses *Syntrichia ruraliformis*, *Hypnum lutescens* and *Rhytidiadelphus squarrosus*. Late October–early March, though most were full-fed in February.

Pupa. In a firm sand cocoon (Stainton, 1866).

Imago. Univoltine; mid-May–late August. Easily disturbed during the day, otherwise nocturnal; comes to light.

Distribution (Map 45)

Widespread in coastal areas, very common in open dune country, scarce inland. Throughout England, Wales and Ireland, more scarce in Scotland, absent from the north coast of Scotland, the Outer Hebrides, and Shetland. Europe.

BRYOTROPHA TERRELLA ([Denis & Schiffermüller])

Tinea terrella [Denis & Schiffermüller], 1775, *Schmett. Wien.*: 140.

Tinea inulella Hübner, [1803], *Samml.eur.Schmett.* **8**: pl.41, fig.286.

Nothris pauperella Hübner, [1825], *Verz.bekannt. Schmett.*: 411.

Recurvaria listeri sensu Haworth, 1828, *Lepid.Br.*: 548.

Gelechia latella Herrich-Schäffer, 1854, *Syst.Bearb. Schmett.Eur.* **5**: 174.

Gelechia lutescens Constant, 1865, *Annls Soc.ent.Fr.* (4)**5**: 196, pl.7, fig.12.

Gelechia suspectella Heinemann, 1870, *Schmett.Dtl. Schweiz* (2)**2**(1): 202.

Type locality: [Austria]; Vienna district.

Description of imago (Pl.2, figs 22,23)

Wingspan 14.0–16.5mm. Head concolorous with forewing; antenna ochreous, ringed fuscous to dark fuscous; labial palpus with segment 3 longer than segment 2 (see figure 52a, p.103), ochreous, weakly to heavily suffused fuscous on outer side, sometimes dark greyish brown all over. Thorax concolorous with forewing. Forewing variable, often dark brown or dark greyish brown, occasionally ferruginous or light brown; discal and plical stigmata blackish brown, moderately well developed, sometimes very diffuse, occasionally prominent; costal and tornal patches fused to form fascia with inner angle of less than 90°, being often accentuated by dark scales both on inner side and especially on outer side, but when these dark scales are absent it is often impossible to discern a fascia since its colour is usually identical with that of forewing; terminal area lined with very dark scales, irroration with dark scales sometimes darkening whole subapical area; cilia concolorous, with several darker and lighter ciliary lines. Hindwing of equal width to forewing, pale fuscous, darker towards apex; cilia variable, concolorous in very dark specimens, ochreous with two dark ciliary lines and pale yellow tips in pale specimens. Legs variably mixed light to dark grey and ochreous. Abdomen ochreous-grey to dark grey above, pale ochreous below; anal tuft in male ochreous. Genitalia, see figures 7e,27b.

Similar species. B. terrella can sometimes be confused with *B. desertella* (Douglas) and *B. politella* (Stainton), qq.v. In dune areas *B. desertella* is clearly smaller and often ochreous, a colour rarely found in *B. terrella*. Inland forms of *B. desertella* are larger and darker than their coastal counterparts and can be difficult to tell apart from small *B. terrella*. In general, *B. desertella* has a slightly narrower forewing, making it a more elegant species, and regularly has an indication of a median streak on the forewing, a feature absent in *B. terrella*. Occasionally it will be necessary to study the genitalia. Females, however, can be recognized by brushing away some scales from the terminal abdominal segment; in *B. terrella* segment 8 is not excavate ventrally while in *B. desertella* the posterior margin of segment 8 is deeply excavate ventrally.

Life history

Based on Heckford (1999c) and Sterling & Heckford (2001).

Ovum. Undescribed.

Larva. Length 10–12mm in final instar. Head black, adfrontal sutures brownish, sometimes a brownish spot anteriorly. Prothoracic plate yellowish brown, marked black, bisected by indistinct whitish median line; body dull reddish brown, dorsal line of same colour but edged slightly paler and therefore discernible, indistinct grey subspiracular line, abdominal segment 8 yellowish dorsally; pinacula minute, black, with short blackish setae; anal plate yellowish or yellowish brown, sometimes with a few small blackish brown spots; thoracic legs black, ringed yellowish brown at joints, with shiny black elongate sclerite just anterior to, and another almost translucent black-marked sclerite just posterior to, base of each leg; prolegs concolorous with body.

Early-instar larvae were found in fairly tough, rather opaque silken tubes attached to moss or, occasionally,

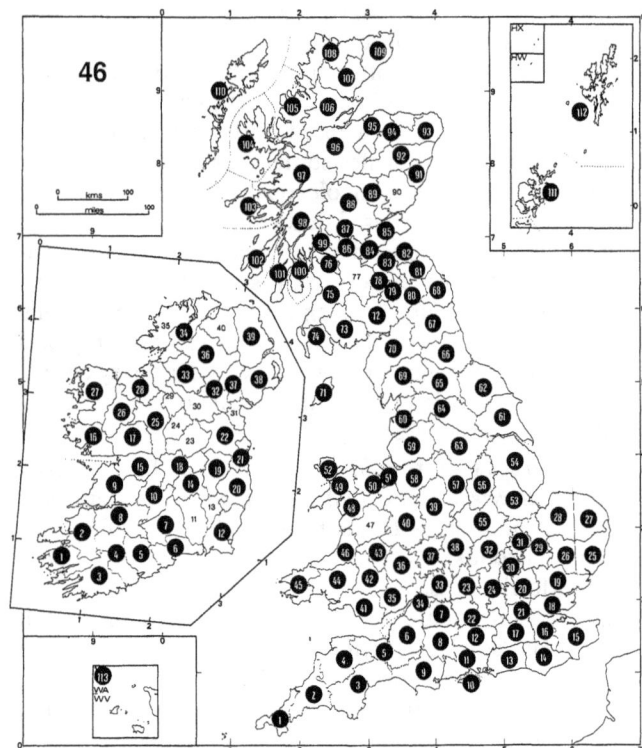

Bryotropha terrella

grass. The lower end of the tube was attached low down amongst the moss or grass and the upper end just below the surface of the moss, the tube being open at each end and covered with chewed fragments of moss and grass. In the final instar the larva constructs a flimsy, more transparent tube which is unadorned. Larvae, observed feeding on common bent (*Agrostis capillaris*) and the moss *Rhytidiadelphus squarrosus*, are able to move very quickly and have a characteristic way of rapidly vibrating the head while the rest of the body is still. Late September–mid-March, though some larvae were fully fed and made cocoons as early as mid-January.

Pupa. Pale reddish brown, within a cocoon covered with fragments of grass, sometimes incorporating moss; in the larval feeding-place.

Imago. Univoltine; late May–late August. The moth becomes active near dusk and comes to light.

Distribution. (Map 46)
Widespread and very common, especially on sandy ground, throughout the whole of the British Isles. Occurs throughout Europe; Asia Minor.

BRYOTROPHA POLITELLA (Stainton)

Gelechia politella Stainton, 1851, *Syst.Cat.Br.Tineidae & Pterophoridae* Suppl: 4.
Gelechia expolitella Doubleday, 1859, *Zoologist synonymic List Br.Lepid.*: 30.
Type locality: England; Skiddaw, Cumbria.

Description of imago (Pl.2, figs 24,25)
Sexual dimorphism pronounced. *Male* (fig.24). Wingspan 12–16mm. Head concolorous with forewing; antenna dark fuscous; labial palpus with segment 3 longer than segment 2, yellowish grey on inner side, heavily suffused fuscous on outer side, sometimes fuscous all over. Thorax concolorous with forewing. Forewing shiny brown-grey; plical and discal stigmata blackish brown, usually very small and occasionally barely visible; costal and tornal patches often absent, when present fused to form very ill-defined fascia with inner angle of less than 90°, subapical area of same colour as rest of wing; termen often lined with very dark scales set singly or in very small clusters; cilia concolorous, with two dark ciliary lines. Hindwing of equal width to forewing, pale greyish, slightly darker towards apex; cilia concolorous. Legs variably mixed dark grey-brown and ochreous. Abdomen greyish brown above, dark greyish brown below; anal tuft concolorous. *Female* (fig.25). Differs from male as follows. Wingspan 13–15mm. Antenna ochreous, ringed fuscous; labial palpus ochreous, weakly to heavily suffused fuscous on outer side. Forewing shiny ochreous to ochreous-grey; plical and discal stigmata blackish brown, rather prominent, first discal usually elongate; costal and tornal patches fused to form fascia with inner angle of less than 90°, fascia, however, very ill defined and sometimes not visible; subapical area slightly darkened owing to irroration with brownish scales; termen lined with small groups of dark scales; cilia yellow to ochreous with dark ciliary line and yellow tips. Hindwing often with ciliary line. Abdomen pale yellowish grey above, darker yellowish grey below. Genitalia, see figures 7c,27d.

Similar species. The male can be confused with *B. terrella* ([Denis & Schiffermüller]). The male of *B. politella*, however, has a smooth greyish brown forewing whereas the forewing of *B. terrella* is brownish and often coarser due to suffusion with scales of different colour. Furthermore in *B. politella* the fascia is absent or very ill-defined, there is no conspicuous darkening of the subapical region, and no ciliary lines in the hindwing cilia. On the whole, *B. politella* has a thinner body, making *B. terrella* appear more robust. The female

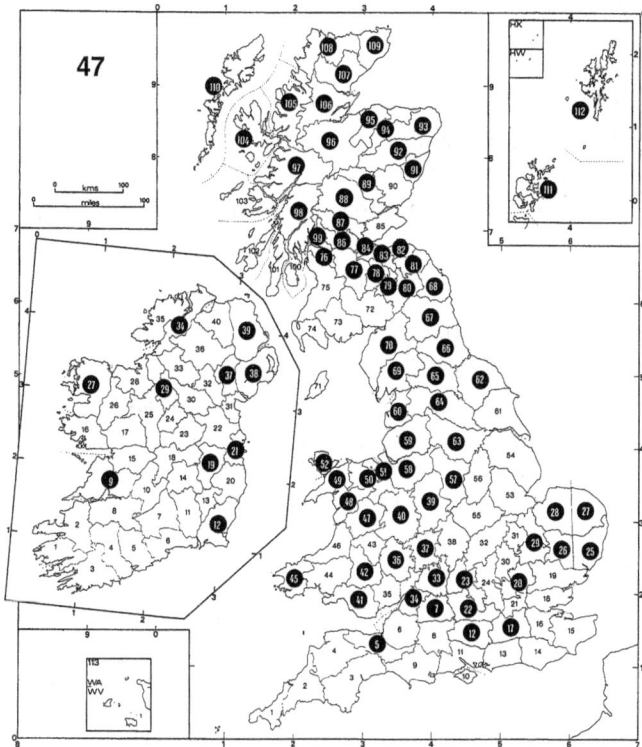

Bryotropha politella

resembles light coastal forms of *B. desertella* (Douglas); however, the female of *B. politella* is larger, the wings are more pointed and the ochreous colour slightly more greyish, while the hindwing is greyish and not ochreous as in light forms of *B. desertella*, and the cilia in both fore- and hindwings have only one ciliary line.

Life history
The early stages are undescribed.

Imago. Univoltine; late May–late July. Comes to light.

Distribution (Map 47)
Occurs mainly in dry grassland. Widespread throughout the British Isles; local and rare in the south, more common in Wales, northern England and especially Scotland, extending to Shetland. On mainland Europe known only from France.

BRYOTROPHA DOMESTICA (Haworth)
Tinea domestica Haworth, 1828, *Lepid.Br.*: 551.
Gelechia domesticella Doubleday, 1859, *Zoologist synonymic List Br.Lepid.*: 30.
Type locality: Great Britain.

Description of imago (Pl.2, fig.26)
Wingspan 12–13mm. Head with vertex yellow to ochreous, often speckled with ochreous-grey, frons yellowish; antenna yellow, ringed fuscous; labial palpus with segment 3 longer than segment 2, pale yellow on inner side, weakly to heavily suffused fuscous on outer side, segment 3 darker than segment 2. Thorax yellow to ochreous, often speckled with ochreous-grey. Forewing greyish ochreous to dark greyish ochreous, the prominent yellow to ochreous colour of bases of scales often visible, costa usually slightly darkened; discal and plical stigmata blackish brown, very prominent against light ground colour, first discal above smaller plical, often separate, sometimes fused, ochreous scales being usually present beyond plical stigma and between discal stigmata; costal and tornal patches ochreous, sometimes fused to form irregular fascia, conspicuous group of dark scales often present on inner side of costal patch; subapical area irrorate with both dark and light scales, occasionally darkening this area slightly; cilia concolorous, with pale yellow tips. Hindwing of equal width to forewing, pale greyish, darker towards apex; cilia concolorous. Legs variably mixed dark brown and ochreous. Abdomen grey above, slightly paler below; anal tuft of male ochreous. Genitalia, see figures 7f,27f.

Life history
Larva. Head and prothoracic plate dark brown. Body brown with conspicuous dark brown spots. Feeds on mosses growing on walls, the larva living in a silken gallery. At the end of March and in early April the then full-grown larva can often be found feeding outside the gallery, especially early on wet or dewy mornings. A heap of yellowish grey frass at the entrance of its gallery indicates its presence (Stainton, 1865b).

Pupa. Light brown; anal segment with hooked spines. Normally in an open network cocoon made in the gallery in which the larva was living.

Imago. Univoltine; mid-May–early August. A nocturnal species which comes to light.

Distribution (Map 48)
Most common in urban areas. Throughout England and Wales, often very common in the south but more

117

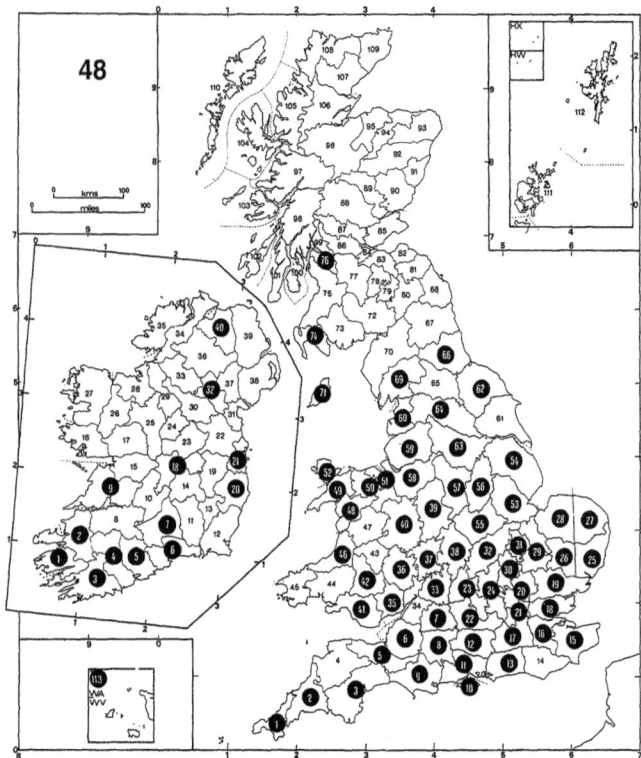

Bryotropha domestica

scarce towards the north; in Scotland known only from the Glasgow area and Wigtownshire; widespread in southern Ireland, rare in the north; the Channel Islands; the Isle of Man. Southern and central Europe; only a single record from The Netherlands; absent from Scandinavia.

Gelechiinae

STENOLECHIA Meyrick

Poecilia Heinemann, 1870, *Schmett.Dtl.Schweiz* (2)**2**(1): 281, *nec* Schneider, 1801.

Stenolechia Meyrick, 1894, *Entomologist's mon.Mag.* **30**: 230.

Gibbosa Omelko, 1988, *Ent.Obozr.* **67**: 152.

A very small Palaearctic genus with a single European species, which occurs in Britain.

Imago. Small to very small gelechiids with labial palpus slender and segment 2 rough-scaled ventrally; antenna without pecten. Forewing lanceolate and elongate, with small tufts of erect scales. Hindwing, narrower than forewing, trapezoidal; termen clearly emarginate before acutely produced apex. Abdomen with tergites 1–3 buff-coloured. Valva of male genitalia separated into spinose costa and digitate sacculus, aedeagus strongly curved; signum of female genitalia in form of two strong thorns, ostium with pronounced dorsal flap.

Larva. Bores into young shoots of oak (*Quercus* spp.) or possibly other trees.

STENOLECHIA GEMMELLA (Linnaeus)

Phalaena (*Tinea*) *gemmella* Linnaeus, 1758, *Syst.Nat.* (Edn 10) **1**: 539.

Recurvaria nivea Haworth, 1828, *Lepid.Br.*: 554.

Type locality: not stated.

Description of imago (Pl.2, fig.27)

Wingspan 10–11mm. Head white, occasionally with brownish fuscous spot on vertex; antenna brownish fuscous, banded lighter; labial palpus white, segment 3 with dark fuscous rings near base and just before apex. Thorax and tegulae white, mixed dark fuscous to a greater or lesser extent. Forewing white, suffused to a greater or lesser extent with dark fuscous-, pale fuscous- and sometimes ochreous-tipped scales, to produce dark fuscous pattern of large dorsal and smaller costal spot at base, almost complete outwardly oblique fascia at one-half of costa, spot on costa between this and base, and a spot at tornus; cilia white becoming pale greyish fuscous towards tornus with ciliary line of fuscous-tipped scales. Hindwing pale greyish fuscous becoming darker apically; cilia pale greyish fuscous. Fore- and midlegs dark fuscous, tarsi banded whitish; hindleg ochreous whitish. Abdomen light greyish fusc-

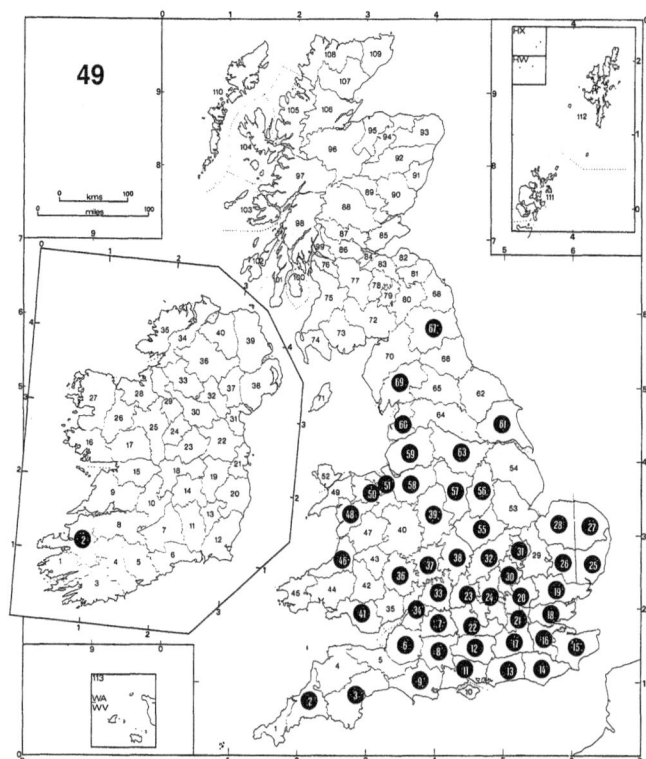

Stenolechia gemmella

ous, posterior three segments often edged whitish caudally. Genitalia, see figures 8a,28a.

Life history

Ovum. Laid on a twig of oak (*Quercus* spp.), probably late July–early September.

Larva. Head and undivided elliptical prothoracic plate yellowish-brown, latter grey-speckled. Body whitish, pinacula large and pale grey (Meyrick, 1895).

It is unknown when the eggs hatch, but in June the larvae bore into the buds and shoots of the foodplant, sometimes causing a swelling and causing the leaves to die. The larval habits of this species were known to Linnaeus (1758).

Pupa. Undescribed; usually in the larval habitation. July.

Imago. Univoltine; July–early September, but has been found as late as October (Bradford & Sokoloff, 1988). May be found resting in crevices on oak trunks during the day (Barrett, 1865); occasionally comes to light (Huemer & Karsholt, 1999).

Distribution (Map 49)

Widespread in England and Wales as far north as southern Northumberland and Cumbria. Unknown in

Scotland and recorded only from Muckross, North Kerry in Ireland (K. G. M. Bond, pers.comm.). Throughout Europe.

PARACHRONISTIS Meyrick

Parachronistis Meyrick, 1925, *in* Wytsman, *Genera Insect.* **184**: 14 (key), 52.

A very small genus restricted to the Palaearctic region of which only a single species occurs in Europe including Britain.

Imago. Small to very small gelechiids with labial palpus long and slender; antenna without pecten. Forewing lanceolate and elongate, with tufts of partially erect scales. Hindwing less broad than forewing, trapezoidal, apex acutely pointed, termen clearly emarginate before apex. Abdomen with tergites 1–3 lighter-coloured. Valva of male genitalia separated into a strongly spatulate cucullus with the blade subtriangular, and an angulate digitate sacculus, saccus fused to basal part of aedeagus; signum of female genitalia denticulate and irregular in shape, antrum strongly sclerotized, with a series of constrictions.

Larva. Early stages in the buds of various deciduous trees and shrubs.

PARACHRONISTIS ALBICEPS (Zeller)

Gelechia (Brachmia) albiceps Zeller, 1839, *Isis, Leipzig* **1839**: 202.

Gelechia aleella sensu Stephens, 1834, *nec* Fabricius, 1794.

Gelechia albicapitella Doubleday, 1859, *Zoologist synonymic List Br.Lepid.*: 30.

Type locality: Germany; Glogau (now Poland; Glogów).

Description of imago (Pl.2, fig.28)

Wingspan 10–11mm. Head white; antenna white, ringed dark fuscous; labial palpus white, segment 3 with strong fuscous ring near apex and an almost obsolete ring more basally, segment 2 fuscous below. Thorax and tegulae white, irrorate dark fuscous in thoracic midline and anteriorly. Forewing dark fuscous with two poorly developed scale-tufts, one just above fold at one-third and other at two-thirds; clear white markings in the form of scattered small spots in discal area, along base of cilia and three large subtriangular spots evenly spaced along costa, the basal edge of the first of these often extended into an oblique narrow transverse fascia which joins an elongate streak along

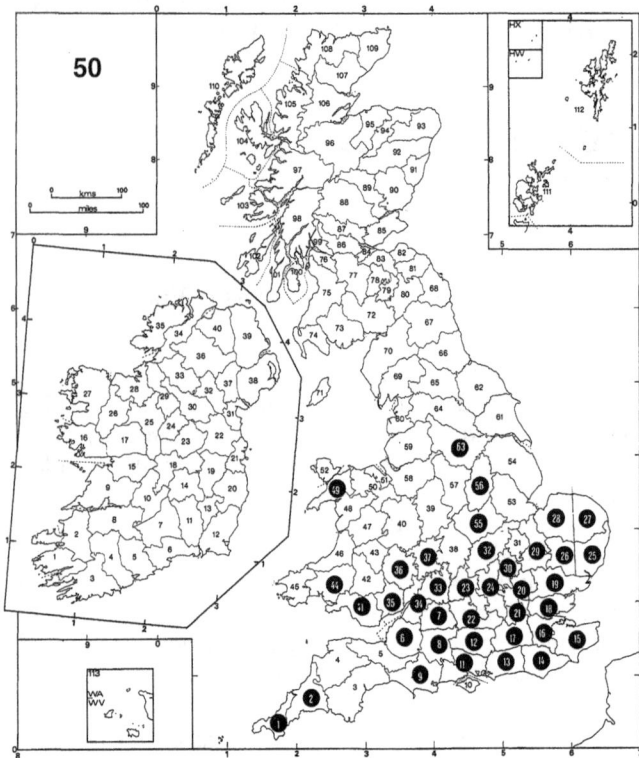

Parachronistis albiceps

dorsum almost to tornus; cilia with ciliary line of dark-tipped scales which becomes double on tornal half of termen. Hindwing pale fuscous; cilia slightly paler. Legs white, barred dark fuscous. Abdomen pale fuscous becoming paler caudally and along caudal edge of each tergite. Genitalia, see figures 8b,28k.

Life history

Ovum. Laid on a twig of hazel (*Corylus avellana*); July–August. Continental authors also give elm (*Ulmus* spp.) and peach (*Prunus persica*) (Huemer & Karsholt, 1999).

Larva. Undescribed. Feeds in a bud of the foodplant during May (Bradford, [1979]).

Pupa. Undescribed. June.

Imago. Univoltine; July–August. Comes to light (Huemer & Karsholt, *loc.cit.*).

Distribution (Map 50)

Widespread in Wales and the southern half of England as far north as southern Yorkshire. Not known from Scotland or Ireland. Throughout Europe.

RECURVARIA Haworth

Recurvaria Haworth, 1828, *Lepid.Br.*: 547.
Lita Kollar, 1832, *Beitr.Landesk.Oest.Enns.Wien* **2**: 95.
Telea Stephens, 1834, *Ill.Br.Ent.* (Haust.) **4**: 244.
Aphanaula Meyrick, 1895, *Handbk Br.Lepid.*: 579.
Hinnebergia Spuler, 1910, *Schmett.Eur.* **2**: 356.
Microlechia Turati, 1924, *Atti Soc.ital.Sci.nat.* **63**: 162.

A small genus with a worldwide distribution but best represented in the Neotropical region. Three species occur in Europe and two of these are found in Britain.

Imago. Small to medium-sized gelechiids with labial palpus long and slender, segment 2 thickened with appressed scales; antenna without pecten. Forewing elongate-subtrapezoidal, with termen rather oblique and tufts of erect scales. Hindwing slightly broader than forewing, trapezoidal with termen slightly emarginate before apex which is rounded. Tergites 1–3 of abdomen concolorous with the rest. Valva of male genitalia flagellate and long with bulbous base; signum of female genitalia subtrapezoid with serrate margins.

Larva. Feeds in buds or spun-leaves of various deciduous trees and shrubs, especially Rosaceae.

Key to species (imagines) of the genus *Recurvaria*

–   Forewing with broad white transverse fascia at one-third .... *leucatella* (p. 121)
–   Forewing with no obvious white fascia ....
                     .... *nanella* (p. 120)

RECURVARIA NANELLA ([Denis & Schiffermüller])

Tinea nanella [Denis & Schiffermüller], 1775, *Schmett. Wien.*: 141.
Tinea pumilella [Denis & Schiffermüller], 1775, *Ibid.*: 142.
Recurvaria nana Haworth, 1828, *Lepid.Br.*: 554.
Type locality: [Austria]; Vienna district.

Description of imago (Pl.2, fig.29)

Wingspan 11-13mm. Head clothed in white, fuscous-tipped scales; antenna white, barred fuscous; labial palpus with segment 2 white above and fuscous below, segment 3 white with two fuscous rings. Thorax and tegulae fuscous with white scale bases showing to varying degrees. Forewing white, irregularly suffused fuscous but leaving a white spot at tornus and one on costa

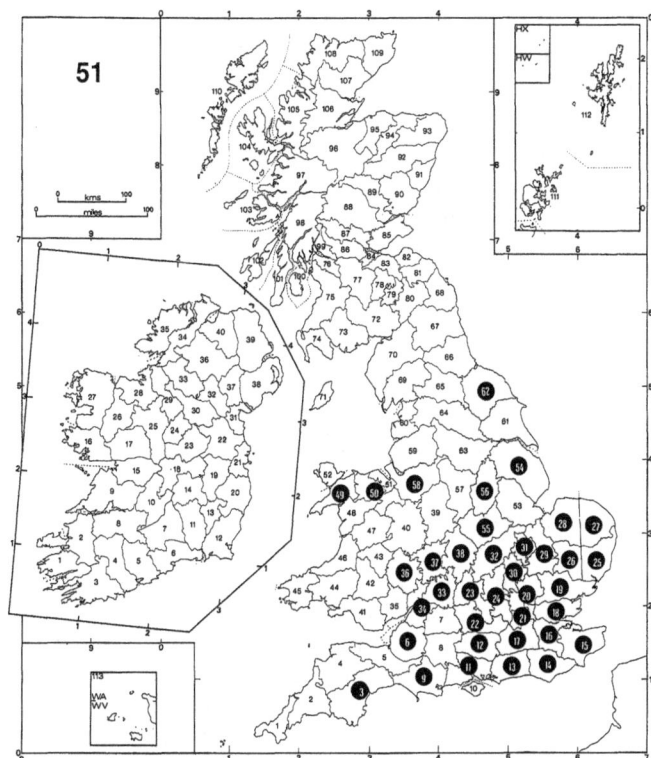

Recurvaria nanella

at one-quarter, often extending obliquely across wing almost to form a transverse fascia; markings of dark blackish fuscous rather variable but usually along basal edge of costal white spots, streak in discal area and another from apical end of discal area to apex; cilia pale ochreous-fuscous in tornal area, on termen becoming white with basal line of dark-tipped scales and a pair of dark ciliary lines close together around apex. Hindwing pale ochreous-fuscous; cilia concolorous. Legs fuscous, banded white, hind tibia clothed in pale fuscous hair-scales. Abdomen fuscous, becoming whitish ventrally. Genitalia, see figures 8d,28b.

Life history

Ovum. Laid in July on apple (*Malus* spp.), pear (*Pyrus* spp.) or species of *Prunus*, especially blackthorn (*P. spinosa*). Continental authors also record it on birch (*Betula* spp.), hazel (*Corylus avellana*), hawthorn (*Crataegus* spp.) and *Sorbus* spp. (Huemer & Karsholt, 1999).

Larva. Head and divided prothoracic plate black. Body reddish brown (Meyrick, 1895). August–May.

When young, the larva mines a leaf of the foodplant. After hibernating from November to March, it feeds in a leaf- or flower-bud and, later, on the early blossom

and leaves until late May. It may sometimes seriously reduce the fruit yield in orchards (Bradford, [1979]).

Pupa. Undescribed. In the bark (Douglas, 1879). June.

Imago. Univoltine; July–August. Flies at night, sometimes coming to light. Rests on a tree-trunk in the daytime.

Distribution (Map 51)

Widespread in England, recorded as far north as north-east Yorkshire; North Wales; not recorded from Scotland or Ireland. There is concern that its range is currently contracting and it has been ascribed Notable B conservation status by Parsons (1995[1996]). Throughout most of Europe but as yet not recorded from Norway; Egypt; Asia Minor; accidentally introduced into North America

RECURVARIA LEUCATELLA (Clerck)

Phalaena (*Tinea*) *leucatella* Clerck, 1759, *Icones Ins.rar.*: pl.2, fig.3.
Erminea leucatea Haworth, 1828, *Lepid.Br.*: 514.
Type locality: [Sweden].

Description of imago (Pl.2, fig.30)

Wingspan 14–15mm. Head white; antenna fuscous, ringed paler; labial palpus white, becoming fuscous basally and usually finely ringed fuscous at extreme apex of segment 3. Thorax and tegulae dark fuscous. Forewing dark fuscous; small scale-tufts just above fold between one-third and one-half and another beyond at two-thirds; white markings in the form of transverse straight-edged fascia at one-quarter from base which is twice as broad on dorsum as on costa, tornal spot, similar spot opposite on costa and several small spots near base of cilia around apex and termen; cilia fuscous, tipped white around apex. Hindwing fuscous; cilia concolorous. Legs dark fuscous, tarsi thinly edged whitish distally. Abdomen dark fuscous becoming paler caudally. Genitalia, see figures 8e,28c.

Life history

Ovum. Laid on hawthorn (*Crataegus* spp.), apple (*Malus* spp.) or, less frequently, on rowan (*Sorbus aucuparia*). Continental authors also record wild plum (*Prunus domestica*), blackthorn (*P. spinosa*) and pear (*Pyrus* spp.) (Huemer & Karsholt, 1999).

Larva. Final instar length 10mm. Head and prothoracic plate black, the latter divided along midline. Body pale brown with a faint rosy tinge; colour somewhat variable ranging from very rosy to whitish-green with only the anterior end tinged rosy; anal segment pale

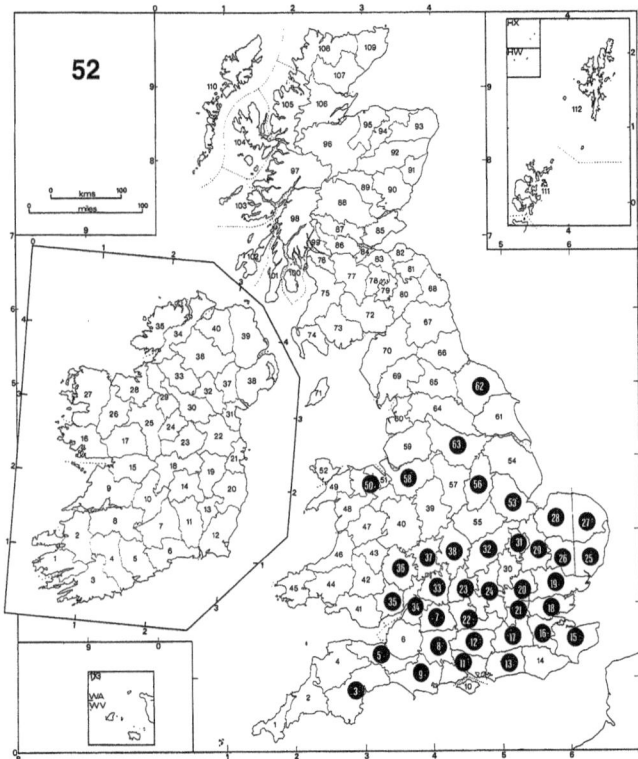

Recurvaria leucatella

COLEOTECHNITES Chambers

Evagora Clemens, 1860, *Proc.Acad.nat.Sci.Philad.* **1860**: 165, *nec* Péron & Lesueur, 1810.

Eidothea Chambers, 1873, *Can.Ent.* **5**: 186,229, *nec* Risso, 1826.

Coleotechnites Chambers, 1880, *in* Comstock, *Rep. Commnr Agric.,Washington (Rep.U.S.Dep.Agric.)* **1879**: 206.

Eucordylea Dietz, 1900, *Ent.News* **11**: 349.

Pulicalvaria Freeman, 1963, *Can.Ent.* **95**: 727.

A small Nearctic genus represented by about 50 species in North America. A single species has been introduced into Europe including Britain, where it is a potential pest species.

Imago. Rather small gelechiids with labial palpus slender; antenna without pecten. Forewing subtrapezoidal with termen rather oblique and tufts of erect scales. Hindwing same breadth as forewing, trapezoidal with termen slightly emarginate before pointed apex; in male a group of ochreous scales at base. Abdomen with tergites 1–3 buff. Valva of male genitalia slender, tapering distally and with bulbous base; vinculum-valva complex very asymmetrical; signum of female genitalia conical and spiny.

Larva. Early stages on Pinaceae.

COLEOTECHNITES PICEAELLA (Kearfott)

Recurvaria piceaella Kearfott, 1903, *Jl N.Y.ent.Soc.* **11**: 155.

Recurvaria niger Kearfott, 1903, *Ibid.* **11**: 156.

Recurvaria obscurella Kearfott, 1907, *Can.Ent.* **39**: 4.

Type locality: U.S.A.; Montclair, New Jersey.

Description of imago (Pl.2, fig.31)

Wingspan 10–12mm. Head, collar and thorax white to creamy white, frons white; antenna dark fuscous, ringed pale brown edged whitish; labial palpus white, segment 2 suffused dark fuscous dorsolaterally and orange-brown ventrolaterally, and segment 3 with orange-fuscous ring basally and dark fuscous ring just before apex. Forewing white, strongly suffused pale brownish; markings dark fuscous consisting of three subtriangular strigulae from costa at base, one-third and two-thirds, each with a pair of spots at its apex, and six vague spots around apex at base of cilia; unsuffused ground colour partly outlining the pairs of spots and forming a transverse chevron-shaped fascia at three-quarters; cilia grey, dark fuscous basally, with a

brown without a black anal plate (Stainton, 1865b).

The larva may be found from late May to mid-June within a pair of spun leaves of its foodplant. Occasionally several leaves may be pulled together, sometimes with a piece of withered brown leaf intermixed.

Pupa. Undescribed. In a whitish cocoon within the larval spinning (Stainton, *loc.cit.*). June–July.

Imago. Univoltine; late June–end of July. In suitable localities can be obtained by beating hawthorn hedges. Occasionally comes to light.

Distribution (Map 52)

Throughout England, as far north as north-east Yorkshire; North Wales. Not recorded from Scotland or Ireland. Occurs throughout most of Europe, but apparently absent from Portugal.

Coleotechnites piceaella

pale silvery fuscous ciliary line and tips whitish. Hind-wing ochreous-grey; cilia concolorous. Legs dark fuscous, ringed white; hind tibial tuft creamy white. Abdomen buff. Genitalia, see figures 8c,28d.

Life history

Ovum. Laid singly or, rarely, in groups of two or three on Norway spruce (*Picea abies*) or other species of spruce; June–July. Eggs are usually located either between the scales in the axils of the current year's foliage, or within damaged foliage and cones, or in spent staminate flowers. Eggs hatch in about ten days.

Larva. Final instar 9mm long. Young larva with head light brown and body pale orange. The coloration changes at onset of dormancy, the head becoming dark brown to black and the body deep rose to brick-red.

The young larva feeds by mining the needles until late September, by which time it has consumed four or five needles. The frass is ejected from the mined needle which has a circular basal entrance hole and is attached to an adjacent leaf or stem by silken threads. Occasionally a silken tube is constructed from one mine-entrance to another. Other feeding modes have also been recorded, especially tunnelling into young

shoots or cones. There is a tendency for larval workings to be associated with foliage damaged by weather or other spruce-feeders. In September the larva becomes dormant inside a silken spinning within the working. Feeding resumes in the same working around late May of the following year but the larva soon transfers to another site lower down the shoot. Occasionally the larva leaves the tree to pupate (McLeod, 1966; Freeman, 1967).

Pupa. Undescribed. Within a silken cell at the base of the mined needles or within a bud. The pupal period lasts about 12 days. June–early July.

Imago. Univoltine; July.

Distribution (Map 53)

First taken in Britain in 1952 (Ellerton, 1970), this native of North America has become established very locally in south-eastern England. Elsewhere in Europe restricted to the central part.

EXOTELEIA Wallengren

Exoteleia Wallengren, 1881, *Ent. Tidskr.* **2**: 94.

Paralechia Busck, [1903] *in* Dyar, *Bull. U.S.natn.Mus.* **52**(1902): 502.

Heringia Spuler, 1910, *Schmett.Eur.* **2**: 357, *nec* Rondani, 1856.

Heringiola Strand,1917, *Int.ent.Z.* **10**: 137.

A small genus with a worldwide distribution. Only two species are represented in Europe; one of these occurs in Britain. It sometimes causes severe damage to pine plantations.

Imago. Small to medium-sized gelechiids with labial palpus slender and rather short; antenna without pecten. Forewing lanceolate, narrow and with tufts of erect scales. Hindwing broader than forewing, trapezoidal without terminal emargination. Abdomen with tergites 1–3 ochreous-grey. Valva of male genitalia filiform with bulbous base; aedeagus curved basally; no signum in female genitalia.

Larva. Mines needles and later shoots of Pinaceae.

EXOTELEIA DODECELLA (Linnaeus)

Phalaena (Tinea) dodecella Linnaeus, 1758, *Syst.Nat.* (Edn 10) **1**: 539.

Recurvaria dodecea Haworth, 1828, *Lepid.Br.*: 549.

Anacampsis annulicornis Stephens, 1834, *Ill.Br.Ent.* (Haust.) **4**: 208.

Type locality: not stated.

Description of imago (Pl.2, fig.32)

Wingspan 12–15mm. Head covered with ochreous-white scales variably tipped pale fuscous; antenna ochreous-white, banded dark fuscous; labial palpus with segment 2 whitish above and fuscous below, segment 3 white with two fuscous bands, which sometimes conjoin into a single broad fuscous band. Thorax and tegulae pale fuscous, sometimes whitish posteriorly. Forewing whitish, variably irrorate pale fuscous and pale brown, except for narrow, angled transverse white fascia just before three-quarters; variable-sized patches of dense fuscous suffused across wing at base of costa, at one-third along costa and on either side of the transverse white fascia; three black spots, sometimes edged whitish, one on fold at one-quarter, one below fold just beyond one-third, and one at one-third across wing from tornus at two-thirds; other spots and dense patches of fuscous suffusion variable; cilia pale fuscous, more ochreous basally and towards tornus, ciliary line of fuscous and whitish-tipped scales variably defined. Hindwing ochreous-fuscous; cilia concolorous. Legs fuscous, banded white. Abdomen fuscous, paler caudally. Genitalia, see figures 8f,28g.

Life history

Ovum. Laid on a shoot of Scots pine (*Pinus sylvestris*), less commonly on larch (*Larix europaeus*) (Styles, 1959); June–July.

Larva. Head black-brown. Prothoracic plate concolorous with body, which is pinkish brown with small black pinacula (Meyrick, 1895).

The larva commences feeding in September by mining the apical half of a pine needle; the mine contains some spinning, has a hole at each end and most of the frass is ejected. After hibernation in the needle, it feeds in shoots and spun needles until mid-May (Bradford & Sokoloff, 1988). The larva of this species was known to Linnaeus (1758). It sometimes causes severe damage to pine plantations.

Pupa. Undescribed. May–June.

Imago. Univoltine; June–July. Occurs in most localities where pine grows; best obtained by beating the lower branches of its foodplant. Occasionally comes to light.

Distribution (Map 54)

Widespread in Britain and Ireland but apparently absent from the Outer Hebrides, Orkney and Shetland. Throughout Europe; accidentally introduced into North America.

ATHRIPS Billberg

Athrips Billberg, 1820, *Enumeratio Insect.Mus.G.J.Billberg*: 93.

Rhynchopacha Staudinger, 1871, *Berl.ent.Z.* **14**: 303.

Epithectis Meyrick, 1895, *Handbk Br.Lepid.*: 580.

Leobatus Walsingham, 1904, *Entomologist's mon.Mag.* **40**: 220.

Ziminiola Gerasimov, 1930, *Dt.ent.Z.Iris* **44**: 72.

Cremona Busck, 1934, *Proc.ent.Soc.Wash.* **36**: 82.

A Holarctic genus with some 10 European species of which three occur in Britain.

Imago. Variably sized gelechiids, ranging from small to large, with labial palpus slender but of variable length; antenna without pecten. Forewing lanceolate and often elongate, sometimes with tufts of erect scales. Hindwing same breadth as forewing, subtrapezoidal, tornus gently rounded, apex acute, degree of emargination of termen variable; abdomen variable, sometimes with tergites 1–3 buff-coloured, sometimes concolorous with the other tergites. Valva of male genitalia with long slender digitate costa and short inwardly curved sacculus; signum of female genitalia a small plate with a transverse crease situated near entrance to corpus bursae.

Larva. Larval stages have been recorded from a wide range of plant families.

Key to species (imagines) of the genus *Athrips*

1 Forewing blackish with no conspicuous markings .... *rancidella* (p. 126)
– Forewing with two or more conspicuous black spots 2

2(1) Large species, wingspan >14mm, forewing pale brownish with four or more conspicuous black spots .... *mouffetella* (p. 127)
– Small species, wingspan <12mm, forewing creamy orange broadly bordered with fuscous and with three conspicuous fine black spots .... *tetrapunctella* (p. 125)

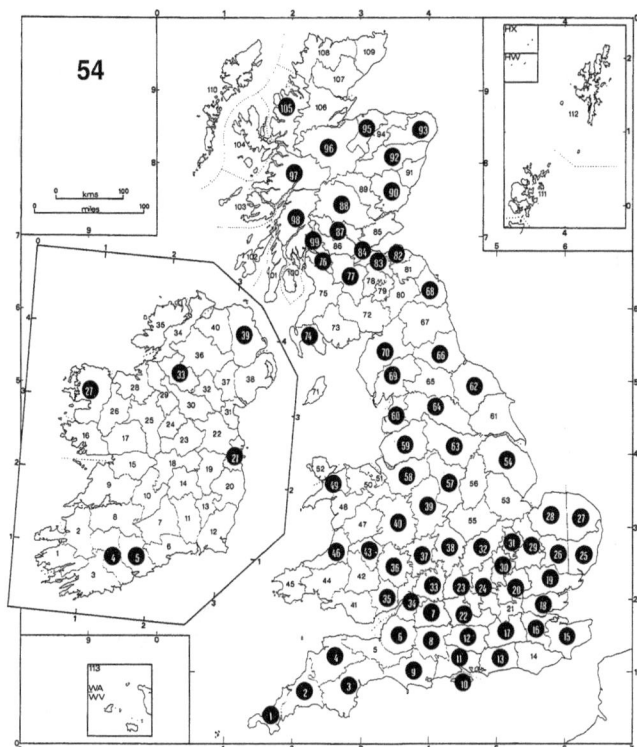

Exoteleia dodecella

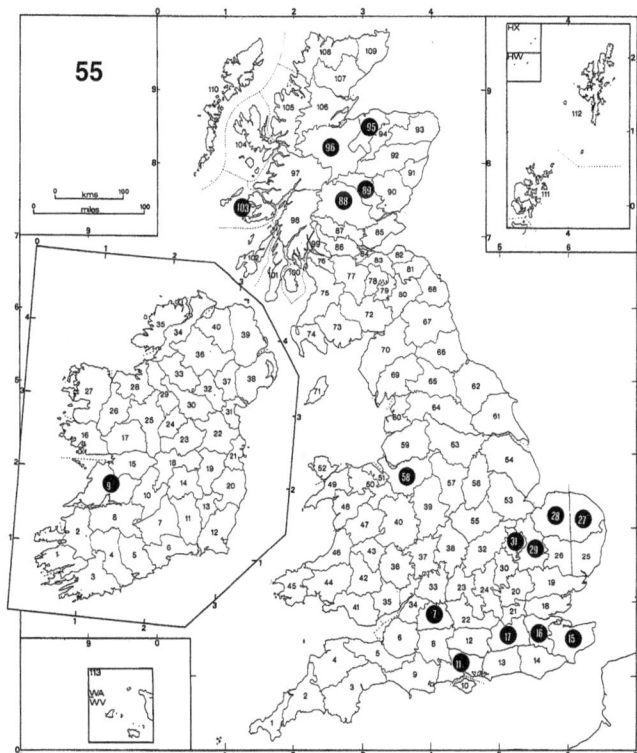

Athrips tetrapunctella

ATHRIPS TETRAPUNCTELLA Thunberg

Tinea tetrapunctella Thunberg, 1794, *D.D.Diss.ent.sistens Insecta suecica* (7): 96.

Gelechia nigricostella sensu Douglas, 1852, *Trans.ent. Soc.Lond.* (N.S.) **1**: 244, *nec* Duponchel, 1842.

Gelechia lathyri Stainton, 1865, *Entomologist's Annu.* **1865**: 130.

Gelechia lathyrella Doubleday, 1866, *Synonymic list Br.Lepid.* (Suppl.2): [2].

Thiotricha subocellea sensu Pierce & Metcalfe, 1935, *Gen.Tineid Fam.Lepid.Br.Is*: 18, pl.10, *nec* Stephens, 1834.

Type locality: Sweden.

Description of imago (Pl.2, fig.33)

Wingspan 9–10mm. Head cream, some scales tipped pale fuscous; antenna dark fuscous, narrowly banded paler; labial palpus white, banded cream. Thorax and tegulae cream to orange-cream with some scales tipped fuscous, especially on tegulae. Forewing cream becoming orange-cream along fold and in discal area, heavily suffused with fuscous-tipped scales broadly along basal half of dorsum excluding base, broadly along costa to two-thirds, a blotch at tornus and whole apical quarter; three black spots, one above fold just before one-quarter, one in discal area at one-third and one halfway across wing above tornus; other spots may variably be present; underside ochreous-fuscous; cilia pale fuscous, yellow-ochreous on costa at three-quarters, irregular ciliary line of fuscous-tipped scales. Hindwing greyish fuscous; cilia concolorous. Legs dark fuscous, banded ochreous. Abdomen dark fuscous, caudal tip ochreous. Genitalia, see figures 8h,28i.

Similar species. In Britain, this species was initially mistaken for the mainland European species *A. nigricostella* (Duponchel) from which it can be separated by its more pointed forewing, the absence of a black spot on the fold before the middle (present in *nigricostella*) and its larval foodplant (in *nigricostella* lucerne (*Medicago sativa*)).

Life history

Ovum. Laid on a shoot of marsh pea (*Lathyrus palustris*) or bitter-vetch (*L. linifolius*); June–early July.

Larva. Final instar length 8mm. Head and undivided prothoracic plate black, the latter edged whitish anteriorly. Body dark dull green with anal segment and membrane between prothorax and mesothorax paler green; pinacula small, dark grey in pale blotches (Stainton, 1867a). The larva feeds from a silken tunnel within a rough untidy spinning of the terminal leaves of the foodplant, June–August. The silk used to hold the leaves together is rather white and conspicuous. The larva is very wriggly when disturbed.

Pupa. Length 4mm. Yellowish brown, evenly tapering caudally; cremaster of about 15 evenly distributed hooked bristles. In a white silken cocoon amongst dead leaves, August–May.

Imago. Univoltine; late May–early July. Best obtained by sweeping the foodplant, although it has been taken on the wing at both dawn and dusk.

Distribution (Map 55)

The two main strongholds of this species in Britain are south-east of a line from Hampshire to The Wash, where the foodplant is marsh pea, and the Highlands of Scotland, where the foodplant is bitter-vetch. There are old records from Cheshire (Meyrick, 1895) but there appear to have been no post-1970 records of this species from any of its English sites. Parsons (1995 [1996]) ascribed the species a RDB I (Indeterminate) conservation status. Also recorded from the Burren, Co. Clare, Ireland. Abroad known from Scandinavia, the Baltic countries of Estonia and Latvia, and France.

ATHRIPS RANCIDELLA (Herrich-Schäffer)
Cotoneaster Webworm

Gelechia rancidella Herrich-Schäffer, 1854, *Syst.Bearb. Schmett.Eur.* **5**: 176.

Type localities: Austria; Vienna, and Germany; Regensburg.

Description of imago (Pl.2, fig.34)

Wingspan 11–12mm. Head dark fuscous; antenna dark fuscous, banded ochreous-fuscous; labial palpus ochreous-fuscous, flecked dark fuscous, especially at apical end of segments 2 and 3. Thorax and tegulae dark fuscous. Forewing dark fuscous, slightly mottled due to pale fuscous scale-bases showing through basally and scales being tipped pale fuscous in apical third; markings pale fuscous consisting of thin transversely elongate spot on costa at two-thirds and another less regular one opposite at tornus, which may meet to form a narrow chevron-shaped fascia; cilia pale fuscous with scattered dark-tipped scales not forming a ciliary line. Hindwing fuscous; cilia ochreous-fuscous. Legs dark fuscous, tarsi edged ochreous-white distally especially above; inner aspect of hindleg ochreous-white. Abdomen dark fuscous. Genitalia, see figures 8j,28h.

Life history

Ovum. Laid on wall cotoneaster (*Cotoneaster horizontalis*) in July or August. In mainland Europe also on

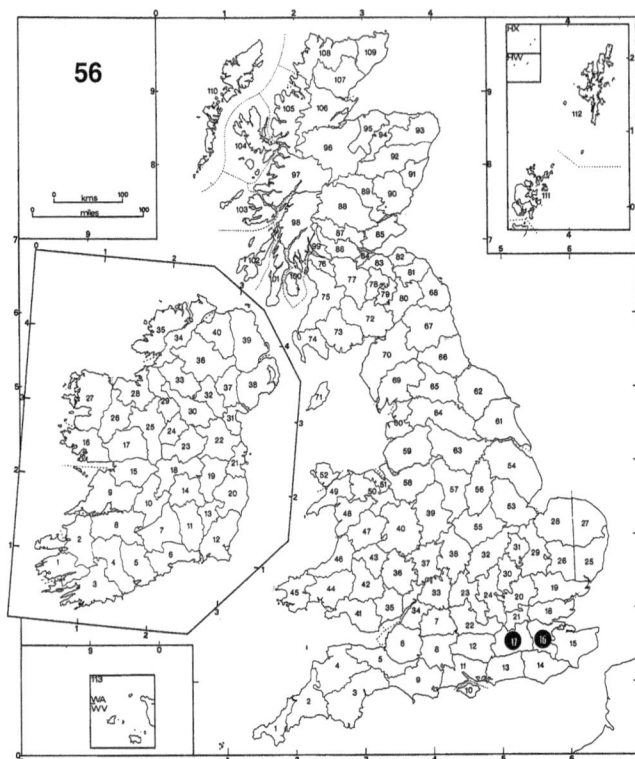

Athrips rancidella

blackthorn (*Prunus spinosa*) and hawthorn (*Crataegus monogyna*) (Huemer & Karsholt, 1999).

Larva. Head black. Prothoracic plate black, brown at front; body varying from dark reddish brown to sooty black; pinacula paler, each with dark central spot.

Whether winter is passed as an egg or larva is unknown, but during May and June the larva lives in an untidy tube of off-white silk, spun along the side of a twig; silk may also extend over leaves and join adjacent twigs. The larva eats the underside of a leaf, leaving only the transparent upper epidermis, which turns slightly brown and withers, disfiguring the appearance of the bush.

Pupa. Undescribed. In a dense white cocoon in the larval habitation, usually along a twig (Sokoloff & Chalmers-Hunt, 1987). June.

Imago. June–August. Comes to light.

Distribution (Map 56)

First recorded in Britain on 7 July 1971 at West Wickham, Kent (Chalmers-Hunt, 1985) and subsequently found to be breeding. Still known only from suburban gardens in west Kent and Surrey. Absent from Scandinavia but widespread throughout the rest of Europe; accidentally introduced into North America.

ATHRIPS MOUFFETELLA (Linnaeus)

Phalaena (Tinea) mouffetella Linnaeus, 1758, *Syst.Nat.* (Edn 10) **1**: 540.

Recurvaria punctifera Haworth, 1828, *Lepid.Br.*: 551.

Type locality: Europe.

Description of imago (Pl.2, fig.35)

Wingspan 15–16mm. Head ochreous, mixed ochreous-fuscous; antenna ochreous- fuscous, ringed dark fuscous; labial palpus ochreous to ochreous-fuscous with dark fuscous band just before apex of segment 3. Thorax and tegulae ochreous, mixed ochreous-fuscous. Forewing ochreous-fuscous with scales narrowly tipped ochreous, slightly more fuscous along costa; markings of fuscous-black scales with partially upturned tips consisting of two spots at one-quarter, one on fold and another larger just beyond it in discal area, two spots in discal area just beyond halfway, the more costal one displaced apically, a row of spots along base of terminal cilia and other variable and smaller spots; underside plain ochreous-fuscous; cilia ochreous-fuscous. Hindwing pale fuscous; cilia concolorous. Legs ochreous, mixed ochreous-fuscous. Abdomen pale fuscous. Genitalia, see figures 8i,28j.

Life history

Ovum. Laid on a twig of honeysuckle (*Lonicera periclymenum*), fly honeysuckle (*L.xylosteum*) or snowberry (*Symphoricarpos rivularis*); August–September.

Larva. Final instar length 11–12 mm. Head and divided prothoracic plate black. Body purplish black or dark grey with white mediodorsal spots between pro- and mesothorax and between meso- and metathorax, these sometimes extending caudally as a white stripe (Michaelis, 1977); spiracular band broad, faint whitish except for thorax where brilliant white but interrupted anteriorly on meso- and metathorax (Stainton, 1865b).

The larva feeds from mid-May to early June. It draws together several terminal leaves of a shoot and spins a dense white silken retreat between them. It quickly withdraws into this spinning if disturbed (Stainton, *loc.cit.*). The larvae of *Ypsolopha dentella* (Fabricius) and *Y. nemorella* (Linnaeus) (Yponomeutidae) feed in a similar way but the former is green with a reddish brown dorsal stripe and the latter is pale reddish brown without a marked midline stripe (see *MBGBI* **3**: 90–91).

Pupa. Length 6.5 mm. Yellow-brown to dark brown, rather broad and parallel-sided anteriorly but tapering strongly from two-thirds to a rounded caudal tip; cremaster comprising six to eight stout hooked hairs. Within a flimsy white silken cocoon spun either in the

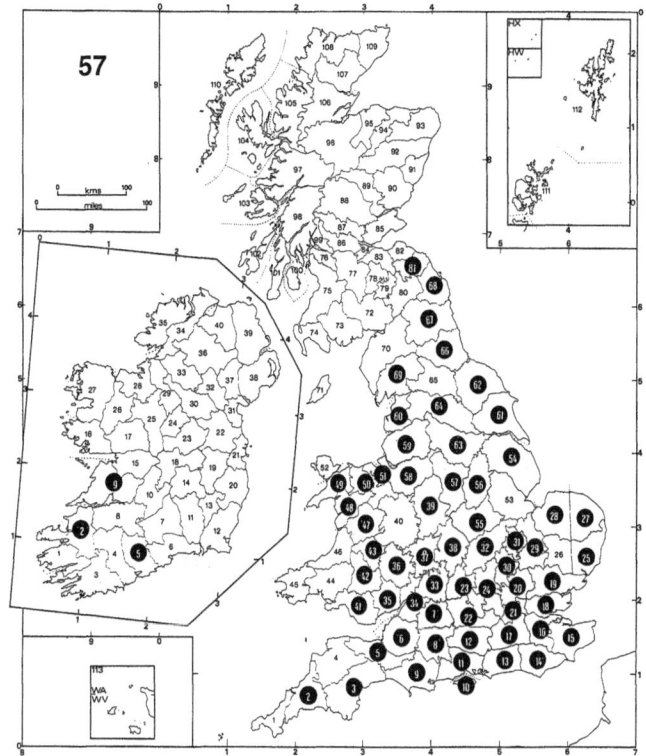

Athrips mouffetella

larval habitation or elsewhere. June–July.

Imago. Univoltine; late July–early September. Owing to its very retiring habits it is rarely encountered in the field but has been found resting on palings next to its foodplants. It comes readily to light.

Distribution. (Map 57)

Throughout England and Wales; in Scotland reported only from Berwickshire. Very local in south-west Ireland. Abroad, widespread throughout Europe; introduced accidentally into North America.

XENOLECHIA Meyrick

Xenolechia Meyrick, 1895, *Handbk Br.Lepid.*: 583.

A small Holarctic genus. Three European species are known but only one of these occurs in Britain.

Imago. Small to quite large gelechiids with labial palpus long and slender, segment 2 prominently rough-scaled ventrally; antenna without pecten. Forewing lanceolate, and often elongate, with tufts of erect scales. Hindwing usually much broader than forewing, trapezoidal, apex pointed, termen slightly emarginate just before apex, tornus gently rounded. Abdomen with tergites concolorous. Valva of male genitalia absent, aedeagus short and broad, uncus bifid with short broad lobes; signum of female genitalia serrate and subrhomboidal with deep transverse equatorial groove, ostium protected by sclerotized tube.

Larva. Known larvae feed on Ericaceae.

XENOLECHIA AETHIOPS (Humphreys & Westwood)

Anacampsis aterrima Edleston, 1844, *Zoologist* **2**: 734 [unused senior synonym].

Anacampsis aethiops Humphreys & Westwood, 1845, *Br. Moths Transform.* **2**: 192.

Gelechia aethiopella Doubleday, 1859, *Zoologist synonymic. List Br.Lepid.*: 30.

Type locality: England; near Manchester.

NOMENCLATURE. The nominal taxon *aterrima* Edleston, 1844, is an unused senior synonym of *A. aethiops* and under Article 23.9.1 of ICZN should be suppressed in the interests of stability.

Description of imago (Pl.2, fig.36)

Wingspan 16–20mm. Head dark fuscous, scales finely tipped whitish especially above eyes; antenna dark fuscous with short white cilia in male; labial palpus dark fuscous with many scales finely tipped whitish. Thorax and tegulae dark fuscous. Forewing dark fuscous, scales finely tipped whitish in apical quarter; markings velvet-black and variable, ranging from scattered spots and streaks especially along veins (most frequent in males) to pairs of transversely placed tufts of raised scales at one-fifth, two-fifths and three-fifths (most frequent in female); cilia fuscous with basal row of white-tipped dark fuscous scales and three faint white ciliary lines. Hindwing pale ochreous-fuscous; cilia concolorous. Legs dark fuscous, inner aspect of hindleg and its tuft pale ochreous-fuscous. Abdomen dark fuscous. Genitalia, see figures 8g,28e.

Similar species. Superficially similar to dark specimens of *Teleiopsis diffinis* (Haworth) and *Carpatolechia decorella* (Haworth). However, the former invariably possesses scattered buff scales on the forewing and the latter retains its yellowish subcostal streak even in almost black specimens. Neither of these features is seen in *X. aethiops*.

Life history

Ovum. Laid on heath (*Erica cinerea*). May–June.

Larva. Final instar length 11mm. Head brown. Prothoracic plate blackish, crescentic, with midline fissure; body dull reddish with intersegmental regions green-ish; pinacula small and black; anal plate blackish (Stainton, 1862).

The larva feeds from mid-June to mid-July, at first mining the leaves and then externally on the leaves from a gallery of silk and frass intermixed with fragments of *Erica* leaves (Stainton, 1865b).

Pupa. Undescribed. July–May.

Imago. Univoltine; mid-May–early June. Often found resting on areas of burnt heather but best obtained by sweeping the foodplant.

Distribution (Map 58)

Widespread on heaths and moors throughout Britain. Not known from the Inner and Outer Hebrides or Shetland. Co. Offaly, Ireland (Bond, 1996). Europe as far north as Denmark; accidentally introduced into North America.

PSEUDOTELPHUSA Janse

Pseudotelphusa Janse, 1958, *Moths S.Africa*, Gelechiidae **6**: 68.

Sattleria Căpuşe, 1968, *nec* Povolný, 1965.

Klaussattleria Căpuşe, 1968, *Ent.Ber.,Amst.* **28**: 80 [replacement name for *Sattleria* Căpuşe, 1968].

A small but worldwide genus with six species in Europe, two of which occur in Britain.

Imago. Small to quite large gelechiids with labial palpus long and slender; antenna without pecten. Forewing lanceolate and with tufts of erect scales. Hindwing as broad as forewing, subtrapezoidal with termen emarginate just before the acute and produced apex, tornus rounded. Valva of male genitalia elongate and spinose with bulbous base, anellus with two long digitate processes; signum of female genitalia rhomboidal with transverse rounded ridge, subgenital plate sometimes elaborately shaped.

Larva. Early stages usually occur in the spun leaves of various deciduous trees and shrubs.

Key to species (imagines) of the genus *Pseudotelphusa*

- Forewing white with black linear markings and spots .... *scalella* (p. 129)
- Forewing brown with six black spots .... *paripunctella* (p. 130)

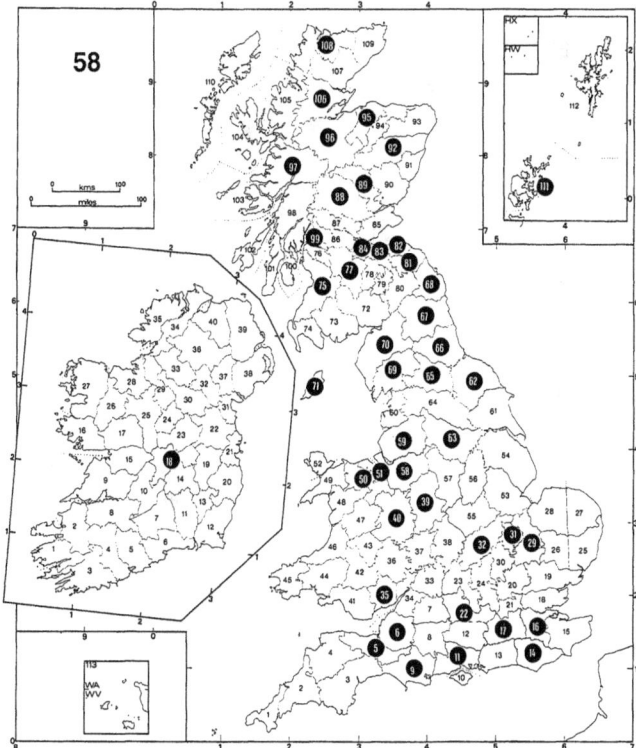

58

Xenolechia aethiops

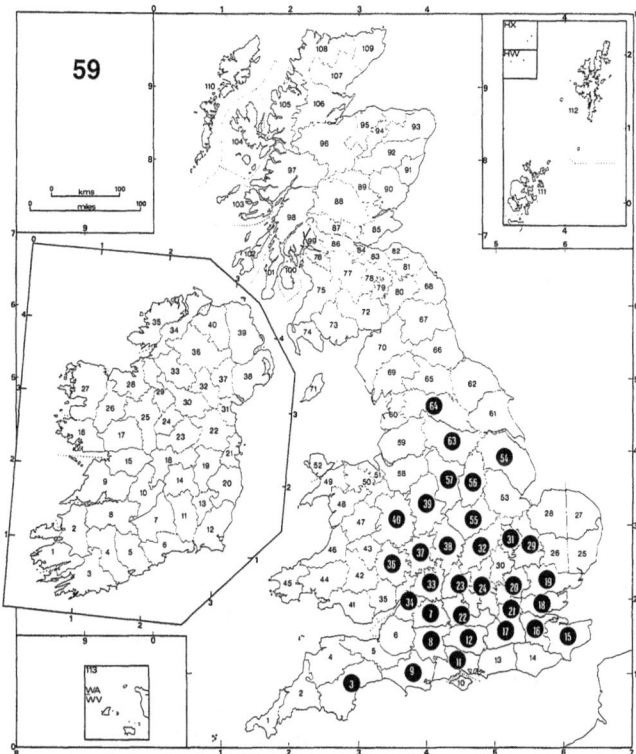

59

Pseudotelphusa scalella

PSEUDOTELPHUSA SCALELLA (Scopoli)

Phalaena scalella Scopoli, 1763, *Ent.Carn.*: 253, fig.654.
Tinea aleella Fabricius, 1794, *Ent.Syst.* **3**(2): 317.
Tinea alternella Hübner, 1796, *Samml.eur.Schmett.* **8**: 22, fig.151, *nec* [Denis & Schiffermüller], 1775.
Type locality: Slovenia (formerly Yugoslavia); 'Carniola' (now area around Idrija and Ljubljana).

Description of imago (Pl.3, fig.1)

Wingspan 11–13mm. Head white; antenna dark fuscous, ringed paler or whitish and finely ciliate below in male; labial palpus white with apical quarter of segment 3 dark fuscous except for extreme apex, a very narrow dark fuscous ring below this and base of palpus also dark fuscous. Thorax white with dark fuscous spot on each side just anterior to scutellum; tegulae white, edged dark fuscous anteriorly. Forewing white with dark fuscous markings consisting of basal spot on costa, outwardly sloping, parallel-sided transverse fascia at one-fifth, triangular costal spot at two-fifths and subquadrate spot at three-fifths; on dorsum, small spot just after base and triangular spot just before tornus, small spot on fold at two-fifths and large spot in apical area with three to five small spots along termen at base of cilia; cilia ochreous-fuscous with scattered dark fuscous and white scales; underside of forewing fuscous. Hindwing fuscous to pale fuscous, often paler between veins; cilia concolorous. Legs banded white and dark fuscous, hindleg tuft pale ochreous-fuscous. Abdomen pale fuscous with tergites edged ochreous-fuscous. Genitalia, see figures 8k,28f.

Life history

Ovum. Site of oviposition not known.

Larva. Undescribed. According to Bradford [1979] the larva feeds from August to April in moss growing on tree-trunks.

Pupa. Undescribed. Within the larval tunnel; April–May. Recent Continental reports record the larva feeding on the leaves of oak (*Quercus* spp.) and probably going into moss only to hibernate (Huemer & Karsholt, 1999).

Imago. Univoltine; May–June. Rests on tree-trunks during the day, and at night comes to light.

Distribution (Map 59)

Widespread in southern England as far north as Yorkshire, but apparently absent from Cornwall and Somerset. Not recorded from Wales, Scotland or Ireland. Abroad, throughout Europe, excepting Norway.

PSEUDOTELPHUSA PARIPUNCTELLA
(Thunberg)

Tinea paripunctella Thunberg, 1794, *D.D.Diss.ent.sistens Insecta suecica* (7): 96.

Gelechia triparella Zeller, 1839, *Isis, Leipzig* **1839**: 201.

Recurvaria dodecea sensu Haworth, 1828, *Lepid.Br.*: 549, nec *dodecella* Linnaeus, 1758.

Type locality: Sweden; Ostergötland.

Description of imago (Pl.3, figs 2,3)

Wingspan 11–12mm. Head concolorous with forewing; antenna fuscous, ringed pale yellowish brown; labial palpus with segments 2 and 3 of equal length, segment 2 thickened with appressed scales, outwardly fuscous, inwardly pale ochreous, segment 3 slender, pale ochreous with two fuscous bands. Thorax concolorous with forewing. Forewing grey, brown or pale yellowish brown; sometimes blackish spot or cloud present at base of costa and two elongate similar costal spots regularly present at two-fifths and three-fifths; three pairs of slightly raised black spots at one-fifth, two-fifths and three-fifths respectively, in first two pairs the more dorsal spot placed more distally, the more costal of first pair sometimes obsolescent; pale patch on costa at three-quarters, sometimes extending as broad, irregular, curved fascia across wing; series of black dots round apex and along termen often present; cilia grey or ochreous-grey, ciliary line darker, becoming obsolescent towards tornus. Hindwing as broad as forewing, light grey; cilia paler. Foreleg fuscous, ringed pale ochreous; mid- and hindlegs with outer sides brown, inner sides ochreous, banded brown. Abdomen fuscous. Genitalia, see figures 81,29a.

The ground colour largely depends on the foodplant. Specimens from larvae that have fed on oak are usually yellowish brown, whereas those on bog-myrtle are darker grey-brown.

Similar species. Teleiodes wagae (Nowicki), *q.v.*

Life history

Ovum. Laid on deciduous oak (*Quercus* spp.) or bog-myrtle (*Myrica gale*). In mainland Europe, beech (*Fagus sylvatica*), birch (*Betula* spp.) and sea-buckthorn (*Hippophae rhamnoides*) are also recorded as foodplants (Sattler, 1980).

Larva. Head, and prothoracic plate with median sulcus, pale brown. Body whitish green in early instars, later yellowish green, frequently developing a pinkish tinge from thoracic segment 2 to abdominal segment 8; pinacula black.

The larva lives between leaves spun flatly together by means of short thongs composed of silk spun upon silk. It feeds by forming large pale green blotches, leaving the upper epidermis of the upper leaf and the lower epidermis of the lower leaf intact. When not feeding, it rests in a chamber constructed of silk and frass. When feeding on oak it is partial to saplings. Late July–September, developing slowly.

The larva of *Strophedra nitidana* (Fabricius) (Tortricidae) feeds on oak in a very similar manner, but its body is more yellow and the pinacula are concolorous.

Pupa. In a flimsy cocoon spun amongst detritus on the ground. October–May.

Imago. Univoltine; May–June. Flies at night and comes to light.

Distribution (Map 60)

Occurs as two ecologically distinct races. That which has oak as its foodplant inhabits young trees growing at the margin of woods, on heaths or in hedgerows; that with larvae on bog-myrtle is found on moorland and in fens, bogs and damp patches of heathland. Throughout the British Isles, but not recorded from the Outer Hebrides, Orkney or Shetland. The race feeding on oak predominates in southern England, that on bog-myrtle in northern England and Scotland. Throughout Europe, extending eastwards to China and Japan. The race feeding on bog-myrtle is known only from western and northern Europe (Sattler, *loc.cit.*).

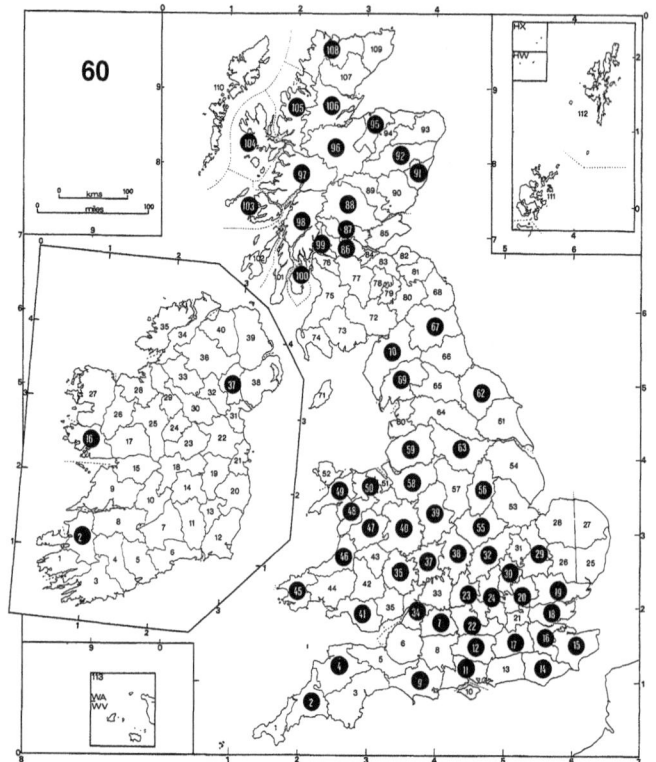

Pseudotelphusa paripunctella

ALTENIA Sattler

Altenia Sattler, 1960, *Dt.ent.Z.* (N.F.) **7**: 58.

A small genus represented by two species in Europe, one of which also occurs in England. This species, *A. scriptella* (Hübner), was until recently placed in *Teleiodes* Sattler, but has been reallocated by Karsholt & Razowski (1996).

ALTENIA SCRIPTELLA (Hübner)

Tinea scriptella Hübner, 1786, *Samml.eur.Schmett.* **8**: 61, pl.22, fig.152.

Type locality: [Europe].

Description of imago (Pl.3, fig.4)

Wingspan 12–13mm. Head white; antenna whitish, ringed fuscous; labial palpus with segment 3 shorter than segment 2, segment 2 thickened with erect scales scarcely enclosing a furrow, fuscous, mixed white, segment 3 rather thick, white with two fuscous bands. Thorax white, mixed ochreous. Forewing greyish white, irrorate darker; ochreous-brown spot on base of costa; large mixed ochreous-brown and blackish blotch from one-fifth to three-fifths extending from dorsum over most of wing but not reaching costa; costa with blackish spot at one-half and larger spot at three-quarters; first discal spot immediately above plical, both composed of raised scales; white-edged black streak in fold distal to plical stigma and similar oblique bar in disc above; second discal stigma two adjacent black spots sometimes joined as black bar and often merging with more distal costal spot; row of indistinct terminal dots formed by scattered dark scales; cilia ochreous-grey. Hindwing as broad as forewing, anal angle pronounced, tornus gently curved, termen straight, then slightly angled before moderately produced apex, grey; cilia grey. Foreleg fuscous, ringed grey, mid- and hindlegs with outer surfaces fuscous, inner surfaces grey, tarsi banded fuscous. Abdomen pale greyish fuscous, ventral surface with whitish stripe. Genitalia, see figures 9a,29b.

Life history

Ovum. Laid on field maple (*Acer campestre*).

Larva. Head yellowish, mouth-parts brown. Body whitish green, paler anteriorly and ventrally; prothoracic plate yellowish, edged black laterally and posteriorly and with two black spots; pinacula black.

The lower branches of saplings growing in hedgerows or at the margin of woods are preferred. The larva folds a leaf flatly, upper side inmost, by means of thongs of silk spun upon silk. It feeds on the

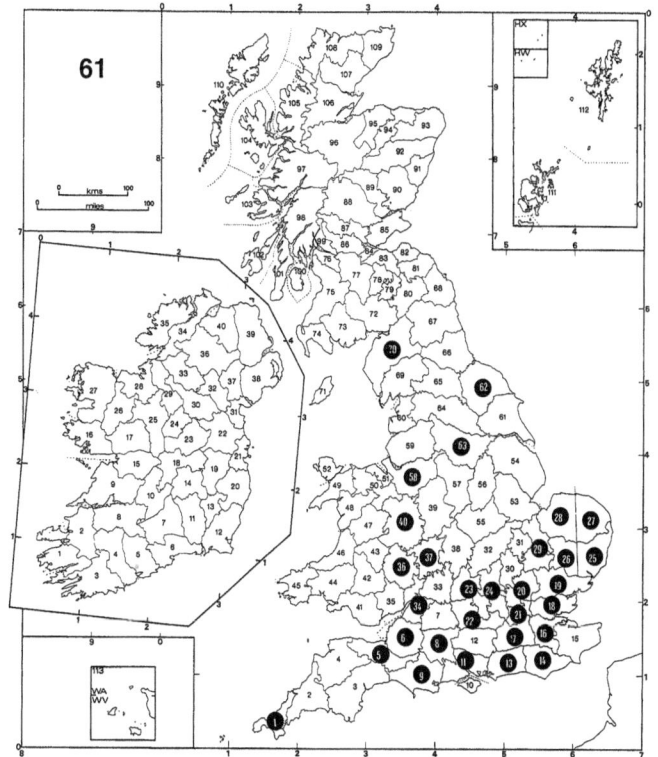

Altenia scriptella

upper epidermis and upper part of the parenchyma without making conspicuous blotches. It readily changes to a fresh leaf. Mid-August–September.

Pupa. In a flimsy cocoon amongst detritus on the ground. October–May.

Imago. Univoltine; June–July. Often rests on a trunk or paling by day; at night has been recorded at light.

Distribution (Map 61)

Occurs mainly along hedgerows in open country. Local in England as far north as Cumberland (Cumbria) but scarce in the north. Possibly declining; Stainton (1865b) described it as common all round London, but that is no longer the case. An old record from Rannoch, Perthshire (Blackburn & Blackburn, 1867; Meyrick [1928]) is almost certainly based on misidentification, since maple does not occur in the area. The moth in question was beaten from birch and was most probably *Carpatolechia alburnella* (Zeller), which had not yet been added to the British list. Not otherwise recorded from Scotland and not recorded from Wales or Ireland. Throughout most of Europe; Asia Minor.

TELEIODES Sattler

Teleiodes Sattler, 1960, *Dt.ent.Z.* (N.F.) **7**: 63.

Teleia Heinemann, 1870, *Schmett.Dtl.Schweiz* (2)**2**(1).: 272, *nec* Hübner, [1825], *Verz.bekannt.Schmett.*: 385.

Telphusa Chambers, 1872, *Can.Ent.* **4**: 132, sensu auctt.

A genus with almost worldwide distribution, having 30 species represented in Europe, one of which is also found in North America; five species occur in Britain and two in Ireland. Karsholt & Razowski (1996) noted that the genus was in need of revision. Since then, Elsner, Huemer & Tokár (1999) and Huemer & Karsholt (1999), followed by Bradley (2000), have reinstated the genus *Carpatolechia* Căpuşe, 1964, and have placed in it species from *Teleiodes* which form 'a strongly monophyletic group', including five occurring in the British Isles. The same authors have transferred *paripunctella* Thunberg from *Teleiodes* to *Pseudotelphusa* Janse.

These changes have been adopted in the present volume, but were made too late for incorporation into the Key to genera in which *Teleiodes* therefore includes *Carpatolechia*.

Imago. Antenna weakly serrate in apical area; labial palpus with segment 3 as long as or slightly shorter than segment 2, segment 2 thickened with appressed or, in a few species, projecting scales; ocelli absent. Forewing generally with tufts of raised scales; veins R_5 (7) and R_4 (8) stalked, R_5 to costa. Hindwing generally as broad as forewing, but broader or narrower in some species; trapezoidal, tornus rounded; termen straight or weakly sinuate, angle with dorsum varying slightly between species; apex moderately or strongly produced; veins Cu_1 (3) to M_2 (5) separate but more or less approximate at base, M_1 (6) and Rs (7) connate or stalked; cilia about one and one-half times width of wing.

All British species are univoltine. Many rest by day on tree-trunks and most are freely attracted to light.

Ovum. Generally undescribed, the oviposition site being unknown. Two or three species probably overwinter in this stage.

Larva. Cylindrical, with the full number of functional prolegs. The larvae feed in spun leaves, all but one of the British species on the foliage of trees or shrubs.

Pupa. Generally in a flimsy cocoon spun amongst detritus on the ground. More than half the British species overwinter as pupae.

Key to species (imagines) of the genus *Teleiodes*

1	Forewing with whitish, yellow or orange costal or subcostal markings before one-half 2
–	Forewing without such costal or subcostal markings 3
2(1)	Forewing with marking whitish, tinged yellow, semicircular, reaching costa and enclosing dark patch .... *luculella* (p. 134)
–	Forewing with marking orange, not semicircular and not reaching costa .... *flavimaculella* (p. 135)
3(1)	Forewing brown 4
–	Forewing white, irrorate grey; an outwardly oblique, blackish fuscous fascia at one-fifth .... *sequax* (p. 136)
4(3)	Forewing with transverse bar of raised black scales at three-quarters .... *vulgella* (p. 132)
–	Forewing without such bar but with four black spots in outwardly oblique row at two-fifths .... *wagae* (p. 133)

TELEIODES VULGELLA ([Denis & Schiffermüller])

Tinea vulgella [Denis & Schiffermüller], 1775, *Schmett. Wien.*: 139.

Recurvaria aspera Haworth, 1828, *Lepid.Br.*: 550.

Type locality: [Austria]; Vienna district.

Description of imago (Pl.3, fig.5)

Wingspan 11–13mm. Head grey, frons paler; antenna fuscous, banded whitish beneath; labial palpus with segment 3 as long as segment 2, segment 2 thickened with appressed scales, outer side fuscous, inner side whitish grey, segment 3 nearly straight, ascending, pale grey with central band and apex fuscous. Thorax grey. Forewing grey, irrorate fuscous; costa with diffused fuscous patches at one-quarter, one-half and three-quarters; stigmata with raised scales, black, two on fold, two in disc, first discal beyond second plical; black bar dorsad of second discal extending to tornus, bar, second discal and costal patch forming interrupted black fascia of raised scales; wing distal of bar slightly paler grey, especially towards costa and dorsum, suggestive of obsolescent, angulate pale fascia; black spots at ends of veins round apex and on termen; cilia grey. Hindwing slightly narrower than forewing, trapezoidal, tornus rounded, termen oblique, slightly sinuate, apex moderately produced; dark grey; cilia equal to width of wing, grey. Foreleg fuscous, ringed

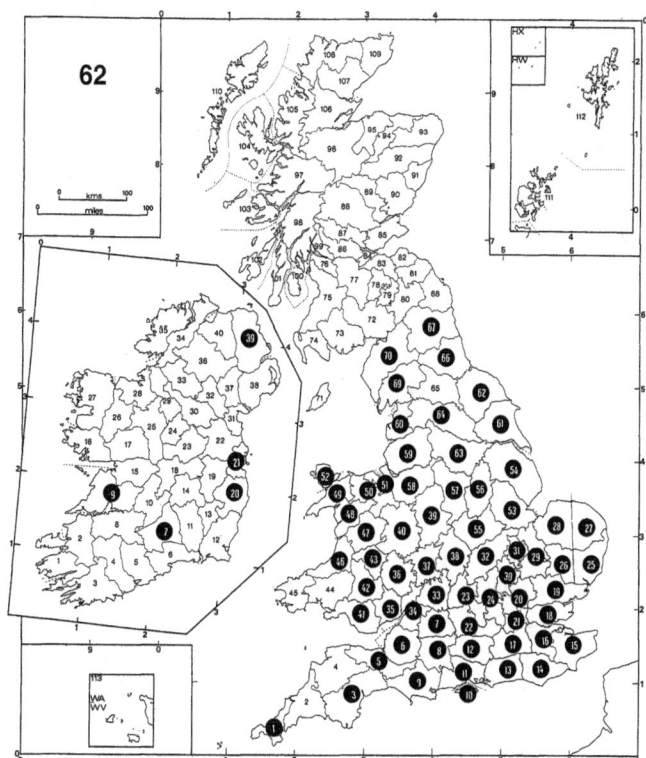

Teleiodes vulgella

whitish; mid- and hindlegs with outer sides fuscous, inner sides whitish grey, tibiae and tarsi ringed whitish. Abdomen fuscous, paler ventrally; anal tuft of male grey; female with ovipositor often protruding in set specimens, slender, pale yellowish. Genitalia, see figures 9d,29c.

Life history

Ovum. Laid on hawthorn (*Crataegus* spp.) or blackthorn (*Prunus spinosa*), occasionally on wall cotoneaster (*Cotoneaster horizontalis*). Has been reared from juniper (*Juniperus communis*) (Sokoloff, 1983). Probably overwinters in this stage.

Larva. Head ochreous-brown. Body greenish grey; prothoracic plate black with broad median sulcus; pinacula small, black; thoracic legs whitish, ringed black.

The larva spins two or more leaves together flatly and grazes through to the upper epidermis, causing conspicuous pale brownish blotches on the uppermost leaves. When not feeding, it rests in a silken tube within the spinning. Late April–May.

Pupa. In the larval habitation or amongst detritus on the ground. Late May–late June.

Imago. Univoltine; late June–August. Rests by day on a tree-trunk or fence; flies by night and comes freely to light.

Distribution (Map 62)

Occurs in gardens, amongst scrub, in open woodland and along hedgerows. Widespread and generally common in England and Wales; local in Ireland; not recorded from Scotland. Northern and central Europe; Spain.

TELEIODES WAGAE (Nowicki)

Gelechia wagae Nowicki, 1860, *Enum.Lepid.Haliciae or.*: 189.
Teleiodes marsata Piskunov, 1973, *Trudȳ vses.ent.Obsch.* **56**: 193, fig.8.
Type locality: Ukraine; Lvov district.

Description of imago (Pl.3, figs 6,7)

Wingspan 12–14mm. Head with grey-edged, brownish fuscous scales, appearing mottled; antenna grey, ringed fuscous; labial palpus with segments 2 and 3 of equal length, segment 2 brownish fuscous, thickened with appressed scales, segment 3 slender, grey with three fuscous bands, third not quite reaching apex. Thorax brown; scales of tegulae grey-edged. Forewing brown, irrorate grey; three black spots, on costa, in disc and on fold, forming outwardly oblique line at one-fifth; four similar spots, two of them in disc, at two-fifths, parallel except spot on fold slightly proximal; two spots in disc at two-thirds; curved pale fascia at three-quarters, seldom visible in worn specimens; series of terminal black dots, rarely present on costa and round apex and sometimes wholly obsolete; cilia grey with irregular ciliary line of grey-tipped, darker scales. Hindwing equal in width to forewing, trapezoidal, tornus gently curved, apex moderately produced, brownish grey; cilia pale grey. Foreleg fuscous, ringed whitish; mid- and hindlegs grey, tarsi banded brown. Abdomen fuscous, paler ventrally; anal tuft of male ochreous-grey. Genitalia, see figures 9b,29f.

Similar species. *Carpatolechia notatella* (Hübner), but *T. wagae* has a lighter ground colour and the markings are distinct black dots, whereas in *C. notatella* they are streak-like and diffuse; *T. wagae* has a darker hindwing. *Pseudotelphusa paripunctella* (Thunberg) has only two black dots at two-fifths, whereas *T. wagae* has four black dots in that area.

Life history

Based mainly on Langmaid (1980).

Ovum. Laid on hazel (*Corylus avellana*), birch (*Betula*

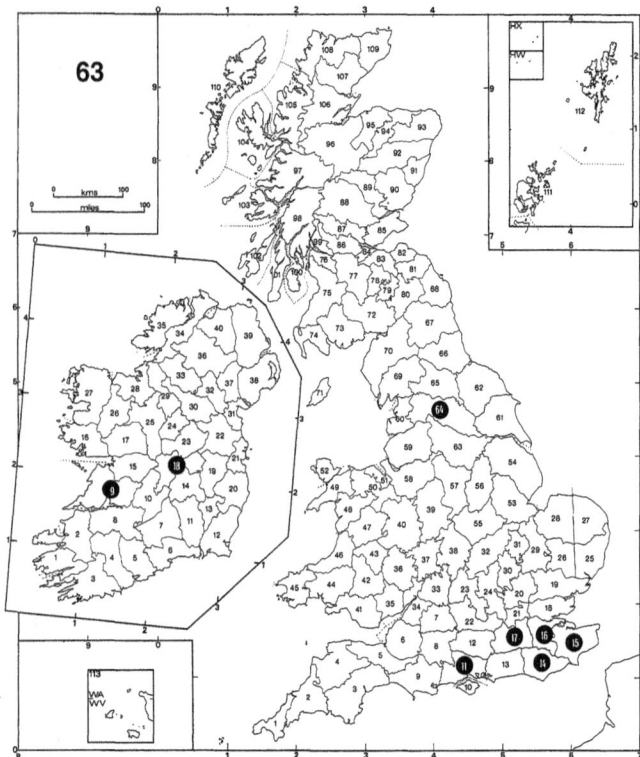

Teleiodes wagae

spp.) (Emmet, 1988a) or sweet chestnut (*Castanea sativa*) (Langmaid & Sattler, 1999).

Larva. Fully fed 11mm long. Head honey-coloured. Prothoracic plate greenish yellow; body pale green, gut darker; pinacula dark grey; male gonads visible on abdominal segment 5 as brown patch. Larva turns pink prior to pupation.

The larva generally feeds between two leaves, one spun flatly over the other, less often in a folded leaf. It lives under a silken pad attached to the upper leaf. When young it makes irregular blotches in both leaves, when larger it makes holes in the leaves. August–September.

Pupa. Uniform light chestnut-brown, enclosed in a flimsy silken cocoon amongst detritus on the ground. October–May, occasionally overwintering twice (Emmet, *loc.cit.*).

Imago. Univoltine; May–June. Rests on a tree-trunk by day and at night occasionally comes to light.

History and distribution (Map 63)

Placed on the British list by Sattler (1980) on the evidence of specimens captured and reared from hazel by J. D. Bradley in the Burren, Co. Clare, western Ireland

from 1961 onwards. Pierce erroneously figured the male genitalia as those of *C. notatella* (Pierce & Metcalfe, 1935a), but it is possible that he used a Continental specimen. The first confirmed English specimen was taken on 30 May 1976 in south Hampshire (Langmaid, *loc.cit.*). The species is currently known from Hampshire, Sussex, Surrey and Kent; a single specimen was taken in a Rothamsted trap at Harrogate, Yorkshire, on 11 August 1978, an unusually late date. In Ireland it is locally fairly common in the Burren, Co. Clare, and has also been recorded from Co. Offaly (K. G. M. Bond, pers.comm.). It is possibly under-recorded through having been mistaken for *C. notatella* or *P. paripunctella*. Northern and central Europe, but Sattler (*loc.cit.*) considers that many of the records require confirmation.

TELEIODES LUCULELLA (Hübner)

Tinea luculella Hübner, [1813], *Samml.eur.Schmett.* **8**: pl.59, fig.397.

Recurvaria subrosea Haworth, 1828, *Lepid.Br.*: 553.

Anacampsis luctuella sensu Stephens, 1834, *Ill.Br.Ent.* (Haust.) **4**: 212, *nec* Hübner, 1793.

Type locality: [Europe].

Description of imago (Pl.3, fig.8)

Wingspan 10–12mm. Head white or grey, sometimes mixed with fuscous scales; antenna whitish, ringed fuscous; labial palpus with segments 2 and 3 of equal length, segment 2 thickened with appressed scales, segment 3 slender, both segments white, ringed fuscous. Thorax grey. Forewing dark grey; large, semi-circular white costal blotch from one-quarter to one-half, not quite reaching dorsum, generally with some yellow scales near distal margin, enclosing blackish fuscous costal bar; prominent subapical white spot with variable white scaling below sometimes to dorsum, indicating an interrupted, straight white subterminal fascia; white markings edged proximally by slightly raised black scales; some white scales at apex and on termen; cilia grey. Hindwing slightly narrower than forewing, trapezoidal, tornus obtusely angled, apex strongly produced; pale grey; cilia paler. Legs fuscous, tarsi banded whitish. Abdomen fuscous; anal tuft of male ochreous-grey. Genitalia, see figures 9c,29d.

Similar species. T. flavimaculella (Herrich-Schäffer), *q.v.*

Life history

Ovum. Laid on deciduous oak (*Quercus* spp.).

Larva. Head yellowish brown. Body whitish green; prothoracic plate yellowish brown marked with black;

pinacula black. Feeds in spun leaves. September.

Pupa. In a loosely spun cocoon amongst debris on the ground. October–May.

Imago. Univoltine; May–June. Rests on a tree-trunk by day; at night comes to light.

Distribution (Map 64)

Widely distributed and fairly common in oak woodland in England and Wales as far north as Co. Durham; not recorded from Scotland or Ireland. Northern and central parts of Europe; Spain; Sardinia.

TELEIODES FLAVIMACULELLA (Herrich-Schäffer)

Gelechia flavimaculella Herrich-Schäffer, 1854, *Syst. Bearb.Schmett.Eur.* **5**: 167, pl.67, fig.497.

Type locality: [Europe].

Account based throughout on Langmaid & Sattler (1999).

Description of imago (Pl.3, fig.9)

Wingspan 10–12mm. Head mixed greyish white and fuscous; antenna shining dark fuscous, indistinctly annulate black; labial palpus with segment 3 slightly shorter than segment 2, dark fuscous, segment 2 inner side with rows of overlapping white-tipped scales forming thin transverse stripes, segment 3 with two indistinct rings and apex whitish. Thorax and tegulae mixed greyish white and fuscous. Forewing elongate, dark fuscous, scales black-tipped giving slightly speckled appearance; outwardly oblique row of three black spots, first on costa near base, second in disc, third on fold, latter two almost coalesced; slender longitudinal orange streak above discal spot widening abruptly into round orange spot just before one-half; above orange spot, small greyish white subcostal spot with a few orange scales; black vertically elongate discocellular spot; indistinct greyish white preapical spot on costa; cilia concolorous with wing. Hindwing grey; cilia concolorous, with faint rusty-orange sheen. Fore- and midlegs dark fuscous, ringed whitish at joints; hindleg fuscous above, greyish ochreous beneath, ringed darker at joints. Abdomen grey, ventral surface lighter. Genitalia, see figures 9e,29e.

Similar species. T. luculella (Hübner), which has the forewing broader distally, the discal marking yellow rather than orange and curving in a semicircle so that both ends touch the costa.

Life history

The early stages have not been described. According to

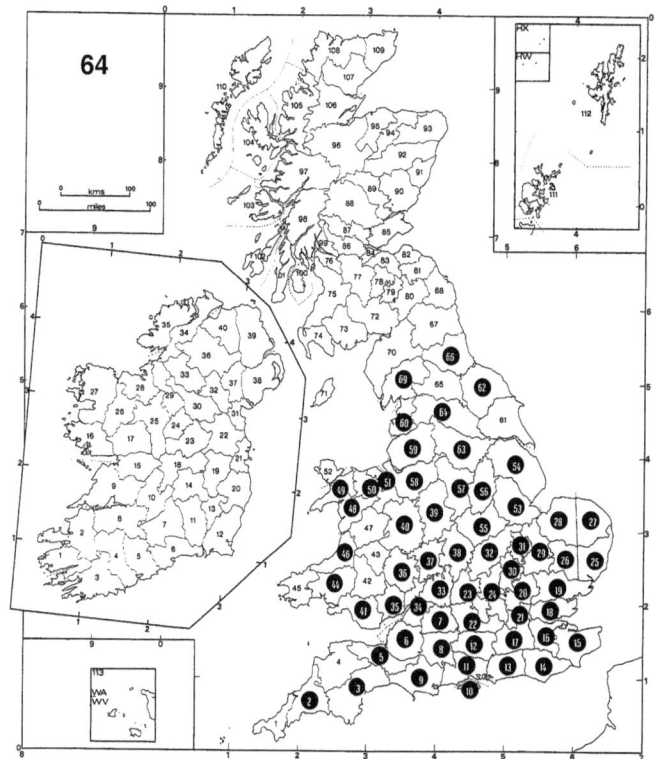

Teleiodes luculella

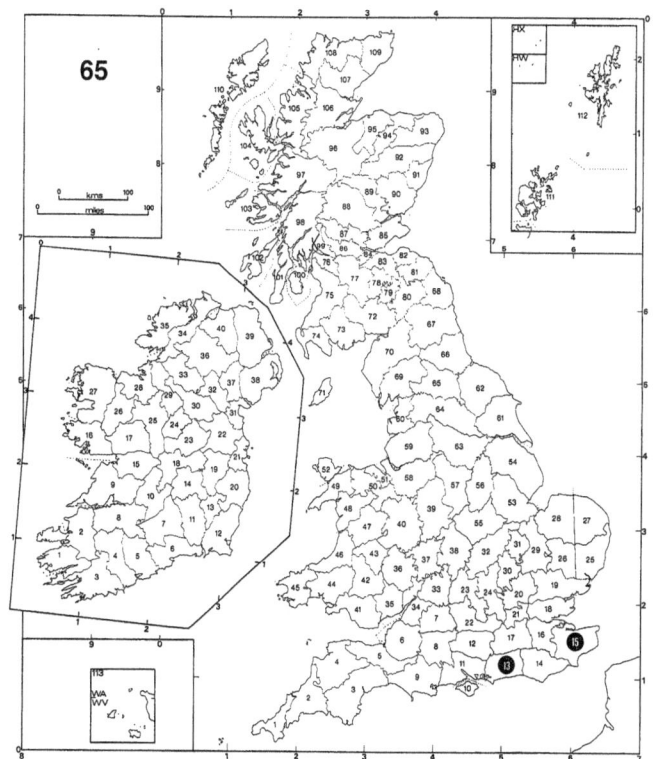

Teleiodes flavimaculella

Svensson (1976), Klimesch reared the species from sweet chestnut (*Castanea sativa*) in Austria; other Continental lepidopterists have taken the adult where oak (*Quercus* spp.) was present and suspected that to be the foodplant. However, Svensson (*loc.cit.*) states that the moth occurs in Sweden in localities where neither of these trees is present.

Imago. Univoltine; June.

Distribution (Map 65)

Occurs in woodland. Known in Britain from only two specimens, the first taken by A. P. Wickham in Blean Wood, Kent, on 6 June 1935, the specimen now being in BMNH, and the second by J. T. Radford at m.v. light in West Sussex on 2 June 1995. Local and rare in Europe from The Netherlands to Russia, and from Sweden and Finland to Italy and the former Yugoslavia.

TELEIODES SEQUAX (Haworth)

Recurvaria sequax Haworth, 1828, *Lepid.Br.*: 552.
Gelechia sequacella Doubleday, 1859, *Zoologist synonymic. List Br.Lepid.*: 30.
Type locality: [England].

Description of imago (Pl.3, fig.10)

Wingspan 11–14mm. Head white, vertex often mixed grey; antenna white, barred fuscous above, almost wholly fuscous below; labial palpus with segments 2 and 3 of equal length, segment 2 thickened with appressed scales, whitish, segment 3 slender, white with black medial and apical bands. Thorax grey; tegulae tipped white. Forewing white, irrorate grey and yellowish ochreous; markings blackish fuscous; spot at base of costa; outwardly oblique fascia at one-fifth, rarely reaching dorsum but ending subdorsally in small patch of raised scales; large costal blotch from one-third to two-thirds, also not reaching dorsum; stigmata, when visible in this blotch, with first discal above plical; discal margin of blotch blacker and excavate; terminal area more heavily mixed ochreous with one or more black wedges between veins; cilia ochreous, spotted fuscous, with three obscure darker ciliary lines. Hindwing as broad as forewing, trapezoidal, tornus gently curved, termen leading almost without angle into moderately produced apex; light grey; cilia paler. Foreleg fuscous, tarsus banded whitish; mid- and hindlegs with outer surfaces fuscous, inner surfaces ochreous, tarsi barred fuscous. Abdomen ochreous on dorsal surface, whitish on ventral surface. Genitalia, see figures 9f,29g.

Specimens from Freshwater Down, Isle of Wight, are

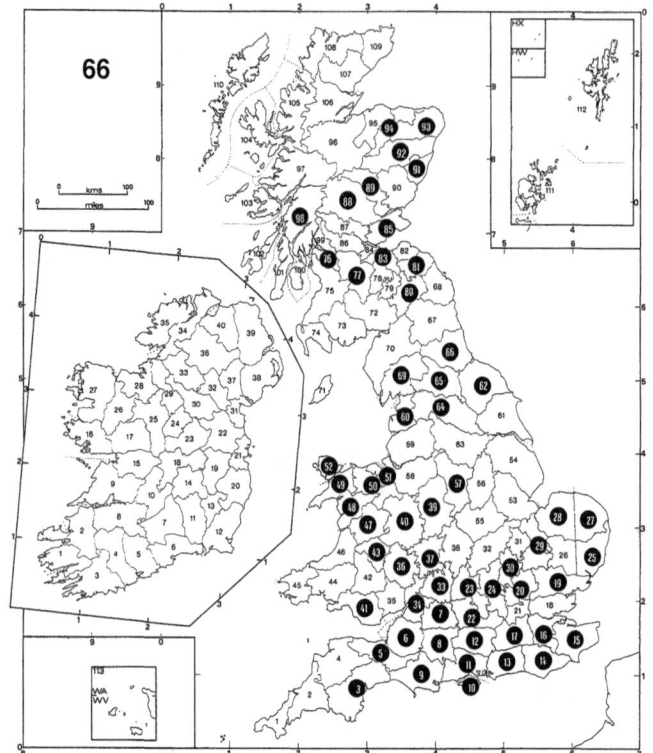

Teleiodes sequax

paler, with the ground colour clear white without irroration (J. R. Langmaid, pers.comm.).

Life history

Ovum. Laid on common rock-rose (*Helianthemum nummularium*) or hoary rock-rose (*H. canum*). Probably overwinters in this stage.

Larva. Head brown. Body greenish or greyish white, dorsal line darker; prothoracic plate brown with a small posterolateral darker spot; pinacula small, grey; thoracic legs annulate brown.

The larva feeds in a tight, ball-like spinning in a terminal shoot, sometimes changing to another shoot. May–early June.

Pupa. In a flimsy cocoon spun amongst detritus on the ground. June.

Imago. Univoltine; July. Can be obtained by sweeping the foodplant; time of flight apparently unrecorded.

Distribution (Map 66)

Widespread and common on chalk and limestone downland in England and Wales and on acid soils in Scotland as far north as Banffshire; not recorded from Ireland. Throughout most of Europe, but absent from Norway, Denmark, The Netherlands and much of the Mediterranean region; North America.

CARPATOLECHIA Căpuşe

Carpatolechia Căpuşe, 1964, *Ent.Tidskr.* **85**: 12.

NOMENCLATURE. Both Elsner, Huemer & Tokar (1999) and Huemer & Karsholt (1999), followed by Bradley (2000) have accepted *Carpatolechia* as a valid generic name for certain species formerly placed in *Teleiodes* Sattler on the ground that they form a clearly monophyletic group. Five of these species occur in Britain and four in Ireland.

All the British species are univoltine; one overwinters as an adult. The larvae feed in spun leaves of trees.

Key to species (imagines) of the genus *Carpatolechia*

1   Forewing brown, with black bar at base of costa .... *decorella* (part) (p. 137)

–   Forewing white or grey, often heavily suffused fuscous 2

2(1)   Forewing white with dark streak on fold or patch of raised scales below fold, or heavily suffused fuscous, obscuring white ground colour .... *fugitivella* (p. 140)

–   Forewing without this dark streak or patch of raised scales and not heavily suffused fuscous 3

3(2)   Forewing with black bar at base of costa .... *decorella* (part) (p. 137)

–   Forewing without black bar, but sometimes with subbasal spot on costa 4

4(3)   Labial palpus segment 2 clear white .... *alburnella* (p. 139)

–   Labial palpus segment 2 with dark markings 5

5(4)   Forewing with black spot below second discal stigma .... *proximella* (p. 139)

–   Forewing without black spot below second discal stigma .... *notatella* (p. 138)

CARPATOLECHIA DECORELLA (Haworth)

Tinea decorella Haworth, 1812, *Trans.ent.Soc.Lond.* 1[1]: 338.

Gelechia humeralis Zeller, 1839, *Isis, Leipzig* **1839**: 200.

Anacampsis lyellella Humphreys & Westwood, 1845, *Br.Moths Transform.*: 190.

Type locality: [England].

Description of imago (Pl.3, figs 11–13)

Wingspan 13–14mm. Head varying from white through grey to dark fuscous; antenna fuscous above, pale grey, spotted fuscous beneath; labial palpus with segments 2 and 3 of equal length, segment 2 thickened with short, erect scales beneath forming shallow furrow, fuscous, apex white, segment 3 gently curved, fuscous with pale, often obscure, spots or bands at base and two-thirds. Thorax concolorous with head. Forewing very variable; ground colour white, grey, pale brown or fuscous; black, longitudinal bar on costa at base, visible in all but the darkest specimens, a diagnostic character; in lightly coloured specimens three stigmata present, consisting of black, slightly raised scales, plical and first discal elongate, first discal slightly beyond plical; two diffuse dark blotches on costa; frequently diffuse black patches in disc, in extreme examples reaching dorsum and forming large elongate dorsal blotch, in pale specimens edged above by yellowish streak; fuscous spots on distal quarter of costa round apex and on termen, sometimes expanding into blotch at tornus; darkest forms almost unicolorous; cilia whitish grey to fuscous, depending on ground colour. Hindwing pale grey; costal and apical cilia sometimes darker than wing, otherwise concolorous. Legs with tibiae and tarsi fuscous, even in pale forms, tarsal joints ringed whitish, femora silvery, spotted or streaked fuscous. Abdomen fuscous, paler and sometimes silvery on ventral surface. Genitalia, see figures 9g,30b.

Similar species. C. alburnella (Zeller), which has segment 2 of the labial palpus clear white, whereas in *C. decorella* it is fuscous with only the apex white.

Polymorphism is rare in Gelechiidae and may be explained in this species by the fact that it is one of the very few members of the family to overwinter as an adult; it is therefore exposed to predation, especially by birds, over a period of up to nine months. Chances of survival are enhanced if predators cannot follow a common search pattern; cf. *Acleris cristana* ([Denis & Schiffermüller]) (Tortricidae) and congeneric species that overwinter as adults and likewise exhibit apostatic polymorphism.

Life history

Ovum. Laid on deciduous oak (*Quercus* spp.) or dogwood (*Cornus* (=*Swida*) *sanguinea*).

Larva. Head and prothoracic plate pale brown. Body light green (Meyrick, [1928]). The larva feeds in a folded leaf. May–June.

Imago. Univoltine; July, overwintering until April or May. Occasionally comes to light, but rarely seen after hibernation.

Distribution (Map 67)

Frequents woods and scrub. Widespread at low density

in the British Isles as far north as Moray; the Channel Islands. Europe; Asia Minor; North Africa.

CARPATOLECHIA NOTATELLA (Hübner)

Tinea notatella Hübner, [1813], *Samml.eur.Schmett.* **8**: pl.50, fig.344.

Type locality: [Europe].

Description of imago (Pl.3, fig.14)

Wingspan 12–15mm. Head grey or fuscous; antenna whitish grey, ringed fuscous; labial palpus with segments 2 and 3 of equal length, segment 2 thickened with appressed scales, segment 3 slender, gently curved, both greyish white and each with two fuscous bands. Thorax grey. Forewing grey, suffused, often heavily, with fuscous tending to obscure pattern; black stigmata, when visible, placed as follows: basal dot; two transversely placed spots at one-fifth; black costal streak at one-quarter; two streaks at one-half, the dorsal member of pair more distal; two transversely placed spots at two-thirds; all spots often edged grey distally; apical one-quarter generally paler with series of black apical and terminal dots; cilia grey, with three indistinct darker ciliary lines Hindwing as broad as forewing, trapezoidal, tornus gently curved, termen not sinuate, apex not strongly produced, rather dark grey; cilia pale grey. Foreleg fuscous, tibia and tarsus ringed whitish grey; mid- and hindlegs whitish grey, tarsi ringed fuscous. Abdomen with dorsal surface greyish fuscous, ventral surface grey. Genitalia, see figures 9h,30e. The male genitalia figured by Pierce & Metcalfe (1935a) and purporting to be of this species are those of *Teleiodes wagae* (Nowicki).

Similar species. Differs from *C. fugitivella* (Zeller) in the absence of raised scales and the dark streak on the fold of the forewing; in the hindwing, the termen is not sinuate and the apex is less strongly produced. *T. wagae, q.v.*

Life history

Ovum. Laid on great sallow (*Salix caprea*) and probably other *Salix* spp.

Larva. Head and prothoracic plate black posteriorly. Body pale grey-green, pinacula black.

Feeds on the parenchyma under the down on the underside of a leaf, often curling the edge, but sometimes spinning two leaves together. August–September.

Pupa. In a flimsy cocoon amongst detritus on the ground. October–May.

Imago. Univoltine; May–June. May be disturbed from its foodplant by day; at night is sometimes attracted to light.

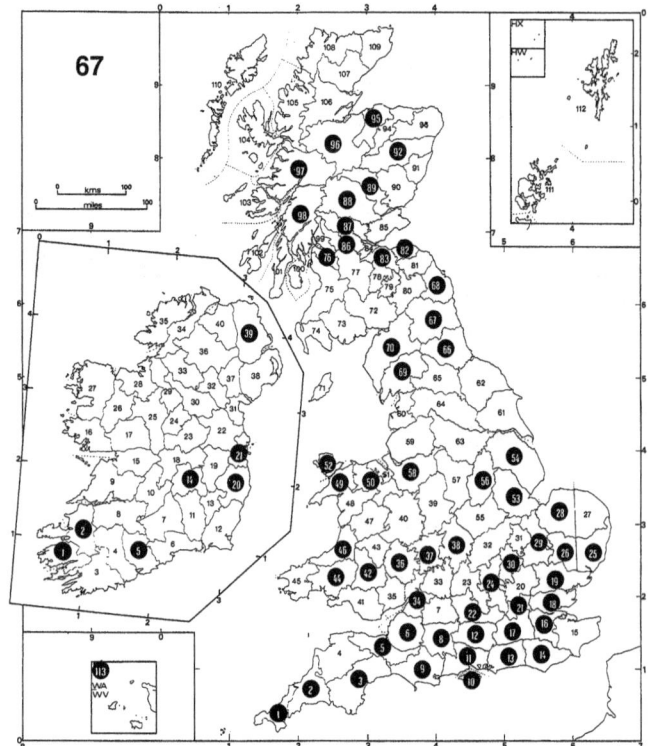

Carpatolechia decorella

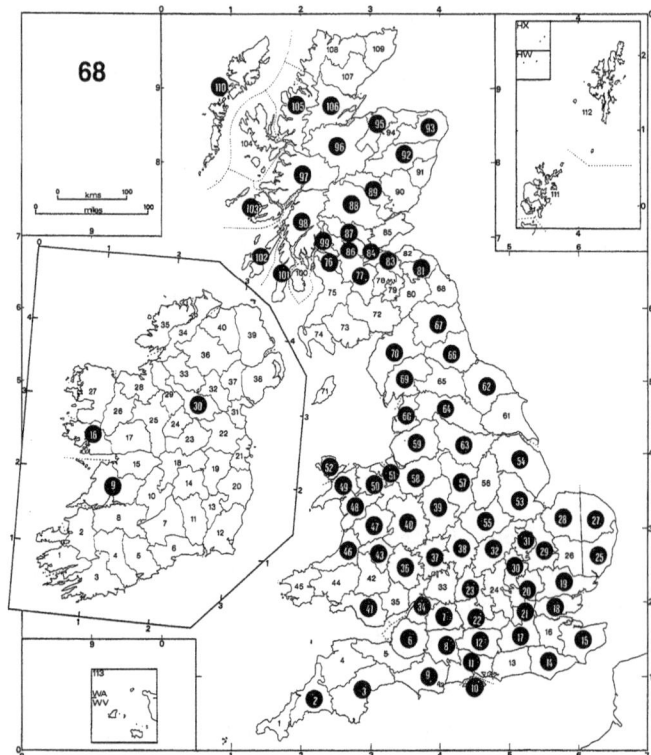

Carpatolechia notatella

Distribution (Map 68)

Occurs in woodland, heathland, scrub and fens. Widespread and fairly common throughout the British Isles except the far north of Scotland, Orkney and Shetland. Northern and central Europe.

CARPATOLECHIA PROXIMELLA (Hübner)

Tinea proximella Hübner, 1796, *Samml.eur.Schmett.* **8**: pl.33, fig.228.

Type locality: [Europe].

Description of imago. (Pl.3, fig.17)

Wingspan 13–16mm. Head white or pale grey; antenna grey, ringed fuscous; labial palpus with segments 2 and 3 of equal length, segment 2 thickened with appressed scales, white, outer side spotted fuscous at base, centre and apex, segment 3 slender, almost straight, white with fuscous bands before and after centre, apex white. Thorax grey. Forewing white, variably irrorate with black and grey scales; base of costa black; outwardly oblique row of two or three black spots at one-fifth; plical stigma present; first discal beyond, but overlapping; second discal with another stigma immediately dorsad; blackish suffusion on costa at two-thirds; reduced irroration often resulting in broad, irregular, angulate white fascia at three-quarters; row of terminal black dots from preapical area to tornus, that at apex often ringed white, spots sometimes extending to form terminal fuscous suffusion, sometimes, however, obsolete; cilia whitish to grey, with three very obscure darker ciliary lines. Hindwing leading almost without angle to weakly produced apex; grey; cilia pale grey. Foreleg dark fuscous, thinly banded grey; mid- and hindlegs grey, tarsi banded slightly darker. Abdomen grey, ventral surface paler. Genitalia, see figures 9i,30d.

Similar species. C. *alburnella* (Zeller), *q.v.*

Life history

Ovum. Laid on birch (*Betula* spp.) or alder (*Alnus glutinosa*); also on Italian alder (*A. cordata*) (Langmaid, 1978b).

Larva. Head yellowish. Body pale green, sometimes reddish-tinged; pinacula black. Feeds in a folded leaf. August–September.

Pupa. In a flimsy cocoon amongst detritus on the ground. October–May.

Imago. Univoltine; May–June. Rests on a tree-trunk during the day; flies by night and comes to light.

Distribution (Map 69)

Frequents all types of terrain where the foodplants occur. Widespread and common throughout the

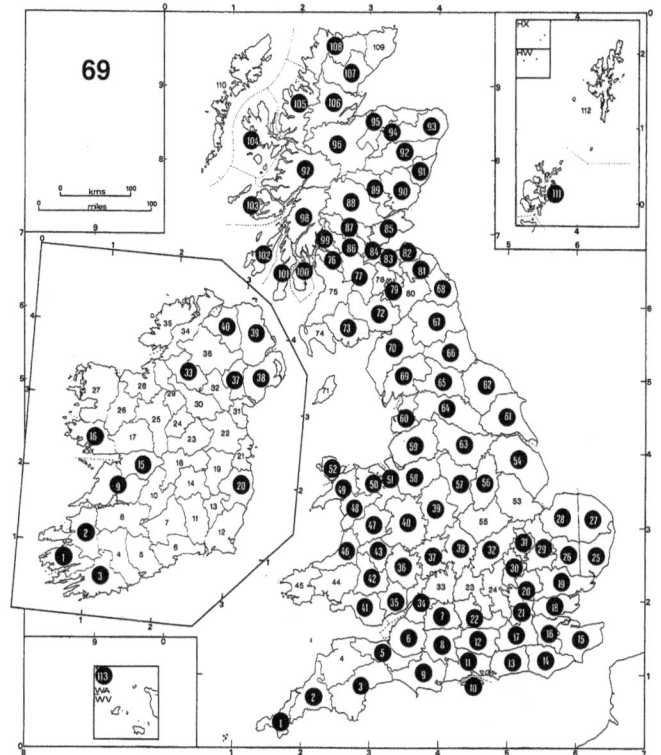

69

Carpatolechia proximella

British Isles, including the Channel Islands, but not recorded from the Outer Hebrides or Shetland. Northern and central Europe; Corsica.

CARPATOLECHIA ALBURNELLA (Zeller)

Gelechia alburnella Zeller, 1839, *Isis, Leipzig* **1839**: 200.

Type localities: Germany; Frankfurt and Glogau (now Poland; Głogów).

Description of imago (Pl.3, figs 18,19)

Wingspan 12–14mm. Head white; antenna white, ringed fuscous; labial palpus with segment 3 slightly shorter than segment 2, segment 2 thickened with appressed scales, clear white, segment 3 slender, nearly straight, greyish white with central and subapical fuscous bands of variable width. Thorax white, generally with some grey scales. Forewing white, variably irrorate grey; variably expressed black spots on costa near base and before and beyond middle, the last generally largest; plical stigma a tuft of raised white or yellowish scales; discal stigmata generally absent; sometimes heavier grey irroration on dorsal half of wing from base to midway; terminal area sometimes with scattered black scales; series of terminal black spots; cilia grey with three indistinct darker ciliary lines. Hindwing as broad as forewing, trapezoidal, tornus gently curved, termen slightly sinuate, apex moderately produced, grey; cilia pale grey. Foreleg fuscous, tarsus ringed white; mid- and hindlegs greyish white, tarsi ringed fuscous. Abdomen grey, ventral surface whitish grey. Genitalia, see figures 9j,30a.

Specimens from the Scottish Highlands (fig.18) have a broad streak of black or dark fuscous irroration in disc, extending from plical stigma to beyond middle, often divided by one or two pale longitudinal streaks, the subcostal dark branch merging with black costal spot beyond middle.

Similar species. *C. proximella* (Hübner) and white forms of *C. decorella* (Haworth), from both of which *C. alburnella* may be distinguished by the clear white coloration of segment 2 of the labial palpus.

Life history

Ovum. Laid on birch (*Betula* spp.), probably overwintering in this stage (Sokoloff, 1995a).

Larva. Head and prothoracic plate yellow. Body light green; pinacula black. Feeds in a folded leaf or spun leaves. May–June.

Pupa. Undescribed. June.

Imago. Univoltine; late June–August. Rests by day on a birch trunk, generally in a crevice of rough bark. Flies at night and comes to light.

History and distribution (Map 70)

First taken by A. Smith at Strensall, Yorkshire, in 1927 and subsequently in the same locality. In 1935, S. Wakely saw the specimens and, suspecting that they represented a species new to Britain, sent one to Pierce for dissection. Pierce confirmed that it was new and Pierce & Metcalfe (1935a) mentioned it as such in their introduction and figured the male genitalia. In his presidential address to the South London Entomological & Natural History Society, delivered in January 1948, L. T. Ford included it , but without description, among the species recently added to the British list (Ford, 1949a), and later gave notes on its life history (Ford, 1949b). However, no illustration or description of the adult appeared in the British literature until 1995, when it was figured by Bradford and described by Sokoloff (*loc.cit.*); it is also figured on the jacket of *MBGBI* 3 (1996).

Widespread at low density in England (though apparently absent from the south-west) as far north as Yorkshire, North Wales, the Scottish Highlands, and Ireland. Northern and central Europe.

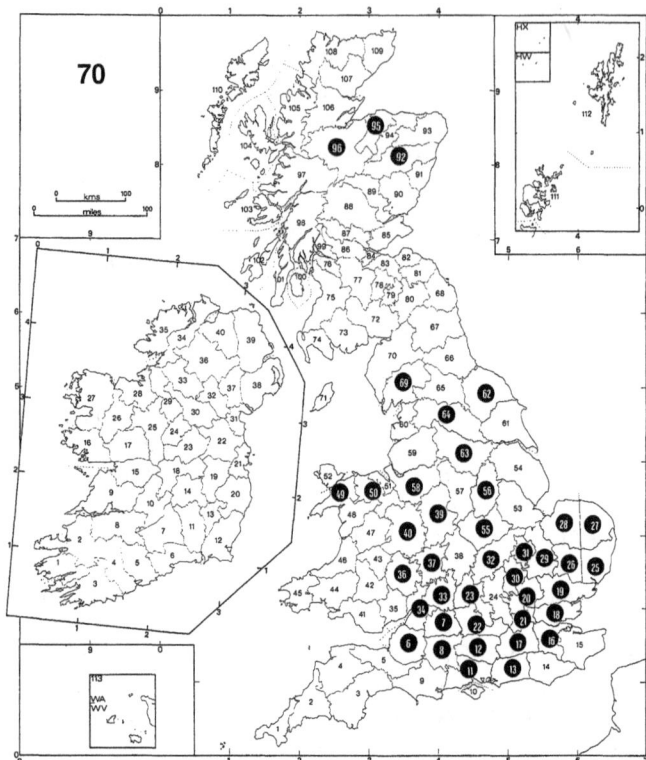

Carpatolechia alburnella

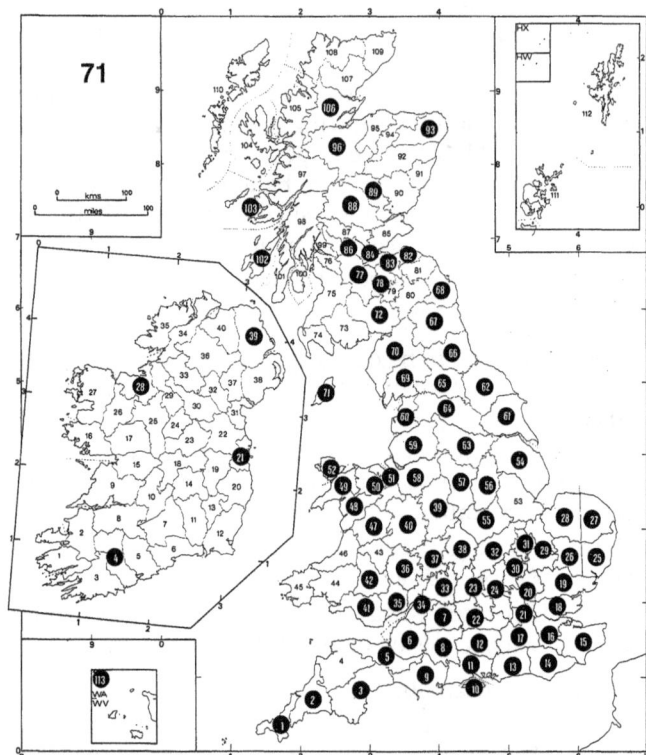

Carpatolechia fugitivella

CARPATOLECHIA FUGITIVELLA (Zeller)

Gelechia fugitivella Zeller, 1839, *Isis, Leipzig* **1839**: 200.
Type localities: Germany; Glogau (now Poland; Glogów) and Berlin.

Description of imago (Pl.3, figs 15,16)

Wingspan 12–15mm. Head white, generally mixed grey; antenna fuscous, ringed whitish; labial palpus with segments 2 and 3 of equal length, segment 2 thickened with appressed scales, segment 3 slender and gently curved, both segments fuscous, banded whitish. Thorax whitish grey to grey. Forewing grey, variably irrorate fuscous, sometimes so heavily as to obscure markings; in pale specimens suffused blackish costal spots near base and before and beyond middle; black, sometimes interrupted, streak along fold, beneath which a small patch of raised, sometimes whitish, scales; above distal end of streak two black spots, sometimes united to form parallel streak; second discal stigma usually present, often with another spot below, both of slightly raised scales; terminal area paler, even in melanic forms (fig.16); series of terminal black spots or streaks; cilia grey with indistinct darker ciliary line. Hindwing of equal width to forewing, trapezoidal, tornus gently curved, apex strongly produced; grey; cilia pale grey. Foreleg fuscous, ringed grey; mid- and hindlegs grey, spotted fuscous. Abdomen fuscous, ventral surface paler. Genitalia, see figures 9k,30c.

Similar species. C. notatella (Hübner), *q.v.*

Life history

Ovum. Laid on elm (*Ulmus* spp.), including wych elm (*U. glabra*). Probably overwinters in this stage.

Larva. Head and prothoracic plate light brown. Body light green, tinged pink on dorsum; pinacula black. Feeds in a folded leaf or spun leaves, readily dropping to the ground by a silken thread if disturbed. May.

Pupa. Undescribed. Late May–June.

Imago. Univoltine; late June–July. Rests on a tree-trunk, generally in a fissure of bark, by day and flies at night, when it readily comes to light.

Distribution (Map 71)

Occurs in woods, parks, gardens and along hedgerows, wherever its foodplants grow. Widespread and common throughout Britain, including the Channel Islands and the Isle of Man, as far north as Easter Ross, but local in the Scottish Highlands; also local in Ireland. Throughout most of Europe, but scarce in the Mediterranean region.

TELEIOPSIS Sattler

Teleiopsis Sattler, 1960, *Dt.ent.Z.* (N.F.) **7**: 66.

A Holarctic genus represented by six species in Europe and one in North America. One species occurs in the British Isles.

TELEIOPSIS DIFFINIS (Haworth)

Recurvaria diffinis Haworth, 1828, *Lepid.Br.*: 551.
Gelechia diffinella Doubleday, 1859, *Zoologist Synonymic List Br.Lepid.*: 30.
Type locality: [England].

Description of imago (Pl.3, fig.20)

Wingspan 14–16mm, moths of the second generation tending to be slightly smaller. Head dark grey; antenna fuscous, obscurely ringed ochreous; labial palpus segment 3 as long as segment 2, segment 2 thickened with short erect scales enclosing slight furrow, segment 3 slender, pale fuscous with two darker fuscous bands. Thorax greyish brown. Forewing ochreous-brown, sometimes with faint violet gloss, irrorate darker and scales tipped whitish; outwardly oblique fascia formed by three black spots of raised scales at one-fifth; variably expressed black streak on fold with plical stigma at its centre; first discal stigma immediately above, second forming transverse bar or two approximate spots; subdorsal area below plical streak paler; obsolescent angulate whitish fascia beyond second plical stigma; sometimes black apical dot; cilia pale grey, mixed darker, with three rather broad bars in costal cilia and two obscure slightly darker ciliary lines in terminal cilia. Hindwing slightly broader than forewing, anal angle prominent, tornus smoothly curved, termen leading without sinuation into weakly produced apex; grey; cilia paler. Foreleg fuscous, ringed ochreous-white; mid- and hindlegs ochreous, tarsi banded darker. Abdomen pale fuscous; anal tuft of male ochreous-fuscous. Genitalia, see figures 9l,30f.

Life history

Ovum. Laid on sheep's sorrel (*Rumex acetosella*).

Larva. Head and prothoracic plate yellow-brown. Body brownish green, marbled reddish. Feeds from a silken gallery spun on the upper part of the roots, along the stem or occasionally among the seeds of the foodplant. August–May; in most years also July–August.

Pupa. Undescribed.

Imago. In most years bivoltine or even continuously brooded, adults having been taken in every month

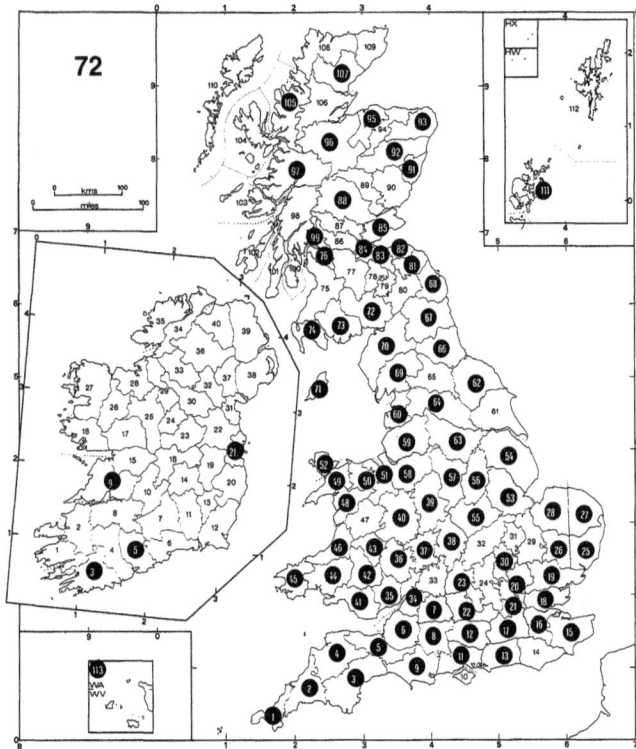

Teleiopsis diffinis

from May to October, the peak periods being May–June and September. Sokoloff (1979) found adults and young larvae on the same day in late September. It is possible that some adults emerge late in the year and then overwinter, since Sokoloff (*loc. cit.*) records a moth taken on 17 March 1977. Adults may be obtained by sweeping the foodplant by day and at night they come readily to light.

Distribution (Map 72)

Occurs on grassland and heaths, especially on acid soil. Widespread and common throughout Britain to Orkney, including the Channel Islands and the Isle of Man, but not recorded from the Outer Hebrides or Shetland; local or under-recorded in Ireland. Throughout Europe.

CHIONODES Hübner

Chionodes Hübner, [1825], *Verz. bekannt. Schmett.*: 420.

A large genus widespread throughout the Holarctic and Neotropical regions with the greatest diversity in North America and the alpine zones of the Palaearctic region. There are 22 European species but only two of these occur in Britain.

Imago. Medium-sized to large gelechiids with labial palpus long and slender, segment 2 rough-scaled beneath. Antenna without pecten. Forewing elongate, tending to lanceolate, without scale-tufts. Hindwing, broader than forewing, subtrapezoidal, very slight to no emargination of termen, tornus gently rounded. Abdomen with tergites 1–3 buff. Valva of male genitalia with long slender spinose costa and short slender digitate sacculus; signum of female genitalia a small serrate plate often reflexed along one margin.

Larva. Larval stages are reported from a wide range of plant families.

Key to species (imagines) of the genus *Chionodes*

Due to the range of variation in each of the two British species of *Chionodes* they are best separated by genitalic characters:

MALES (figures 10a,b, p.30)

–   Uncus with distinct short sclerotized apical thorn*... .... *fumatella* (p. 143)

–   Uncus rounded, at most slightly angled, at apex* .... *distinctella* (p. 143)

FEMALES (figures 31a,b, p.51)

–   Apophyses anteriores modified into a pair of flat sclerotized plate-like extensions of segment 8; apophyseal plates almost spatulate, broad at the apex and with basal two-thirds fused .... *fumatella* (p. 143)

–   Apophyseal plates subtriangular and evenly tapered to apex .... *distinctella* (p. 143)

---

* Can be seen under a binocular microscope without dissection by cleaning away the scales at the tip of the uncus.

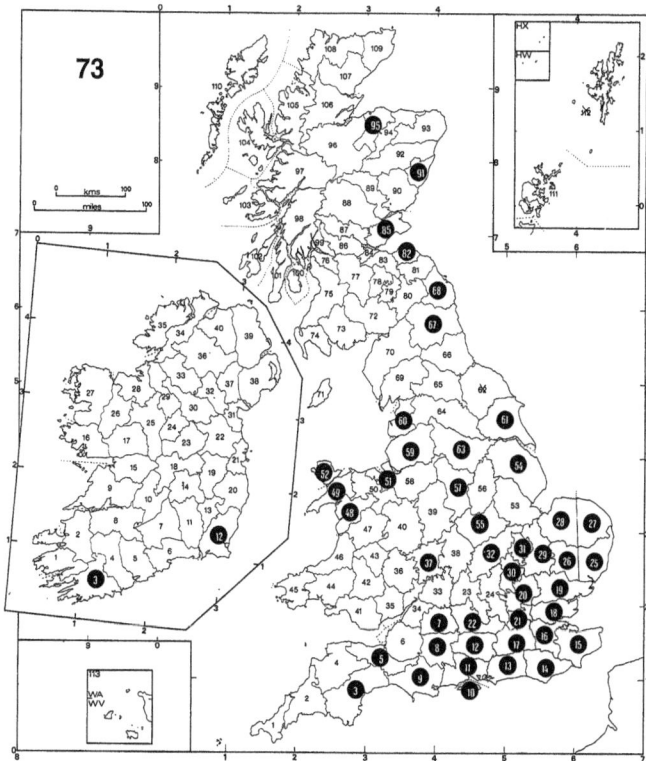

Chionodes fumatella

CHIONODES FUMATELLA (Douglas)

Gelechia fumatella Douglas, 1850, *Trans. ent. Soc. Lond.* (N.S.) **1**: 67.

Gelechia celerella Stainton, 1851, *Suppl. Cat. Br. Tineidae & Pterophoridae*: 5.

Gelechia oppletella Herrich-Schäffer, 1854, *Syst. Bearb. Schmett. Eur.* **5**: 180.

Type locality: England; New Brighton, Cheshire.

Description of imago (Pl.3, fig.21)

Wingspan 13–16mm. Head brownish mixed with pale fuscous, becoming paler on frons; antenna fuscous; labial palpus brownish fuscous becoming more whitish above in basal half of segments 2 and 3. Thorax and tegulae dark fuscous. Forewing dark fuscous mixed with brownish fuscous and scattered black scales; only consistent markings a black spot halfway along fold, usually with a few whitish scales either side, and a blackish spot in the discal area at one-half, usually preceded by a few white scales; cilia pale fuscous with ciliary line of dark-tipped scales. Hindwing pale ochreous-fuscous; cilia concolorous. Fore- and midlegs dark fuscous, tarsi edged paler distally; hindleg pale ochreous-fuscous, tarsi darker basally. Abdomen fuscous. Genitalia, see figures 10b,31b.

Life history

Ovum. Laid on moss.

Larva. Undescribed. According to Continental authors the larva feeds on moss in May (Bradford, [1979]). However Michaelis (1977) reared the species from bird's-foot-trefoil (*Lotus corniculatus*) collected to feed other larvae, so this may be its foodplant.

Pupa. Undescribed.

Imago. Univoltine; late June–August. At night it comes to light.

Distribution (Map 73)

Formerly restricted mainly to sandy coasts, but has recently occurred in increasing numbers well inland (Bradford & Sokoloff, 1988). Widespread in England and Wales, also occurring up the east coast of Scotland as far as Elgin and in the extreme south of Ireland. Throughout Europe.

CHIONODES DISTINCTELLA (Zeller)

Gelechia distinctella Zeller, 1839, *Isis, Leipzig* **1839**: 199.

Type locality: Germany; Glogau (now Poland; Glogów).

Description of imago (Pl.3, fig.22)

Wingspan 14–15mm. Head ochreous-brown; antenna dark fuscous; labial palpus fuscous-brown mixed pale ochreous-fuscous, especially on inner side and above on segment 2. Thorax and tegulae brownish fuscous. Forewing brownish fuscous; fuscous spots at base of fold, at halfway along fold, in discal area just distal to the latter and at halfway across the wing just before two-thirds; cilia pale fuscous with two poorly defined ciliary lines of fuscous-tipped scales. Hindwing pale ochreous-fuscous; cilia concolorous. Fore- and midlegs dark brownish fuscous, hindleg and tuft paler. Abdomen ochreous-fuscous. Genitalia, see figures 10a,31a.

Life history

Ovum. Undescribed. There are no recent observations on the biology. Supposed hostplants such as moss, greenweed (*Genista* spp.) and field wormwood (*Artemisia campestris*), mentioned by various Continental authors in the nineteenth and early twentieth centuries, are extensively discussed by Huemer & Sattler (1995), who suggest that the species may occupy a niche, such as grass roots, which is rarely explored by microlepidopterists.

Larva. Final instar length 10mm. Head, prothoracic and anal plates mid-brown. Body creamy buff with two

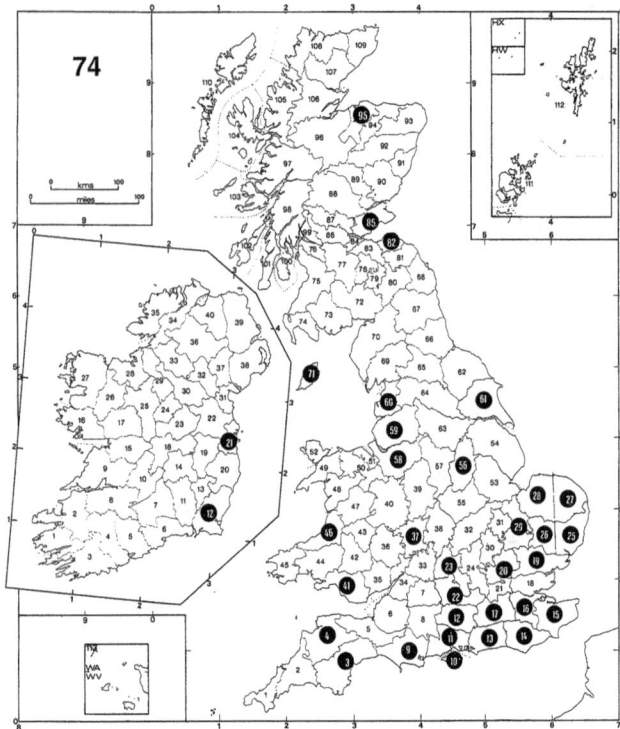

Chionodes distinctella

dorsal and two lateral broad longitudinal creamy purple lines (description taken from a preserved specimen in BMNH). In mainland Europe it is said to make silken tubes in the sand under its foodplant from which it feeds on the dead leaves (Bradford, [1979]).

Pupa. Undescribed.

Imago. Univoltine; end June–August. Has been taken both at light and at sugar (Chalmers-Hunt & Wakely, 1964).

Distribution (Map 74)

Widespread but local in England and Wales, extending up the east coast of Scotland to Elgin. In Ireland known only from Cos Dublin and Wexford (K. G. M. Bond, pers.comm.). In view of the local nature of its occurrence, Parsons (1995[1996]) ascribed the species a Notable B conservation status. Throughout Europe.

PSORICOPTERA Stainton

Psoricoptera Stainton, 1854, *Ins.Br.Lepid.*: 76 (key), 100.

A genus represented by two species in Europe, one of which also occurs in England.

Imago. Head with appressed scales; ocelli present; antenna three-quarters length of forewing, apical two-thirds serrate, scape without pecten; labial palpus with projecting scales on segments 2 and 3. Forewing elongate-ovate; with scale-tufts; veins M_3 (4) and CuA_1 (3) approximate or connate. Hindwing broader than forewing, subtrapezoidal; apex slightly produced; termen sinuate, merging imperceptibly into dorsum, with tornus not distinguishable; anal angle slightly expanded; veins Rs (7) and M_1 (4) closely approximate or connate.

PSORICOPTERA GIBBOSELLA (Zeller)

Gelechia gibbosella Zeller, 1839, *Isis, Leipzig* **1839**: 202.
Type locality: Germany; Berlin.

Description of imago (Pl.3, fig.23)

Wingspan 15–17mm. Head with appressed whitish, brown-tipped scales giving speckled appearance; antenna greyish brown, annulate dark fuscous; labial palpus mixed brownish fuscous, pinkish ochreous and whitish, segments 2 and 3 of equal length, segment 2 with furrowed brush of scales beneath, segment 3 with forward-projecting scales above. Thorax and tegulae concolorous with head. Forewing mixed brownish fuscous, whitish and pinkish ochreous; scale-tufts, two along fold in basal half and two, vertically adjacent, in disc at end of cell, brownish to fuscous; extreme base of dorsum white, bordered by oblique black streak from base of costa; costa with three ill-defined brownish fuscous marks in middle one-third; ill-defined acutely angled whitish fascia at three-quarters; cilia greyish ochreous distal to dark ciliary line, proximally concolorous with wing. Hindwing fuscous, paler toward base; cilia ochreous-grey. Foreleg brownish fuscous above, tarsus marked paler at joints and tip, mixed fuscous and whitish beneath; mid- and hindlegs speckled whitish and fuscous, ringed whitish at joints. Abdomen speckled brownish fuscous and whitish dorsally, venter with whitish median line. Genitalia, see figures 12c,31c.

Life history

Ovum. Laid on deciduous oak (*Quercus* spp.), probably on a twig and overwintering.

Larva. Head and prothoracic plate black. Body grey,

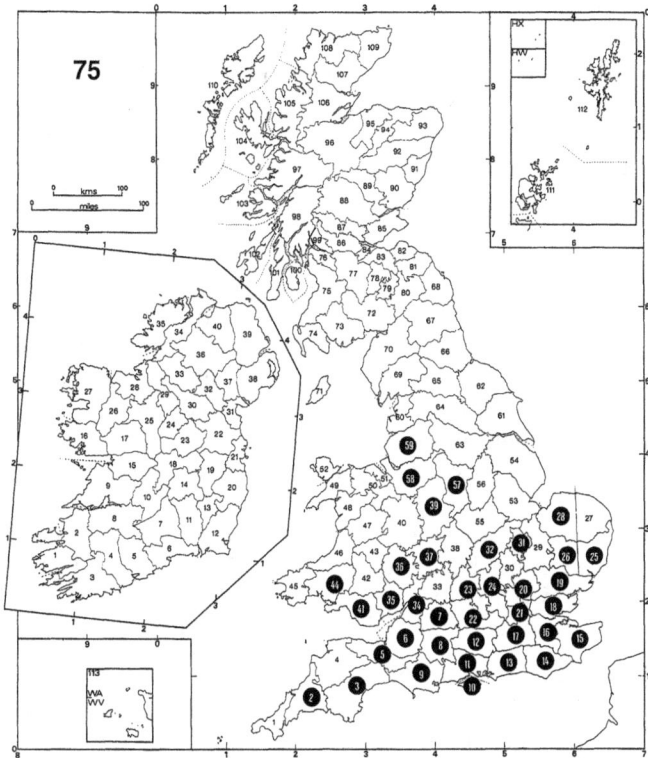

Psoricoptera gibbosella

dorsal line slightly darker, subdorsal lines distinctly darker; pinacula black (Meyrick, [1928]).

Feeds in a longitudinally rolled leaf. May–June.

Pupa. Undescribed. June–July.

Imago. Univoltine; July–early October. Rests by day on a tree-trunk, often in a crevice in the bark, and can be dislodged by jarring the trunk with a stout stick. Flies at night and comes to light.

Distribution (Map 75)

Inhabits mature oak woodland. Local and seldom common in England as far north as Cheshire; South Wales. Throughout Europe except the Mediterranean islands.

MIRIFICARMA Gozmány

Mirificarma Gozmány, 1955, *Annls hist.-nat.Mus. natn.hung.* (S.N.) **6**: 308, 309 [keys], 313.

Helina Guenée, 1849, *nec* Robineau-Desvoidy, 1830.

A genus of 21 species occurring in Europe and the Mediterranean region, with one introduced to the U.S.A. where it has become a pest of clover crops. Two

species are resident in Britain, one of which is also in Ireland.

Imago. Head without frontal modifications; antenna without pecten on scape; ocelli present; labial palpus recurved, segment 1 much shorter than segment 2, segment 3 of similar length to segment 2 or slightly shorter, segment 2 usually without brush below; maxillary palpus with four segments; haustellum well developed. Metascutum with paired group of narrow hair-like scales. Forewing with veins R_5 (7) and R_4 (8) on long common stalk; distance R_2 (10) to R_1 (11) ranging from slightly greater than to four times distance R_3 (9) to R_2 (10); CuA_1 (3) and M_3 (4) separate. Hindwing with veins M_1 (6) and Rs (7) on common stalk or separate; shape of hindwing typical for subfamily Gelechiinae; frenulum of female with three or sometimes two long setae. Distinguished from other genera by the presence, in the male genitalia, of a filament arising from a pair of supporting sclerites extending from the base of the valva and sacculus.

Ovum. Unknown.

Larva. Makes a spinning amongst the leaves of the foodplant or feeds in the shoots; in plants with small leaves, such as gorse (*Ulex europaeus*), it usually feeds in the flowers.

Pupa. See under species.

Key to species (imagines) of the genus *Mirificarma*

NOTE: The wing pattern of *M. mulinella* (Zeller) is variable and occasionally approaches that of *M. lentiginosella* (Zeller). Genitalic examination will confirm in instances of doubt (see *Description of imago* and figures 10c,d (p.30) and 31d,e (p.51)).

– Forewing predominantly dark brown, unstriped .... *lentiginosella* (p. 146)
– Forewing mottled brown, not dark, often with broken or unbroken median longitudinal stripe .... *mulinella* (p. 145)

MIRIFICARMA MULINELLA (Zeller)

Gelechia mulinella Zeller, 1839, *Isis, Leipzig* **1839**: 199.

Gelechia interruptella sensu auctt., *nec* Goeze, 1783.

Type locality: Germany; Dresden.

Description of imago (Pl.3, figs 24,25)

Wingspan 11–15mm. Head cream to light brown; antenna brown, dark basally alternating with paler patches towards apex; labial palpus mottled brown and cream with apex dark brown, apex of segment 2 usu-

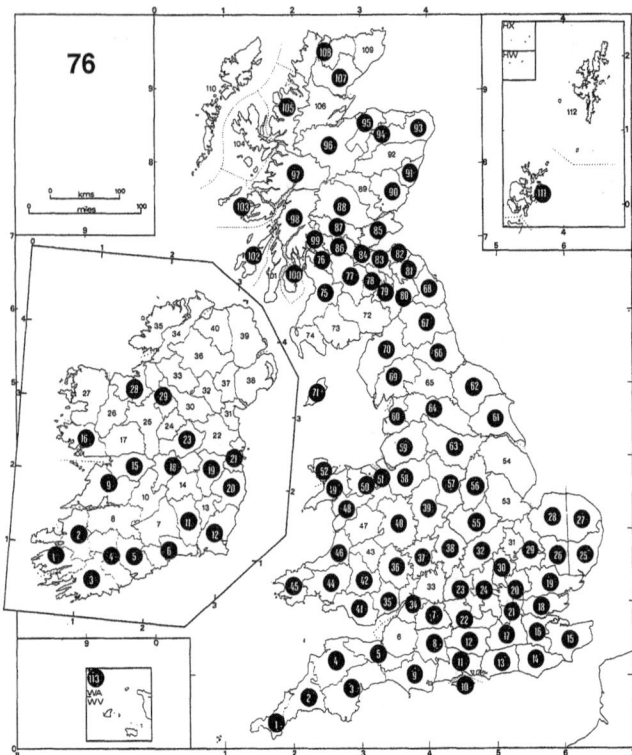

Mirificarma mulinella

ally mostly cream. Thorax and tegulae light to mid-brown, often mottled with dark brown; tegulae slightly darker towards base. Forewing mottled brown, mixed with cream or light brown along costal side and occasionally also near dorsal margin; dark brown spot usually present at end of cell; smaller dark brown spots sometimes present at one-third, one on fold, one in cell; dark brown median longitudinal stripe sometimes present from one-third to apex, another on fold line, these stripes merging to form single diffuse band, or slightly separated by pale area. Hindwing and cilia cream to pale buff. Legs dark brown to buff flecked with cream, tibiae and tarsal segments cream at apex; inner side of hindleg cream. Abdomen cream to pale buff. Male genitalia (figure 10c) with filament straight or gently curved, not reaching hind edge of vinculum; female genitalia (figure 31d) with antrum not more than twice length of apophysis anterior.

Life history

Ovum. Oviposition site not observed. On gorse (*Ulex europaeus*) or broom (*Cytisus scoparius*) (Stainton, 1865b). This species has also been recorded as having been reared from tree lupin (*Lupinus arboreus*) and from a cultivar of dyer's greenweed (*Genista tinctoria*)

(Pitkin, 1984). Abroad, foodplants recorded include black broom (*Cytisus nigricans*) (Hartig, 1964), German greenweed (*Genista germanica*) (Sorhagen, 1886), and *Calicotome spinosa* (Lhomme, [1946–49]).

Larva. Full-grown 10–12mm. Head black. Prothoracic plate black, divided by fine median white line; body dull green; pinacula small, black (Stainton, *loc.cit.*).

The larva makes a small hole in a bud that is not fully open and feeds on the interior of the flower, before repeating the process in another flower. A different strategy is adopted on tree lupin, which does not have reduced leaves; there, the larva is found on the leaves (Pitkin, *loc.cit.*), spinning a leaflet into a pod (J. R. Langmaid, pers.comm.). April to early May; also in June (Sorhagen, *loc.cit.*).

Pupa. In a flimsy cocoon amongst leaves on the ground (Stainton, *loc.cit.*). May–June (Bradford, [1979]).

Imago. In Britain flies from July to September; elsewhere, it has been found also in February, June, October and November. At night it is attracted to light. It runs with its wings slightly elevated (Stainton, *loc.cit.*).

Distribution (Map 76)

On heathland and wasteland. Widespread and common throughout the British Isles, including the Channel Islands and the Isle of Man, as far north as Orkney, but not recorded from the Inner and Outer Hebrides or Shetland. It is the most northerly occurring species of *Mirificarma*; widespread in Europe west of 20°E, north to Norway (Opheim, 1978), extending south to North Africa (Algeria and Tunisia), and known from Russia (Piskunov, 1981).

MIRIFICARMA LENTIGINOSELLA (Zeller)

Gelechia lentiginosella Zeller, 1839, *Isis, Leipzig* **1839**: 198.
Type locality: Germany; Dresden.

Description of imago (Pl.3, fig.26)

Wingspan 12–17mm. Head mid- to dark brown; antenna dark brown; labial palpus dark brown with scattered cream scales, paler on inner surface and at apices of segments. Thorax and tegulae mid- to dark brown. Forewing dark brown with scattered pinkish buff scales; small pinkish spots on costa at two-thirds to three-quarters, sometimes extended to dorsal margin; three very small darker brown spots with ochreous surround, one on fold and one in cell, both approximately at one-third, one at end of cell, all spots indistinct. Hindwing and cilia greyish brown, paler than forewing. Legs dark brown flecked with cream, tibiae and tarsal segments cream at apices; inner side of hindleg cream

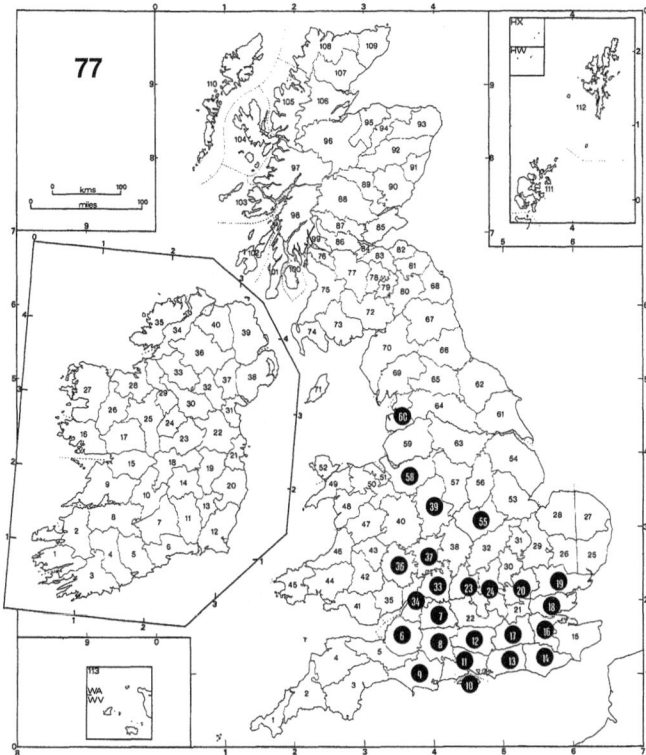

Mirificarma lentiginosella

to buff. Abdomen mid-brown. Male genitalia (figure 10d) with filament almost straight basally but with strong kink at apex, reaching hind edge of vinculum or extending slightly beyond; female genitalia (figure 31e) with extremely long antrum, more than twice to several times length of apophysis anterior.

Life history

Ovum. Oviposition site not observed. On dyer's greenweed (*Genista tinctoria*) (Stainton, 1865b). Outside Britain, foodplants recorded include petty whin (*G. anglica*), German greenweed (*G. germanica*), winged broom (*G. sagittalis*) (Sorhagen, 1886); also laburnum (*Laburnum anagyroides*) (Müller-Rutz, 1913–14).

Larva. Full-grown 10–12mm. Head and prothoracic plate black. Body yellowish green; pinacula minute, black.

Spins together the young terminal shoots of the foodplant (Stainton, *loc.cit.*). May and June.

Pupa. In a cocoon amongst leaves on the ground (Stainton, *loc.cit.*). June–July (Bradford, [1979]).

Imago. Flies from July to August in Britain. It is recorded elsewhere also in May (Mariani, 1943) and in June (Pitkin, 1984).

Distribution (Map 77)

On heathland and downland (Parsons, 1995[1996]). Fairly widespread but very local, occurring only in England. It is found in most southern counties from Somerset to Kent, thence through the Midlands to Staffordshire and Derbyshire, with an extension northwards on the west coast in Cheshire and Lancashire. It is distributed from southern Europe north to Sweden (Krogerus *et al.*, 1971), its range extending east to Turkey (Klimesch, 1961), Armenia (Rebel, 1901) and Russia (Piskunov, 1981).

PROLITA Leraut

Prolita Leraut, 1993, *Alexanor* **18**: 182 [replacement name for *Lita* Treitschke, 1833].

Lita Treitschke, 1833, *Schmett.Eur.* **9**(2): 76, *nec* Kollar, 1832 (praeocc.).

A Holarctic genus with over 20 species occurring in North America and two in Europe, both of which are found in the British Isles.

Imago. Head with appressed scales; ocelli present; antenna three-quarters to seven-eighths length of forewing, apical one-quarter slightly serrate in male, scape without pecten; labial palpus with segment 3 slightly shorter than segment 2 in *P. sexpunctella* (Fabricius), and slightly longer than segment 2 in *P. solutella* (Zeller), segment 2 thickened with scales, somewhat appressed in *P. sexpunctella*, roughened and furrowed beneath in *P. solutella*. Forewing elongate; vein R_3 (9) connate with joint stalk of veins R_4 (8) and R_5 (7). Hindwing broader than forewing, trapezoidal; apex slightly produced; termen sinuate; tornus gently rounded; cilia longer than one-half breadth of wing; veins R_5 (7) and M_1 (6) separate, M_3 (4) and CuA_1 (3) stalked.

Larva. One British species feeds on *Genista*, the other on *Calluna*.

Key to species (imagines) of the genus *Prolita*

–   Forewing with five irregularly shaped white fasciae, that at three-quarters angled outwards toward apex .... *sexpunctella* (p. 148)

–   Forewing with single indistinct whitish fascia at three-quarters perpendicular to costa .... *solutella* (p. 148)

PROLITA SEXPUNCTELLA (Fabricius)

Tinea sexpunctella Fabricius, 1794, *Ent.syst.* **3**(2): 313.

Tinea virgella Thunberg, 1794, *D.D.Diss.ent.sistens Insecta suecica* (7): 92, fig.10.

Anacampsis longicornis Curtis, 1827, *Br.Ent.* **4**: 189.

Gelechia longicornella Doubleday, 1859, *Zoologist synonymic List Br.Lepid.*: 30.

Type locality: Italy.

Description of imago (Pl.3, fig.27)

Wingspan 15–19mm. Head dark shining fuscous, irrorate with whitish scales at base of frons, collar fuscous; antenna dark fuscous, narrowly ringed buffish; labial palpus dark fuscous, irrorate with white scales, more densely at base. Thorax and tegulae dark fuscous, irrorate pale brown, sometimes with band of scattered whitish scales anteriorly. Forewing dark fuscous, more or less suffused pale reddish fuscous; five irregularly shaped white fasciae, first basal extending along dorsum to one-fifth, second oblique at one-quarter sometimes reaching dorsum, third at one-half, fourth somewhat incurved from apex to tornus, fifth slender, terminal; rather elongate blackish spots on fold at one-fifth and two-fifths sometimes confluent; blackish fuscous discal spots at two-fifths and three-fifths; broken line of dark fuscous scales at termen; cilia pale silvery grey. Hindwing greyish fuscous; cilia concolorous, with darker basal ciliary line. Legs dark fuscous, sprinkled white, more densely towards bases of femora; tibiae and tarsi ringed whitish. Abdomen fuscous to greyish fuscous with five transverse creamy white lines and scattered creamy white scales laterally. Genitalia, see figures 10f,32a.

Life history

Ovum. Laid on heather (*Calluna vulgaris*). In mainland Europe also on heath (*Erica* spp.), bilberry (*Vaccinium myrtillus*), cowberry (*V. vitis-idaea*) (Lhomme, [1946–49]), crowberry (*Empetrum nigrum*) (Benander, 1965) and possibly mosses (Piskunov, 1981).

Larva. Head, prothoracic plate and thoracic legs brown, heavily mottled black. Body greyish rust-brown with pale greyish dorsal and lateral lines; pinacula black; anal plate brown, finely dotted black; prolegs translucent, ringed pale brown.

Feeds in spun leaves and overwinters in a dense spinning. August–May.

Pupa. Undescribed. May.

Imago. Univoltine; May–June. Flies over heather on warm sunny days (Poynton, 1996). Observed flying over burnt patches among regenerating heather following a fire, as well as in dry rocky areas among short

herbage (Parsons, 1995[1996]). On one Lancashire moss the moth is restricted to the damper areas, particularly where cranberry (*Vaccinium oxycoccus*) grows.

Distribution (Map 78)

Occurs on heaths, moors and mosses and has been found at an altitude of 1200m (4000ft) (MacAlpine, 1979); occasionally numerous (Poynton, *loc.cit.*), possibly declining in some areas (Parsons, *loc.cit.*). In Herefordshire and on the Sheffield moors, Yorkshire, from which old records exist, the moth has not been found for many years (M. W. Harper and H. E. Beaumont, pers.comm.). Widely distributed in Wales, central and northern England, the Isle of Man and Scotland; discovered new to Ireland in 1977 (Dowling, 1979). Europe, except for Portugal and the Mediterranean islands; North America.

PROLITA SOLUTELLA (Zeller)

Gelechia solutella Zeller, 1839, *Isis, Leipzig* **1839**: 199.

Gelechia fumosella Douglas, 1852, *Trans.ent.Soc.Lond.* (N.S.) **1**: 241.

Type localities: Germany; Dresden, and Hungary.

Description of imago (Pl.3, figs 28,29).

Wingspan 16–21mm. Head, antenna and labial palpus dark fuscous, sprinkled whitish on frons and labial palpus. Thorax and tegulae dark fuscous, thorax variably irrorate with scattered greyish scales. Forewing fuscous, sprinkled greyish white from base to four-fifths, more densely in costal half; blackish spots on fold near base and at one-quarter, in disc at one-third, sometimes coalescing with that on fold at one-quarter, and just beyond one-half; indistinct blackish spots on costa and dorsum near base, that on dorsum sometimes extending as streak to one-fifth; sometimes dark fuscous mark on costa at one-half; whitish, frequently interrupted and often indistinct, fascia at three-quarters; scattered small blackish spots, sometimes interspersed with white scales, around termen sometimes forming indistinct line; cilia pale greyish, sprinkled with fuscous to blackish scales in basal one-third. Hindwing pale greyish fuscous; cilia concolorous. Legs fuscous to dark fuscous, sparsely ringed greyish fuscous. Abdomen dark reddish fuscous, each segment with scattered greyish scales posteriorly. Genitalia, see figures 10g, 32b.

Variation in forewing ground colour from pale greyish fuscous to dark greyish fuscous occurs within populations and between the two separate geographical populations. Specimens from Cornwall are generally paler than those from Scotland, where darker specimens predominate (fig.28).

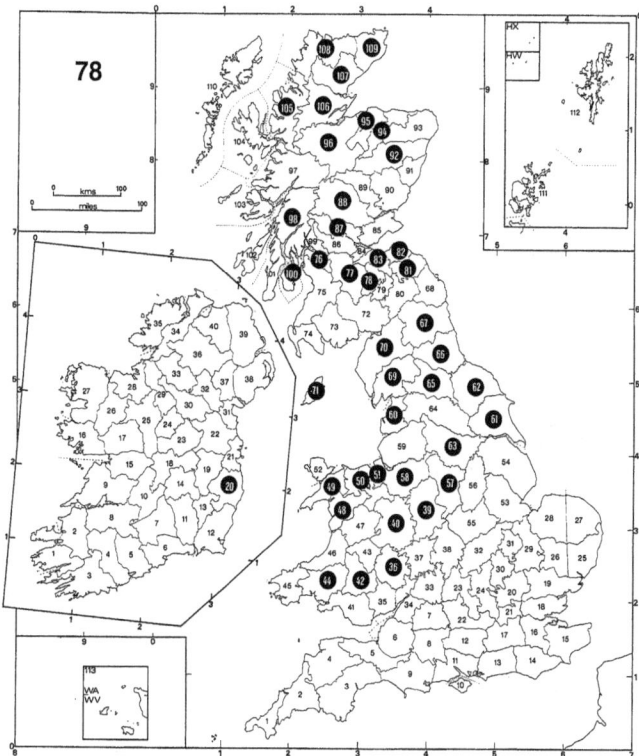

Prolita sexpunctella

Prolita solutella

Life history

Ovum. Laid on recumbent plants of petty whin (*Genista anglica*) or hairy greenweed (*G. pilosa*). Dyer's greenweed (*G. tinctoria*) has also been cited as a food-plant (Bradford [1979]) but probably in error (Heckford, 1999a). In mainland Europe noted on broom (*Cytisus scoparius*).

Larva. Head pale brown, dotted dark brown. Prothoracic plate pale greenish grey, dotted blackish; body pale greenish grey with broad, irregularly shaped, reddish brown longitudinal stripes; pinacula black; anal plate pale greenish grey finely dotted black; all legs pale greenish grey, thoracic legs finely dotted blackish.

On *Genista anglica*, the larva feeds from a dense vertical silken tube constructed in moss and dead plant material situated around the base of the foodplant, a finer silk tube extending upwards to the stem and leaves. It is found in August and September but it is not known whether the species overwinters in this stage or as a pupa.

Pupa. Undescribed.

Imago. Univoltine; May–July. Flies in sunshine (M. W. Harper, pers.comm.), making short flights close to the ground. Can be disturbed during mid-morning (R. J. Heckford, pers.comm.); at night has also been recorded at light (Parsons, 1995[1996]).

Distribution (Map 79)

Very local, occurring in dry, herb-rich, cattle-grazed pasture, dry grassland heath, and known also from a small heather-filled clearing (Parsons, *loc.cit.*). It has a disjunct distribution, occurring on the Lizard Peninsula in Cornwall, in northern England, although apparently not recorded in recent years, and in the central Highlands and north-eastern Scotland. Denmark, Sweden, Belgium, France, Spain, Portugal, south-eastern Europe, western Russia and Asia Minor.

AROGA Busck

Aroga Busck, 1914, *Proc. U.S.natn.Mus.* **47**: 13.

A Holarctic genus with over 20 species occurring in North America and five in Europe, one of which is found in Britain.

Imago. Head with appressed scales; ocelli present; antenna three-quarters length of forewing, apical two-fifths serrate, scape without pecten; labial palpus with segment 3 slightly longer than segment 2, segment 2 with furrowed brush of long scales beneath. Forewing elongate. Hindwing trapezoidal, slightly broader than forewing; apex slightly produced; termen nearly straight; tornus gently rounded; cilia about three-quarters breadth of wing; veins Rs (7) and M_1 (6) separate, M_3 (4) and CuA_1 (3) connate.

Larva. The British species feeds on sheep's sorrel (*Rumex acetosella*).

AROGA VELOCELLA (Zeller)

Gelechia velocella Zeller, 1839, *Isis, Leipzig* **1839**: 198.

Type localities: Germany; Glogau (now Poland; Glogów), Frankfurt and Berlin.

Description of imago (Pl.3, fig.30)

Wingspan 14–19 mm. Head ochreous to fuscous, paler on frons; antenna fuscous; labial palpus ochreous to fuscous, whitish ochreous above. Thorax and tegulae fuscous to dark fuscous, more or less sprinkled with ochreous to dark ochreous scales. Forewing with ground colour fuscous to dark fuscous, sometimes heavily overlaid with pale ochreous and sprinkled sandy brown and black, the latter particularly in apical one-third; veins overlaid whitish in paler specimens, darker specimens only showing this feature adjacent to tornal spot; tapered whitish streak on dorsum from base to tornus, almost obsolete in darker specimens; blackish spots on dorsum at base, in disc at two-fifths and three-fifths and on fold at one-third; scattered blackish scales on fold near base sometimes forming streak extending to one-third; whitish costal and tornal spots; sometimes distinct line of black spots on termen from apex to tornus; cilia grey to creamy white, sometimes with indistinct ochreous to fuscous ciliary line from below apex to tornus. Hindwing grey; cilia concolorous, with two indistinct darker lines in basal half, sometimes coalescent. Legs greyish fuscous, ringed dull creamy white. Abdomen greyish fuscous, ringed dull pale ochreous. Genitalia, see figures 10e,31f.

Life history

Ovum. Laid on sheep's sorrel (*Rumex acetosella*).

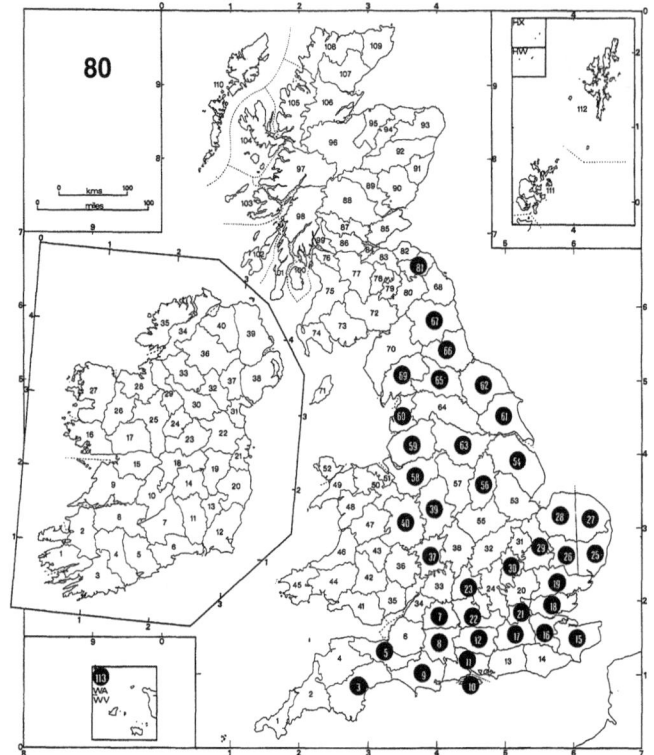

Aroga velocella

Larva. Head and prothoracic plate black. Body brownish yellow with several fine reddish longitudinal stripes; pinacula black (Spuler, 1910).

Feeds in a silken tube at the base of the stems under the lowest leaves (Snellen, 1882), sometimes in a silken tube leading up the stem to higher leaves (Tutt, 1905), the tube extending downwards into the soil (Piskunov, 1981). September–April; late May–early July.

Pupa. In the larval tube, under the ground (Piskunov, *loc.cit.*). April–May; July.

Imago. Bivoltine; May; August. Flies freely in warm weather during the afternoon making short darting flights close to the ground; later comes to light (Emmet, 1988b), occasionally several kilometres from the nearest suitable habitat.

Distribution (Map 80)

Found on breckland, heathland, moorland and in larger woodland clearings, often commonly for several years in areas where the foodplant proliferates following clear-felling. Widely distributed but local throughout England; in Scotland recorded only from Berwickshire; the Channel Islands. Unknown from Wales or Ireland. Throughout mainland Europe and Asia Minor.

NEOFRISERIA Sattler

Neofriseria Sattler, 1960, *Dt.ent.Z.* (N.F.) 7: 16–17 (keys), 48.

A small Palaearctic genus represented by three species in Europe, of which two are found in England.

Imago. Head with appressed scales; ocelli present; antenna three-quarters length of forewing in male, four-fifths in female, apical two-fifths to one-half serrate, scape without pecten; labial palpus with segment 3 slightly longer than segment 2. Forewing elongate. Hindwing trapezoidal, slightly broader than forewing; apex moderately produced; termen slightly sinuate; tornus gently rounded; cilia length equal to breadth of wing; veins Rs (7) and M$_1$ (6) stalked, M$_3$ (4) and CuA$_1$ (3) connate.

Larva. The British species both feed on sheep's sorrel (*Rumex acetosella*).

Key to species (imagines) of the genus *Neofriseria*

– Forewing blackish fuscous, with black spot in disc at two-fifths slightly distal to that on fold .... *peliella* (p. 151)
– Forewing greyish fuscous dusted yellowish white, with black spots at two-fifths vertically one above the other .... *singula* (p. 152)

NEOFRISERIA PELIELLA (Treitschke)

Lita peliella Treitschke, 1835, *Schmett.Eur.* 10(3): 198.
Type locality: Germany; Dresden.

Description of imago (Pl.4, fig.1)
Wingspan 13–16 mm. Head with vertex dark fuscous, frons pale ochreous at base, greyish fuscous above becoming dark fuscous towards vertex; antenna dark fuscous; labial palpus with segments 1 and 2 dark fuscous, sprinkled with pale ochreous more densely at base and on inner side, segment 2 narrowly pale above, segment 3 dark fuscous, apex whitish ochreous. Thorax and tegulae dark fuscous. Forewing blackish fuscous, each scale more or less greyish at base giving slightly mottled appearance; indistinct black spot on fold near base; two black spots at two-fifths, that in disc slightly distal to that on fold, more or less strongly edged distally with whitish scales; blackish spot in disc just beyond middle, usually edged with whitish scales; whitish spots on costa at four-fifths and at tornus, more or less extended to form slightly broken subterminal fascia; tornal spot slightly beyond that before

Neofriseria peliella

apex; cilia greyish fuscous, paler at tips on termen, darkened at three-quarters forming broken and indistinct ciliary line from apex to below tornus; underside with whitish ochreous spot on costa just before apex. Hindwing greyish fuscous, darker along margin and at apex; cilia greyish fuscous, pale-tipped at termen. Legs dark fuscous, more or less heavily sprinkled with pale whitish ochreous scales, tarsi ringed whitish ochreous; hind femur pale ochreous on outer side with concolorous long hairs above. Abdomen greyish fuscous, whitish ochreous on underside. Genitalia, see figures 10h,32c.

Similar species. N. singula (Staudinger) has the black spots at two-fifths vertically one above the other. In *N. peliella* the lower spot is usually slightly basad of the upper spot. *N. singula* has a dusting of dull yellowish white scales on the more greyish fuscous-tinged forewing, making it paler than *N. peliella*.

Life history
Ovum. Laid on sheep's sorrel (*Rumex acetosella*).

Larva. Head and prothoracic plate black. Body chocolate-brown; pinacula minute, black; anal plate dirty greyish buff with darker patches anterolaterally; thoracic legs black.

Feeds in a white silken tube at the base of the stems,

the lower leaves sometimes being drawn together around the stem. The larva retreats rapidly down its tube when disturbed. May–early June.

Pupa. Pale brown, turning dark brown before emergence of the moth, in a loosely spun cocoon amongst the lower leaves of the foodplant, under a stone or in debris on the ground. June.

Imago. Univoltine; July. Can be obtained by sweeping; has been taken at light (Parsons, 1995[1996]).

Distribution (Map 81)

A species confined to dry, open habitats on coastal shingle sites in south-east England, where it can be locally common. Mainland Europe; North Africa.

NEOFRISERIA SINGULA (Staudinger)

Lita singula Staudinger, 1876, *Stettin.ent.Ztg* **37**: 145.
Gelechia suppeliella Walsingham, 1896, *Entomologist's mon.Mag.* **32**: 250.
Lita peliella sensu Meyrick, 1895, *Handbk.Br.Lepid.*: 602, *nec* Treitschke, 1835.
Type locality: Sicily.

Description of imago (Pl.4, fig.2)

Wingspan 13–15 mm. Head greyish white with scales more or less tipped fuscous to dark fuscous, frons dull whitish, sprinkled fuscous near vertex; antenna dark fuscous, indistinctly ringed greyish fuscous; labial palpus dull whitish, sprinkled dark fuscous, particularly on segment 3. Thorax and tegulae fuscous to dark fuscous more or less tinged greyish. Forewing with ground colour fuscous to dark greyish fuscous, sprinkled dull yellowish white; indistinct dark fuscous spots at base of costa, along dorsum from base to one-fifth, near base on fold, and at one-third on costa; blackish fuscous spots in disc and on fold at one-third, one directly above the other, more or less encircled by pale yellowish to whitish scales; blackish fuscous spot in disc at two-thirds; indistinct dark fuscous line between discal spots at one-third and two-thirds, and sometimes another from spot on fold at one-third to tornus; dull yellowish white to whitish spots on costa at four-fifths and at tornus, sometimes extended to form broken angulate fascia, tornal spot slightly beyond costal; two small and indistinct dark spots on termen below apex with several dull yellowish white to white scales proximally; cilia greyish, paler terminally. Hindwing greyish fuscous, paler basally; cilia greyish to greyish fuscous. Legs dark fuscous, evenly ringed whitish. Abdomen shining pale greyish fuscous; anal tuft of male pale yellowish white. Genitalia, see figures 10i,32d.

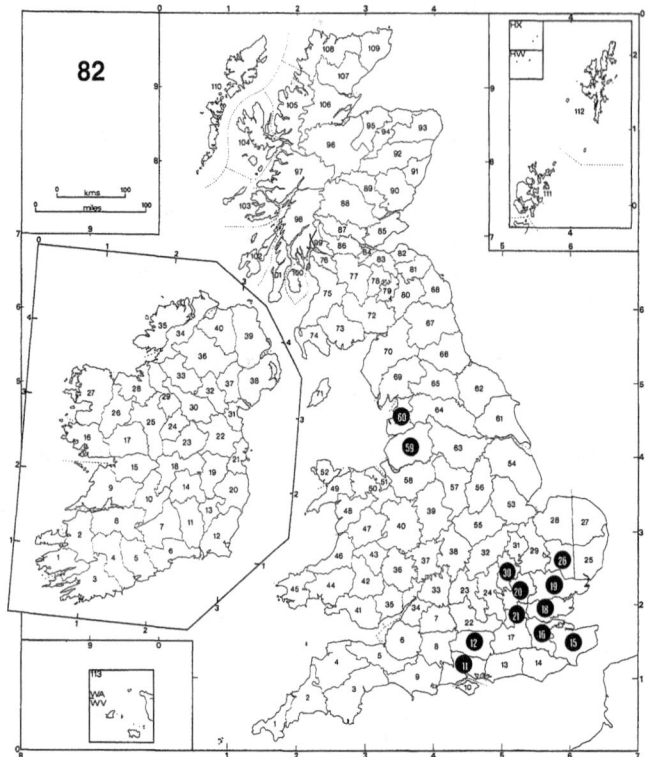

Neofriseria singula

Similar species. *N. peliella* (Treitschke), *q.v.Life history*

Ovum. Laid on sheep's sorrel (*Rumex acetosella*).

Larva. Head, prothoracic plate and first two pairs of thoracic legs black. Body pale dull green; pinacula minute and grey; anal plate concolorous with body, with a few dark fuscous markings.

Feeds in a silken tube at the base of the stems. A search at a site in 1995 revealed larvae only on plants growing on anthills (J. R. Langmaid, pers.comm.). May.

Pupa. Probably among detritus or in the soil. June.

Imago. Univoltine; June–July. Many moths were observed flying around and inside the mouths of rabbit-burrows at Lakenheath Warren, Suffolk (Parry, 1980); at night it has been taken at light (Parsons, 1995[1996]).

Distribution (Map 82)

Found on heathland and breckland, occurring mainly inland where it can sometimes be locally common. Although noted in the past from north-western England (Parsons, *loc.cit.*), all more recent records have been in the south and south-east from Hampshire and Kent northwards to Suffolk. Western Europe from Spain to Sweden; Austria, Hungary and Sardinia.

GELECHIA Hübner

Gelechia Hübner, [1825], *Verz. bekannt. Schmett.*: 415.

As originally described this genus contained a large number of species grouped together on the basis of similar wing venation and external characteristics. Most of these have now been placed in many other genera after discovery of genitalic differences. The genus (*sensu stricto*) contains ten British species, with ten others occurring in the western Palaearctic region. Further species are found in North America and southern Africa.

Imago. Labial palpus segment 2 with a brush of thickened scales beneath, in some species developed into a large tuft, often with distinct markings. Forewing with vein R_5 (7) to costa, R_4 (8) stalked with R_5 (7), M_3 (4) and CuA_1 (3) also stalked. Hindwing with veins Rs (7) and M_1 (6) stalked, M_3 (4) and CuA_1 (3) connate. Male genitalia with uncus hood-shaped, apex sometimes indented or bifid; alimentary canal with tegumen supported by two flattened rods. Female genitalia with short lateral lobes on genital plates, signum quadrangular or absent.

Larva. In spun shoots or between joined leaves. Five British species feed on Salicaceae, two on Rosaceae, one on Elaeagnaceae, and two, probably introduced, on horticultural species of *Juniperus* and *Chamaecyparis*.

Most species are local or rare, and although the imagines can be found on tree-trunks, beaten from the foodplants, or occasionally taken at light, they are seldom seen and are therefore probably under-recorded.

Key to species (imagines) of the genus *Gelechia*

1	Labial palpus segment 2 with large tuft of scales beneath	2
–	Labial palpus segment 2 with ventral scales not forming large tuft	4
2(1)	Labial palpus segment 2 outer surface divided into black basal and white distal halves .... *sabinellus* (p. 156)	
–	Labial palpus segment 2 outer surface not divided into black basal and white distal halves	3
3(2)	Labial palpus segment 2 outer surface dark brown with central white, sometimes fuscous-speckled, mark	*cuneatella* (p. 158)
–	Labial palpus segment 2 outer surface reddish brown with white central mark	*senticetella* (p. 155)
4(1)	Forewing with distinct black mark at base of costa	5
–	Forewing without distinct black mark at base of costa	6
5(4)	Forewing with stigmata large and black, second discal forming curved mark	*rhombella* (p. 153)
–	Forewing with stigmata small and indistinct	*hippophaella* (p. 159)
6(4)	Abdominal segments 1–3 dorsally ochreous-yellow	7
–	Abdominal segments 1–3 not dorsally ochreous-yellow	9
7(6)	Forewing densely suffused black	*nigra* (p. 160)
–	Forewing not densely suffused black	8
8(7)	Forewing with first discal stigma narrowly white-ringed; smaller grey species, wingspan 13–16mm	*muscosella* (p. 157)
–	Forewing with first discal stigma not white-ringed, although followed by white mark; larger browner species, wingspan 17–20mm	*turpella* (p. 161)
9(6)	Forewing with first discal stigma distinctly white-ringed	*sororculella* (p. 157)
–	Forewing with first discal stigma not white-ringed	*scotinella* (p. 154)

GELECHIA RHOMBELLA ([Denis & Schiffermüller])

Tinea rhombella [Denis & Schiffermüller], 1775, *Schmett. Wien.*: 139.

Recurvaria rhombea Haworth, 1828, *Lepid. Br.*: 549.

Type locality: [Austria]; Vienna district.

Description of imago (Pl.4, fig.3)

Wingspan 12–16mm. Head ochreous-grey; antenna ochreous, ringed black; labial palpus segment 2 with outer surface black, inner surface white, ventrally black, segment 3 ochreous, marked black. Thorax ochreous-grey. Forewing pale ochreous-grey with scattered pale brown scales; large black mark along base of costa; dark brown mark in disc near base; discal stigmata large and black with second discal forming a hook-shaped mark; plical stigma a tiny black spot; faint curved ochreous postmedian fascia; some black subterminal dots; cilia ochreous-white. Hindwing grey, darker towards apex; cilia grey. Legs pale ochreous, marked with brown. Abdomen ochreous-grey. Genitalia, see figures 11a,33a.

Life history

Ovum. Laid on apple (*Malus* spp.) or pear (*Pyrus communis*).

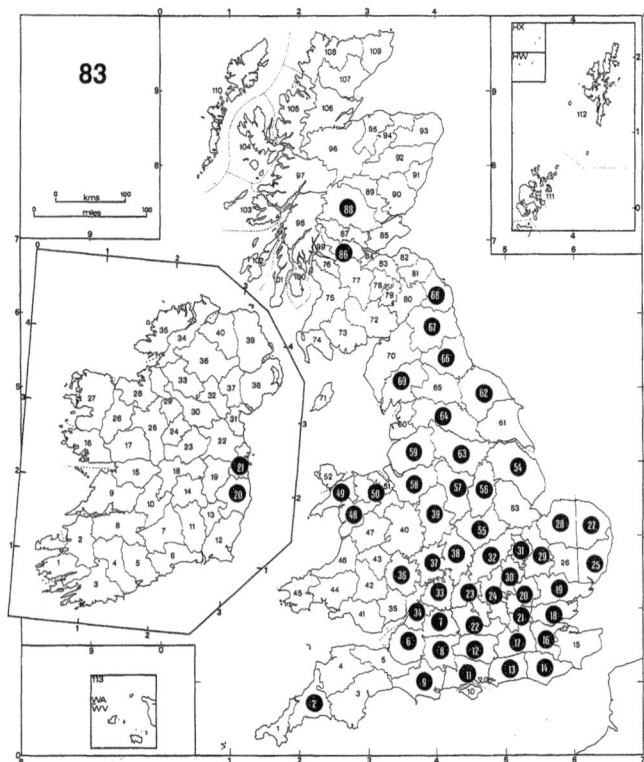

Gelechia rhombella

Larva. Head dark brown. Prothoracic plate black, divided in centre; body dark green, paler below, becoming tinged pink with paler subdorsal and spiracular lines.

In a turned-down leaf-edge, or flat spinning between two leaves. May–June.

Pupa. Pale brown, in a spinning between leaves, or in tissue in captivity. June–July.

Imago. Univoltine; July–August. Can be beaten from the foodplant but is more often
recorded at light.

Distribution (Map 83)

Uncommon but widespread, especially around old orchards, in southern and central England north to Perthshire in Scotland; North Wales; eastern Ireland (Beirne, 1941). Throughout most of Europe to Russia.

GELECHIA SCOTINELLA Herrich-Schäffer

Gelechia scotinella Herrich-Schäffer, 1854, *Syst.Bearb. Schmett.Eur.* **5**: 181; 1853, *Ibid.*: pl.68, fig.505 [non-binominal].

Type locality: Austria.

Description of imago (Pl.4, fig.4)

Wingspan 10–12mm. Head whitish grey with violet-tipped scales; antenna grey, ringed black; labial palpus segment 2 black on outer and ventral surfaces, white on inner surface, segment 3 ochreous-white, marked black. Thorax pale grey with violet tinge and dark, grey-tipped scales. Forewing pale grey, irrorate dark brown and violet; small black mark at base of costa; first discal stigma large, black, connected to second by large, quadrate pale mark; indistinct, slightly curved, white postmedian fascia; cilia grey. Hindwing grey with scattered darker scales; cilia grey. Legs black, marked white. Abdomen grey. Genitalia, see figures 11b,32f.

Life history

Ovum. Laid on blackthorn (*Prunus spinosa*).

Larva. Length 5mm. Head pale brown. Prothoracic plate greenish yellow, posteriorly edged dark brown; body pale green, pinacula concolorous; anal plate pale green.

According to Meyrick ([1928]), the larva feeds in spun flowers; however, searching and beating flowers was unsuccessful, but the species was twice reared from larvae found in spinnings in the tips of blackthorn shoots in late April and early May, just after the flowers had faded (A. N. B. Simpson, pers.obs.).

Pupa. Pale brown. In captivity in a spinning in tissue. May–June.

Imago. Univoltine; June–August. It has been beaten from the foodplant but most records are at light.

Distribution (Map 84)

Very local and rare, although possibly overlooked as it is not easily recorded. It is found in old uncut hedges and in blackthorn scrub. The first British specimens seem to have been reared from larvae found in shoots of blackthorn in a lane at Powick, near Worcester, in 1873 (specimens in Worcester museum collected by J. E. Fletcher). In 1976 a larva was found in the same lane, and another was found at Old Hills, Callow End, in 1996 (A. N. B. Simpson, pers.obs.); also in Worcestershire, single moths occurred at light at Bransford in 1986 and 1995, and another was beaten from blackthorn in 1993. Elsewhere it has been recorded from Borehamwood, Hertfordshire, in 1966, 1969 and 1971 (Bradford, 1978); from Laindon, south Essex, in 1975

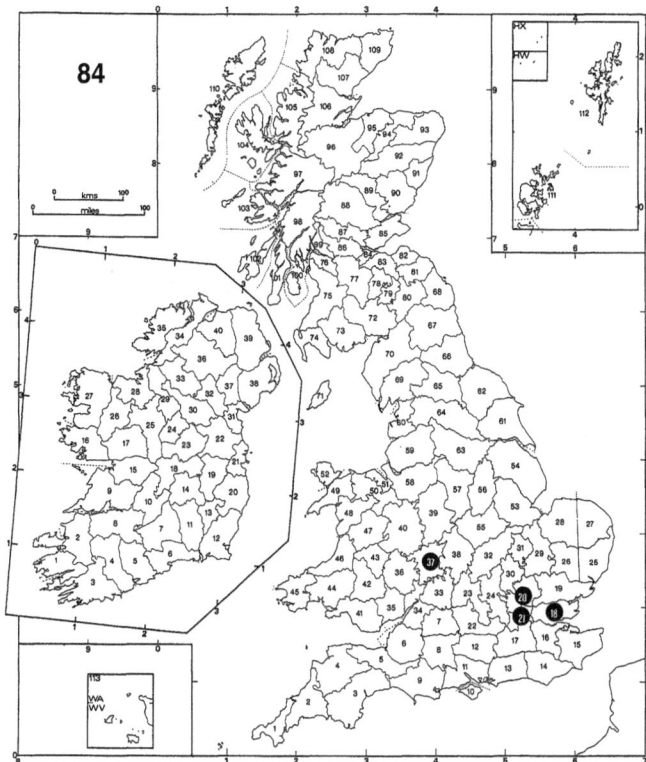

Gelechia scotinella

(Emmet, 1981); and from South Kensington, London, in 1999 (M. R. Honey, pers.comm). Records from Dorset, north Essex, Lancashire and Perthshire are unconfirmed. Europe, from France and Sweden to Greece and Russia.

GELECHIA SENTICETELLA (Staudinger)

Nothris senticetella Staudinger, 1859, *Stettin.ent.Ztg* **20**: 238.

Type locality: Spain; Chiclana, Andalusia.

Description of imago (Pl.4, fig.5)

Wingspan 10–13mm. Head red-brown with a few paler scales, frons creamy white; antenna dark brown, slightly serrate; labial palpus with segment 2 tufted beneath, outer surface red-brown, with white mark in middle superiorly, inner surface white, segment 3 red-brown with a few white scales. Thorax red-brown, mottled darker. Forewing ochreous, densely stippled red-brown; darker marks on costa and in disc at base; discal stigmata elongate, dark brown; dark brown dash in disc between stigmata, connected to them by white scales; plical stigma an ill-defined brown streak along fold with a few white scales in middle; some dark brown streaks between veins distally; termen ochreous beyond irregular line of red-brown scales; cilia ochreous-grey. Hindwing ochreous-grey; cilia ochreous. Abdomen red-brown with first three segments dorsally ochreous-yellow. Genitalia, see figures 11c,32c.

Life history

Ovum. Laid on juniper (*Juniperus* spp.), or cypress (*Chamaecyparis* and *Cupressus* spp.). In southern Europe recorded on *Juniperus excelsa*, *J. sabina*, *J. phoenicea*, *J. oxycedrus* and *J. thurifera* (Burmann, 1950). In France on *Cupressus sempervirens*, in Belgium on *Thuja* spp., and in England on *Chamaecyparis lawsoniana* (Heckford, 1999c).

Larva. Length 8mm. Head brown. Prothoracic plate green, with posterior margin black; body dark green, with paler intersegmental divisions; pinacula small, black and conspicuous; ill-defined paler green dorsal, subdorsal, and lateral lines; anal plate green; thoracic legs pale green.

Lhomme ([1946–49]) quotes Chrétien as having reared moths from the cones of cypresses, collected from the cemetery at Bize, France, in April and May. He was uncertain whether they had fed in the cones, or just pupated there. Almost certainly the latter, as larvae were found feeding on the foliage of junipers in Spain, in November 1997, and in spinnings in Lawson's cypress in south-east England in November 1997 and March 1998 (R. J. Heckford, *loc.cit.*). From Heckford's observations it would appear that the larva feeds slowly through the winter until the end of March or beginning of April.

The larva at first mines the leaves and later feeds in a tight silken tube between leaflets or developing flower buds which turn brown. It deposits greenish brown frass, which turns yellowish brown, on the edges of the tube and in the vicinity. The spinning is conspicuous, but the larva very cryptic (Heckford, *loc.cit.*).

Pupa. Length 5mm. Head brown; wings pale green, tips brown; thorax and abdomen yellow. In captivity in a strong white silken cocoon in tissue. April–May.

Imago. Univoltine; July–August. At night it comes to light in parks and gardens.

Distribution (Map 85)

Adventive; now resident. A southern European species probably introduced with the foodplants by the horticultural trade. The first record was at light at Grays, south Essex, on 6 August 1988 (Agassiz, 1989). There were further records on 23 July 1992 at Petts Wood, Kent (O'Keeffe, 1993); 3 August 1993 at Southsea, Hampshire (Langmaid, 1994); 13 July and 3 August 1994 at Raynes Park, London (Parsons, 1995); in 1996

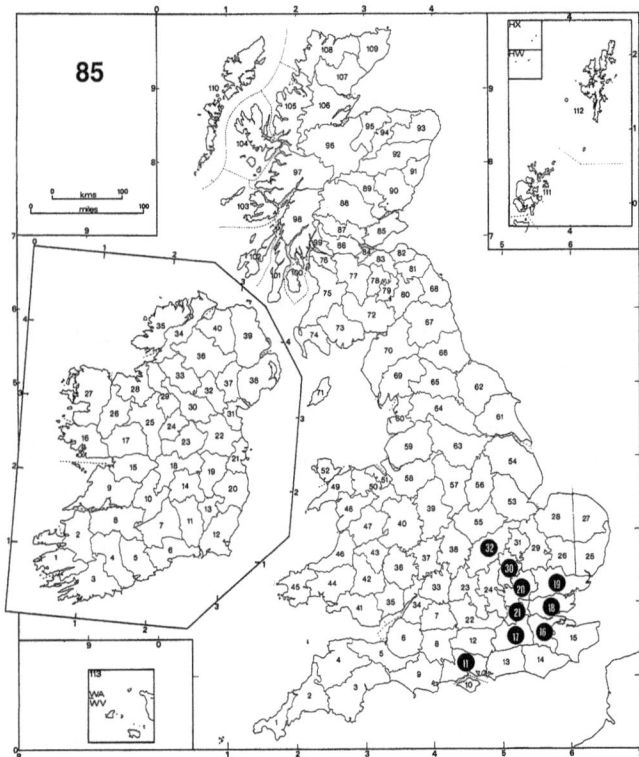

Gelechia senticetella

Gelechia sabinellus

and 1997 at Saffron Walden, north Essex (Emmet, 1997; 1998) and from Bedfordshire and Northampton-shire in 1997 and 1999 respectively (D. V. Manning, pers.comm.). In November 1997 larvae were found at Canvey Island, south Essex, and on 21 March 1998 at Dartford and Crayford, Kent (R. J. Heckford, *loc.cit.*). Spain, France, Belgium, Italy, the Balkans; Morocco.

GELECHIA SABINELLUS (Zeller)

Ypsolophus sabinellus Zeller, 1839, *Isis, Leipzig* **1839**: 190.
Type locality: Switzerland.

Description of imago (Pl.4, fig.6)

Wingspan 15–18mm. Head dark grey, mottled lighter; antenna grey, faintly ringed darker grey; labial palpus with segment 2 tufted beneath, outer surface divided into black basal and white distal halves, inner surface white, segment 3 dark grey with a few white scales. Thorax grey, mottled lighter. Forewing grey, darkened dorsally with fuscous scales; some scattered white scales in costal half; small black basal costal mark; long black streak in disc including discal stigmata, which are joined by a white mark; elongate plical stigma included in black streak along fold; some small black subterminal dots; cilia grey beyond line of fuscous-tipped scales. Hindwing pale grey, mottled darker; cilia pale grey. Legs pale grey with some fuscous scales. Abdomen grey. Genitalia, see figures 11e,33c.

Life history

Ovum. Laid on juniper (*Juniperus* spp.); probably on garden junipers in the British Isles.
Larva. Not recorded in the British Isles. 'Head honey-yellow with fine dark spots. Body green with dorsal and lateral lines darker or red. Feeding on *Juniperus sabina* in June' (Spuler, 1910).
Pupa. Not described.
Imago. Univoltine; August. Likely to be found in parks and gardens.

Distribution (Map 86)

Adventive; possibly now resident. Probably introduced on garden junipers by the horticultural trade. First recorded at light at Winchmore Hill, north London, on 28 August 1971 (Agassiz, 1978). A second specimen was found at Colletts Green, Worcestershire, on 30

August 1978, resting on an inside wall near a window, close to a *Juniperus squamata* bush in the garden, which was searched unsuccessfully for larvae the following year (Simpson, 1979). Spain, France, Belgium, The Netherlands, Scandinavia, Germany, Switzerland; Austria as subsp. *hofmanniella* Strand, and Corsica as subsp. *corsella* Rebel.

GELECHIA SORORCULELLA (Hübner)

Tinea sororculella Hübner, [1817], *Samml.eur.Schmett.* **8**: pl.68, fig.440.

Type locality; Germany, Augsburg.

Description of imago (Pl.4, fig.7)

Wingspan 13–15mm. Head brown; antenna dark brown, faintly ringed black; labial palpus segment 2 with outer surface dark brown with a few white scales, inner surface white, segment 3 dark brown with pale tip. Thorax dark brown. Forewing dark brown with a few scattered white scales; white mark in disc near base; stigmata black, first discal distinctly white-ringed and connected to second discal by black streak, white-edged below; plical stigma white-edged, included in a black streak along fold; slightly angulate white postmedian fascia; black subterminal dots inwardly edged white; cilia brown with darker tips. Hindwing grey; cilia pale grey. Legs dark brown, marked white. Abdomen fuscous, paler ventrally. Genitalia, see figures 11f,33b.

Life history

Ovum. Laid on *Salix* spp., usually goat willow (*S. caprea*) or grey willow (*S. cinerea*).

Larva. Head brown. Prothoracic plate pale brown, anteriorly pale grey; body varies from pale grey to pale green; a narrow brown dorsal line; broad reddish brown subdorsal lines; pinacula dark brown, ringed white; anal plate yellowish, marked black.

In a spinning among the leaves and sometimes in female catkins. May.

Pupa. Pale brown; in a spinning among leaves, or in tissue in captivity. May–June.

Imago. Univoltine; July–August. Can be beaten from sallow bushes or found on the trunks of large old sallows.

Distribution (Map 87)

Fairly common and widespread throughout Britain to Sutherland; the Channel Islands; the Isle of Man; local or under-recorded in Ireland. Throughout central and northern Europe to Poland and the Balkans.

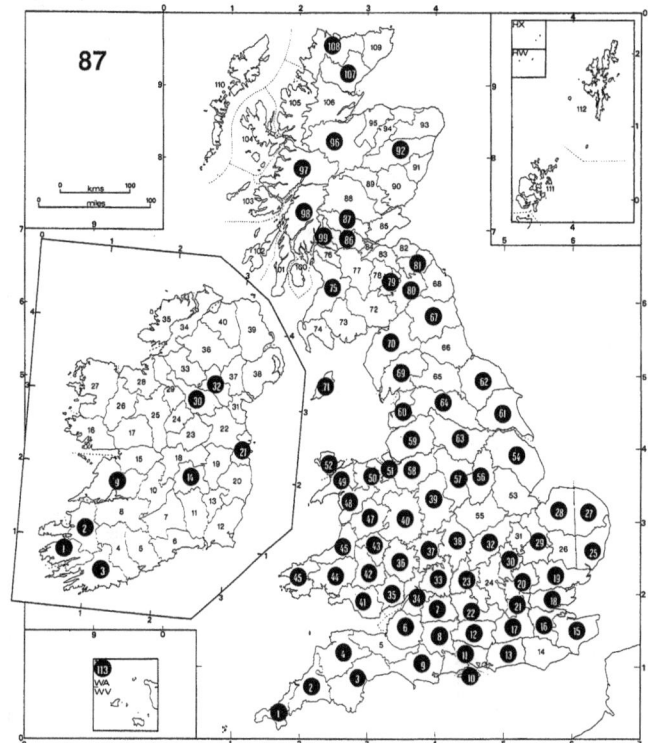

Gelechia sororculella

GELECHIA MUSCOSELLA Zeller

Gelechia muscosella Zeller, 1839, *Isis, Leipzig* **1839**: 197.

Type locality: Germany; Glogau (now Poland; Głogów).

Description of imago (Pl.4, fig.8)

Wingspan 12–17mm. Head dark grey with paler scales, frons white; antenna dark brown, ringed pale brown; labial palpus segment 2 with outer surface dark brown speckled with white, inner surface white, segment 3 dark grey with white scales. Thorax pale grey, mottled darker. Forewing white, densely suffused fuscous and black; indistinct dark costal marks at base, one-third and two-thirds; black subcostal dash near base; first discal stigma small, black ringed with white, connected to second discal stigma by black streak; plical stigma included in black streak along fold, interrupted by white mark; indistinct angulate white postmedian fascia; some subterminal black spots; cilia white, beyond dark ciliary line. Hindwing grey, mottled; cilia pale grey. Legs fuscous with white markings. Abdomen fuscous, segments 1–3 dorsally ochreous-yellow. Genitalia, see figures 11d,33f.

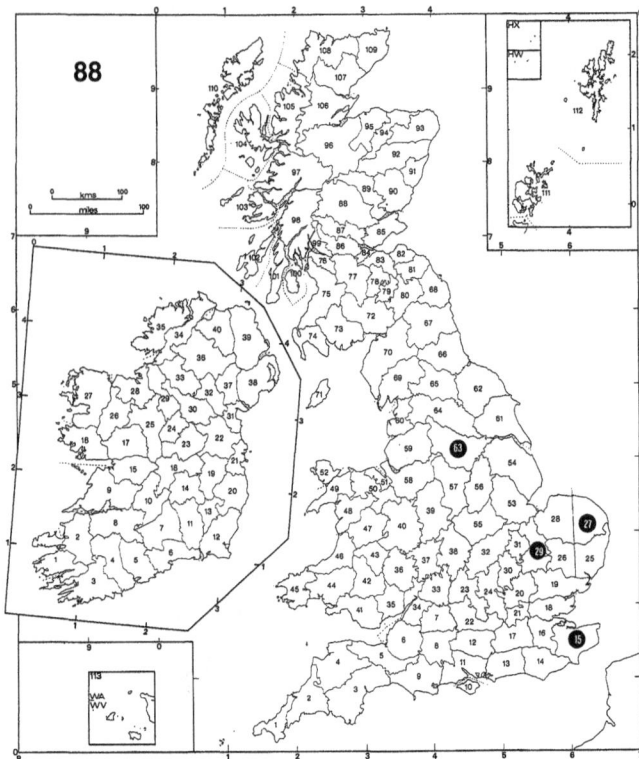

Gelechia muscosella

Life history

Ovum. Laid on sallows (*Salix* spp.) or, possibly, poplars (*Populus* spp.)

Larva. According to Spuler (1910), the larva is whitish, having an indistinct brownish dorsal line, with prothoracic plate dark brown and feeds in the catkins of goat willow (*Salix caprea*) in the spring. Meyrick ([1928]) repeats this description and states that it feeds in the catkins of poplars and sallows in May. Lhomme ([1946–49]) states that it rolls leaves of sallows after leaving the catkins. Large amounts of female sallow catkins and spinnings in sallow were collected at Dungeness in May 1989, but no moths were reared. Male sallow and aspen catkins collected in April 1996 were similarly unproductive. On 22 May 1997, a larva which matched Spuler's description was found at the same site in a terminal spinning of grey willow (*S. cinerea*). Unfortunately it was parasitized and therefore identification could not be confirmed but a description of it follows. Length 9mm. Head brown. Prothoracic plate pale brown, with black spots; body whitish, faintly tinged green, with indistinct pale brown, slightly interrupted, dorsal and subdorsal lines; anal plate whitish; thoracic legs pale brown.

Pupa. Not described.

Imago. Univoltine; June–July. Comes to light. In July 1997 at Dungeness many were found resting on the trunks of sallows, whence they were easily disturbed but very elusive and difficult to catch (A. N. B. Simpson, pers.obs.).

Distribution (Map 88)

Very local and rare. At the BMNH there are nine specimens in the Bankes collection labelled 'coll. by W. Farren Wicken 1889 & 1890', and one in the Stainton collection labelled 'coll. by W. Farren Cambridge 1889' which may refer to the same site (Wicken Fen, Cambridgshire). There are also old records from Ranworth, east Norfolk (Barrett, 1901). In recent years it has been recorded only from Wicken Fen, Cambridgeshire (Fairclough & Fairclough, 1980), Dungeness and Whitstable, east Kent (Agassiz, 1984, 1990; A. N. B. Simpson, pers.obs.) and at Blacktofts Sands Nature Reserve, south-west Yorkshire in 1998 (H. E. Beaumont, pers.comm.). France, Germany, Poland, Romania.

GELECHIA CUNEATELLA Douglas

Gelechia cuneatella Douglas, 1852, *Trans.ent.Soc.Lond.* (N.S.) **1**: 242.

Type locality: England; London.

Description of imago (Pl.4, fig.9)

Wingspan 13–16mm. Head pale grey with fuscous-tipped scales; antenna dark brown, faintly ringed pale brown; labial palpus tufted beneath, segment 2 with outer surface blackish brown with white central mark, inner surface white. Thorax pale grey with fuscous-tipped scales. Forewing pale grey with scattered dark brown and ochreous scales; black spot at base of costa; some indistinct black costal marks; black mark in disc near base; discal stigmata black joined by a white mark; plical stigma a black dash; slender angulate white postmedian fascia; some black subterminal spots; cilia ochreous-white. Hindwing grey; cilia ochreous-white. Abdomen fuscous, segments 1–3 dorsally ochreous-yellow. Genitalia, see figures 11h,33d.

Life history

Ovum. Laid on willows (*Salix* spp.); white willow (*S. alba*) and crack willow (*S. fragilis*) have been recorded in the past.

Larva. Undescribed. June and July. Two were reared from 'willow' collected on Hackney Marshes, east London (Emmet, 1981).

Pupa. Not described.

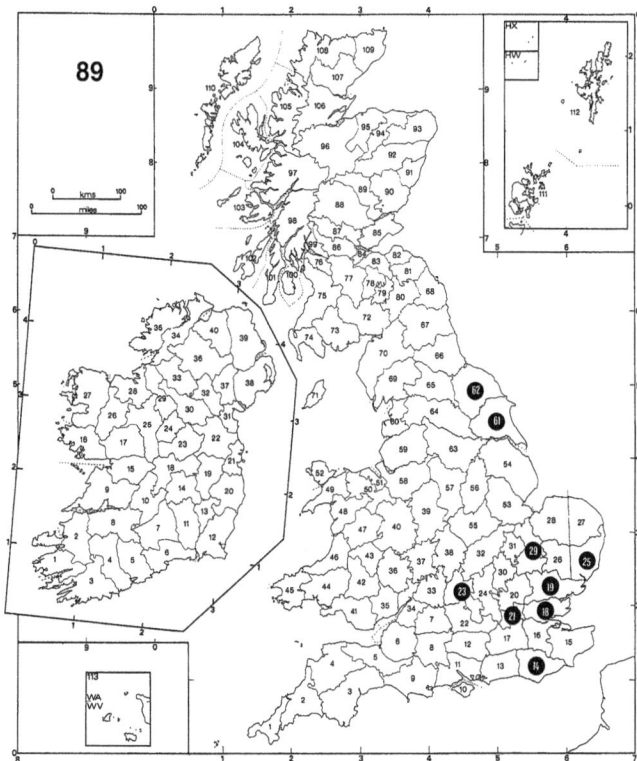

Gelechia cuneatella

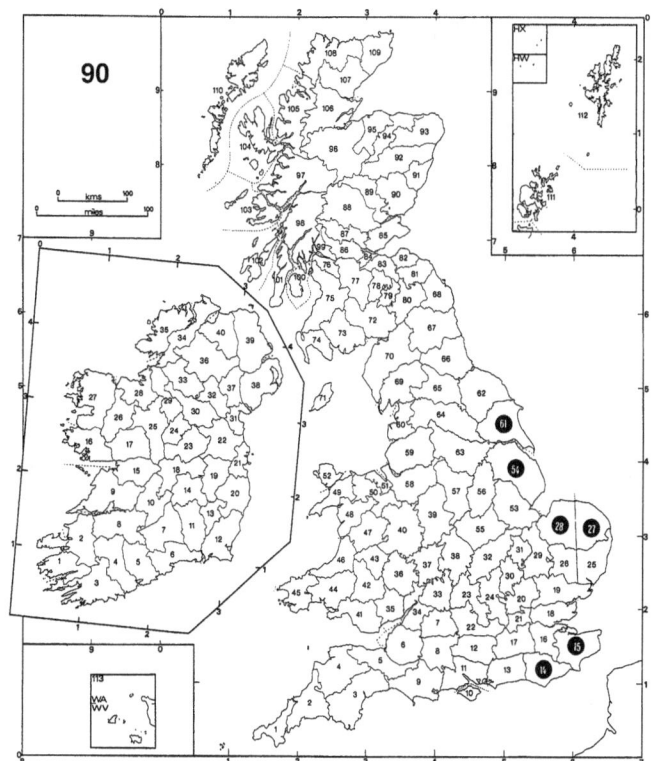

Gelechia hippophaella

Imago. Univoltine; August–September. Has been found resting on a willow trunk.

Distribution (Map 89)

Very rare. No recent record. First found in 1853 in south Essex on Hackney Marshes and by the banks of the river Lea (Stainton, 1854), and there are other old records from Colchester, north Essex, in 1911 (Emmet, *loc.cit.*), Oxfordshire (Waters, 1929), Beccles, east Suffolk in 1905 (Morley, 1937), Cambridgeshire (Fryer & Edelsten, 1938), Allerthorpe, south-east Yorkshire, in 1948, and Strensall, north-east Yorkshire, in 1950 (Sutton & Beaumont, 1989). Might be found again if looked for on willows in the eastern counties. France, Denmark, Germany, Austria and Russia.

GELECHIA HIPPOPHAELLA (Schrank)

Tinea hippophaella Schrank, 1802, *Fauna Boica* **2**(2): 115.

Tinea basalis Stainton, 1854, *Ins.Br.Lepid.*: 105.

Type locality: Germany; Bavaria.

Description of imago (Pl.4, fig.10)

Wingspan 15–18mm. Head ochreous-grey; antenna dark brown, indistinctly ringed pale brown; labial palpus segment 2 with outer surface ochreous-grey, speckled dark brown, inner surface ochreous-white, segment 3 ochreous, ringed pale brown. Thorax ochreous-grey. Forewing pale ochreous-grey with scattered pale brown and a few dark brown scales; distinct large black spot along base of costa; small black spot in disc at base; stigmata small, black and indistinct; faint pale ochreous postmedian fascia; some small black subterminal spots; cilia ochreous-grey. Hindwing and cilia ochreous-grey. Legs dark brown, marked darker brown. Abdomen pale grey with segments 1–3 dorsally ochreous-yellow. Genitalia, see figures 11g,33e.

Life history

Ovum. Laid on sea-buckthorn (*Hippophae rhamnoides*).

Larva. Head pale brown, speckled darker brown. Prothoracic plate greenish grey, speckled pale brown; body pale greenish grey, darker dorsally, with pale spiracular line; pinacula small, black and distinct; anal plate greenish grey, marked darker green. Larvae become tinged with pink in last instar.

Feeds between leaves of the terminal shoots spun tightly and flatly together. End of June and July.

Pupa. Pale brown. Between spun leaves of foodplant;

in captivity, in tissue. July–August

Imago. Univoltine; late August–September. Occasionally comes to light. When disturbed conceals itself on the ground amongst herbage.

Distribution (Map 90)

Rare and very local. Occurs amongst the foodplant on sandhills by the sea. Recorded only from Camber, east Sussex (Wakely, 1956; A. N. B. Simpson in 1989, pers.obs.), Deal, Folkestone and Sandwich, east Kent (Parsons, 1995[1996]), Hemsby, east Norfolk (Wakely, 1951), Holme, west Norfolk (Parsons, *loc.cit.*), Gibraltar Point to Saltfleetby, north Lincolnshire (Parsons, *loc.cit.*), and Spurn, south-east Yorkshire (Sutton & Beaumont, 1989). Denmark, Germany, Austria.

GELECHIA NIGRA (Haworth)

Recurvaria nigra Haworth, 1828, *Lepid.Br.*: 550.
Ypsolophus cautella Zeller, 1839, *Isis, Leipzig* **1839**: 200.
Type locality: England.

Description of imago (Pl.4, fig.11)

Wingspan 13–18mm. Head mixed grey and black, frons white; antenna dark fuscous, faintly ringed darker; labial palpus segment 2 black with a few white scales, segment 3 black. Thorax blackish grey. Forewing fuscous, densely suffused black and with a few white scales; stigmata indistinct, included in black longitudinal streak in disc; black plical streak; faint angulate white postmedian fascia; some white subterminal dots; cilia pale grey. Hindwing grey; cilia pale grey. Legs fuscous with white markings. Abdomen dark fuscous with segments 1–3 dorsally ochreous-yellow. Genitalia, see figures 12b,34a.

Life history

Ovum. Laid on aspen (*Populus tremula*), white poplar (*P. alba*) or grey poplar (*P. canescens*).

Larva. Head brown. Body green (Meyrick, [1928]).

Feeds in a flat spinning between leaves of the foodplant. May and June.

Pupa. Undescribed.

Imago. Univoltine; June–July. The moths come to light and have been found resting on aspen trunks.

Distribution (Map 91)

Uncommon and local. Larvae have been found on aspen in woods, and on white poplar along river banks. Recorded from south-east England north to Yorkshire and west to Somerset, Worcestershire and Cornwall (Smith, 1997). North-western and central Europe.

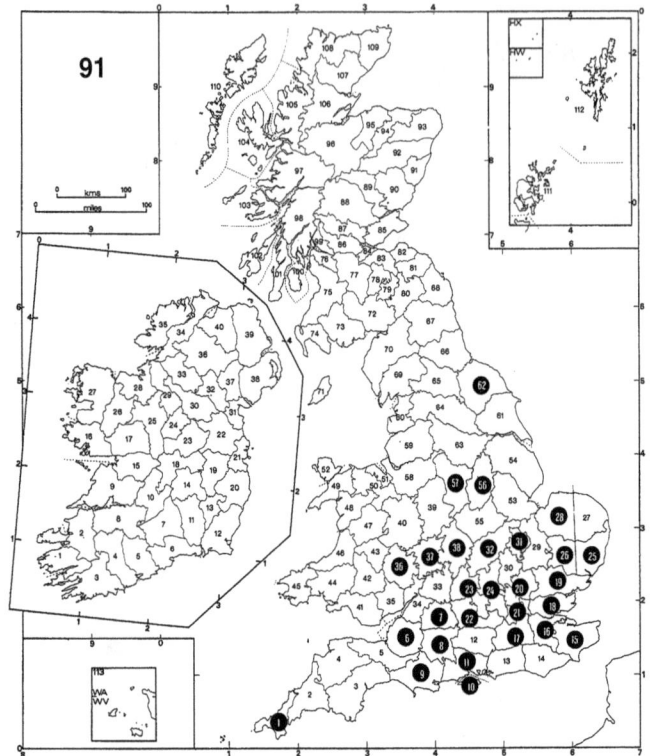

Gelechia nigra

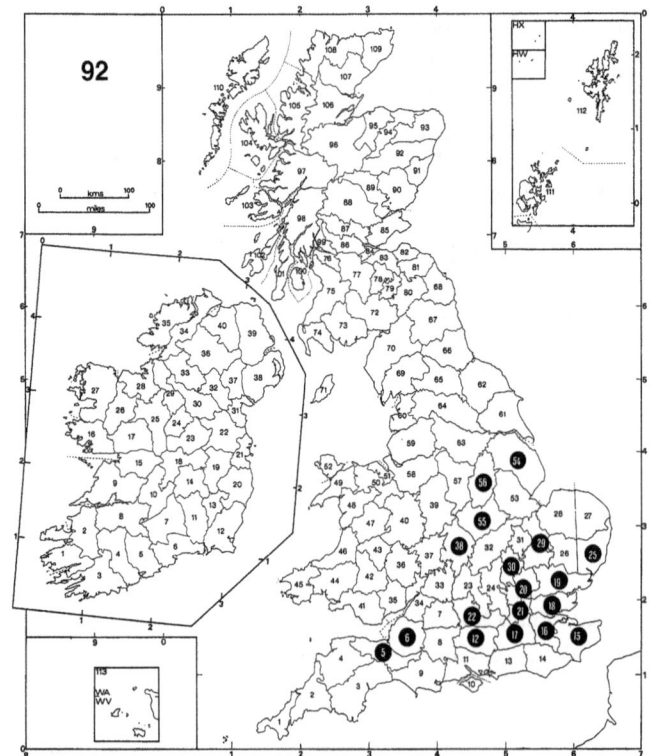

Gelechia turpella

GELECHIA TURPELLA [Denis & Schiffermüller]

Tinea turpella ([Denis & Schiffermüller]), 1775, *Schmett. Wien.*: 139.
Haemylis pinguinella, Treitschke, 1832, *Schmett.Eur.* **9**(1): 244.
Type locality: [Austria]; Vienna district.

Description of imago (Pl.4, fig.12)

Wingspan 17–20mm. Head brown with some paler brown scales; antenna brown, ringed black; labial palpus segment 2 with outer surface dark brown with a few ochreous scales, inner surface pale ochreous-brown, segment 3 brown, marked white. Thorax mixed brown and ochreous. Forewing dark brown with many scattered ochreous-brown scales; dark marks at base and at middle of costa; stigmata blackish brown; black dash between discal stigmata joined to them by ochreous scales; pale ochreous angulate postmedian fascia; cilia pale brown. Hindwing fuscous; cilia ochreous-grey. Legs dark brown, marked white. Abdomen fuscous, segments 1–3 dorsally ochreous-yellow. Genitalia, see figures 12a,34b.

Life history

Ovum. Laid on black poplar (*Populus nigra*) or Lombardy poplar (*P. nigra* var. *italica*); it may also feed on *Salix* spp., as moths have recently been found resting on a willow trunk.

Larva. Length 10–11mm. Head dark brown, mottled black. Prothoracic plate pale brown, divided in centre, with black lateral marks; body plump, pale green with paler green subdorsal lines; pinacula small, black and inconspicuous; anal plate pale green, marked black; thoracic legs black with brown tips.

According to Lhomme ([1946–49]), it feeds in May to June between joined leaves of poplar. On 19 May 1997, numbers of larvae were found under the bark of old Lombardy poplars at Bexley, Kent, having apparently just moved there to pupate. On the lower branches of the trees were many leaves which had been spun upwards into pods, with only the inner surfaces eaten. These seemed very likely to have been the feeding sites, but they had all just been vacated (A. N. B. Simpson, pers.obs.).

Pupa. Pale brown with long tapering abdomen. Most easily found under loose bark. Many were found thus on old Lombardy poplars on 29 June 1989 at Bexley, spun together gregariously in dense white silken cocoons incorporating black debris. There was no frass to suggest they had fed *in situ*, and it would appear that they descended the tree to pupate together *en masse*.

Imago. Univoltine; July. Can be found resting on the trunks of old poplars but is very difficult to catch as it runs up the trunk with great rapidity. At night the moth has been taken at light.

Distribution (Map 92)

Rare and very local in south-eastern England. In recent years recorded regularly only near Bexley, west Kent, but, in the past, its distribution extended north to Nottinghamshire and west to Somerset. Isolated specimens have been recorded from Berkshire (Baker, 1994), Bedfordshire (Arnold *et al.*, 1997) and, in west London, from the gardens of Buckingham Palace (Bradley & Mere, 1964) and the garden of the Natural History Museum, South Kensington (Parsons, 1995[1996]). It was also found at Petts Wood, Kent, in 1996 on the trunk of an old crack willow (*Salix fragilis*) (D. O'Keeffe, pers.comm.). Spain, France, Belgium to central and eastern Europe.

SCROBIPALPA Janse

Scrobipalpa Janse, 1951, *Moths of South Africa* **5**(3): 199.
Ilseopsis Povolný, 1965, *Acta ent.bohemoslovaka* **62**: 481.

A large genus containing over 250 species, with representatives in North America, Africa and Australia. The genus is particularly diverse in Asia Minor. Over 60 species have been recorded in Europe (Povolný, 1996), of which fifteen species are known from the British Isles, with a further species dubiously recorded from the Channel Islands. Several are widespread and many are distinctly coastal in distribution. A few are very local or recorded only from a few examples in the British Isles. The species are here arranged in a different order from that of Bradley (1998) to reflect affinities in life history.

Imago. Head smooth-scaled; ocelli present; antenna simple, scape without pecten; labial palpus about twice diameter of eye, upturned, segment 2 roughly scaled below, so as to form two ridges, with groove in between, segment 3 about as long as segment 2, smooth-scaled (figure 56). Forewing broadly lanceolate, width about one-quarter of length; costa more or less evenly curved; termen evenly curved, no pronounced tornus; cell about two-thirds length of wing,

Figure 56 *Scrobipalpa obsoletella* (Fischer von Röslerstamm), side view of head showing palpus

closed; venation standard. Hindwing subtrapezoidal, width about one-fifth length; termen oblique, moderately excavated beneath apex; tornus broadly rounded; cell about one-half length of wing; venation standard, R_s (7) and M_1 (6) separate, M_3 (4) connate or slightly stalked with CuA_1 (3). Hind tibia fringed with hairs above, outer spurs about half length of inner.

Ovum. The oviposition site has been recorded for only two of the species occurring in the British Isles; however, oviposition probably takes place on the hostplant in all cases.

Larva. Generally green or greenish with varying degrees of rosy to reddish brown suffusion. Larvae of different species have different habits; many mine leaves, particularly in the early instars, several loosely spin leaves and seeds together to form a larval habitation, others feed in stems, whilst the larva of one species feeds from within the root of the foodplant. The larvae of some species can be locally abundant.

Pupa. The pupation site varies from species to species. Some pupate on the hostplant, either in the stem or within the larval habitation, others leave the larval feeding site and pupate amongst litter or on the ground.

Imago. Some species are attracted to light and some have been found with the aid of a bee-smoker. Several are bivoltine. The forewing pattern of many species is very variable. If disturbed adults will often run to find a hiding-place rather than fly.

Key to species (imagines) of the genus *Scrobipalpa*

NOTE. The adults of many species of *Scrobipalpa* are very variable and can be difficult to determine, particularly if worn. The key is meant as an aid to the identification of the more frequently encountered forms, but reference should also be made to the key to genitalia, the description, illustrations, the habitat and, where known, the hostplant as an aid to identification. Although typical habitats are given in some sections of the key, it must be borne in mind that individuals can wander some distance from where they emerged.

1   Forewing with substantial subtriangular costal blotch, visible even in darker examples .... *costella** (p. 182)
–   Forewing without subtriangular costal blotch 2

---

* *S. hyoscyamella* (Stainton) would also key out at this point but, as this species is only very doubtfully thought to occur on the Channel Islands, it has not been included in the key. Differences between *S. hyoscyamella* and *S. costella* (Humphreys & Westwood) are given in the text for *S. hyoscyamella*.

2(1)   Wingspan 10mm or less. Forewing grey with dense fuscous scales at apex; stigmata large .... *murinella* (p. 168)
–   Wingspan usually more than 10mm. If ground colour of forewing grey then fuscous scales at apex, if present, only sparse 3

3(2)   Forewing relatively uniform, brown; if appearing speckled, then speckling generally evenly distributed across forewing or becoming more dense towards apex; stigmata frequently obscure or small 4
–   Forewing variegated, appearing mottled, marbled or speckled, or greyish; if speckled, then this unevenly distributed across forewing 8

4(3)   Forewing without reddish brown coloration, veins usually distinctly paler. A coastal species of sandy or shingle beaches .... *clintoni* (p. 181)
–   Forewing reddish brown or with reddish brown streaks; if not then veins not paler 5

5(4)   Larger species, wingspan 14–16mm. Forewing with stigmata often forming dark streaks, obscure dark streaks sometimes present at apex. A coastal species .... *stangei* (p. 180)
–   Smaller species, wingspan less than 14mm 6

6(5)   Wingspan 10–11mm. Forewing usually with stigmata clearly visible, and often with dark streak at apex; ground colour brownish with greyish white scaling more distinct and appearing heavier towards apex. A species of sandhill or limestone habitats .... *artemisiella* (part) (p. 179)
–   Wingspan 10–14mm. Stigmata of forewing usually obscure, no apical streak present; ground colour dark greyish brown with grey scaling finely scattered over wing 7

7(6)   Wingspan 10–14mm. See genitalia differences .... *acuminatella* (p. 166)
–   Wingspan 10–13mm. See genitalia differences .... *pauperella* (p. 167)

8(3)   Wingspan 10–11mm. Forewing usually with stigmata clearly visible; brown ground colour, which is not variable in shade, evenly distributed over wing, with greyish white scales more distinct and appearing heavier towards apex; often a dark streak at apex. A species of sandhill or limestone habitats .... *artemisiella* (part) (p. 179)
–   Wingspan more than 11mm. Forewing with ground colour brownish, not evenly distributed across forewing, or grey 9

9(8)   Forewing appearing pale; ground colour uniformly pale greyish buff, stigmata usually contrasting well with ground colour. Foreleg usually appearing darker than ground colour of forewing. Predominantly a coastal species .... *obsoletella* (p. 175)

– Forewing ground colour not pale greyish buff; if greyish then with ochreous and reddish brown markings 10

10(9) Forewing with distinct pale area along dorsum, with or without slight fuscous speckling within pale area 11

– Forewing with or without pale area along dorsum; if with such a marking, then with distinct fuscous speckling within pale area 12

It can be particularly difficult to identify some specimens from this point. Again, reference to the genitalia and taking into account the description, illustrations, the hostplant and the habitat will aid determination.

11(10) Dorsum of forewing contrasting strongly with general wing coloration; stigmata usually distinct. A salt-marsh species .... *suaedella* (part) (p. 169)

– Pale area of dorsum rather obscure, not contrasting strongly with forewing coloration; stigmata obscure or distinct. A coastal species .... *samadensis* (part) (p. 179)

12(10) Forewing with pale angulate fascia at about three-quarters, contrasting distinctly with rest of forewing. A species of coastal shingle and the upper margins of salt-marshes .... *ocellatella* (p. 177)

– Forewing without such distinctly contrasting marking 13

13(12) Forewing mottled, dark grey to brownish with fuscous scales, these usually more dense at wing apex; stigmata usually indistinct because of variable fuscous scaling. A species of disturbed habitats .... *atriplicella* (p. 176)

– Forewing not usually mottled; stigmata usually distinct, or forewing with large reddish brown blotches 14

14(13) Forewing with reddish brown frequently large blotches, giving a marbled appearance; stigmata obscure. A salt-marsh species ....*salinella* (part) (p. 170)

– Forewing with stigmata usually distinct, if not, then without large reddish brown blotches 15

15(14) Ground colour of forewing dark greyish buff, evenly speckled with fuscous scaling; stigmata well marked; fine reddish brown streaks over wing, this coloration sometimes encircling stigmata. A species of salt-marshes .... *nitentella* (p. 173)

– Ground colour of forewing not usually dark greyish buff, if so, then quite heavily marked reddish brown 16

16(15) Forewing heavily marked with dark fuscous scales and reddish brown streaks, giving a variegated appearance. A coastal species .... *samadensis* (part) (p. 179)

– Forewing not usually with combination of dark fuscous scales and reddish brown streaks; if so, then of a mottled appearance and reddish brown streaks usually short 17

17(16) Forewing with reddish brown streaks; ground colour greyish buff, wing evenly and finely speckled fuscous; stigmata usually small. A salt-marsh species .... *suaedella* (part) (p. 169)

– Forewing with or without reddish brown streaks; if ground colour greyish buff, then wing not finely and evenly speckled fuscous 18

18(17) Forewing with reddish brown streaks; ground colour greyish buff. A coastal species .... *samadensis* (part) (p. 179)

– Forewing without reddish brown streaks, if present then streaks short 19

19(18) Forewing with mottled appearance, sometimes with large dark fuscous markings. A species of salt-marshes and saline situations .... *instabilella* (p. 171)

– Ground colour of forewing usually grey or greyish brown. A salt-marsh species ... *salinella* (part) (p. 170)

Key to genitalia of species (imagines) of the genus *Scrobipalpa*

NOTE. Figures 57, 58 (p.164) are generalized diagrams of the male and female genitalia. The terms used in the keys below are shown on these figures. In *Scrobipalpa* the structure of the genitalia is rather uniform and the differences between species are sometimes slight. Moreover, individual variation in some of the important characters can further obscure these differences. The keys are therefore based on typical examples and are intended as an aid that should be used in conjunction with the description, illustrations, habitat and hostplants. *S. hyoscyamella* is not treated in either of the following keys.

Great care should be taken in the preparation of the genitalia to avoid distortion of the structures.

MALES (figures 12d–f; 13a–h; 14a–d, pp. 32–34)

Caution is needed in interpreting the first couplet of this key as the coremata (a paired bundle of long hair-like scales at the base of abdominal tergite 8) are easily lost during preparation, although their sockets can usually still be seen.

1 Coremata present 2

– Coremata absent 8

2(1) Posterior margin of uncus notched .... *suaedella* (p. 169)

– Posterior margin of uncus rounded or flat (seldom weakly concave) 3

uncus

valva

culcitula

gnathos

gnathos hook

sacculus

posterior
process of
vinculum

median
emargination
of vinculum

vinculum

pendunculus

saccus

(**a**) ♂ genitalia capsule

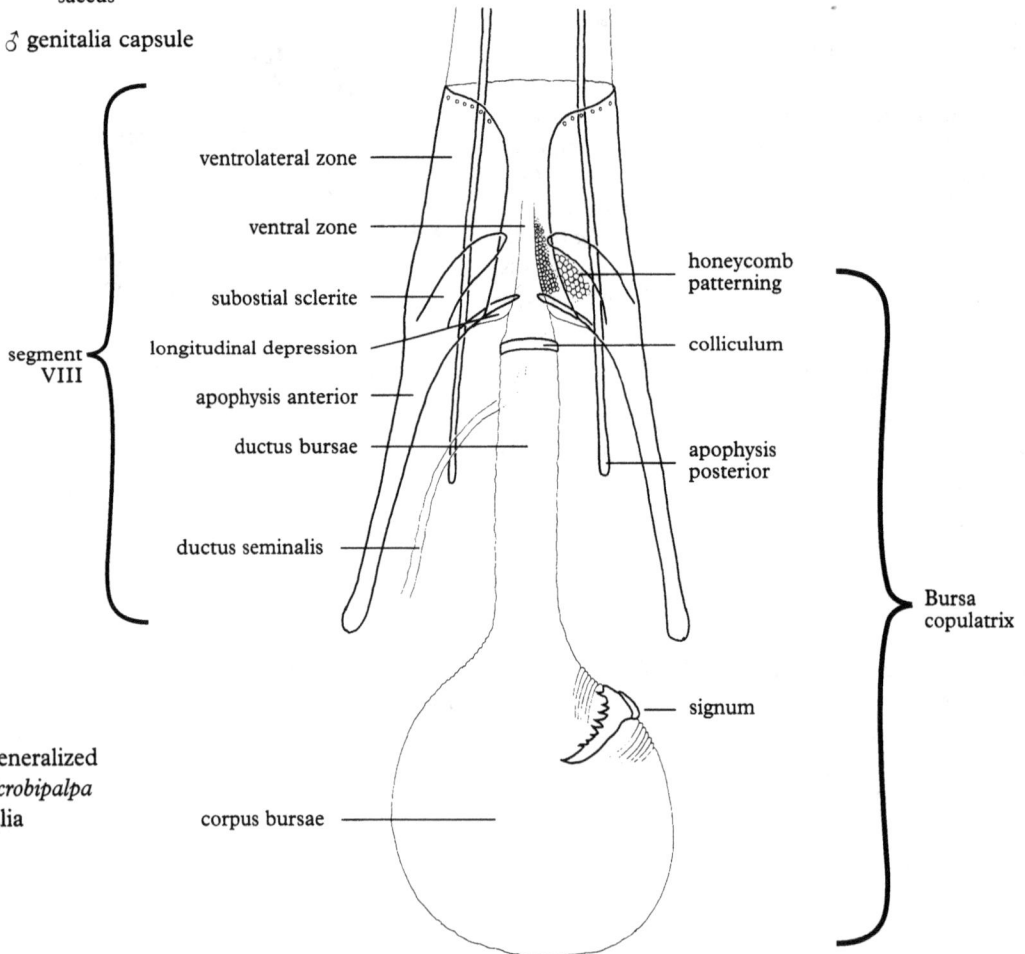

subapical arc

apical arm

coecum

bulbus
ejaculatorius

(**b**) Aedeagus
seen from left side

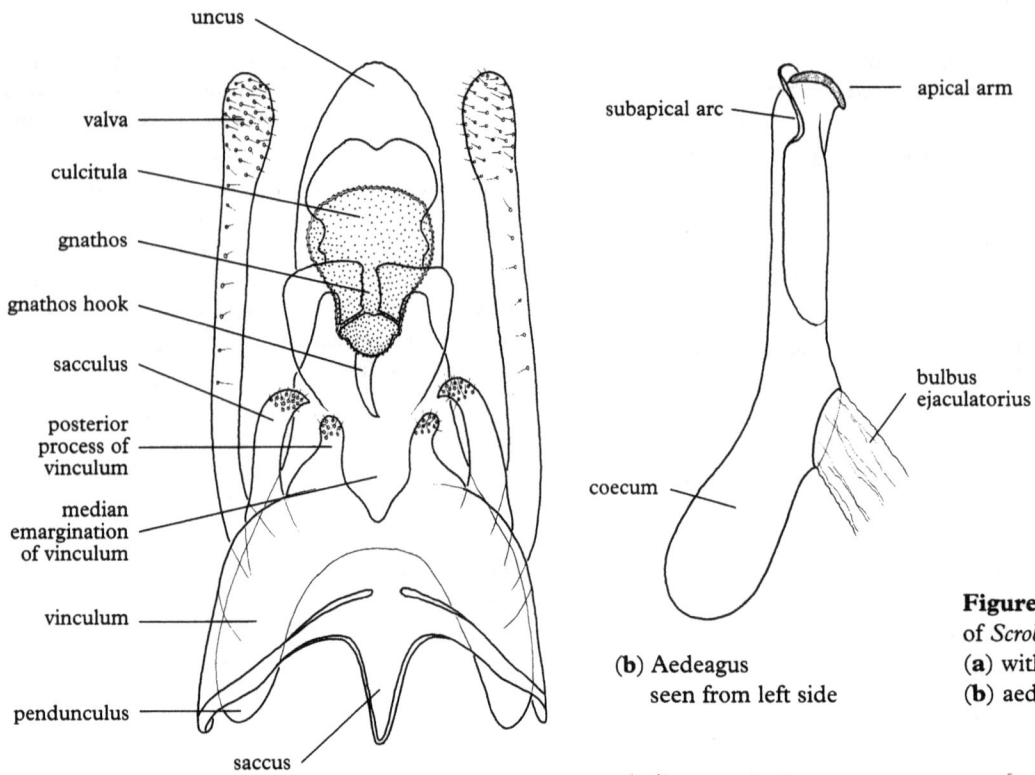

Figure 57 Generalized diagram
of *Scrobipalpa* male genitalia
(**a**) with aedeagus removed;
(**b**) aedeagus

ventrolateral zone

ventral zone

subostial sclerite

longitudinal depression

apophysis anterior

ductus bursae

ductus seminalis

segment
VIII

honeycomb
patterning

colliculum

apophysis
posterior

Bursa
copulatrix

signum

corpus bursae

Figure 58 Generalized
diagram of *Scrobipalpa*
female genitalia

3(2) Uncus and saccus distinctly exceeding valva and pedunculus respectively .... *costella* (p. 182)
– Uncus and saccus not significantly exceeding valva and pedunculus respectively 4

4(3) Posterior process of vinculum distinctly shorter than sacculus 5
– Posterior process of vinculum hardly shorter than sacculus 6

5(4) Aedeagus stout, almost straight, subapical arc broad .... *instabilella* (p. 171)
– Aedeagus slimmer, distinctly curved, subapical arc narrow .... *salinella* (p. 170)

6(4) Posterior process of vinculum squat; uncus slightly exceeding valva .... *samadensis* (p. 179)
– Posterior process of vinculum slender; uncus normally exceeded by valva, rarely longer than valva ... 7

7(6) Small species (wingspan 10–12mm) .... *artemisiella* (p. 179)
– Large species (wingspan 14–16mm) .... *stangei* (p. 180)

8(1) Posterior margin of uncus broadly truncate to concave 9
– Posterior margin of uncus rounded 10

9(8) Valva clearly exceeding uncus ...... *atriplicella* (p. 176)
– Valva to posterior margin of uncus or shorter .... *ocellatella* (p. 177)

10(8) Posterior process of vinculum small, remote from median emargination .... *acuminatella* (p. 166)
– Posterior process of vinculum continuous with median emargination 11

11(10) Posterior process of vinculum noticeably shorter than sacculus 12
– Posterior process of vinculum as long as sacculus or slightly shorter 14

12(11) Uncus exceeding valva .... *clintoni** (p. 181)
– Uncus not exceeding valva 13

13(12) Posterior processes of vinculum inclined towards each other .... *nitentella* (p. 173)
– Posterior processes of vinculum diverging or parallel .... *murinella* (p. 168)

14(11) Uncus slightly exceeding valva ...... *obsoletella* (p. 175)
– Valva slightly exceeding uncus ...... *pauperella* (p. 167)

---

* The posterior margin of the uncus can vary, sometimes being flat with a slight indentation.

FEMALES (figures 34c,d; 35a–e; 36a–d; 37a–d, pp. 54–57)

1 Transition from ductus to corpus bursae gradual ... 2
– Transition from ductus to corpus bursae abrupt; corpus bursae clearly defined 5

2(1) Segment 8 with characteristic honeycomb pattern .. 3
– Segment 8 without honeycomb pattern ....*acuminatella* (p. 166)

3(2) Ventral zone of segment 8 without honeycomb pattern .... *samadensis* (part) (p. 179)
– Ventro-lateral and ventral zones of segment 8 with honeycomb pattern 4

4(3) Honeycomb pattern of ventral zone very dense in anterior third .... *ocellatella* (p. 177)
– Honeycomb pattern of ventral zone evenly distributed ....*pauperella* (p. 167)

5(1) Signum clearly in corpus bursae 6
– Signum in transition zone between ductus and corpus bursae 7

6(5) Longitudinal depressions of ventral zone anteriorly broad; signum long, penetrating three-quarters of corpus bursae .... *clintoni* (p. 181)
– Longitudinal depressions of ventral zone anteriorly tapered; signum penetrating less than half of corpus bursae .... *obsoletella* (p. 175)

7(5) Corpus bursae oval, longer than wide 8
– Corpus bursae spherical 11

8(7) Ventral zone densely set with microtrichia; concave side of signum with strong teeth almost to apex .... *nitentella* (p. 173)
– Ventral zone bare or at most with weak patches of microtrichia; concave side of signum with few teeth at base or without 9

9(8) Ventral zone posteriorly constricted, at narrowest point narrower than ductus bursae at colliculum .... *salinella* (p. 170)
– Ventral zone hardly constricted, wider than ductus bursae at colliculum 10

10(9) Ductus bursae straight ...... *samadensis* (part) (p. 179)
– Ductus bursae distinctly curved near corpus .... *instabilella* (p. 171)

11(7) Ventral zone of segment 8 with coarse honeycomb pattern; apophyses posteriores distinctly shorter than ductus bursae .... *murinella* (p. 168)
– Honeycomb pattern of ventral zone at best weak; apophyses posteriores at least as long as ductus bursae 12

12(11) Longitudinal depressions of ventral zone anteriorly with honeycomb pattern or strong microtrichia ... 13
– Longitudinal depressions of ventral zone at least anteriorly free from honeycomb pattern and microtrichia 14

SCROBIPALPA ACUMINATELLA (Sircom)

Gelechia acuminatella Sircom, 1850, *Zoologist* **8** (Appendix): 72.

?*Gelechia pulliginella* Sircom, 1850, *Ibid.* **8** (Appendix): 72.

?*Gelechia cirsiella* Stainton, 1851, *Suppl.Cat.Br.Tineidae & Pterophoridae*: 4.

?*Gelechia gracilella* Stainton, 1871, *Entomologist's Annu.* **1871**: 97.

Type locality: England; Brislington, Bristol.

Description of imago (Pl.4, figs 13,14)

Wingspan 10–14mm. Head ochreous-grey to fuscous; antenna ochreous-grey partly ringed fuscous, in male appearing serrate in apical half; labial palpus with segment 3 slightly shorter than segment 2, ochreous-brown with apex buff, segment 2 ochreous-brown sparsely irrorate fuscous, inner side whitish ochreous. Thorax greyish brown to dark brown; tegulae sometimes paler. Forewing with apex acute, especially in female, greyish brown to ochreous-brown, variably irrorate fuscous; stigmata absent or weak if present, blackish, first discal beyond plical; veins sometimes with reddish brown colouring, particularly at base of wing, in one form extending over entire wing along and between veins; cilia brownish grey, basally irrorate fuscous, two obscure darker ciliary lines sometimes present. Female often smaller than male and more distinctly brownish, whereas male can have greyish hue. Hindwing brownish grey, veins and margins generally darker; cilia concolorous with wing, paler at base. Legs pale fuscous or fuscous, ringed ochreous at intersegmental joints. Abdomen brownish fuscous above and below; anal tuft of male brownish. Genitalia, see figures 14d,37d.

A form from the Scottish Highlands has a pale grey-brown forewing, pale orange-brown streaks at base of wing; veins, particularly towards apex, faintly lined with darker scales; plical and discal stigmata dark, second discal forming short streak.

Similar species. Can be confused with *S. pauperella* (Heinemann) but both are reliably distinguished by their genitalia. In the male of *S. acuminatella* the posterior process of the vinculum is much smaller than the sacculus and arises remote from the median emargination; in *S. pauperella* the posterior process is broader than the sacculus and its edge is continuous with the median emargination. In the female genitalia, *S. acuminatella* is the only species in Britain to combine the absence of the honeycomb pattern on segment 8 with an undifferentiated bursa copulatrix. The female genitalia of *S. pauperella* are immediately distinguished by the presence of honeycomb pattern. *S. acuminatella* is also superficially similar to *S. clintoni* Povolný, although they can usually be separated on external characters.

Life history

Ovum. Oviposition site not observed. On thistles (*Carduus* spp. and *Cirsium* spp.) and once reported from colt's-foot (*Tussilago farfara*) (Bland, 1992). On the Continent it has been reared from carline thistle (*Carlina vulgaris*) (O. Karsholt, pers.comm.); records of knapweed (*Centaurea* spp.) (Schmid, 1886) and saw-wort (*Serratula* spp.) (Povolný, 1980) probably refer to *S. pauperella* (= *klimeschi* auctt.) (Sattler, 1987b, 1989). Those for blessed thistle (*Cnicus benedictus*) and cotton thistle (*Onopordum* spp.) (Hering, 1957) and for field wormwood (*Artemisia campestris*) (Povolný, *loc.cit.*) are unconfirmed (Sattler, 1987b), and a record for red clover (*Trifolium pratense*) (Lhomme, [1946–49]) is clearly erroneous.

Larva. Full-grown *c.*9–10mm. Head brown. Prothoracic plate pale brown with two blackish spots at posterior margin; body greenish yellow, with yellowish, pale brick-red or pinkish red on anterior part of each segment; pinacula black, faintly ringed dark grey; anal plate brown; thoracic legs of body colour, with black basal markings; prolegs of body colour, sometimes partly ringed black.

The larva feeds on a lower leaf of the foodplant in a brownish, sometimes branched, mine extending along the midrib. The mine usually starts at the base; older mines are brownish and can be slightly inflated and blotch the leaf. The larva rests along the midrib in the blotch mine, with frass congregated in a very few areas. On coltsfoot, the larva was found feeding in an irregular blotch mine in the upper surface of a leaf (Bland, *loc.cit.*). July–September.

93

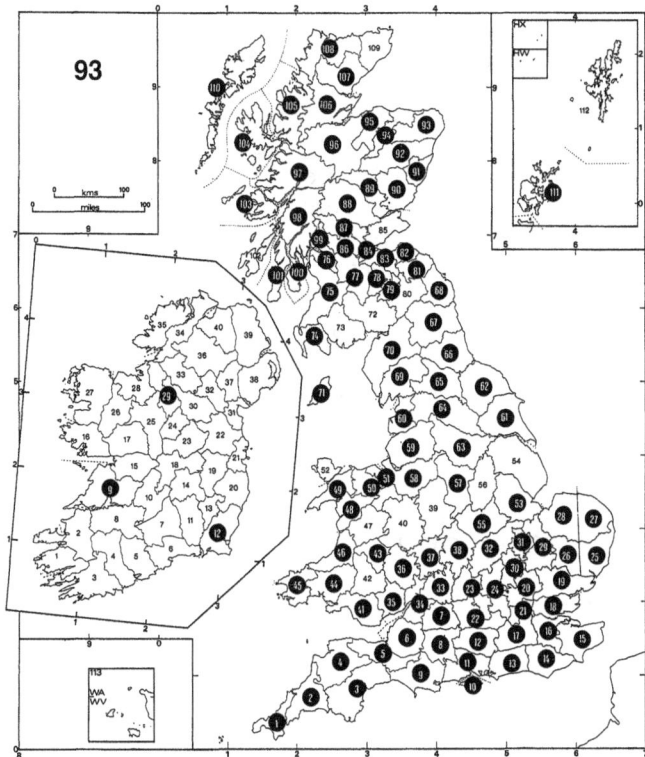

Scrobipalpa acuminatella

Pupa. In litter on the ground. July–August and October–May.

Imago. Bivoltine; late April–June and late July–early September. There is possibly only one generation a year in northern Britain (unpublished observations by E. C. Pelham-Clinton). The moth has been recorded at light.

Distribution (Map 93)

Frequents rough meadows, edges of fields, broken ground on chalk downland and parkland. Widely distributed in Britain from southern England to Orkney. Seemingly very local in Ireland, but possibly underrecorded. Distributed throughout Europe; recorded from Asia Minor.

SCROBIPALPA PAUPERELLA (Heinemann)

Lita pauperella Heinemann, 1870, *Schmett.Dtl.Schweiz* (2)**2**(1): 256.
Scrobipalpa klimeschi sensu auctt., *nec* Povolný, 1967.
Type locality: Germany; Regensburg, Bavaria.

Description of imago (Pl.4, figs 16,17)

Wingspan 10–13mm. Head with frons greyish fuscous; antenna dark fuscous, annulate ochreous; labial palpus with segment 3 as long as segment 2, greyish fuscous, paler on inner side. Thorax greyish fuscous. Forewing dark fuscous, mixed whitish and brown, some faint reddish brown scaling following veins, particularly at base; costa dark greyish fuscous. Hindwing pale grey; cilia concolorous, slightly paler at base. Foreleg dark fuscous, ringed pale greyish ochreous at intersegmental joints; mid- and hindlegs fuscous, ringed pale greyish ochreous, hairs pale ochreous. Abdomen greyish fuscous. Genitalia, see figures 13c,35e.

Similar species. The few British specimens known to date closely resemble typical dark *S. acuminatella* (Sircom), *q.v.*

Life history

Ovum. Biology in Britain not known. In continental Europe, this species is associated with various species of Asteraceae (Compositae), such as marsh thistle (*Cirsium palustre*) (Krogerus & von Schantz, 1970), greater knapweed (*Centaurea scabiosa*) and white butterbur (*Petasites albus*) (Sattler, 1989). It is possibly also associated with saw-wort (*Serratula tinctoria*) (Sattler, *loc.cit.*).

Larva. Undescribed.

Larvae occur in July and, probably, September. In Finland, the larva has been found in August mining along the midrib of a leaf of marsh thistle, the mine being slightly different from that of *S. acuminatella* (Kaitila, 1996). It is not known if *S. pauperella* overwinters as a mature larva or as a pupa (Sattler, *loc.cit.*). The biology of related species would suggest that the larva mines the lower leaves of the hostplant. Label data on old German specimens (in BMNH) record it as living in the stem of greater knapweed.

Pupa. Undescribed.

Imago. In Britain, the adult has been found only in June. In mainland Europe, the species is seemingly bivoltine, April–early June and from July–August. Adults have been noted flying about marsh thistle.

Distribution (Map 94).

In Britain, this species is known only from seven examples recorded in 1972 and 1973, all from Chippenham

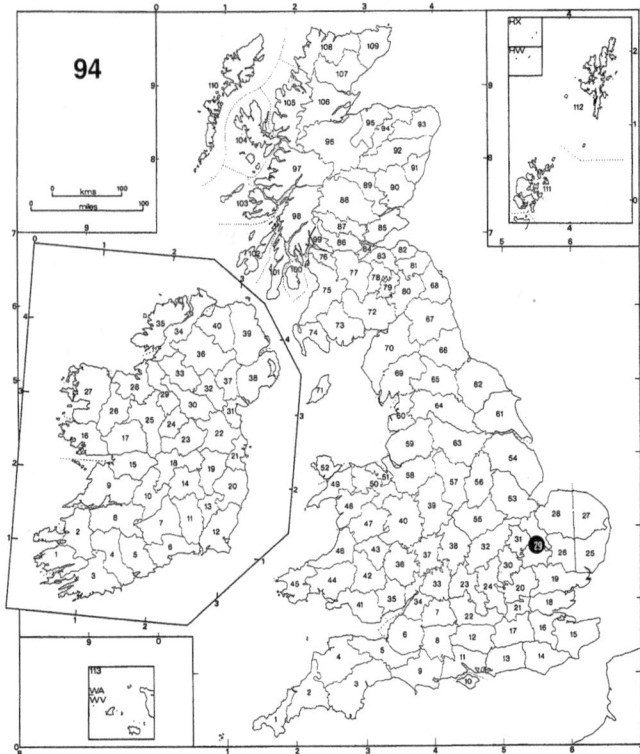

Scrobipalpa pauperella

Fen NNR, Cambridgeshire. It was added to the British list as *Scrobipalpa klimeschi* Povolný by Agassiz (1986), a name which was shown to be based on a misidentification (Sattler, *loc.cit.*). Recent searches have failed to refind the species (D. J. L. Agassiz, pers.comm.). Treated as RDB K (Insufficiently Known) by Parsons (1995[1996]). This moth has been recorded in continental Europe from Finland, France, Switzerland, Italy, Germany, Austria, Hungary, Czech Republic, Slovakia, Latvia, and western Russia (Povolný, 1996); Afghanistan.

SCROBIPALPA MURINELLA (Duponchel)

Lita murinella Duponchel, 1843, *in* Godart & Duponchel, *Hist.nat.Lépid.Fr.* Suppl. **4**: 458, pl.85, fig.7.

Gelechia murinella Herrich-Schäffer, 1854, *Syst.Bearb. Schmett.Eur.* **5**: 162 [key], 178; 1853, *Ibid.* **5**: pl.71, fig.535 [non-binominal].

Gelechia excelsa Frey, 1880, *Lepid.Schweiz*: 363.

Type locality: [France].

Description of imago (Pl.4, fig.15)

Wingspan 9–10mm. Head with vertex grey-brown,

frons often paler; antenna dark fuscous, obscurely ringed grey-brown; labial palpus with segment 3 as long as segment 2, segment 3 fuscous below, above pale brownish fuscous with apex fuscous, segment 2 fuscous-brown on outer side, whitish on inner side. Thorax grey or brownish grey. Forewing whitish, irrorate greyish brown and dark ochreous, sometimes so heavily that wing appears uniform brown; stigmata black, first discal, sometimes divided in two, beyond plical; subapical area more heavily irrorate fuscous, tending to form blackish apical and interneural spots; cilia greyish brown with an obscure darker ciliary line. Hindwing grey to pale grey; cilia concolorous. Foreleg pale fuscous, ringed buff at intersegmental joints; mid- and hindlegs buff, irrorate fuscous on outer sides, hairs pale buff. Abdomen greyish fuscous or brownish fuscous, somewhat paler below; anal tuft of male buff. Genitalia, see figures 14c,37b.

Similar species. S. *murinella* is the smallest *Scrobipalpa* in western and northern Europe. On external characters it may be mistaken for a pale or worn example of *S. artemisiella* (Treitschke) (Bradley, 1964), although they can be distinguished on genitalia differences.

Life history

Ovum. Oviposition site not observed. The hostplant is mountain everlasting (*Antennaria dioica*). Continental records of *Leontopodium* and *Anthemis cretica* (Povolný, 1980) are unconfirmed and that of dwarf cudweed (*Gnaphalium supinum*) (Povolný, *loc.cit.*) refers to another species. *Gnaphalium* has been recorded as a hostplant by Hering (1957), but no voucher material exists in his herbarium to support this.

Larva. Head black in early instars, yellow in final instar, with a black mark on the ocelli and one behind. Prothoracic plate at first blackish brown, later yellow and often with brown posterior and ventral margins; body bluish white with black pinacula, a reddish medio-dorsal line from abdominal segment 1 to penultimate segment, a similar much broader lateral line somewhat interrupted between segments producing a sinuous effect, encircling upper pinacula, followed by a mark a little below and behind the end; in early instars these lines are yellow; anal plate brown to black (Bradley, *loc.cit.*).

The larva feeds in the stems and in blotch-mines in leaves of the foodplant, and later in the flowers and seedheads. The larva, which mines several leaves, moves from one leaf to another, spinning a slight web as it progresses. July–September.

Pupa. Overwinters as a pupa amongst spun parts of the hostplant.

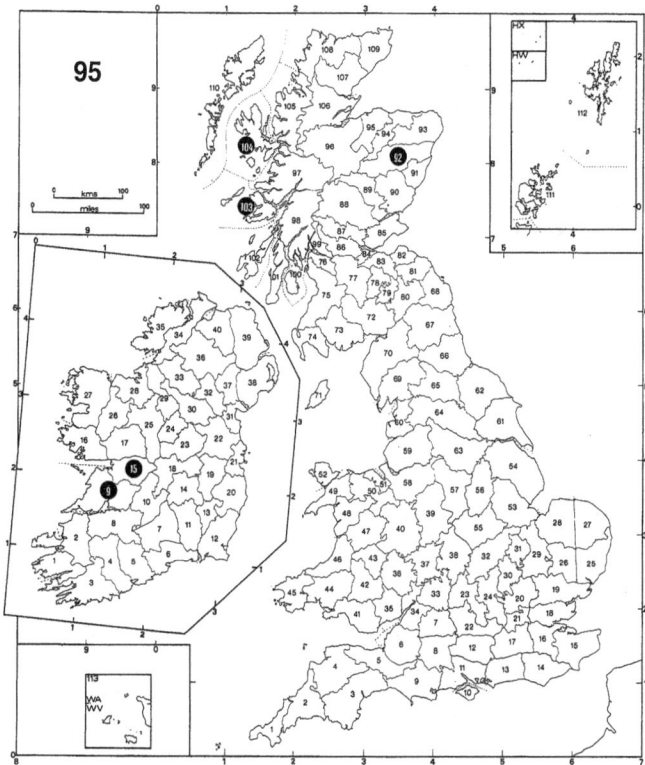

95

Scrobipalpa murinella

Imago. April–June and has been found amongst patches of the hostplant.

Distribution (Map 95)

The species was added to the British list by Bradley (*loc.cit.*), based on specimens collected in the Burren, Co. Clare, Ireland, in 1962 (Pelham-Clinton, 1964). In Ireland, it has been found in areas of limestone pavement. *S. murinella* was first recorded in Britain in 1969 on Rum (Wormell, 1982). Treated as RDB I (Indeterminate) by Parsons (1995[1996]). Extremely local in Britain and recorded only from Rum, North Ebudes; one small area in the northern part of the Isle of Coll, Mid Ebudes, despite the foodplant being widespread in rocky areas of the island (Bland *et al.*, 1987a); and near Braemar, Aberdeenshire, in 2000 (Heckford, 2001). In Ireland, it has been found only in Co. Clare and south-east Galway. In continental Europe, the species has been recorded from France, The Netherlands, Norway, Sweden, Finland, Germany, Switzerland, Italy, Austria, Hungary, Czech Republic, Slovakia, Romania, Poland, Latvia, Lithuania and western Russia (Povolný, 1996).

SCROBIPALPA SUAEDELLA (Richardson)

Lita suaedella Richardson, 1893, *Entomologist's mon. Mag.* **29**: 241.

Type locality: England; near Weymouth, Dorset.

Description of imago (Pl.4, figs 18,19)

Wingspan 13–15mm. Head ochreous-grey, frons whitish grey; antenna fuscous, obscurely ringed ochreous, apical one-half slightly serrate in male; labial palpus with segment 3 as long as segment 2, segment 3 ochreous-grey with median band and apex fuscous, segment 2 whitish ochreous, sometimes grey irrorate fuscous on outer side, greyish white on inner side. Thorax greyish ochreous to ochreous, tegulae slightly darker. Forewing whitish ochreous, densely irrorate grey; discal area and veins ochreous; usually with dorsal area pale ochreous with reduced grey irroration occupying about one-third of breadth of wing becoming less distinct to termen, often edged above in proximal half of wing with diffused greyish streak; two diffused blackish subbasal spots on costa and two on fold; stigmata black, first discal beyond plical and often weakly expressed, often with second plical stigma between discal stigmata, second discal elongate or occasionally divided into two spots; occasionally destitute of pale ochreous dorsal marking, with reduced black spots and with orange-brown markings along veins; blackish subapical and subterminal spots between veins; cilia whitish grey, irrorate grey, with three indistinct ciliary lines; underside grey, subcostal area paler. Hindwing whitish grey, veins, costal and terminal areas darker; cilia whitish, darker on costa and at apex, paler at base. Foreleg fuscous, narrowly banded ochreous at intersegmental joints; mid- and hindlegs whitish ochreous, variably irrorate fuscous and banded whitish ochreous. Abdomen above grey to greyish fuscous, segment 1 and often anterior half of segment 2 pale ochreous; below whitish ochreous with fuscous lateral stripe; anal tuft of male whitish ochreous. Genitalia, see figures 12e,34c.

Similar species. Although it can vary considerably, this is one of the more easily recognizable *Scrobipalpa* species. The female genitalia are very similar to those of *S. artemisiella* (Treitschke) and *S. stangei* (Hering). Confusion with *S. artemisiella* is unlikely because the latter is much smaller and occurs in a different habitat. An infrequent form of *S. suaedella* is superficially similar to *S. instabilella* (Douglas), *S. nitentella* (Fuchs) and *S. salinella* (Zeller), *qq.v.*

Life history

Ovum. Oviposition site not observed. On shrubby sea-

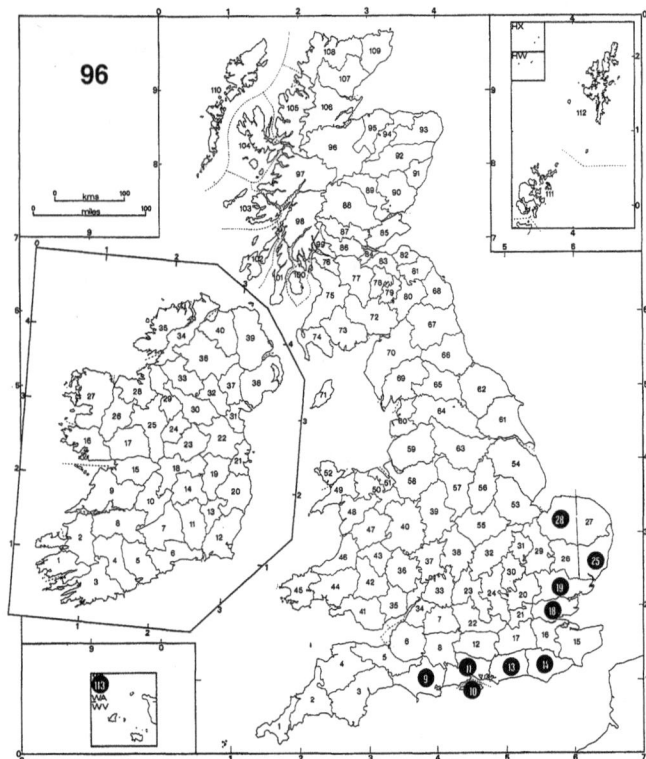

Scrobipalpa suaedella

blite (*Suaeda vera*) or occasionally on annual sea-blite (*S. maritima*) (Bankes, 1894).

Larva. Full-grown *c.*10mm. Head black. Prothoracic plate black with a thin lighter median stripe; body pale greenish grey, posteriorly darker; brownish red dorsal stripe almost continuous with a more irregularly shaped and broken subdorsal stripe; spiracular line slightly paler; pinacula appearing as black dots but brownish under magnification; anal plate black; thoracic legs black; some scattered pale brick-red coloration on prolegs.

The larva feeds in a silken gallery amongst the leaves and blossom of the foodplant, generally at the tips of the shoots (Richardson, 1893). Occasionally two shoots are spun together. On shrubby sea-blite the larval feeding causes the tips of the shoots to wither and turn brown. Larvae can occasionally be abundant where they are found. May–early June. The larva has been recorded on annual sea-blite in August, a plant that is hardly visible in May and June and is unlikely to be available as a hostplant at that time (Richardson, *loc.cit.*).

Pupa. In a tough cocoon, often coated with mud, on the ground (Bankes, *loc.cit.*). May–June.

Imago. June–July.

Distribution (Map 96)

Associated with coastal salt-marshes. Very local, but often abundant where it occurs. Recorded along the coast from Dorset to west Norfolk, although seemingly not reported in Kent and only twice recorded in east Sussex; the Channel Islands. Abroad, this species has been recorded from Portugal, Spain, France, Germany, Sardinia (a separate subspecies) and Italy (Povolný, 1996); Asia Minor.

SCROBIPALPA SALINELLA (Zeller)

Gelechia salinella Zeller, 1847, *Isis, Leipzig* **1847**: 853.
Lita salicorniae E. Hering, 1889, *Stettin.ent.Ztg* **50**: 302.
Type locality: Sicily; Syracuse.

Description of imago (Pl.4, figs 20,21)

Wingspan 11–15mm. Head with vertex greyish ochreous, irrorate fuscous, frons pale ochreous; antenna fuscous, indistinctly annulate pale ochreous, apical half serrate in male; labial palpus with segment 3 as long as segment 2, segment 3 fuscous with obscure ochreous median band, segment 2 ochreous, irrorate fuscous on outer side, whitish ochreous on inner side. Thorax ochreous-fuscous. Forewing whitish ochreous, heavily irrorate fuscous and variably suffused and streaked orange-ochreous; subbasal area sometimes with one or two fuscous spots on costa and one on fold; stigmata large, either diffused black variably mixed with orange-ochreous, or with no black, appearing as orange-ochreous blotches; first discal beyond plical; often whitish ochreous subapical spot on costa and similar spot on tornus, sometimes joined to form weak outwardly angulate fascia; cilia whitish ochreous, irrorate fuscous at apex, two very obscure darker ciliary lines; underside fuscous to pale fuscous. Hindwing greyish white; cilia whitish ochreous. Foreleg fuscous, banded ochreous at intersegmental joints; mid- and hindlegs pale ochreous, irrorate fuscous on outer sides, banded pale ochreous at intersegmental joints, hairs whitish ochreous. Abdomen fuscous above, segments 1–3 sometimes ochreous, underside ochreous-fuscous; anal tuft of male ochreous-fuscous. Genitalia, see figures 13a,35c.

Similar species. Most likely to be confused with *S. instabilella* (Douglas), *q.v.*, but some examples can be mistaken for *S. atriplicella* (Fischer von Röslerstamm), *q.v.*, or *S. obsoletella* (Fischer von Röslerstamm), *q.v.* An unusual form of *S. suaedella* (Richardson) is similar to *S. salinella*, but the stigmata are usually much smaller and more indistinct in *S. suaedella*.

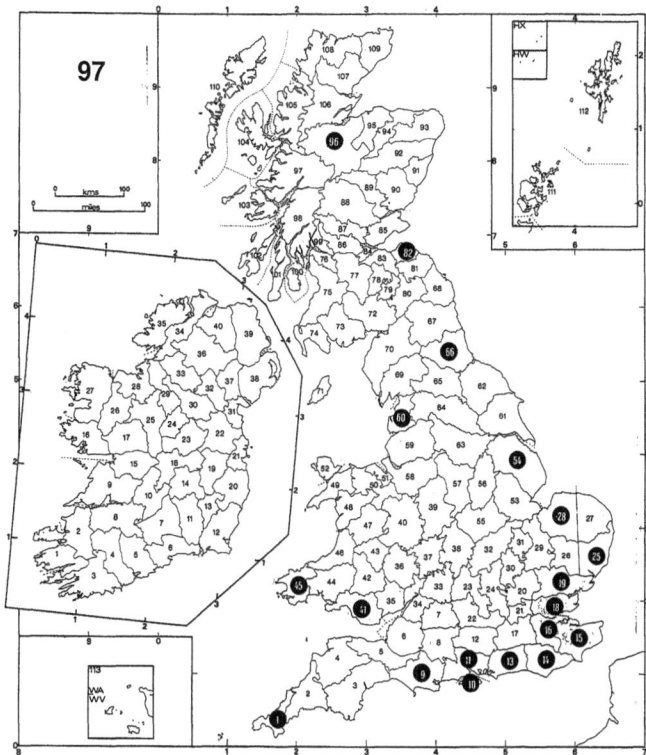

Scrobipalpa salinella

Life history

Ovum. Oviposition site not recorded. Possibly overwintering in this stage. On glassworts (*Salicornia* spp.). Sea aster (*Aster tripolium*), sea-blite (*Suaeda* spp.) including annual sea-blite (*S. maritima*), sand-spurrey (*Spergularia* spp.) including greater sea-spurrey (*S. media*) (Bankes, 1894), and common sea-lavender (*Limonium vulgare*) (BMNH material) have all been recorded as hostplants, but it is thought unlikely that these are regular hosts. In Germany, pedunculate sea-purslane (*Atriplex pedunculata*) has been recorded as a foodplant but only where it was growing in close proximity to glasswort (Martini, 1916). It is probable that the species will accept most if not all glasswort species within its region.

Larva. Full-grown *c.*12mm. Head honey-brown with black speckling at posterior edge, mandibles slightly darker. Prothoracic plate black; body dull greenish grey with a pale brick-red dorsal line; subdorsal and subspiracular stripes blotch-like, reddish; red markings variable and sometimes obsolete; pinacula black, ringed brown; anal plate blackish; thoracic legs black; prolegs of ground colour, sometimes with slight pale brick-red markings.

From Continental observations on glasswort, the species feeds within a tent of several spun shoots or branches on strong erect plants; where the plants are small and thinly spread it lives often in a dense silken tunnel on or in the ground and extends a loose web over nearby prostrate branches to feed on the surface of glasswort segments. On sea aster, the larva makes a series of galleries in the leaves, similar to those of *Bucculatrix maritima* Stainton (Bucculatricidae) (*MBGBI* 2: 231) and is most easily found on this foodplant (A. M. Emmet, pers.comm.). On sea-blite, the larva spins together several leaves or shoots and eats green parts from within this shelter, leaving only the colourless remains (Bankes, *loc.cit.*). The larva feeds from April to June.

Pupa. Amongst detritus; June–August. Cocoons of this species have been found in the dead stems of sea aster which must have been regularly submerged at high tide (Heckford, 1997).

Imago. Probably univoltine, although Bankes (*loc.cit.*) states that the species has two or more broods, apparently in succession, but with emergence spread over three months. June–September. On the Continent, the adult has been recorded as occurring also in May (Bankes, *loc.cit.*), but no voucher specimens exist to support this. The moth has been recorded at light.

Distribution (Map 97)

Associated with coastal and estuarine salt-marshes, although occasionally found inland. In mainland Europe, this species has been recorded from inland saline habitats. This local and difficult-to-record species has historically been found on the southern and eastern coasts of England from Cornwall to Lincolnshire; Co. Durham; Lancashire; Pembrokeshire, south Wales; in Scotland, East Lothian and Nairn. Not recorded from Ireland. Elsewhere, it has been recorded through Europe to western Russia (Povolný, 1996).

SCROBIPALPA INSTABILELLA (Douglas)

Anacampsis instabilella Douglas, 1846, *Zoologist* 4: 1270, fig.10.

Type locality: England, St Osyth, Essex.

Description of imago (Pl.4, figs 22,23)

Wingspan 12–14mm. Head with vertex whitish, irrorate ochreous and greyish fuscous, frons whitish; antenna fuscous, narrowly ringed ochreous; labial palpus with segment 3 as long as segment 2, segment 3 ochreous-fuscous, segment 2 ochreous, fuscous on outer side, whitish on inner side. Thorax as vertex of head. Forewing variable, whitish, heavily irrorate fuscous and suffused orange-brown; stigmata often black, sometimes ringed with orange-brown or completely orange-brown, first discal beyond plical, discal spots often divided into two diagonally, in rare form black spots confluent, forming irregular longitudinal mark; subbasal area often with one or two suffused black spots on costa; whitish ochreous subapical patch on costa and similar spot on tornus, both sometimes weakly expressed, occasionally joining to form sharply outwardly angulate fascia; blackish subterminal spots between veins, often weakly expressed or absent; cilia ochreous, irrorate fuscous at apex, two or three indistinct darker ciliary lines; underside fuscous, sometimes paler between veins towards apex. Hindwing whitish grey, darker on costa and at apex; cilia ochreous-white, darker at apex. Foreleg pale fuscous, banded ochreous at intersegmental joints; mid- and hindlegs whitish ochreous, outwardly irrorate fuscous. Abdomen greyish fuscous above, segments 1 and 2 ochreous, whitish ochreous below; anal tuft of male greyish ochreous. Genitalia, see figures 12f,35b.

Similar species. S. instabilella can be confused with *S. salinella* (Zeller), although they can be distinguished by small but constant differences in their genitalia. The male genitalia of *S. instabilella* and *S. salinella* are characterized by the short, very broad posterior processes of the vinculum. Both species vary in the shape of the uncus, length of the valva, width of the median emargination of the vinculum and shape of the saccus. They can best be distinguished from each other by characters of the aedeagus. In *S. instabilella* the aedeagus is almost straight and the apical arm (see figure 12f) is relatively broad, curved, with the apex blunt, whereas in *S. salinella* the aedeagus is slightly curved and the apical arm (see figure 13a) is narrow and almost straight, with the apex sharply bent; the subapical arc of the aedeagus is broad in *S. instabilella* and narrow in *S. salinella*. The female genitalia of *S. instabilella* differ from those of *S. salinella* by the shorter and broader segment 8 (widest at the base of the apophyses anteriores and noticeably tapered posteriorly), the shorter apophyses anteriores and the reduced area of honeycombing on the ventrolateral zone. The forewing of *S. salinella* is often more marbled in appearance and the stigmata of this species can also be large blotches. Some examples of *S. instabilella* have dark fuscous markings on the forewings which are not usual in *S. salinella*. An infrequent form of *S. suaedella* (Richardson) closely resembles some forms of *S. instabilella*. In the male genitalia the posterior margin of the uncus is notched in *S. suaedella* but not in *S. instabilella*. The females can be separated by the more spherical corpus

bursae in *S. suaedella*; this species also has a straight ductus bursae. *S. nitentella* (Fuchs) is superficially similar, but can appear more heavily irrorate fuscous. The male genitalia of *S. nitentella* have the valva generally exceeding the uncus and a broad and comparatively shallow median emargination of the vinculum. In *S. instabilella* the valva rarely reaches the posterior margin of the uncus and the median emargination of the vinculum is more or less V-shaped. In the female the ventral zone of *S. nitentella* is densely set with microtrichia and the signum has teeth almost to the apex. In *S. instabilella* the ventral zone is without microtrichia and the signum at most has a small number of teeth at the base. Some examples of *S. samadensis* (Pfaffenzeller) can also be similar, but *S. instabilella* generally has more heavily marked stigmata. In the male of *S. instabilella* the posterior process of the vinculum is distinctly shorter than the sacculus, in *S. samadensis* it is hardly shorter than the sacculus. In the former the apex of the posterior process of the vinculum has a short point facing the sacculus whereas in *S. samadensis* the posterior process is broad and strongly curved inwards. The females of *S. instabilella* and *S. samadensis* can be distinguished by the ductus bursae, which is curved in the former and straight in the latter. *S. instabilella* and *S. atriplicella* (Fischer von Röslerstamm), *q.v.* can be distinguished by the genitalia and some confusion may arise between *S. instabilella* and *S. obsoletella* (Fischer von Röslerstamm), although the latter is more obscurely marked.

Life history

Ovum. Oviposition site not observed. On sea-purslane (*Atriplex portulacoides*). Other recorded hostplants, such as grass-leaved orache (*A. littoralis*), buck's-horn plantain (*Plantago coronopus*), sea aster (*Aster tripolium*) (BMNH material), common glasswort (*Salicornia europaea*) (Stainton, 1883) and Duke of Argyll's teaplant (*Lycium barbarum*) (M. S. Parsons, pers.obs.) are probably accidental. Threlfall (1878) records the 'roots of sea plantain' (*Plantago maritima*) as a larval pabulum; this is almost certainly in error for *S. samadensis*. Stainton (*loc.cit.*) refers to the 'true *instabilella*' reared from shrubby sea-blite (*Suaeda vera*), but this is likely to be in error for *S. suaedella*. Annual sea-blite (*Suaeda maritima*) and sea beet (*Beta vulgaris* ssp. *maritima*) were also reported as hostplants by Stainton (*loc.cit.*). The former is probably based on misidentification of *S. salinella* and the latter refers to *S. ocellatella* (Boyd). *Chenopodium* has been recorded as a hostplant (Hering, 1957), but no voucher material exists in his herbarium to support this. Other Chenopodiaceae have been recorded, although these are based on observations in

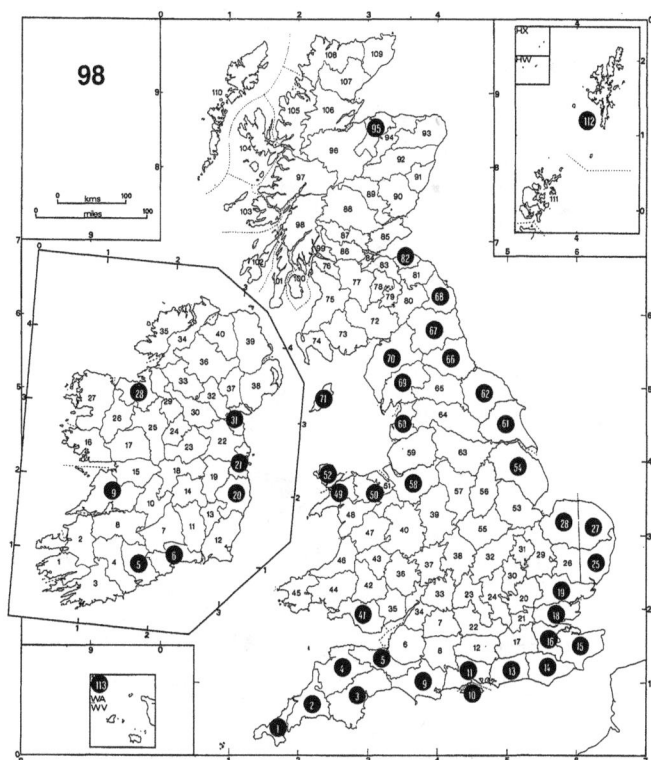

Scrobipalpa instabilella

north-west Africa (Povolný, 1979).

Larva. Full-grown *c*.12mm. Head brown or light brown, mouth-parts darker, with a dark spot laterally, or black. Prothoracic plate light brown, slightly lighter than head and sparsely mixed with black at posterior edge; body greenish grey to yellowish green; three reddish brown lines along dorsum, the subdorsal lines being broader, and a much fainter spiracular line; pinacula dark brownish under magnification, appearing as black dots; anal plate light brown or black; thoracic legs blackish; prolegs green.

On sea-purslane the larva feeds in the leaves of the foodplant causing an almost bladdered effect. The larva makes a small round hole in the skin of the leaf through which it ejects its frass, the mine being quite clean (Richardson, 1893). Larvae within mines are seemingly unaffected by periodic submergence of their habitat. Some larvae spin leaves together in an untidy spinning. These spinnings possibly occur more frequently on plants growing amongst salt-marsh grasses (M. S. Parsons, pers.obs.). Larvae feeding on orache, glasswort and Duke of Argyll's teaplant are found in loose spinnings. The larvae can be locally abundant. March–May, although Bankes (1894) suggested that on sea-purslane it possibly overwinters as a young

larva in the evergreen leaves.

Pupa. In a mud-covered cocoon on the ground, in which the larva remains for weeks before pupating (Bankes, *loc.cit.*). Pupae have been found in the larval habitation and in mud-covered cocoons on the stems of sea-purslane.

Imago. May–July; it can be disturbed easily from the foodplant by day and is also attracted to light.

Distribution (Map 98)

Associated with salt-marshes, mudflats and the banks of tidal rivers. Widely distributed around the coasts of England. Local on the coast of Wales and reported from a very few localities in Scotland north to the Shetland Islands. Seemingly local in Ireland, although possibly under-recorded, and reported from the Channel Islands. Abroad, this species has been widely recorded from western Europe, Poland, Bulgaria and western Russia (Povolný, 1996). Because of taxonomic problems the distribution of *S. instabilella* in southern Europe and the Mediterranean region is uncertain. Also recorded from north-west Africa and the Canary Islands.

SCROBIPALPA NITENTELLA (Fuchs)

Lita nitentella Fuchs, 1902, *Stettin.ent.Ztg* **63**: 324.
Phthorimaea seminella Pierce & Metcalfe, 1935, *Entomologist* **68**: 97.

Type locality: Germany; Artern, Thuringia.

Description of imago (Pl.4, fig.24)

Wingspan 12–15mm. Head with vertex pale ochreous heavily irrorate fuscous, frons greyish ochreous; antenna ochreous, annulate fuscous, in male apical half slightly serrate; labial palpus segment 3 as long as segment 2, segment 2 on outer side ochreous-fuscous, on inner side whitish ochreous. Thorax greyish ochreous, irrorate darker. Forewing greyish buff heavily irrorate fuscous and orange-brown, subbasal area sometimes with cloudy fuscous spots on costa and fold; stigmata black, first discal slightly beyond plical, often elongate and sometimes, especially second discal, divided into two, sometimes additional spot costad to second discal; occasionally a pale subterminal costal spot, sometimes similar spot on tornus, the two occasionally joined to form pale fascia; cilia whitish ochreous, irrorate fuscous, three slightly darker ciliary lines sometimes present; underside greyish fuscous, veins darker. Hindwing whitish grey, veins and apex darker; cilia whitish ochreous. Fore- and midlegs fuscous, annulate pale ochreous at intersegmental joints; hindleg pale ochreous, irrorate fuscous on outer side, ochreous on inner side. Abdomen greyish fuscous above, segments 1–3 ochreous-yellow, ochreous below with lateral line of fuscous spots. Genitalia, see figures 13e,36a.

Similar species. S. nitentella can be confused with several other members of the genus. *S. obsoletella* (Fischer von Röslerstamm) is generally more obscurely marked and has less fuscous irroration, but examination of the genitalia is often required for confirmation of determination. The male genitalia of *S. nitentella* differ from those of *S. obsoletella* in that those of the latter species have shorter valvae, parallel sacculi and a deep V-shaped posterior emargination on the vinculum. The female genitalia of *S. obsoletella* differ from those of *S. nitentella* by the much reduced scobination on segment 8, the shape of the corpus bursae and the position of the signum in the corpus bursae rather than the transition zone. Some forms of *S. samadensis* (Pfaffenzeller) approach *S. nitentella*, although the latter species usually appears more grey. Similar forms of *S. salinella* (Zeller) are usually less heavily irrorate and have more distinct stigmata. One form of *S. suaedella* (Richardson) is similar, but this form has the stigmata less well marked and with less fuscous irroration than *S. nitentella. S. nitentella* and *S. instabilella* (Douglas), *qq.v. S. samadensis, S. salinella* and *S. obsoletella* can all be distinguished by examination of the genitalia. In the male *S. nitentella* the valva slightly exceeds the uncus, in *S. samadensis* and *S. obsoletella* the valva does not, whereas in *S. salinella* the valva can just reach the posterior margin of the uncus. The females of *S. nitentella* and *S. samadensis* can be separated by the former having a curved ductus bursae and a signum with teeth almost to the apex. In the latter the ductus bursae is straight and the signum has only a small number of teeth at the base. The female of *S. nitentella* has a ventral zone with densely set microtrichia, unlike that in *S. salinella*. In *S. nitentella* the signum is in the transition zone of the ductus bursae and corpus bursae, in *S. obsoletella* it is in the corpus bursae.

Life history

Ovum. Laid on the foliage of the hostplant (Stüning, 1988). On goosefoot (*Chenopodium* spp.) or annual sea-blite (*Suaeda maritima*). Also recorded from sea rocket (*Cakile maritima*), although it is possible that the plant was misidentified; sea-purslane (*Atriplex portulacoides*); among the seeds and flowers of sea beet (*Beta vulgaris* ssp. *maritima*); grass-leaved orache (*Atriplex littoralis*); hastate orache (*A. hastata*) (Michaelis, 1977); and perennial glasswort (*Sarcocornia perennis*) (BMNH material). A record for common orache (*Atriplex patula*) (Newton, 1985) requires con-

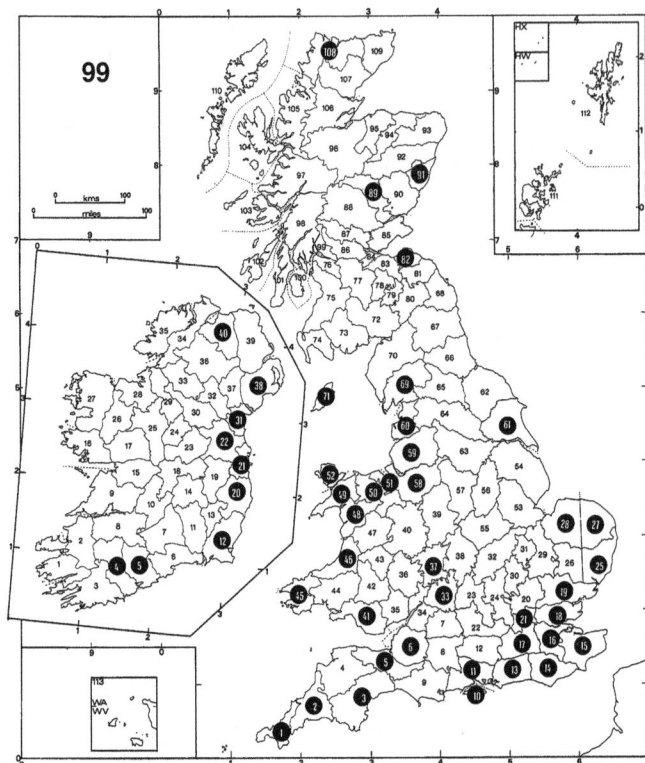

Scrobipalpa nitentella

firmation. In continental Europe, common glasswort (*Salicornia europaea*) (BMNH material) has been recorded as a hostplant.

Larva. The following description is largely based on that of Pierce & Metcalfe (1935b). Full-grown 8–9mm. Head pale brown, mouth darker; a large black spot adjacent to the mouth on each side and a second a little behind it. Prothoracic plate pale yellowish; body pale yellowish green, posteriorly pinkish; dorsal line brick-red, barely interrupted at segmental divisions; dorsal area suffused whitish, subdorsal lines reddish, paler than dorsal line; a transverse pinkish crescentic mark on each segment below the subdorsal line; reddish markings becoming darker and denser toward posterior end; each segment with a transverse series of four (or perhaps more) small black dots, except segments with plates; anal plate pale yellowish; thoracic legs and prolegs yellowish green.

The young larvae mine the leaves. First instar mines are typically U-shaped whilst later mines are irregular and even branched. The penultimate instar larva leaves the mine and lives until maturity in a silken tube amongst the seeds, feeding mostly on the unripe seeds. Can be numerous where it is found. September–October.

Pupa. Overwinters as a pupa in a cocoon made of detritus and sand grains spun together.

Imago. July–August. Can be disturbed from amongst its foodplants by day; flies at night and comes to light, often some distance from its nearest foodplant (J. R. Langmaid, pers.comm.).

Distribution (Map 99)

Not recognized in Britain until 1935 when it was separated from *S. obsoletella* and described as *Phthorimaea seminella* (Pierce & Metcalfe). Possibly our commonest salt-marsh gelechiid (Michaelis, *loc.cit.*), it has a widely scattered distribution around the coasts of Britain and has been recorded as far north as Sutherland. Records for west Gloucestershire and south-west Yorkshire require confirmation. It is widely distributed in Ireland, particularly along the eastern coast. Abroad, this moth has been recorded throughout northern Europe; also Mongolia.

SCROBIPALPA OBSOLETELLA (Fischer von Röslerstamm)

Lita obsoletella Fischer von Röslerstamm, [1841] 1834, *Abbild.Bericht.Ergänzung Schmettkde*: 225, pl.79, figs a–k.

Type locality: Austria; Vienna.

Description of imago (Pl.4, fig.25)

Wingspan 12–14mm. Head buff, irrorate grey; antenna ochreous, annulate dark fuscous; labial palpus with segment 3 as long as segment 2 (see figure 56, p.161), segment 3 buff on underside, irrorate fuscous, especially towards apex on upperside, segment 2 buff on outer side, pale buff on inner side. Thorax concolorous with head. Forewing whitish, irrorate with mixed grey and ochreous-yellow scales, the latter colour almost absent in some specimens; subbasal area unspotted; stigmata black, first discal beyond plical, second discal elongate and sometimes forming two spots, the second component distad and dorsad of first; an ill-defined, suffused, angulate pale fascia at three-quarters; apical area often paler and more sparsely irrorate fuscous; fuscous interneural spots sometimes present round apex and on termen; cilia ochreous-white with three slightly darker ciliary lines sometimes visible. Hindwing whitish grey, costa, apex and veins darker; cilia concolorous. Foreleg dark fuscous, annulate buff at intersegmental joints; midleg buff, irrorate fuscous especially on outer side; hindleg buff, intersegmental joints annulate pale fuscous. Abdomen greyish fuscous, segments 1–3 ochreous-yellow, below buff, anterior edge of segments banded fuscous; anal tuft of

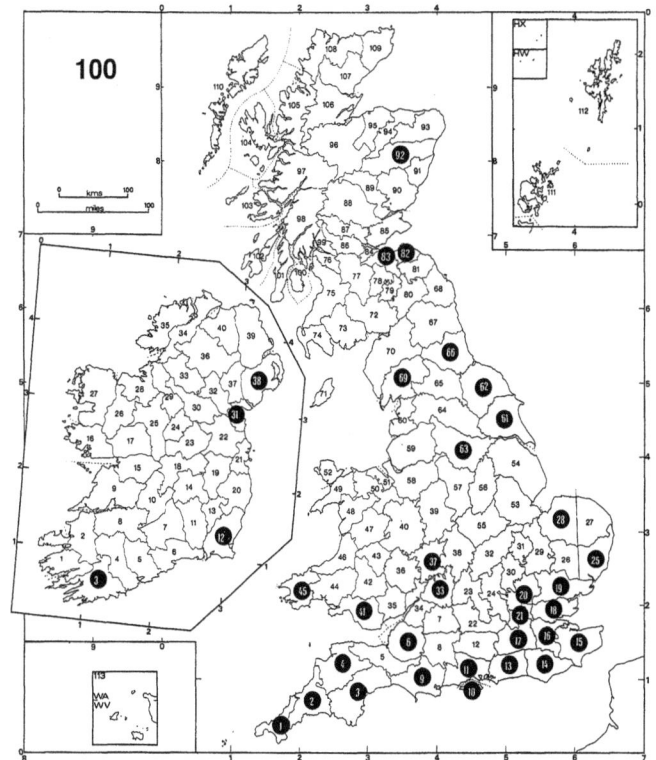

Scrobipalpa obsoletella

male buff. Genitalia, see figures 13f,36d.

Similar species. The palest of our *Scrobipalpa* species. Other species with which it can be confused, for example *S. nitentella* (Fuchs), *q.v.*, and *S. atriplicella* (Fischer von Röslerstamm), *q.v.*, usually have much darker forewings with often bigger or more distinct dark spots. Some forms of *S. salinella* (Zeller) can be superficially similar, but *S. obsoletella* is greyer and without any red-brown coloration on the wings.

Life history

Ovum. Oviposition site not observed. On orache (*Atriplex* spp.) and supposedly on goosefoot (*Chenopodium* spp.). It has been recorded from Babington's orache (*Atriplex glabriuscula*) (Langmaid, 1995).

Larva. Full-grown 9–10mm. Head brown with a dark brown spot at the basal angles, mandibles darker. Prothoracic plate blackish brown with brown median line, sometimes pale anteriorly; body pale green; pinacula small, dark greyish to black; anal plate and legs concolorous with body. Benander (1965) states 'when full-grown with reddish dorsal and dorsolateral lines'.

Larvae can be found from May to October in the stems of the foodplant, eating the pith. Dry whitish frass exuding from a small hole in the stem can betray the presence of a larva. Langmaid (*loc.cit.*) found larvae feeding within the fruits of Babington's orache on the strand-line. In some instances the larva was in a single fruit, but more usually two were spun together. In captivity the larvae left the hostplant and spun up in tissue.

Pupa. Pupates in the stem of the hostplant, a small exit hole betraying its presence. Overwinters as a pupa. June–September and October–May.

Imago. Two or more generations. The adult has been found from May to September. It flies at night and comes to light, often some miles from its nearest food-plant site (J. R. Langmaid, pers.comm.).

Distribution (Map 100)

S. nitentella (Fuchs) was distinguished from *S. obsoletella* in Britain only in 1935 by Pierce & Metcalfe (1935b) and as a consequence there is some uncertainty as to the true identity of some of the older records. Seemingly distributed widely in coastal situations, with a few inland records, in southern England, becoming more local northwards and recorded as far north as south Aberdeenshire, Scotland; South Wales. Recorded from very few sites in Ireland. This species has been widely recorded through Europe and has been found in the U.S.A. (probably as an introduction from Europe).

SCROBIPALPA ATRIPLICELLA (Fischer von Röslerstamm)

Lita atriplicella Fischer von Röslerstamm, [1841] 1834, *Abbild.Bericht.Ergänzung Schmettkde*: 223, pl.78, figs a–h.

Tinea atrella Thunberg, 1788, *D.D.Mus.nat.Acad. upsal.* (6): 78 [Junior primary homonym of *Tinea atrella* [Denis & Schiffermüller], 1775].

Type locality: Austria; Vienna.

Description of imago (Pl.4, fig.26)

Wingspan 13–15mm. Head fuscous; antenna buff, annulate fuscous; labial palpus with segment 3 as long as segment 2, segment 3 fuscous below, buff above, barred black at base and just below apex, segment 2 fuscous on outer side, buff on inner side. Thorax varying from pale to dark fuscous. Forewing with ground colour whitish in pale specimens but generally so heavily irrorate with blackish fuscous as to appear that colour; often ochreous-yellow suffusion; subbasal area sometimes with suffused dark spots on costa and fold, but often obscured by dark suffusion; stigmata gener-

ally large, suffused, ill-defined, first discal beyond plical; additional spots often present on fold and distal to second discal; suffused indistinct angulate pale fascia at three-quarters; terminal area often suffused fuscous; dorsal area sometimes less heavily suffused fuscous; cilia pale ochreous-fuscous with three obscure darker ciliary lines. Underside of forewing often with pale interneural streaks in apical area. Hindwing pale grey with veins and margins darker; cilia pale grey to pale ochreous. Fore- and midlegs fuscous, banded ochreous at intersegmental joints; hindleg dark ochreous to fuscous on outer side, pale buff on inner side, annulate pale ochreous at intersegmental joints; hairs whitish buff. Abdomen fuscous above with segments 1–3 often ochreous-yellow, buff below with fuscous lateral line. Genitalia, see figures 13h,36b.

Similar species. *S. atriplicella* has been confused with *S. nitentella* (Fuchs). Dark specimens of the latter can usually be distinguished from *S. atriplicella* by the greyish buff rather than blackish grey appearance. Some forms of *S. instabilella* (Douglas) and *S. samadensis* (Pfaffenzeller) approach *S. atriplicella*, but the stigmata are usually more distinct in the former two species, whereas *S. samadensis* does not have the mottled appearance of *S. atriplicella*. *S. atriplicella* can be distinguished from *S. nitentella*, *S. instabilella* and *S. samadensis* on genitalia characters. The male genitalia of *S. atriplicella* differ from those of all other species by the very long valvae which significantly exceed the uncus. In the female genitalia the apophyses anteriores usually diverge strongly and the signum is situated on the left side of the bursa copulatrix. *S. obsoletella* (Fischer von Röslerstamm) and some forms of *S. salinella* (Zeller) can seem superficially similar, but the former appears greyer and paler in comparison with *S. atriplicella* and the latter is generally less heavily suffused fuscous and appears less variegated in wing pattern.

Life history

Ovum. Oviposition site not observed. On orache (*Atriplex* spp.), including common orache (*A. patula*), and goosefoot (*Chenopodium* spp.). Bankes (1894) records the species from annual sea-blite (*Suaeda maritima*) and perennial glasswort (*Sarcocornia perennis*), but these hostplants almost certainly apply to *S. nitentella*. In continental Europe, hostplants recorded include frosted orache (*Atriplex laciniata*) (Fischer von Röslerstamm, [1841]), fat-hen (*Chenopodium album*); nettle-leaved goosefoot (*C. murale*), maple-leaved goosefoot (*C. hybridum*) (Povolný, 1990) and *Atriplex portulacoides* (Hering, 1957). A Continental record of spear thistle (*Cirsium vulgare*) (Povolný, 1980) must be considered as erroneous.

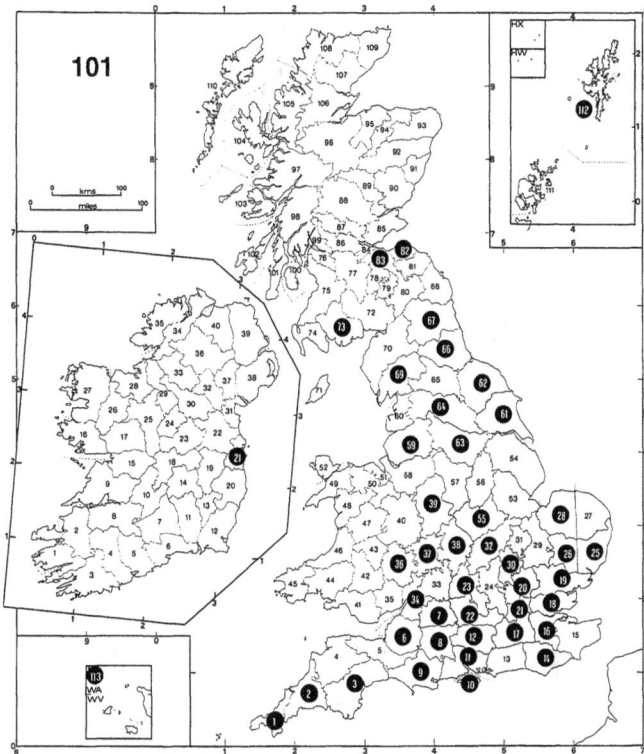

Scrobipalpa atriplicella

Larva. Full-grown *c.*13–14mm. Head honey-yellow with four brown 'side spots'. Prothoracic plate green, sparsely dotted brown laterally; body greenish yellow to grass-green, sometimes with dark to light rose-red suffusion on dorsum; pinacula black; anal plate green with slight black speckling; thoracic legs pale green to light brown, tipped blackish brown; prolegs concolorous with body.

The larva mines the leaves or feeds in spun leaves and on flowers and seeds of the foodplant. May–July and in October.

Pupa. June–July, pupal period lasting 10–12 days, and October–May.

Imago. May and July–August. Bankes (*loc. cit.*) gives May–September and suggests that *S. atriplicella* may have more than two generations a year. Karsholt (pers.comm.) notes that there is a partial third brood from September–October in Denmark.

Distribution (Map 101)

Generally associated with areas of disturbed habitat. Widely distributed in England. Local in Scotland, recorded from near the Borders and Shetland. There is some doubt over the true identity of the record from east Kent; consequently this should be treated as unconfirmed. Seemingly extremely local in Ireland, being recorded only from Co. Dublin. Also recorded from the Channel Islands. Throughout continental Europe to Turkey (Povolný, 1996); Turkestan. Established in the U.S.A. (Pennsylvania; California) and Canada as an introduction from Europe.

SCROBIPALPA OCELLATELLA (Boyd)
Beet Moth

Gelechia ocellatella Boyd, 1858, *Entomologist's wkly Intell.* **4**: 143.

Type locality: England; The Lizard, Cornwall.

Description of imago (Pl.4, fig.27)

Wingspan 11–15mm. Head with vertex varying from ochreous to pale fuscous, frons whitish ochreous; antenna fuscous, ringed ochreous, apical half serrate in male; labial palpus with segment 3 as long as segment 2, segment 3 whitish ochreous, banded fuscous just below tip and at base, segment 2 ochreous, partly irrorate fuscous on outer side, whitish ochreous on inner side. Thorax ochreous, sometimes partly grey-fuscous. Forewing pale ochreous, sometimes buff-tinged, and suffused fuscous, especially on basal half of costa and in cell; dorsal area usually without such irroration; subbasal area with suffused fuscous spots on costa and on fold; stigmata black, generally ringed with ground colour, first discal beyond plical, discal stigmata often divided into two; subapical patch of ground colour on costa and similar patch on tornus opposite, generally joined to form a fascia, though often interspersed or made zigzag by wedges of fuscous irroration; terminal area with fuscous irroration, sometimes tending to form subterminal interneural spots; cilia ochreous-white, an indistinct fuscous basal line more strongly expressed at apex; underside pale fuscous, ochreous between veins towards apex. Hindwing whitish grey, slightly darker at apex; cilia concolorous. Foreleg fuscous, ringed ochreous at intersegmental joints; mid- and hindlegs ochreous, ringed fuscous. Abdomen fuscous to grey-fuscous above, pale ochreous below, partly banded fuscous; anal tuft of male concolorous. Genitalia, see figures 13b,35d.

Similar species. One of the more distinctive of our *Scrobipalpa* species because of the impression of an angulate fascia at about three-quarters. Some forms approach *S. atriplicella* (Fischer von Röslerstamm) but the stigmata and other darker markings usually contrast more with the ground colour in *S. ocellatella*. Some forms of *S. instabilella* (Douglas) are similar, but this species appears darker overall, whereas *S. nitentella* (Fuchs) is greyer in general coloration.

Life history

Ovum. Oval, slightly flattened; white, becoming yellowish brown. Inserted singly or in groups in young, curled heart leaves, in dried petioles, in cracks in the skin of the root or on the soil (Carter, 1984). On sea beet (*Beta vulgaris* ssp. *maritima*). A record of sea aster (*Aster tripolium*) (Threlfall, 1878) is erroneous and refers to *S. salinella* (Bankes, 1894). Abroad, goosefoot (*Chenopodium*), sea-blite (*Suaeda*) and glasswort (*Salicornia*) have been cited as foodplants by Povolný (1980), although all seem to be unlikely regular host-plants, at least in Britain. It can be a pest of sugar beet and beetroot (*Beta vulgaris* ssp. *vulgaris*) in parts of southern Europe, the Middle East and North Africa.

Larva. Full-grown *c.*10mm. Head brown to pale brown, mouth-parts blackish. Prothoracic plate pale brown with black speckling, particularly to the sides and posterior edge; body greenish grey to yellowish green, markings varying from a finer reddish brown transverse series of blotches on each segment forming vague stripes, one dorsal, one subdorsal and two further weaker stripes on the side, to broad dorsal markings pinkish red in colour with a stripe-like marking along the side and a much weaker stripe above the legs; pinacula dark brownish black under magnification, appearing as black dots; anal plate brownish; thoracic legs black.

The larva feeds on the buds and stems and in spun or mined leaves of the foodplant, with a preference for plants in the open. The larval spinnings can be very untidy. Larvae can be found in numbers where they occur. January–May, although possibly overwintering from October (Bankes, *loc.cit.*), and June–August.

Pupa. Light brown, becoming darker; cremaster blunt, without hooked spines (Carter, *loc.cit.*). In captivity the larva pupates in the earth in a flimsy cocoon or amongst detritus (Richardson, 1893). May–June and July–September.

Imago. Bivoltine; May–July and August–October. The adult occasionally wanders a little from its breeding grounds and at night has been recorded at light. The species has been observed flying freely over 'wild beet' (Boyd, 1858).

Distribution (Map 102)

Frequents vegetated coastal shingle and the upper margins of salt-marshes. Although primarily found along the coast of southern England from west Cornwall to east Suffolk, this local moth has also been recorded in Pembrokeshire. In Ireland it is known only from east Cork; it has also been reported from the Channel Islands. Widely distributed in Europe; Asia Minor; North Africa.

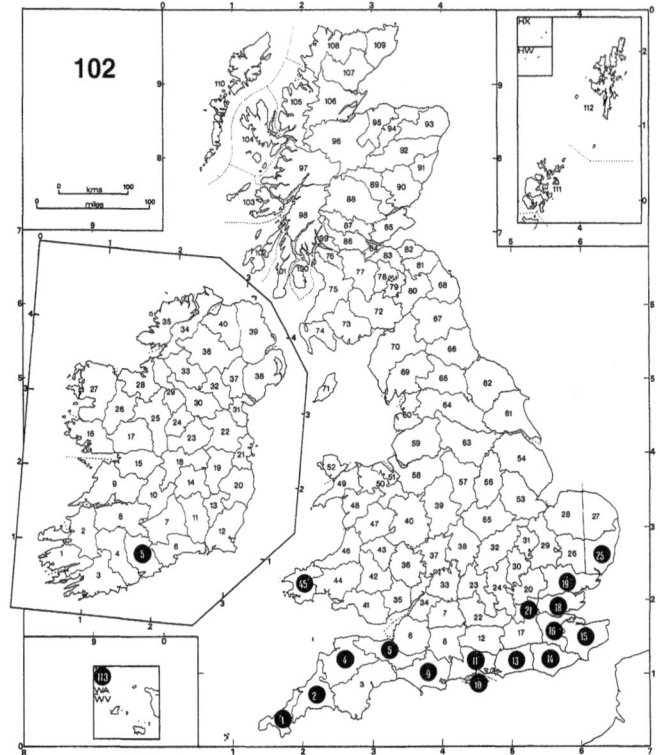

Scrobipalpa ocellatella

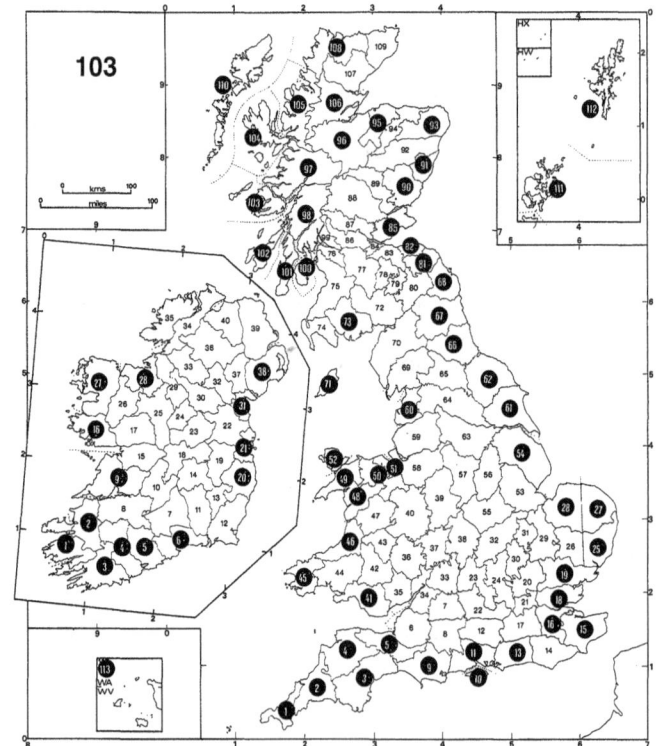

Scrobipalpa samadensis

SCROBIPALPA SAMADENSIS (Pfaffenzeller)

Gelechia samadensis Pfaffenzeller, 1870, *Stettin.ent.Ztg* **31**: 321.

Gelechia plantaginella Stainton, 1883, *Entomologist's mon.Mag.* **19**: 253.

Type locality: Switzerland; Samedan ('Samaden'), Piz Padella (Voralpe Schafberg), Upper Engadin.

Description of imago (Pl.4, fig.28)

Wingspan 11–15mm. Head with vertex whitish ochreous, irrorate pale fuscous, frons whitish; antenna ochreous, ringed fuscous, apical one-half slightly serrate in male; labial palpus with segment 3 as long as segment 2, segment 3 whitish, irrorate fuscous, with apical area black, segment 2 whitish ochreous, irrorate fuscous on outer side, whitish ochreous on inner side. Thorax greyish ochreous. Forewing whitish, irrorate fuscous; costa and veins sometimes broadly streaked orange-brown; four fuscous plical spots, first and second discal stigmata elongate or divided into two spots; apex and termen with blackish spots between veins; cilia greyish ochreous, darker at apex, with two indistinct ciliary lines; underside greyish ochreous. Hindwing whitish grey, slightly darker at apex; cilia concolorous. Foreleg grey, irrorate fuscous, banded ochreous at intersegmental joints; mid- and hindlegs ochreous, darker on outer sides, banded pale ochreous at intersegmental joints. Abdomen greyish above, irrorate fuscous, segments 1 and 2, sometimes also 3, yellowish or blackish ochreous; underside whitish ochreous, lateral line blackish; anal tuft of male greyish ochreous. Genitalia, see figures 12d,34d.

Similar species. *S. atriplicella* (Fischer von Röslerstamm), *S. instabilella* (Douglas) and *S. nitentella* (Fuchs), *qq.v.* *S. stangei* (Hering) has been found among specimens of *S. samadensis* and Karsholt (pers.comm.) notes that the two can be indistinguishable on external characters. In the male genitalia *S. samadensis* differs from *S. stangei* by the squat posterior process of the vinculum and in the female by the oval corpus bursae.

Life history

Ovum. Oviposition site not observed. On buck's-horn plantain (*Plantago coronopus*) or sea plantain (*P. maritima*) (Bankes, 1894). Also reported from ribwort plantain (*P. lanceolata*) (Richardson, 1893).

Larva. Full-grown *c.*10mm. Head brown, appearing black in earlier instar larvae. Prothoracic plate black with a thin lighter line down middle; body dull creamy yellow; dorsal, subdorsal and spiracular lines pale salmon-pink; pinacula black, being slightly larger and appearing dark brown on last segment; anal plate pale brown, lighter than head; thoracic legs black.

The larva feeds in the root and in blotches in the leaves. Wilting and greying leaves can indicate the presence of a larva in the root as does frass in the centre of a plant. Young larvae have been found on sea plantain lightly spinning two leaves together and eating the parenchyma. Occasionally the early instars probably mine the leaves (Bankes, *loc.cit.*). April–August.

Pupa. Pupates in the root of the foodplant. May–August.

Imago. Probably univoltine, but emerging over a long period; June–September. The adult comes to light (Bland *et al.*, 1987a).

Distribution (Map 103)

Associated with salt-marshes but also recorded from sandhills and on vegetated shingle. Povolný (1967) considers *S. samadensis plantaginella* to be the subspecies of European coasts and inland saline areas, differing from the nominotypical subspecies mainly by its ecological dependence on halophilous hostplants. However, it is indistinguishable from the nominotypical subspecies on external characters. Widely distributed around the coast of Britain and Ireland and found as far north as Shetland; the Channel Islands. Widely recorded in Europe.

SCROBIPALPA ARTEMISIELLA (Treitschke)
Thyme Moth

Lita artemisiella Treitschke, 1833, *Schmett.Eur.* **9**(2): 97.

Type locality: Germany; Dresden.

Description of imago (Pl.4, fig.29)

Wingspan 10–12mm. Head with vertex dark fuscous, frons paler; antenna dark fuscous, thinly ringed buff; labial palpus with segment 3 as long as segment 2, both fuscous, inner margin of segment 2 paler. Thorax dark fuscous. Variable in wing markings and coloration. Forewing dark brown, sometimes irrorate whitish and with ferruginous-orange longitudinal streaks; stigmata elongate, black, first discal beyond plical; often black streak on fold distal to plical stigma; dorsal area sometimes paler; sometimes traces of black interneural spots in terminal area and occasionally short dark streak to apex; cilia greyish brown with scattered dark scales at base, two faint dark ciliary lines present in some examples. Hindwing pale grey, darker terminally; cilia brownish grey. Abdomen fuscous above, pale fuscous below. Foreleg fuscous, annulate buff at intersegmental joints; mid- and hindlegs pale fuscous. Genitalia, see figures 14b,37c.

Similar species. Being one of the smallest members of the British *Scrobipalpa*, *S. artemisiella*, although variable in wing markings, can usually be identified without reference to genitalia. The genitalia are indistinguishable from those of *S. stangei* (Hering) except for their generally smaller size. The female genitalia of *S. artemisiella* are also similar to those of *S. suaedella* (Richardson), but this latter species can usually be separated on external characters and occurs in a different habitat.

Life history

Ovum. Oviposition site not observed. On wild thyme (*Thymus polytrichus britannicus*). The record from breckland thyme (*T. serpyllum*) (Stainton, 1865b) is incorrect. In The Netherlands this species is associated with large thyme (*T. pulegioides*) (M. Jansen, pers.comm.) and in southern France it has been reared from winter savory (*Satureja montana*) (BMNH material). Mint (*Mentha* sp.) has been given as a foodplant in Continental literature (Nickerl, 1908). Field wormwood (*Artemisia campestris*) (Treitschke, 1833) and Jersey knapweed (*Centaurea paniculata*) (Nickerl, *loc.cit.*) have been listed as foodplants, but these are probably in error, the former probably referring to *Scrobipalpula psilella* (Herrich-Schäffer) (Stainton, *loc.cit.*).

Larva. Head and prothoracic plate black. Body dull brownish green, dorsal, subdorsal and lateral lines purplish brown, mesothoracic segment purplish brown; pinacula small, black; anal plate blackish; thoracic legs black; prolegs concolorous with body (Heckford, 1995a).

The young larva can mine the foodplant (Hering, 1957), but in later instars it feeds within a silken web on the underside of a stem. May–July.

Pupa. In Ireland, pupae have been found under hoary rock-rose (*Helianthemum canum*) (Heckford & Langmaid, 1991). June–July.

Imago. Univoltine; June–July. The adult can be disturbed from the foodplant.

Distribution (Map 104)

A species of sandhills and limestone habitats. Stainton (1854) gives it as occurring among short grass in sandy and gravelly places. Bland *et al.* (1987a) consider it to be a species characteristic of the machair. Widely distributed but local in Britain, occurring as far north as the Outer Hebrides, and in Ireland. Distributed throughout Europe.

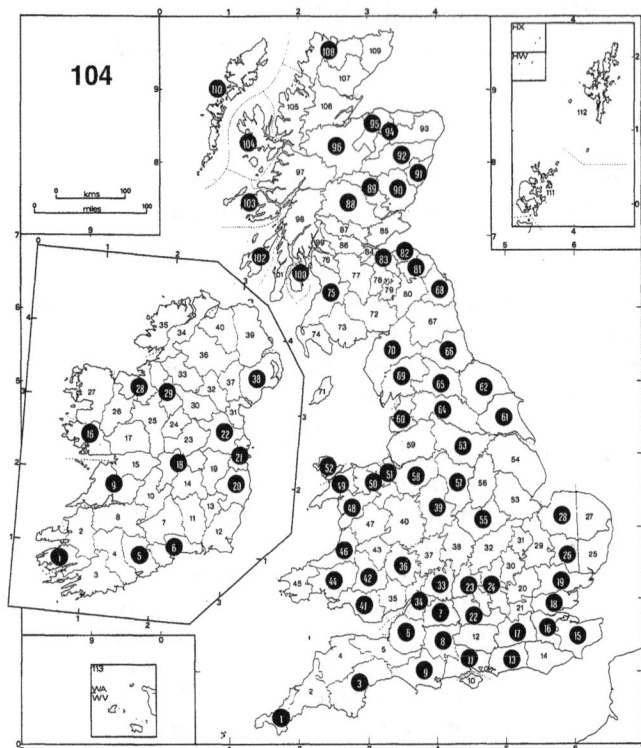

Scrobipalpa artemisiella

SCROBIPALPA STANGEI (E. Hering)

Gelechia stangei E. Hering, 1889, *Stettin.ent.Ztg* **50**: 299.
Type locality: Germany; Friedland, Mecklenburg.

Description of imago (Pl.4, fig.30)

Wingspan 14–16mm. Head grey with brown irroration; antenna brown, ringed dark fuscous; labial palpus with segment 2 as long as segment 3, ochreous, irrorate brown, paler on inner side. Thorax grey with brown irroration. Forewing very variable; greyish brown, dorsum often lighter, sometimes rusty brown dashes on dorsum, in cell and along veins; often indistinct black streak at base of wing near dorsum; plical spots usually distinct, discal and discocellular rarely so; variably developed dark longitudinal streak through cell to apex. Hindwing pale grey, darker at margins and apex; cilia greyish. Legs fuscous, irrorate ochreous, annulate ochreous at intersegmental joints. Abdomen fuscous on upperside, sometimes segments 1–4 with brownish colouring; underside pale ochreous. Genitalia, see figures 13d,35a.

Similar species. Can be confused with *S. samadensis* (Pfaffenzeller), *q.v.* The genitalia are similar to those of *S. artemisiella* (Treitschke), but can be distinguished by their larger size. The female genitalia are similar to those of *S. suaedella* (Richardson) but both can usually be distinguished on external characters.

Life history

Ovum. Biology in Britain not known. On the Continent this species is associated with sea arrowgrass (*Triglochin maritima*) or marsh arrowgrass (*T. palustris*) (Sattler, 1987a).

Larva. Head and prothoracic plate black. Body green with reddish dorsal, dorsolateral and lateral lines; thoracic segments 1 and 2 reddish brown; pinacula dark green; legs black (Benander, 1965). Larsen (1927) records the larva as yellowish or greenish with a weak longitudinal stripe from segment 3; thoracic segments 1 and 2 dark yellowish brown; head and thoracic legs blackish brown.

The larva feeds in the roots and leaves of the food-plant. Autumn–June, the young larva probably mining the leaves during the autumn and overwintering in the rootstock; after hibernation it mines the leaves or feeds in the stem (Hering, 1957). Larsen (*loc.cit.*) notes that feeding in sea arrowgrass can be detected by a drooping leaf, the angle of the droop supporting the larva.

Pupa. In the larval mine. June–July.

Imago. June–July. In Denmark, the adult has been recorded at light (O. Karsholt, pers.comm.).

Distribution (Map 105)

On the Continent, this species has been found mostly but not exclusively in coastal and inland saline habitats (Sattler, *loc.cit.*). Not yet known from localities with extreme levels of salinity. It has also been collected from along a dry footpath that crossed a boggy meadow. This species was added to the British list in 1986 following the discovery in the BMNH collections of old specimens which had previously been misidentified as *S. samadensis plantaginella* (Stainton) and *S. instabilella* (Douglas) (Sattler, *loc.cit.*;1986). Most of the specimens were not labelled, but two had data which indicated that they were found at Yarmouth, Isle of Wight, in July 1882. Searches in 1985 and 1986 failed to rediscover this species (Sattler, *loc.cit.*). In mainland Europe, it has been recorded from Norway, Sweden, Denmark, Finland, Latvia, Germany and Austria (Povolný, 1996).

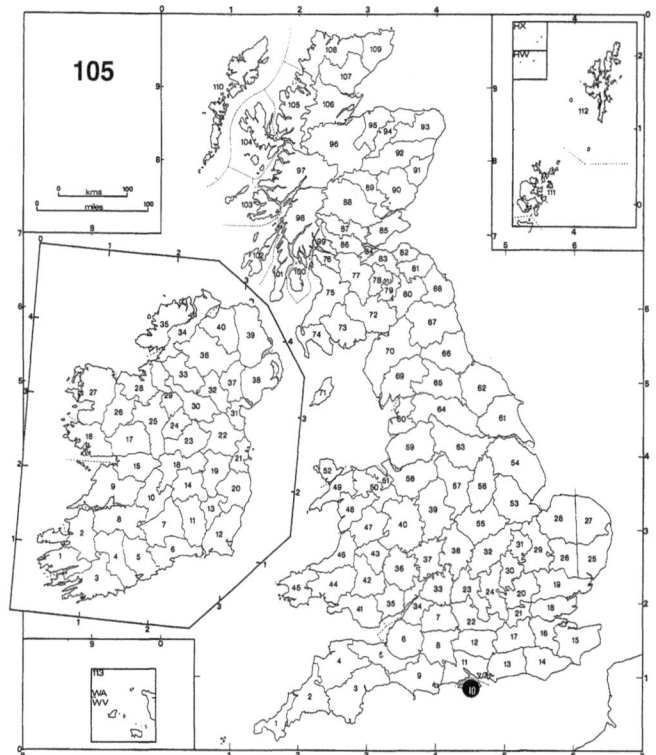

Scrobipalpa stangei

SCROBIPALPA CLINTONI Povolný

Scrobipalpa clintoni Povolný, 1968, *Entomologist's Gaz.* **19**: 113, figs 1–5.

Type locality: Scotland; Glenborrodale, Ardnamurchan, Argyll.

Description of imago (Pl.4, fig.31)

Wingspan 13–15mm. Head with vertex greyish fuscous, frons pale fuscous; antenna fuscous, obscurely ringed ochreous; labial palpus with segment 3 as long as segment 2, fuscous, inner side of segment 2 paler. Thorax dark greyish fuscous, sometimes paler posteriorly. Forewing dark brownish fuscous, streaked greyish ochreous along veins; stigmata absent although second discal stigma visible in some examples; sometimes weak marginal spots at base of fringe from costa to tornus; cilia pale greyish ochreous, often with three faint darker ciliary lines. Underside of forewing with ochreous streaks in apical area. Hindwing whitish grey; cilia whitish. Foreleg fuscous, ringed ochreous at intersegmental joints; mid- and hindlegs ochreous-fuscous, annulate buff. Abdomen fuscous above, fuscous banded buff below; anal tuft of male fuscous-buff. Genitalia, see figures 13g,36b.

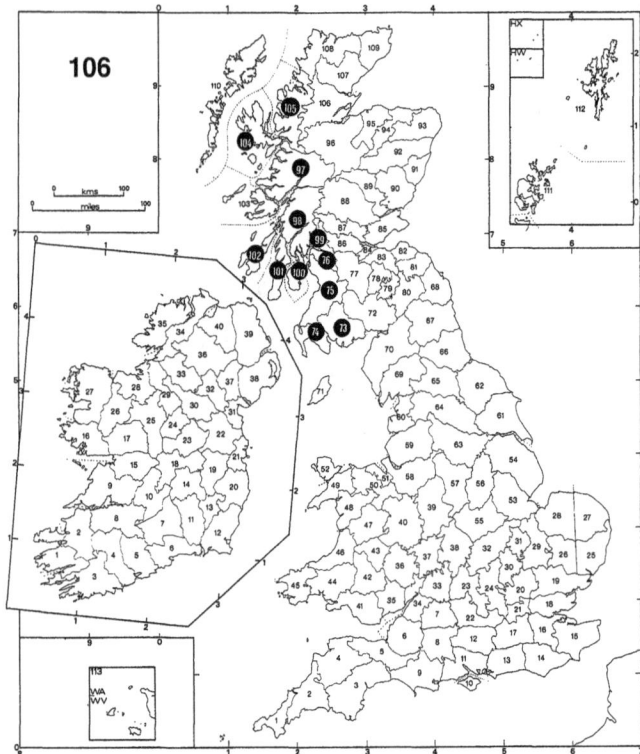

Scrobipalpa clintoni

Similar species. Superficially, *S. clintoni* can resemble *S. pauperella* (Heinemann) and some forms of *S. acuminatella* (Sircom). The pale markings along the veins and the lack of any red-brown coloration normally serve to distinguish *S. clintoni* from these species.

Life history

Ovum. Oviposition site not observed. The hostplant is curled dock (*Rumex crispus*).

Larva. Apparently undescribed.

The larva feeds in July within the stems of the food-plant, ejecting quantities of pale brown frass from holes situated mostly at the nodes. All parts of the stem can be tenanted including the bases of the flower stems and petioles. Larvae can be abundant (Pelham-Clinton, 1971).

Pupa. Within the stem of the foodplant where it over-winters. A small hole in the stem betrays the presence of the pupa.

Imago. April, May and June. The species is bivoltine in Denmark, the second generation flying in July and early August (O. Karsholt, pers.comm.). The adult is seemingly rarely observed in the wild.

Distribution (Map 106)

This species was described from examples collected at Ardnamurchan, Argyll, in 1965. It is local on sandy and shingle beaches and has been found at or a little above high-water mark in south-facing bays (Povolný, 1968). In the British Isles, it has been recorded only on the western coast of Scotland from Kirkcudbrightshire north to North Ebudes and Wester Ross. It has been looked for, without success, in Ireland and Orkney (Pelham-Clinton, *loc.cit.*). In mainland Europe, it is known from Norway, Sweden, Denmark, Estonia, Latvia, western Russia and Germany (Povolný, 1996).

SCROBIPALPA COSTELLA (Humphreys & Westwood)

Anacampsis costella Humphreys & Westwood, 1845, *Br. Moths Transform.* **2**: 192, pl. 107, fig.15.

Type locality: England; Camberwell, London.

Description of imago (Pl.5, figs 1,2)

Wingspan 12–14mm. Head with vertex brown, frons sometimes paler; antenna ochreous-brown, annulate fuscous; labial palpus with segment 3 slightly shorter than segment 2, segment 3 ochreous with subbasal and subapical fuscous bands, segment 2 on outer side brown, irrorate fuscous, on inner side whitish ochreous. Thorax brown, variably irrorate fuscous. Forewing brownish ochreous, variably irrorate fuscous; subtriangular fuscous blotch extending from costa to fold at about one-quarter, broadly extended along costa to two-thirds, often completely obscuring stigmata; discal stigma sometimes ringed with ground colour; first discal sometimes followed by black dot obliquely beneath it; in darker examples subtriangular blotch merging with ground colour but proximal edge still remaining traceable as oblique line; generally an irregular, sometimes angled, fascia of ground colour at three-quarters; apical area with series of fuscous interneural spots or streaks; cilia brownish ochreous, faint ciliary lines sometimes visible. Sexually dimorphic, subtriangular blotch of female contrasts distinctly with ground colour of forewing, more weakly marked in the male. Underside grey. Hindwing greyish brown; cilia concolorous, paler at base on dorsum. Fore- and midlegs fuscous, banded ochreous (reddish in fresh specimens) at intersegmental joints; hindleg pale ochreous, outwardly irrorate brownish fuscous. Abdomen fuscous above, segments 1 and 2 more smoothly scaled, pale ochreous below with lateral row of fuscous spots; anal tuft of male ochreous. Genitalia, see figures 14a,37a.

Similar species. Unlikely to be confused with any other

native species of *Scrobipalpa*. Even in very dark examples the proximal edge of the large subtriangular costal marking remains traceable. Superficially similar to *S. hyoscyamella* (Stainton), *q.v.*, a species not recorded in the British Isles except for a dubious record from the Channel Islands.

Life history

Ovum. Oviposition site not known. On bittersweet (*Solanum dulcamara*). A Continental record for potato (*S. tuberosum*) (Lhomme, [1946–49]) requires confirmation.

Larva. Full-grown *c.*9–10mm. Head black to brown with black speckling, particularly at sides and anterior margin. Prothoracic plate black, brownish anteriorly and at margins, thoracic segments 2 and 3 dull purplish brown; body grey to greyish green; dorsal and subdorsal lines sometimes faintly pink or darker than body colour; pinacula black, faintly margined brown; anal plate brownish grey; thoracic legs black; prolegs concolorous with body.

The young larva lives in the midrib and from there produces irregular blotch-mines which sometimes contain frass. Older mines, which can extend to the leaf margin, are brownish and often inflated (figure 59). The larva can live on other parts of the plant, for

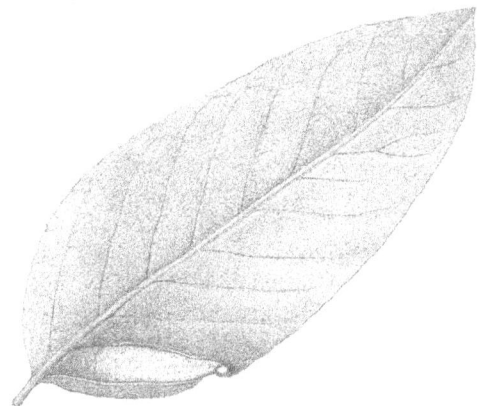

Figure 59 *Scrobipalpa costella* (Humphreys & Westwood), larval mine on *Solanum dulcamara* (×5)

example in a berry or the stem or between lightly spun leaves without mining. May (Bland, 1987), August, September and the beginning of December (J. R. Langmaid, pers.comm.). It has also been found in the stem 'hybernating' (Anonymous, 1852). In The Netherlands this species hibernates as a larva (M. Jansen,

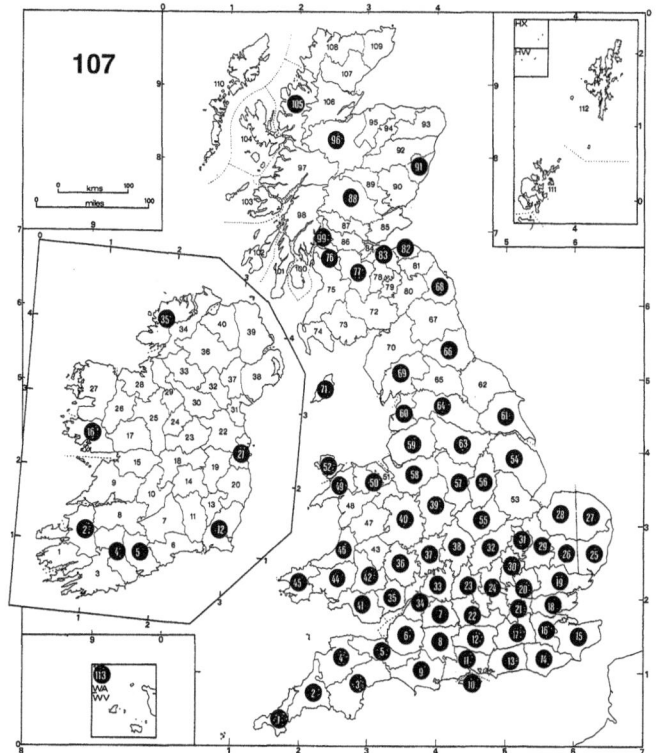

Scrobipalpa costella

pers.comm.). The moth has emerged from spongy oak galls collected in December and from thistle heads collected in winter (J. R. Langmaid, pers.comm.), suggesting that the larva can wander some distance from the hostplant in search of a suitable pupation site.

Pupa. Pupation takes place outside the larval mine. April (R. J. Heckford, pers.comm.), June and September (Bradford & Sokoloff, 1988). In Denmark, the pupa has been found in the stem of bittersweet in early May (Buhl *et al.*, 1996).

Imago. Possibly more than one brood, emerging over a long period. This species was formerly considered to hibernate as an adult. Recorded from late March to mid-November and noted on one occasion in mid-February (J. R. Langmaid, pers.comm.). Occasionally found in houses. Comes to light.

Distribution (Map 107)

Recorded from hedgerows, coastal shingle and gardens. Widely distributed over much of England and Wales, becoming local in northern England and southern and central Scotland. Seemingly local, but perhaps under-recorded, in Ireland. Recorded from the Isle of Man and the Channel Islands. In mainland Europe, this species has been found in France, The Netherlands, Denmark, Austria and doubtfully recorded from Romania (Povolný, 1996). Erroneously listed by Povolný (*loc.cit.*) as being found in the Czech Republic (O. Karsholt, pers.comm.).

SCROBIPALPA HYOSCYAMELLA (Stainton)

Gelechia hyoscyamella Stainton, 1869, *The Tineina of southern Europe*: 233.
Type locality: France; Île St. Honoré.

Recorded from the Channel Islands by Povolný (1996). The record is based on the determination of two females from Jersey in the Zoologisches Museum der Humboldt-Universität, Berlin (D. Povolný, pers. comm.). However, inquiries to the Museum have failed to locate any appropriately labelled specimen. A genitalia slide labelled Jersey does exist, but this relates to an unlabelled specimen. Therefore, in the absence of any further information, the occurrence of this species on the Channel Islands must remain as unconfirmed. In mainland Europe, it has otherwise been recorded only from France and Spain (Povolný, *loc.cit.*). Stainton (1869a) gives a larval description and records henbane (*Hyoscyamus*) as a hostplant.

Similar species. S. *hyoscyamella* is superficially similar to S. *costella* (Humphreys & Westwood) but the former is generally paler. The costal marking in S. *hyoscyamella* just reaches half-way to the dorsum of the wing whereas in S. *costella* it regularly reaches two-thirds. This costal marking is more diffuse towards the termen in S. *costella* and the stigmata in S. *hyoscyamella* are usually more distinct.

SCROBIPALPULA Povolný

Scrobipalpula Povolný, 1964, *Cas.Csl.Spol.ent.* **61**: 339.

A genus of about 40 species, mainly from the New World, represented by four species in central and northern Europe, two of which occur in southern England.

Imago. Head without ocelli; scape of antenna without pecten; labial palpus long and slender, segments 2 and 3 of approximately equal length, segment 2 slightly thickened with appressed scales. Forewing with veins R_4 (8) and R_5 (7) stalked, M_3 (4), CuA_1 (3) and CuA_2 (2) separate. Male genitalia with spatulate gnathos, a feature not found in other Palaearctic Gelechiidae. Female genitalia with a corniform signum in corpus bursae.

Larva. Known larva mines a leaf of colt's-foot (*Tussilago farfara*).

Key to species (imagines) of the genus *Scrobipalpula*

– Wingspan 11–12mm; forewing greyish fuscous, speckled dark fuscous with some ferruginous scales; foodplant unknown in the British Isles .... *diffluella* (p. 184)
– Wingspan 13–14mm; forewing more uniform paler grey, with fewer ferruginous scales; foodplant *Tussilago farfara* .... *tussilaginis* (p. 186)

Unless reared, these species must be dissected to confirm identification.

Key to genitalia of species (imagines) of the genus *Scrobipalpula* (figures14g–i, p. 34)

(Males only, there being no known British females of S. *diffluella*.)
– Saccus subrectangular .... *diffluella* (p. 184)
– Saccus subtriangular .... *tussilaginis* (p. 186)

SCROBIPALPULA DIFFLUELLA (Frey)

Gelechia diffluella Frey, 1870, *Mitt.schweiz.ent.Ges.* **3**: 252.
Scrobipalpa psilella sensu Povolný & Bradley, 1965, *Entomologist's Gaz.* **16**: 9, *nec* Herrich-Schäffer, 1854.
Type locality: Switzerland; Zermatt.

Description of imago (Pl.5, fig.3)
This description is based on the three known British specimens and not on undissected Continental material standing under this taxon in the Natural History Museum, London (BMNH). Wingspan 11–12mm. Head with greyish fuscous scales tipped light grey; antenna dark fuscous; labial palpus whitish, segment 2 suffused fuscous beneath and on outer edge, segment 3 ringed fuscous basally and at apex. Thorax and tegulae greyish fuscous. Forewing greyish fuscous, speckled dark fuscous; stigmata blackish brown, first discal obliquely beyond plical, second discal elongate, with ferruginous scaling between the two discal stigmata and from base to one third of costa and towards apex on one specimen, but such scaling (on base and near apex) reduced or obsolete in the other two specimens; blackish scaling at apex; cilia light fuscous with darker ciliary line. Hindwing light fuscous; cilia concolorous. Fore- and midlegs greyish, speckled whitish, tibiae and tarsal segments ringed whitish; hindleg femur whitish, speckled greyish, tibia and tarsal segments ringed whitish. Abdomen not available for description. Male genitalia, see figures 14g,h.

Similar species. Scrobipalpa artemisiella (Treitschke) has a darker brown forewing, often becoming paler dorsally, and segment 3 of the labial palpus is generally dark fuscous with some buffish scaling; *S. instabilella* (Douglas) usually has a greater wingspan, the labial palpus is buff and segment 3 is usually ringed fuscous both basally and apically; *S. acuminatella* (Sircom) usually has the ferruginous scaling forming rather diffuse longitudinal streaks, often obsolete, the labial palpus is buff and segment 3 is not ringed fuscous basally, but is ringed fuscous apically; *Scrobipalpula diffluella* also differs from these by the more speckled appearance of the thorax and forewing. *S. tussilaginis* (Stainton) is very similar to *S. diffluella*, but generally has a slightly larger wingspan, more obvious in series, a more uniform paler grey appearance, with fewer ferruginous scales; but, unless reared, these two species can be separated reliably only by examination of the genitalia.

Life history

Early stages. Unknown in the British Isles, though there is a specimen in the BMNH among a series of *Scrobipalpa acuminatella* (Sircom) taken by L. T. Ford in 1926 and labelled 'ex l., brown blotches, thistle, early July'. However, as thistles (*Carduus* spp., *Cirsium* spp.) are not reliably recorded elsewhere as hostplants, Ford's record requires confirmation. Klimesch (1951) notes that the larva is similar to *Scrobipalpa psilella* (Herrich-Schäffer) and causes blotch-like leaf-mines. According to Huemer & Karsholt (1998), it has been reared from alpine aster (*Aster alpinus*) and blue flea-bane (*Erigeron acre* [sic]), and in Scandinavia the hostplant is *Erigeron politus*, considered by some to be a subspecies of the latter. Klimesch (1958) and Povolný (1964) record *Erigeron* spp., purple colt's-foot (*Homogyne alpina*) and false aster (*Bellidiastrum michelii* (=*Aster bellidiastrum*)). Blue fleabane alone of these occurs in the British Isles.

Imago. Late June–early July.

History and distribution (Map 108)

Added to the British list as *Scrobipalpa psilella* by Povolný & Bradley (1965) following the genitalic determination of three males in the BMNH. They had been taken at Dartford Heath, Kent, on 24 June 1850; Benfleet, Essex, on 3 July 1899; and Erith Marshes, Kent, at the end of July 1926, and were misidentified as *S. artemisiella*, *S. instabilella* and *S. acuminatella* respectively. No other specimen has yet been recorded. Because the uncus, gnathos and saccus differ slightly in the two dissected specimens for which slides were made, the third having been mounted on a piece of card, the genitalia of both specimens have been illustrated.

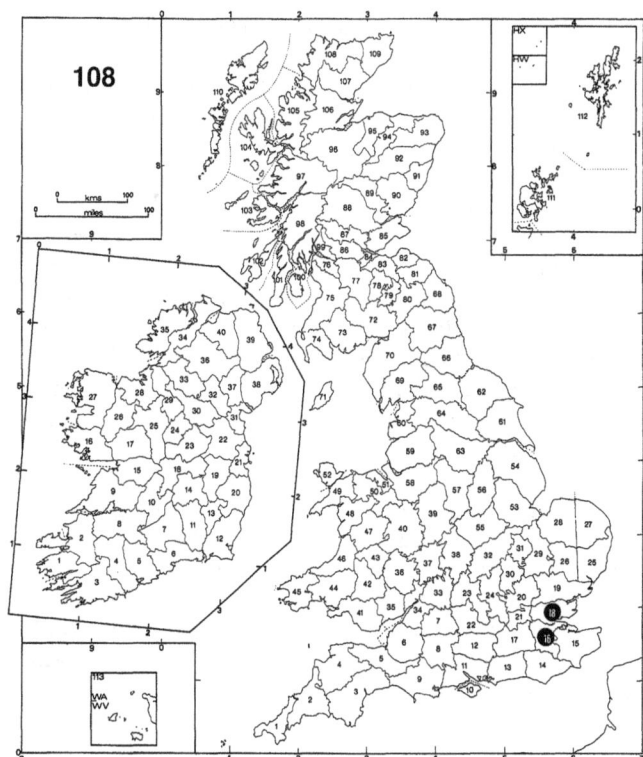

Scrobipalpula diffluella

As the female is, as yet, unknown in the British Isles, the female genitalia of a Continental *diffluella* have not been illustrated. The figure of the female genitalia in Povolný & Bradley (*loc.cit.*) is, in fact, that of *psilella*.

In their review of Old World *Scrobipalpula* spp., Huemer & Karsholt (*loc.cit.*) recognize four species from central and northern Europe: *S. psilella*, *S. ramosella* (Müller-Rutz) sp.rev., *S. diffluella* and *S. tussilaginis*. The present author was aware of this paper before it was finalized and had drawn attention to the fact that the British specimens did not agree with the key genitalic character proposed for separating male *S. psilella* from the other three species, namely that the apex of the valva exceeded the uncus. In the English specimens they do not, the genitalia more closely resembling *S. diffluella*, which Huemer & Karsholt (*loc.cit.*) consider a mainly montane species. It was following these comments and examination of the English specimens and slides that they stated that the English records of *S. psilella* were dubious. Until more material is obtained and the life history in Britain is discovered, their status remains problematic. It has been decided to assign the British specimens to *S. diffluella*, unless further investigation shows that the British taxon is undescribed, as they best fit this species as understood by Huemer & Karsholt (*loc.cit.*) (Heckford & Sattler, 1999).

It seems unlikely that the British moths were immigrants or accidental importations and probable that the species is resident in the extreme south-east of England. On 11 June 1996, at the time when they were believed to be *S. psilella*, the present author, with D. J. L. Agassiz and J. R. Langmaid, visited Dartford Heath and an area near Erith Marshes but no evidence of larvae was found on mugwort (*Artemisia vulgaris*), one of its foodplants. The author also visited Benfleet on 30 May 1998 with the same negative result. Benfleet has a range of habitats and it is not known exactly where the specimen was taken but, as it was misidentified as *Scrobipalpa instabilella*, it might have been found on or near what remains of the salt-marsh bordering the Thames. However, if the specimen from Dartford Heath had bred locally and not near the Thames, the pabulum would not have been a salt-marsh plant. Blue fleabane has been recorded from the Essex tetrad which includes the Benfleet area and the Kent tetrads which encompass Erith Marshes and Dartford Heath, but the author has not observed it at these localities.

SCROBIPALPULA TUSSILAGINIS (Stainton)

Gelechia tussilaginis Stainton, 1867, *Nat.Hist.Tineina* **10**: 14.

Lita tussilaginella Heinemann, 1870, *Schmett.Dtl.Schweiz* (2)**2**(1): 251.

Type locality: Switzerland; Zurich.

Description of imago (Pl.5, fig.4)

Wingspan 13–14mm. Thorax, forewing and legs with a speckled appearance, caused by dark scales having pale bases and tips. Head varying from dark fuscous to ochreous; antenna dark fuscous, ringed ochreous, in female lighter and more distinctly ringed than in male, scape dark fuscous; labial palpus ochreous-white, often variably mixed dark fuscous, segment 3 marked dark fuscous below near base and ringed dark fuscous subapically. Forewing very similar to *S. diffluella* (Frey), but having a more uniform paler grey appearance, with fewer ferruginous scales and plical stigma sometimes obsolete; cilia light fuscous, with an indistinct line of dark scales near base. Hindwing pale fuscous; cilia concolorous. Fore- and midlegs dark fuscous, speckled ochreous, tibiae and tarsal segments ringed ochreous; hindleg femur ochreous, speckled dark fuscous, tibia and tarsal segments ringed ochreous. Abdomen dark greyish fuscous above, mixed dull reddish towards base, light ochreous towards apex beneath; anal tuft light ochreous. Genitalia, see figures 14i,37f.

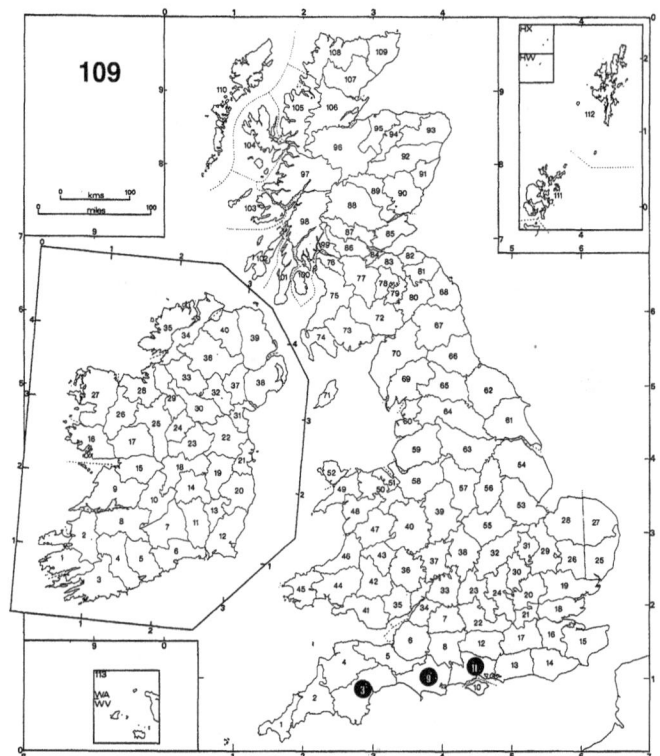

Scrobipalpula tussilaginis

Similar species. *S. diffluella*, q.v.

Life history

Ovum. Undescribed. Oviposition site unknown, but presumably on the foodplant, colt's-foot (*Tussilago farfara*).

Larva. Head dark ochreous-brown, darker at sides and spotted darker posteriorly. Prothoracic plate coloured as head, darker posteriorly, bisected; body bright apple-green, dorsal surface matt; pinacula small and indistinct with short setae; small anal plate concolorous with body. Immediately prior to vacating the mine the body becomes suffused with pink, especially posteriorly and between segments.

The larva mines the upper surface of a leaf, usually causing the surrounding area to turn purple. The frass is usually deposited in one part beneath which the larva may retreat; consequently the mine may appear to be vacated. It apparently occurs only on plants on bare soil with a preference for those with small leaves close to the ground. July–August; October–November.

Pupa. Undescribed; in a cocoon. In captivity larvae spun cryptic cocoons incorporating soil particles on the underside of leaves (R. J. Heckford & J. R. Langmaid, pers.obs.).

Imago. Bivoltine; June –July; August–September.

Distribution (Map 109)

It occurs only near the shore on bare sand or clay soil where there is little vegetation. First found in 1983 (Pelham-Clinton, 1989) at Axmouth in south-east Devon. It has since been recorded in several localities in Dorset and one in Hampshire. It has not been found to the west of Axmouth despite searches in apparently suitable localities in Devon and Cornwall (R. J. Heckford, pers.obs.). France, The Netherlands, Germany, Switzerland, Italy, Austria, Poland, Hungary, Greece and Turkey.

GNORIMOSCHEMA Busck

Gnorimoschema Busck, 1900, *Proc.U.S.natn.Mus.* **25**: 405.

A Holarctic genus of over 50 species, represented in Great Britain by a single species recorded only from Aviemore, Scotland and not observed since 1909.

Imago. Head with small ocelli immediately above and posterior to eye; scape of antenna without pecten; labial palpus recurved, long and slender, segment 3 shorter than segment 2, segment 2 much thickened with dense scales, rough and furrowed beneath. Forewing with veins R_4 (8) and R_5 (7) stalked, R_5 to costa, M_3 (4) and CuA_1 (3) connate, M_2 (5) somewhat approximated, M_1 (6) and R_5 parallel.

Larva. Unknown.

GNORIMOSCHEMA STRELICIELLA (Herrich-Schäffer)

Gelechia streliciella Herrich-Schäffer, 1854, *Syst.Bearb. Schmett.Eur.* **5**: 171; 1853, *Ibid*: pl.67, fig. 495 (non-binominal).

Phthorimaea strelitziella Heinemann, 1870, *Schmett.Dtl. Schweiz* (2)**2**(I): 245 [incorrect subsequent spelling].

Type locality: Germany; Neustrelitz.

Description of imago (Pl.5, fig.5)

Wingspan 11–14mm. Head brownish fuscous, irrorate white, frons white; antenna dark fuscous, ringed whitish; labial palpus segments 1 and 2 brownish fuscous, mixed whitish, inner surface whitish, segment 3 brownish fuscous, mixed whitish, sometimes with sub-basal whitish band. Thorax and tegulae dark fuscous, mixed whitish grey. Forewing dark fuscous, irregularly sprinkled whitish grey; stigmata black, usually edged brown and usually whitish grey between discal spots,

plical before first discal; brown spot on costa at base; slender oblique whitish grey fascia at one-quarter reaching fold; black oblique fascia immediately beyond also reaching fold; angulate slender whitish grey fascia at three-quarters, forming a spot on costa; some whitish grey scales towards apex; cilia light fuscous with black ciliary line from apex fading to tornus. Hindwing light fuscous; cilia concolorous. Legs dark fuscous, mixed whitish grey, tarsal segments banded whitish distally; hindleg with whitish band at tibial spurs. Abdomen dark greyish, sometimes with brown scaling along dorsum, often becoming lighter grey posteriorly. Genitalia, see figures 14f,38b.

Wolff (1971) stated that Pierce & Metcalfe (1935a, pl.X) apparently illustrated the male genitalia of *G. valesiella* (Staudinger) (then *valesiellum*) as *streliciella*. Although their figure of the male genitalia resembles the photograph of *G. valesiella* in Wolff's work, comparison of Pierce's slide in BMNH with Wolff's photographs of *streliciella* and *valesiella* shows them all to be of *streliciella*. The illustration in Pierce & Metcalfe (*loc. cit.*) of the female genitalia is in fact of *Exoteleia dodecella* (Linnaeus) (Heckford, 1995b).

Life history

Early stages. Undescribed. Some Continental authors have suggested an association with thyme (*Thymus* spp.), especially wild thyme (*Thymus serpyllum*) (Piskunov, 1981), but, in his diary for 1909 (in Entomology Library at BMNH; unpublished), Bankes speculated that it might be associated with mountain everlasting (*Antennaria dioica*), a view supported from Finnish observations by Kaitila (1996), who considers *Thymus* spp. to be unlikely hostplants but that *Antennaria* is probable, as adults have been netted in numbers over clumps of *A. dioica*, or possibly *Solidago virgaurea*.

Imago. Univoltine. Recorded in the British Isles only between 14–29 June.

History and distribution (Map 110)

Known with certainty only from Aviemore, Inverness-shire, Scotland. Parsons (1995[1996]) refers to an undated, but presumed old, record from Westmorland which has not been traced and so is not included on the map. Stainton (1872) added it to the British list from two specimens taken amongst marram grass by E. N. Bloomfield at Lowestoft, Suffolk, on 28 July 1871. Bankes (1893) showed this to be a misidentification of *Chionodes fumatella* (Douglas), but *G. streliciella* was restored to the British list after the discovery of previously misidentified specimens taken by C. T. Cruttwell at Aviemore in 1905 (Bankes, 1907). The

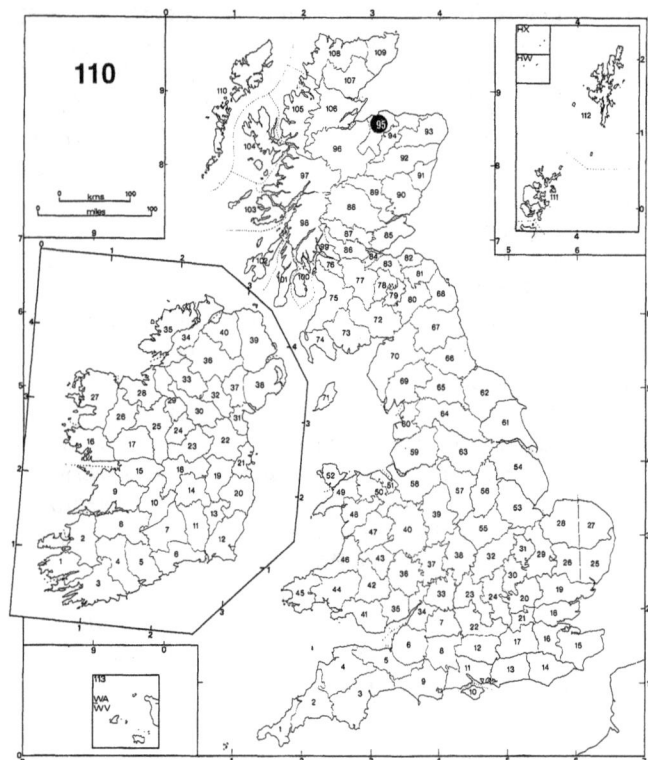

Gnorimoschema streliciella

BMNH has two specimens taken by Cruttwell, eight specimens without data from the Rait-Smith collection, and several from the Bankes collection, taken at Aviemore in 1909, which appear to be the last specimens recorded in the British Isles. The locality where the species had been found is an area of mixed dry heathland known as Granish Moor, though not marked as such on the current Ordnance Survey 1:50,000 map. It lies within NH9014/9015, immediately to the east of the A95 and to the west of the Strathspey railway line. According to Cruttwell (1907), the moth was to be found among heather which had been burnt the previous year. Bankes thought so too, for he wrote in his diary for 21 June 1909, 'I think it probable that, like G[elechia] solutella [*Prolita solutella* (Zeller)], it is always drawing to the recently burnt patches of heath, thriving there for 2 or 3 years, then gradually disappearing as its foodplant (Can this be *Antennaria dioica* which occurs where I take the insect?) becomes choked by the ling'.

Parsons (*loc.cit.*) presumes the species to be extinct in the British Isles. Granish Moor was visited from 30 June to 2 July 1996 and again from 19–24 June 1997 in an attempt to rediscover it. It is clear from a study of Bankes's diary that since his time a considerable area

has been enclosed and is now cattle-grazed; part of what remains has become a landfill site. There was no evidence that any part had been recently burnt; only a few plants of *Antennaria dioica* were found and no *Solidago virgaurea* (R. J. Heckford, pers.obs.). The species may occur elsewhere in the area but visits made during the same period to nearby sites containing *Antennaria dioica* were unproductive.

Believed to be declining generally in Europe. Recorded from Italy, France, northwards to Scandinavia and east to Poland and the Baltic States, but regarded as doubtful from Sweden (O. Karsholt, pers.comm.).

PHTHORIMAEA Meyrick

Phthorimaea Meyrick, 1902, *Entomologist's mon.Mag.* **38**: 103.

A Neotropical genus of about seven species, represented in Britain by *P. operculella* (Zeller), which is mainly imported with produce but possibly also an immigrant.

Imago. Head with small ocellus immediately above eye; scape of antenna without pecten; labial palpus long and slender, segments 2 and 3 of approximately equal length, segment 2 slightly thickened with appressed scales. Forewing with veins R_5 (7) and M_1 (6) remote, nearly parallel, M_3 (4) and CuA_1 (3) connate and CuA_2 (2) remote. Hindwing of male with brush of long pale scales on costa.

Larva. On Solanaceae.

PHTHORIMAEA OPERCULELLA (Zeller)
Potato Tuber Moth

Gelechia (?*Bryotropha*) *operculella* Zeller, 1873, *Verh.zool.-bot.Ges.Wien* **23**: 262, pl.3, fig.17.
Gelechia terrella Walker, 1864, *List Specimens lepid. Insects Colln Br.Mus.* **30**: 1024, *nec* [Denis & Schiffermüller], 1775.
Gelechia tabacella Ragonot, 1878, *Bull.Soc.ent.Fr.* **1879**: 146.
Gelechia sedata Butler, 1880, *Cistula ent.* **2**: 560.
Type locality: U.S.A.; Texas.

Description of imago (Pl.5, fig.6)
Wingspan 14–15mm. Head greyish ochreous, mottled fuscous; antenna dark fuscous, mottled ochreous; labial palpus segments 1 and 2 greyish ochreous, mottled fuscous, segment 2 inner edge more ochreous, outer edge fuscous, banded ochreous medially and api-

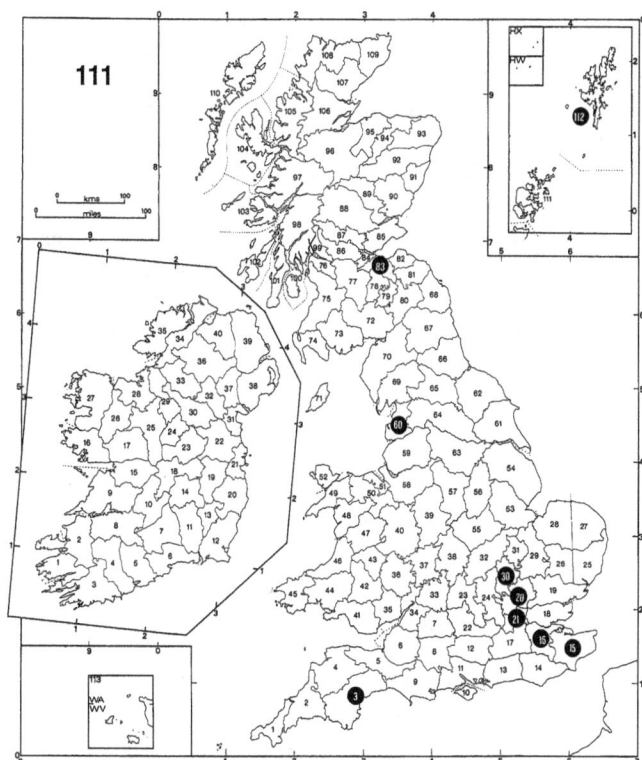

Phthorimaea operculella

cally, segment 3 greyish ochreous, darker towards apex. Thorax and tegulae greyish ochreous, mottled fuscous. Forewing greyish ochreous, mixed light brown, basally with obscure black markings along fold and scattered black scaling over wing and along termen; cilia greyish ochreous with black ciliary line becoming obsolete towards tornus. Hindwing fuscous, in male with brush of long pale scales on costa; cilia concolorous. Legs dark fuscous, tarsi of all legs ringed ochreous. Abdomen dark fuscous, laterally fuscous mottled ochreous, posterior edge of each segment light fuscous, becoming yellowish posteriorly; in the male the posterior three segments have slender semi-erect scales, and segment 8 is greatly elongate and flattened. Genitalia, see figures 14e,38a.

Similar species. Resembles some *Scrobipalpa* spp. Adult *P. operculella* taken in the wild can be reliably identified only by dissection, unless they are male and the brush of long pale scales on the costa of the hindwing is apparent.

Life history

Ovum. Oval and yellowish. Laid on potato (*Solanum tuberosum*) or on other Solanaceae spp. including tomato (*Lycopersicon esculentum*).

Larva. Head dark brown. Prothoracic plate black; body rose-pink, intersegmental areas greenish brown; anal plate greenish brown (Chalmers-Hunt, 1970). On potato, mining the tubers and leaves. Larvae have not been found in the wild in the British Isles. July–September.

Pupa. Yellowish or reddish brown; normally in the soil.

Imago. June, July and September.

Distribution (Map 111)

First recorded in the British Isles in 1935 from the East India Dock, London, by Jacobs (1936; 1958) from potatoes imported from Malta; since then occasionally imported with both potatoes and tomatoes. Has been taken at least three times at light, twice in Kent, on one occasion with immigrants, and once in Devon, so it is possible that it is also migratory. Widespread throughout the tropical, subtropical and Mediterranean regions, and a significant cosmopolitan pest.

CARYOCOLUM Gregor & Povolný

Caryocolum Gregor & Povolný, 1954, *Zool.ent.Listy* **3**: 87.

A Holarctic genus represented by approximately 70 species from western Europe to the Middle East and occurring mainly in montane areas. There are 13 species known from Great Britain and Ireland, the type localities of ten of which are in the British Isles.

Imago. Differs from related genera primarily in characters of the genitalia (Huemer, 1988). The antenna has the scape fuscous, shortly tipped white or whitish, and the flagellum ringed. Segment 3 of the labial palpus is never ringed but may have a white tip. This is useful for separating *Caryocolum* from species such as *Parachronistis albiceps* (Zeller), *Recurvaria nanella* ([Denis & Schiffermüller]) and *Teleiodes sequax* (Haworth) which all have dark rings on segment 3 of the labial palpus. The basic forewing pattern varies little throughout the genus; some of the species, however, vary considerably, particularly in the extent of the dark markings, adding to the difficulty of distinguishing the species. The basic forewing pattern consists of four blackish marks, an elongate mark on costa at one-sixth, an elongate mark on fold at one-third, two discal spots, the first at two-fifths, small and only slightly elongate, the second, at three-fifths more elongate, often inverted L-shaped, these marks frequently more or less obscured by patches of dark fuscous scales, the costal and plical marks being very commonly joined by a dark fuscous bar; in addition to the blackish marks, there is a white or pale fascia at four-fifths, usually interrupted in the middle to form separate costal and dorsal spots. One form of *Carpatolechia decorella* (Haworth) has a superficial resemblance to a *Caryocolum* after the scale-tufts are lost, but is immediately separable by its wholly dark fuscous antenna. The species of *Caryocolum*, apart from *C. alsinella* (Zeller) and *C. viscariella* (Stainton), are readily distinguished by their genitalia. Univoltine, a single species hibernating as an adult. Species are attracted to light and can sometimes be found during the day, flying, or resting on a tree-trunk or by sweeping the foodplant.

Early stages. The ovum is not described and nothing is known about the oviposition site. All known European species feed on Caryophyllaceae, mostly on leaves, flower-heads, or seeds, but occasionally as stem-borers or leaf-miners. The larval descriptions are based mainly on those of Stainton (1867a). Patočka (1989) provided a key based on pupal characters which includes five species that occur in Britain.

Key to species (imagines) of the genus *Caryocolum*

1   Thorax white, or creamy white, sometimes with some greyish fuscous or ochreous scales anteriorly, or more general mottling with light fuscous scale-tips, but appearing white or whitish 2
–   Thorax grey, ochreous or fuscous 6

2(1) Head and thorax without fuscous scale-tips; forewing with dark fuscous bar from costa to fold, obscuring spots 3
–   Head and thorax more or less mottled with fuscous scale-tips; forewing without dark fuscous bar from costa to fold, or with bar not dark enough to obscure spots 4

3(2) Wingspan usually over 11mm; forewing with series of terminal fuscous dots; male genitalia with narrow incision in posterior margin of vinculum ....
.... *blandella* (p. 197)
–   Wingspan up to 11mm; forewing without distinct series of fuscous terminal dots; male genitalia with V-shaped incision in posterior margin of vinculum ......
.... *blandulella* (p. 201)

4(2) Wingspan 12mm or more; no fuscous bar from costa to fold .... *kroesmanniella* (p. 202)
–   Wingspan up to 12mm; forewing with weakly marked, ochreous- and fuscous- mixed bar from costa to fold 5

5(4) Forewing with costal and dorsal white spots at four-fifths more or less separated by an ochreous patch ....
.... *blandelloides* (p. 198)
–   Forewing with costal and dorsal white spots forming angulate fascia at four-fifths, not separated by ochreous scales, but sometimes nearly separated by fuscous scales .... *huebneri* (p. 203)

6(1) Forewing blackish fuscous, usually with four white spots, without orange-ochreous markings ....
.... *vicinella* (p. 191)
–   Forewing always with some orange-ochreous markings, occasionally much reduced in dark specimens, but these lack four white spots 7

7(6) Forewing with dark fuscous patch or bar from costa at one-sixth or beyond and extending to fold 8
–   Forewing with costa fuscous or dark fuscous, but without distinct patch or bar from costa at one-sixth to fold 12

8(7) Forewing with proximal margin of dark fuscous patch at angle of 60–80° to costa ....
.... *proxima* (p. 196)
–   Forewing with proximal margin of dark fuscous patch at angle of about 45° to costa 9

9(8) Forewing with large dark fuscous triangle, or two overlapping triangles, extending along large part of costa .... *tricolorella* (p. 200)
–   Forewing with dark fuscous bar meeting costa only for a short distance 10

10(9) Frons silvery; male genitalia with posterior margin of vinculum notched ....*junctella* (p. 199)
–   Frons shining pale ochreous-grey; male genitalia with posterior margin of vinculum with processes
.... 11

11(10) Wingspan 7.5–11.0mm; forewing with mainly white scales forming incomplete fascia preceding, and triangle following, the dark fuscous bar ....
.... *alsinella* (p. 192)
–   Wingspan 10–14mm; forewing with orange-ochreous scales forming incomplete fascia preceding, and triangle following, the dark fuscous bar ....
....*viscariella* (p. 193)

12(7) Forewing with white triangle, often more or less heavily infuscate, then appearing as grey triangle, in mid-wing between fold and discal spots ....
.... *marmorea* (p. 194)
–   Forewing without such triangle ....
....*fraternella* (p. 195)

Provisional key to larvae of *Caryocolum*

Some life histories are very poorly known, with some larvae undescribed. Some species cannot at present be separated as larvae. The discovery of new foodplants could affect the reliability of this key.

CARYOCOLUM VICINELLA (Douglas)

Gelechia vicinella Douglas, 1851, *Trans. ent. Soc. Lond.* (N.S.) **1**: 102.
Lita inflatella Chrétien, 1901, *Naturaliste* **23**: 17.
Gelechia leucomelanella sensu auctt., *nec* Zeller, 1839.
Type locality: Ireland; [near Belfast].

Description of imago (Pl.5, fig.7)

Wingspan 10–15mm. Head greyish fuscous from white scales with variably darkened tips, frons shining whitish; antenna with scape dark fuscous, white at apex, flagellum dark fuscous, narrowly ringed grey, paler beneath; labial palpus fuscous, white on upper and inner sides of segment 2. Thorax fuscous. Forewing fuscous to dark fuscous, paler in dorsal region to mid-wing; four white spots of varying extent, one forming partial fascia from near costa at one-sixth towards dorsum at one-quarter, one in middle of wing closer to dorsum than to costa, usually with incursion of fuscous scales in middle of proximal edge, costal and dorsal spots at about four-fifths narrowly separated by fuscous scales; small blackish fuscous spots, more or less obscured by fuscous ground colour, in middle of base adjacent to first white spot, plical spot elongate between first and second white spots, discal spots before and after costal side of central white spot; apical cilia grey at base, with greyish fuscous ciliary line, greyish buff at apex, tornal cilia grey. Hindwing light grey; cilia pale greyish fuscous. Legs whitish on inner sides, outer sides fuscous, distal ends of tibiae and tarsal segments white, hind tibia with white median band. Abdomen whitish. Genitalia, see figures 15a,38c.

Similar species. No other *Caryocolum* species has four large white spots, but these may be much reduced in some specimens, which resemble the darkest forms of other species. Genitalic examination is desirable for such forms.

Life history

Ovum. British hostplant records are restricted to sea campion (*Silene uniflora* (=*S. maritima*)); on the Continent also recorded from sand spurrey (*Spergularia rubra*), field mouse-ear (*Cerastium arvense*), alpine catchfly (*Lychnis alpina*), Nottingham catchfly (*Silene nutans*) and tunic-flower (*Petrorhagia saxifraga*) (Huemer, 1988, 1993); oviposition site not known.

Larva. Head black. Prothoracic segment reddish brown, with divided black prothoracic plate; body green; pinacula small, black.

It feeds between spun shoots and leaves which become withered. The young larva is possibly a leaf-miner (Stainton, 1867a). Mid-April–early June.

Pupa. Brown; in a whitish cocoon on the ground, frequently attached to a stone (Bradford & Sokoloff, 1988). June.

Imago. Univoltine; late June–early September. It flies by night and comes to light.

Distribution (Map 112)

Occurs locally on coastal shingle and cliffs. In England it is found in the southern and south-eastern counties from Cornwall to Norfolk; in Wales in most coastal counties; in Scotland mostly on the east coast but also in Ayrshire; the Isle of Man; in Ireland in scattered localities on the southern, eastern and northern coasts, the type locality being the Belfast district; the Channel Islands. Throughout Europe, frequently on alpine scree.

CARYOCOLUM ALSINELLA (Zeller)

Gelechia alsinella Zeller, 1868, *Stettin.ent.Ztg* **29**: 145.

Lita albifrontella Heinemann, 1870, *Schmett.Dtl.Schweiz* (2)**2**(1): 266.

Lita tristella Heinemann, 1870, *Ibid.* (2)**2**(1): 267.

Lita semidecandriella Tutt, 1887, *Entomologist* **20**: 29.

Gelechia semidecandrella Threlfall, 1887, *Entomologist's mon.Mag.* **23**: 233.

Type locality: Italy; Raibl (now Cave del Predil).

NOTE. The British populations were described as a valid species under the name *Gelechia semidecandrella* (Threlfall), which was later relegated to a subspecies of *C. alsinella*. However, adults of similar phenotype occur in the Mediterranean region and therefore *semidecandrella* is considered as infrasubspecific.

Description of imago (Pl.5, fig.8)

Wingspan 7.5–11.0mm. Head dark fuscous, frons shining pale ochreous-grey; antenna with scape dark fuscous, whitish at apex, flagellum ringed grey, brown and dark fuscous, paler beneath; labial palpus with segment 2 whitish, light fuscous on outer side, segment 3 dark fuscous, narrowly whitish above. Thorax dark fuscous, posterior part orange-brown. Forewing with base, apex, costal and dorsal areas light fuscous from dark-tipped whitish scales; opposite white costal and dorsal spots at four-fifths; central area with small groups of white and orange-ochreous scales; a larger elongate orange-ochreous spot extending between costal and dorsal white spots which are preceded on costa and dorsum by darker fuscous patches; small elongate blackish spots below costa near base and on costa at

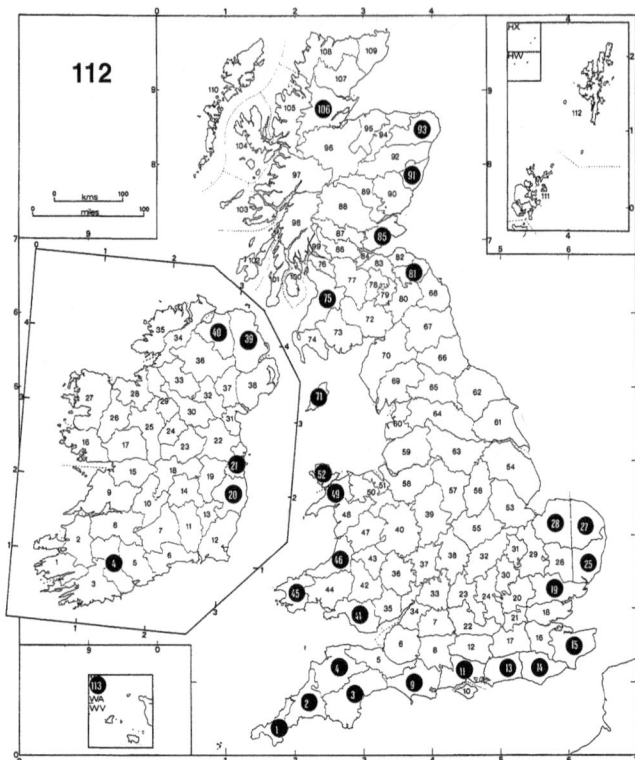

Caryocolum vicinella

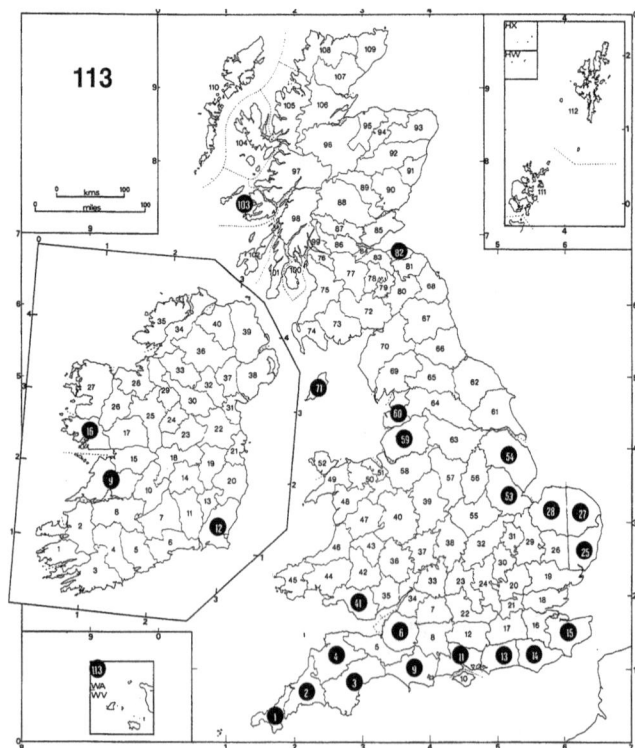

Caryocolum alsinella

one-sixth, one placed medially after basal patch, two in disc, the first slightly elongate at two-fifths, the second at three-fifths, usually inverted L-shaped with distal extension towards dorsum, one on fold at one-third much elongate, and one placed medially beyond white spots at four-fifths, these blackish spots frequently obscured by more extensive dark fuscous or blackish fuscous patches, particularly along posterior margin of basal patch and from costa at one-fifth to plical spot, often enclosing first discal spot; cilia with fuscous ciliary line, buff at apex. Hindwing light grey, fuscous near apex; cilia buff. Legs whitish on inner sides, fuscous on outer sides, with broad whitish bands on tibiae and tarsal segments; Abdomen grey, whitish below. Genitalia, see figures 15b,38e.

Similar species. C. alsinella and C. viscariella (Stainton) can hardly be distinguished by their genitalia, which suggests the possibility that they may be conspecific. However they differ in choice of foodplant, and C. alsinella is nearly always smaller and has far more white coloration on the forewing. C. junctella (Douglas) is similar in pattern (though not in genitalia), but has the first discal spot always included in the bar from costa to fold and the second discal spot not or only slightly L-shaped. C. proxima (Haworth), q.v.

Life history

Ovum. British records are mainly from little mouse-ear (*Cerastium semidecandrum*), but larvae have been found on common mouse-ear (*C. fontanum*) as well as little mouse-ear in a locality where both plants occurred together (Agassiz, 1988); also locally common on sea mouse-ear (*C. diffusum*) (J. R. Langmaid, pers.comm). In the Burren, Co. Clare, recorded from spring sandwort (*Minuartia verna*) (Agassiz et al., 1998); on the Continent also recorded on field mouse-ear (*C. arvense*); oviposition site not observed.

Larva. Head black. Prothoracic segment reddish brown, with black plate; body yellow (Stainton, 1867a, as *maculiferella*), however according to Zeller (1868) and J. R. Langmaid (pers.comm.), light green; pinacula extremely small, black.

Initially the larva is a leaf-miner (Sønderup, 1949), but later it spins the young buds together and feeds on the flowers and seeds. May.

Pupa. In a cocoon amongst leaves on the ground. Late May–June.

Imago. Univoltine; July–mid-August; on the Continent from late June–early October. It flies by night and is attracted to light; it may occasionally be disturbed during the day.

Distribution (Map 113)

Occurs locally on sandy coasts. Recorded in coastal counties of England from Cornwall to Lincolnshire but not from Essex; Lancashire, the type locality of f. *semidecandrella* Threlfall; South Wales (Agassiz, *loc.cit.*); from Scotland there are single records from East Lothian and the Inner Hebrides; the Isle of Man; from Ireland, three records: Slyne Head, west Galway, 14 August 1963, the first Irish record, the specimen now being in BMNH (Emmet, 1968); Co. Wexford (coll. H. G. Heal); and Co. Clare (Agassiz *et. al.*, *loc.cit.*); the Channel Islands. Throughout Europe; Morocco; Central Asia.

CARYOCOLUM VISCARIELLA (Stainton)

Gelechia viscariella Stainton, 1855, *Entomologist's Annu.* **1855**: 43.

Type locality: Great Britain [Scotland; Edinburgh].

Description of imago (Pl.5, fig.9)

Wingspan 10–14mm. Head dark fuscous, sometimes only at sides and posterior margin, frons and sometimes vertex shining pale ochreous-grey; antenna with scape fuscous, whitish ochreous at apex, flagellum ringed grey, brown and dark fuscous, paler beneath; labial palpus with segment 2 whitish ochreous on inner side, fuscous on outer side and below, segment 3 fuscous, whitish above. Thorax fuscous anteriorly, orange-brown posteriorly; tegulae sometimes orange-brown posteriorly. Forewing dark fuscous at extreme base, on costa and at apex; dorsal area light fuscous; narrow white or ochreous-white costal and dorsal spots at four-fifths; central part of wing orange-ochreous between fuscous patches; large orange-ochreous spot between discal spots and plical spot; usually a few white scales in orange-ochreous areas, sometimes forming white spots; blackish spots obscured by larger areas of fuscous scales, extending from costa at one-fifth to plical spot, sometimes to first discal spot, usually joining discal spots to costa; further fuscous scales around basal patch and enclosing an orange-ochreous spot; second discal inverted L-shaped with extension towards dorsum; cilia grey with greyish fuscous ciliary line, buff at tips. Hindwing light grey, fuscous on costa and at apex; cilia buff. Legs whitish on inner sides, fuscous on outer sides, tibiae banded whitish in middle and at apex, tarsal segments whitish at apex. Abdomen greyish fuscous above, whitish below. Genitalia, see figures 15c,38d.

Similar species. C. alsinella (Zeller), q.v., has more white scaling in the central part of the forewing and lacks the

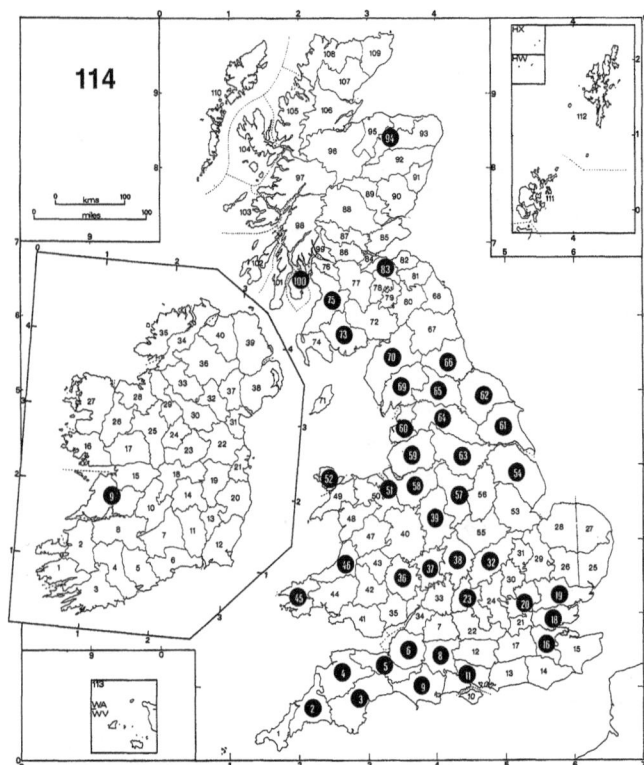

Caryocolum viscariella

small orange-ochreous patch in the basal area; the costal and dorsal spots are wider. It is usually smaller, but there is a wide overlap in size. *C. fraternella* (Douglas) has similar coloration, but the dark bar from the costa to the fold is lost in the general dark coloration extending along the costa.

Life history
Ovum. British records are from red campion (*Silene dioica*), white campion (*S. latifolia*) or sticky catchfly (*Lychnis viscaria*); on the Continent recorded also on berry catchfly (*Cucubalus baccifer*) and probably occurring on bog stitchwort (*Stellaria uliginosa*); oviposition site not known.

Larva. Head black. Prothoracic segment reddish grey, with divided black plate; body dull green, with faint rosy tinge anteriorly and posteriorly.

Feeds on the young shoots and leaves which become contorted. It spins the outer leaves together and feeds inside, leaving a characteristic mass of black frass. Later it frequently bores into the central stem which becomes swollen. April–May.

Pupa. Brown; in a slight cocoon amongst detritus on the ground; in captivity also amongst spun leaves. May–June.

Imago. Univoltine; late June–July. It flies by night when it is attracted to light, but may also be found by sweeping stands of the hostplant during the day.

Distribution (Map 114)
Occurs on dry, open grassland, along hedgerows and in open woodland. Widespread and locally fairly common in England except for East Anglia and adjacent counties; formerly considered to be restricted to the north but it has been taken in many southern counties in recent years; scattered records from Wales; in Scotland mainly in the south, but recorded also from Banffshire; in Ireland recorded only from the Burren, Co. Clare (Bradley & Pelham-Clinton, 1967). Throughout Europe, becoming scarce towards the south.

CARYOCOLUM MARMOREA (Haworth)
Recurvaria marmorea Haworth, 1828, *Lepid.Br.*: 553.
Gelechia manniella Zeller, 1839, *Isis, Leipzig* **1839**: 198.
Gelechia pulchra Wollaston, 1858, *Annl.Mag.nat.Hist.* (3)**1**: 121.
Gelechia marmorella Doubleday, 1859, *Zoologist synonymic List Br.Lepid.*: 30.
Type locality: England.

Description of imago (Pl.5, figs 10,11)
Wingspan 9.5–12.0mm. Head shining pale greyish fuscous or variably tinged ochreous or fuscous, vertex sometimes darker than frons, frons shining pale greyish fuscous; antenna with scape fuscous, apex white or greyish fuscous, flagellum ringed greyish fuscous, fuscous and blackish fuscous, paler beneath; labial palpus with segment 2 white, usually with some fuscous mottling except on dorsal surface, segment 3 mottled dark fuscous, white dorsally. Thorax greyish ochreous to ochreous-fuscous, tinged fuscous anteriorly, sometimes whitish towards posterior margin. Forewing with basal two-fifths of costa and apical area heavily mottled with light to dark fuscous scales; following two-fifths of costa white, speckled with pale ochreous and pale fuscous scales, or more heavily mottled dark fuscous; area between dorsum and fold white, often with a light dusting of mixed ochreous and light fuscous scales, appearing whitish, or more heavily mottled fuscous, appearing fuscous; ochreous streak between fuscous costal stripe and discal spots from near base, sometimes interrupted, expanded beyond second discal spot and extending between costal and dorsal white spots at four-fifths, in darkest specimens only the portion beyond the second discal spot not completely infuscate; costal white spot triangular or subquadrate,

dorsal white spot narrowly triangular, both sometimes obscure; small dark fuscous spots, one subcostal near base, one in mid-wing at one-sixth, one on fold, two discal, the second elongate, one near apex; dark fuscous scales forming oblique bar joining basal costal spot and spot at one-sixth; triangular spot between plical spot and first discal and extending towards or reaching costa at one-quarter; small subcostal spot just before one-half, and oblique patch from second discal spot extending to dorsal white spot at four-fifths and expanded on dorsum; white median spot at one-quarter, another much larger triangular white spot based on fold, between two discal spots, these two white spots much infilled with fuscous scales in specimens with dorsal white area infuscate; cilia light grey with greyish fuscous ciliary line, greyish buff at apex. Hindwing light grey, slightly darker on costa; cilia greyish buff. Legs white on inner sides, outer sides speckled light to dark fuscous, tibiae with two white bands, sometimes obscured by fuscous scales, tarsi whitebanded at distal ends of segments. Abdomen white below, pale grey to greyish fuscous above. Genitalia, see figures 15d,38f.

Similar species. Lightly marked forms are distinctive. Darker forms might be confused with *C. fraternella* (Douglas), *q.v.*

Life history

A fuller account of the early stages is given by Heckford (2000).

Ovum. British hostplant records are restricted to common mouse-ear (*Cerastium fontanum*) and little mouse-ear (*C. semidecandrum*); abroad also recorded on Spanish catchfly (*Silene otites*) (Burmann, 1990).

Larva. Head and prothoracic plate black. Body yellowish grey-green, with dirty green dorsal and subdorsal lines; pinacula small, black; anal plate dark brown.

It lives in a silken tube covered with grains of sand at the base of the hostplant (Stainton, 1867a) and feeds on the basal leaves. On the Continent the larva has been observed between spun terminal shoots of *Silene* spp., later boring into the stem (P. Huemer, pers.obs.). Late March–early May.

Pupa. Brown; in a cocoon in the last larval sand-tube. In captivity, specimens from *Silene* pupated amongst leaf-litter. May.

Imago. Univoltine; late May–August. It flies by night and comes to light, but is also easily disturbed during the day.

Distribution (Map 115)

Widespread and common on sandy coasts throughout

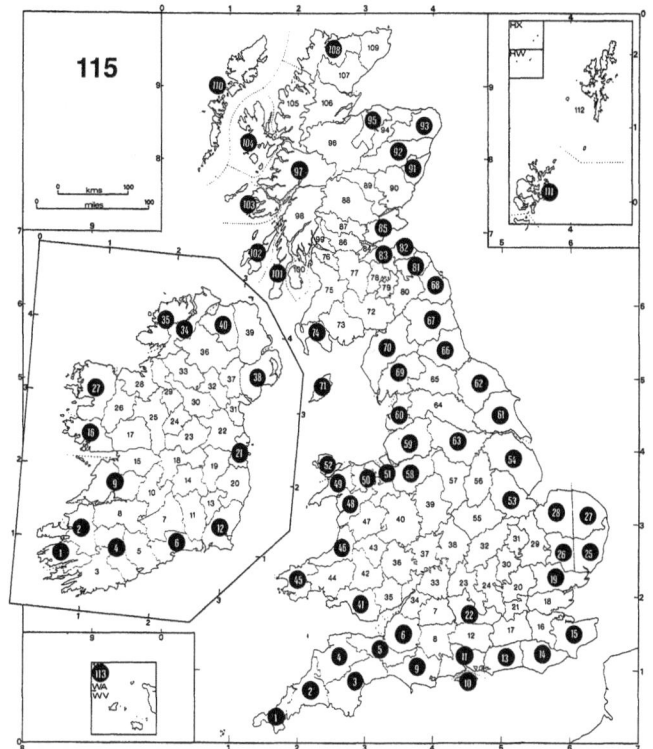

Caryocolum marmorea

the British Isles to Orkney, including the Channel Islands and the Isle of Man. Records inland are rare and some may refer to vagrants, but in the 1920s Waters (1929) recorded a flourishing colony around an old sand-pit at Tubney, Berkshire. Europe, particularly central and southern parts; Morocco; Turkey; Canada.

CARYOCOLUM FRATERNELLA (Douglas)

Gelechia fraternella Douglas, 1851, *Trans.ent.Soc.Lond.* (N.S.) **1**: 101.

Lita intermediella Hodgkinson, 1897, *Entomologist's Rec.J.Var.* **9**: 36.

Type locality: England.

Description of imago (Pl.5, fig.12)

Wingspan 9–13mm. Head ochreous, mottled fuscous, frons whitish, pale greyish ochreous or ochreous; antenna with scape fuscous, whitish at apex, flagellum pale grey, ringed brown and black, often very narrowly, paler beneath; labial palpus with segment 2 white above and upper part of inner side, ochreous below and usually fuscous on outer side, segment 3 fuscous, whitish above and at apex. Thorax ochreous-fuscous to dark fuscous, sometimes paler at posterior margin and at posterior

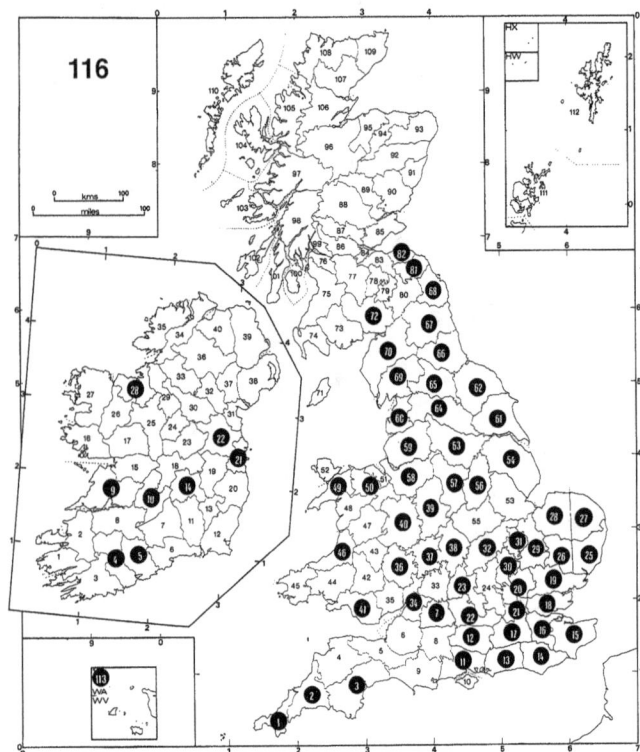

Caryocolum fraternella

ends of tegulae. Forewing with base, costa, dorsum and apical region dark fuscous from dark-tipped scales; central area of wing orange-ochreous with scattered fuscous and white scales except in area beyond second discal spot extending between white costal and discal spots at four-fifths; small blackish fuscous spots in middle of wing-base, in middle of wing at one-sixth, between fold and dorsum at one-sixth, and on fold; two discal spots, the second curved towards tornus; irregularly shaped and poorly delimited fuscous area from spot on fold through first discal towards costa, and another from second discal spot to dorsum; cilia pale buff, with dark fuscous ciliary line, pale buff beyond line. Hindwing light grey, slightly darker on costa; cilia pale buff. Legs whitish on inner sides, outer sides fuscous with two whitish bands on tibiae and narrow whitish bands at distal ends of tarsal segments. Abdomen greyish fuscous above and below. Genitalia, see figures 15f,39a.

Similar species. Only *C. viscariella* (Stainton) and dark forms of *C. marmorea* (Haworth) have similar coloration. In *C. viscariella* the dark bar from the plical spot clearly reaches the costa, whereas in *C. fraternella* it is lost in the fuscous ground colour of the costa. *C. marmorea* has the costal and dorsal white spots much reduced in dark specimens.

Life history

Ovum. British records are from bog stitchwort (*Stellaria uliginosa*), lesser stitchwort (*S. graminea*), field mouse-ear (*Cerastium arvense*) and common mouse-ear (*C. fontanum*); oviposition site not known.

Larva. Head and prothoracic plate black. Body dull reddish brown; pinacula black.

It bores into a young shoot, which is easily recognized by its gall-like abnormal growth. When the spun shoot is excavated, the larva proceeds to a fresh one. April–mid-May.

Pupa. Brown; in a slight cocoon among detritus on the ground. Mid-May–June.

Imago. Univoltine; late June–August. It flies by night, but is rarely found at light.

Distribution (Map 116)

Frequents dry grassland, heaths or damp areas, according to the foodplant. Occurs in Britain as far north as East Lothian, being more common in the south-eastern sector; southern Ireland; the Channel Islands. Throughout northern and central Europe but apparently absent from the Mediterranean region (Huemer, 1989).

CARYOCOLUM PROXIMA (Haworth)

Recurvaria proxima Haworth, 1828, *Lepid.Br.*: 552.
Gelechia maculiferella Douglas, 1851, *Trans.ent.Soc. Lond.* (N.S.) **1**: 102. [unnecessary objective replacement name for *Recurvaria proxima* Haworth, which is not an emendation of *Tinea proximella* Hübner, 1796].
Gelechia horticolla Peyerimhoff, 1871, *Mitt.schweiz.ent. Ges.* **3**: 411.

Type locality: England.

Description of imago (Pl.5, fig.13)

Wingspan 9.0–11.5mm. Head greyish fuscous, frons shining pale ochreous-grey, small pink tuft below eye; antenna with scape fuscous, white at apex, flagellum ringed black and silvery grey; labial palpus with segment 2 whitish ochreous on inner side, fuscous on outer side and below, segment 3 fuscous, whitish above. Thorax grey or ochreous-grey, mottled with fuscous-tipped scales, posteriorly becoming orange-ochreous or whitish ochreous. Forewing whitish ochreous, heavily overlaid with fuscous-tipped scales along costa, narrowly near base, more widely in mid-wing, and in dorsal and apical areas; overlaid with scales tipped orange-ochreous in centre of wing; costal and dorsal white or whitish ochreous spots at four-fifths, narrow and often indistinct, separated by orange-

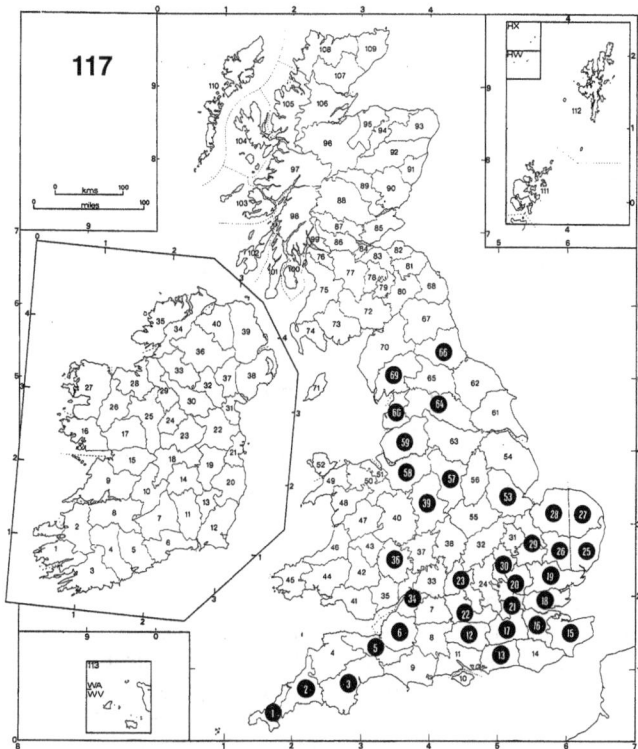

117

Caryocolum proxima

ochreous stripe; small blackish spots in middle of wing-base, on costa at one-fifth, just dorsad of fold at one-fifth, on fold, two in disc, the second elongate, inverted L-shaped, with projection directed towards dorsum; irregularly shaped dark fuscous area joining blackish costal spot at one-fifth, plical and first discal spots, usually preceded by whitish scales; cilia pale grey with greyish fuscous ciliary line and buff apex. Hindwing light grey, greyish fuscous on costa and at apex; cilia pale grey. Legs whitish on inner sides, outer sides with femora light fuscous, tibiae and tarsi dark fuscous, three pink bands on fore and mid tibiae and three white bands on hind tibiae, tarsal segments all tipped whitish. Abdomen greyish fuscous above, white below. Genitalia, see figures 15g,39c.

Similar species. C. viscariella (Stainton), *C. alsinella* (Zeller) and *C. junctella* (Douglas), from all of which it differs in the shape of the fuscous blotch and the angle it makes with the costa.

Life history

Ovum. British records are from common mouse-ear (*Cerastium fontanum*) and common chickweed (*Stellaria media*). Records from little mouse-ear (*Cerastium semidecandrum*) (Stainton, 1867a) are based on

misidentification of *C. alsinella* (Zeller); oviposition site not known.

Larva. Head and prothoracic plate black. Body bright green, matching the colour of the leaves of the food-plant when on common chickweed.

The larva makes an inconspicuous spinning in a shoot and feeds therein on the developing flowers and seeds, or enters a seed-capsule, spinning it loosely to a leaf; this spinning, though difficult to detect, is the only visible sign that the capsule is tenanted (Corley, 1995). May.

Pupa. Brown; in a cocoon among detritus on the ground. June.

Imago. Univoltine; late June–August. On the Continent from late June–early September. It flies at night and comes to light, and may also be found resting on tree-trunks during the day.

Distribution (Map 117)

Occurs mainly in open grassland, but also along hedgerows and in parkland. Widespread but local in England as far north as Co. Durham; not recorded from Wales, Scotland or Ireland. Present, though scarce, in many continental European countries; U.S.A.

CARYOCOLUM BLANDELLA (Douglas)

Gelechia blandella Douglas, 1852, *Trans.ent.Soc.Lond.* (N.S.) **1**: 246.

Recurvaria maculea Haworth, 1828, *Lepid.Br.*: 552 [misidentification and unjustified emendation of *Tinea maculella* Fabricius, 1794].

Anacampsis maculella Stephens, 1834, *Ill.Br.Ent.* (Haust.) **4**: 214, *nec* Fabricius, 1794.

Anacampsis nivella Stephens, 1834, *Ibid.* **4**: 215, *nec* Fabricius, 1794.

Type locality: England.

Description of imago (Pl.5, fig.14)

Wingspan 9.5–14.5mm. Head white, or white mottled with scales tipped greyish buff, frons white; antenna with scape fuscous, white at apex, flagellum white, ringed dark fuscous; labial palpus white, segment 2 fuscous on outer side at base, elsewhere lightly mottled fuscous, segment 3 mottled fuscous. Thorax white, mottled greyish fuscous, pale grey or pale ochreous, particularly anteriorly, often with three blackish spots on mesoscutellum. Forewing white, more or less lightly dusted with greyish fuscous scales, particularly in costal area, and with pale ochreous scales; narrow white costal and dorsal spots at four-fifths usually

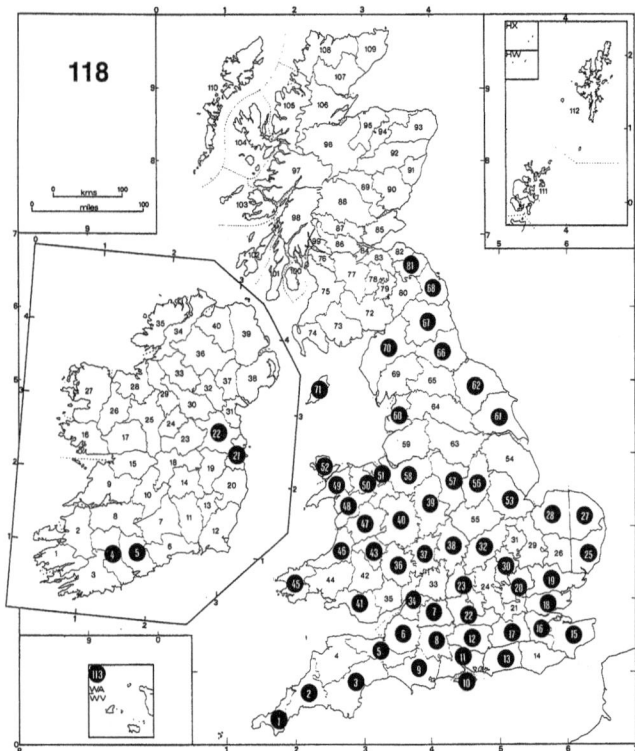

Caryocolum blandella

joined to form curved fascia; dark fuscous-tipped scales concentrated to form indistinct patches on costa and dorsum at three-fifths and in apical area; pale ochreous scales forming patches midway between costa and dorsum at one-fifth and at three-fifths, these more or less mixed with fuscous-tipped scales; small blackish spots on costa at base, sometimes two more near base, one subcostal, one dorsal, one at one-fifth on fold or immediately dorsad of fold, one at apex, with additional spots around termen; larger blackish spots on costa at one-fifth and on fold, these joined to form an oblique blackish bar, two in disc, not connected to bar, the second inverted L-shaped with distal extension directed towards dorsum, sometimes separate; cilia whitish buff, with grey ciliary line, often indistinct. Hindwing pale grey, slightly darker on costa; cilia whitish buff. Legs white, with small areas of light fuscous scales on outer sides. Abdomen pale grey above, white below. Genitalia, see figures 15h,39e.

Similar species. C. blandulella (Tutt) is generally smaller and whiter and lacks a distinct row of terminal spots. *C. blandelloides* Karsholt, *C. kroesmanniella* (Herrich-Schäffer) and *C. huebneri* (Haworth) lack the very dark bar from costa to fold which completely obscures the costal and plical spots.

Life history

Ovum. Restricted to greater stitchwort (*Stellaria holostea*); oviposition site not known.

Larva. Head and prothoracic plate black. Body yellowish green; pinacula small, black.

Initially, in April, it mines a leaf. The mine consists of a long and slender gallery close to the midrib. Later it feeds on young leaves, buds and between spun terminal shoots which become contorted and wrinkled; occasionally it also enters the seed-capsules, which become slate-coloured, and feeds on unripe seeds. April–early June.

Pupa. Pale brown; in a whitish cocoon amongst litter on the ground. June–July.

Imago. Univoltine; July–August. It flies freely at night and comes to light, but has also been recorded by sweeping the hostplant (Stainton, 1867a).

Distribution (Map 118)

Occurs mainly where the foodplant grows in shaded situations such as woodland or along hedgerows. Widespread in England as far north as Northumberland, just reaching Scotland in Berwickshire; the Isle of Man; Cos Cork, Meath and Dublin in Ireland; the Channel Islands. Europe, except the Mediterranean region.

CARYOCOLUM BLANDELLOIDES Karsholt

Caryocolum blandelloides Karsholt, 1981, *Entomologica scand.* **12**: 252.

Type locality: Denmark; Læsø; north-east Jutland.

Description of imago (Pl.5, fig.15)

Wingspan 10–12mm. Head light greyish fuscous, frons white; antenna with scape fuscous, white at apex, flagellum whitish, barred fuscous, whitish beneath; labial palpus white on inner side and dorsally, segment 2 light fuscous on outer side and beneath, segment 3 darker fuscous. Thorax whitish, anteriorly greyish fuscous, sometimes tinged ochreous, mesoscutellum with one or three dark fuscous spots. Forewing white, with a few or, in some specimens, many scattered ochreous and greyish fuscous scales, especially at base, on costa and in terminal area, in darker specimens also along dorsum; costa and dorsum with more heavily mottled fuscous scaling immediately before costal and dorsal white spots at four-fifths, which are sometimes separated by an ochreous patch; small dark fuscous dots at wing-base in middle and on costa and dorsum, in midwing and on dorsum at one-sixth, dorsad of fold at one-fifth; larger spots, one on costa at one-fifth, an elongate spot on fold and two discal spots, the second

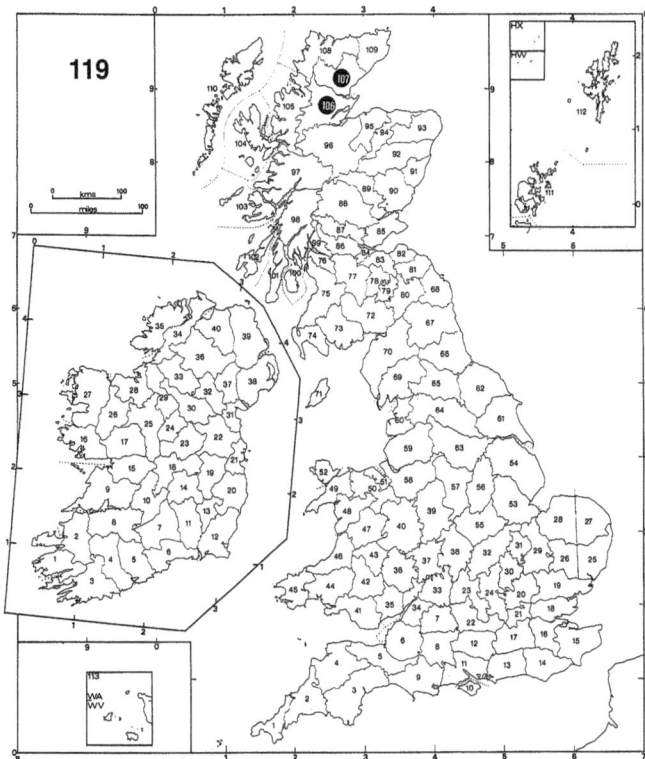

119

Caryocolum blandelloides

inverted L-shaped; costal spot at one-fifth more or less connected to plical spot by indistinct, oblique, fuscous mixed ochreous bar not obscuring spots; more or less extensive ochreous scaling on fold and between second discal spot and costa; cilia greyish buff, with fuscous or greyish fuscous ciliary line, apex pale buff. Hindwing pale grey, light greyish fuscous towards costa; cilia pale greyish buff. Legs white on inner sides, outer sides fuscous, tibiae with two white bars, tarsi with distal ends of segments whitish. Abdomen grey. Genitalia, see figures 15i,39d.

Similar species. C. huebneri (Haworth) and *C. kroesmanniella* (Herrich-Schäffer), for differences see under the former (p.203).

Life history

Ovum. Laid on common mouse-ear (*Cerastium fontanum*) (Hoare *et al.*, 1999). In Austria also on field mouse-ear (*C. arvense strictum*) (Huemer, 1987). Karsholt (1981) suspects that the larva also feeds on little mouse-ear (*C. semidecandrum*), but this is not the case at the Scottish localities.

Larva. Head and prothoracic plate black. Body whitish to pale grey-green.

Feeds, without spinning, in the flowers and on unripe seeds, occasionally also on the capsule wall. June.

Pupa. Yellowish brown; in a cocoon amongst litter on the ground. Late June–mid-July.

Imago. Univoltine; mid-July–late August. It flies freely at night and comes to light.

Distribution (Map 119)

Known only from three Scottish localities and restricted to sand-dunes. It was first discovered by R. J. B. Hoare who took three specimens on Coull Links, Loch Fleet, east Sutherland on 23 August 1994. Two further examples were collected by M. R. Young at the same locality on 19 August 1996 (Agassiz, *in litt.*). In June 1998, larvae were found in the same locality and another nearby, and also on the Black Isle in Easter Ross (Hoare *et al., loc.cit.*). In coastal localities of southern Scandinavia and the Baltic states; in mountains of central and south-western Europe.

CARYOCOLUM JUNCTELLA (Douglas)

Gelechia junctella Douglas, 1851, *Trans.ent.Soc.Lond.* (N.S.) **1**: 103.

Phthorimaea aganocarpa Meyrick, 1935, *Exotic Microlep.* **4**: 585.

Type locality: England.

Description of imago (Pl.5, fig.16)

Wingspan 9.5–11.5mm. Head pale greyish fuscous, frons silvery grey, small pink tuft below eye; antenna with scape fuscous, white at apex, flagellum grey, ringed dark fuscous, paler beneath; labial palpus with segment 2 white on inner side and above, light fuscous on outer side and beneath, segment 3 dark fuscous, white above. Thorax greyish fuscous, sometimes tinged ochreous. Forewing white with scattered ochreous and greyish fuscous scales; wing-base, costa and dorsum more heavily dusted with greyish fuscous scales; terminal area dark fuscous; dark fuscous patches on costa and dorsum immediately before white costal and dorsal spots at four-fifths, these occasionally forming complete fascia but usually separated by orange-ochreous patch extending from second dorsal spot; narrow orange-brown patch between this and fuscous patch on costa; small dark or blackish fuscous spots at wing-base in middle and on costa, at one-sixth in mid-wing and just dorsad of fold at one-fifth; slightly larger blackish spots at one-fifth on costa, on fold, and two discal spots, the second linear, oblique, and a small fuscous second plical spot; poorly defined fuscous bar, more or less mixed orange-ochreous, from costal spot

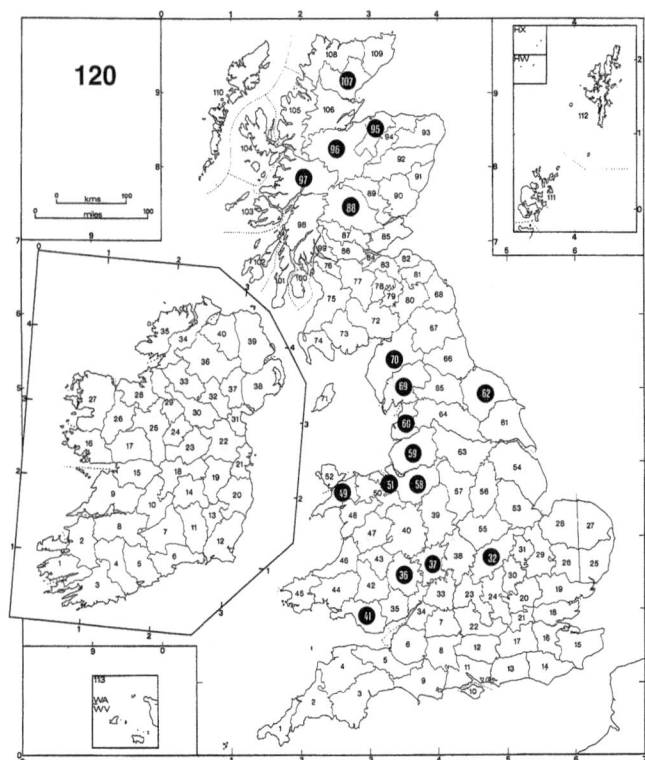

Caryocolum junctella

at one-fifth to plical spot, usually extending to include first discal spot; cilia pale grey, with greyish fuscous ciliary line, pale buff beyond line. Hindwing grey, darker towards costa; cilia pale grey. Legs white on inner sides, fuscous on outer sides, tibiae with three white or pink bars, tarsi with segments white-tipped. Abdomen greyish buff, white below. Genitalia, see figures 15k,40d.

Similar species. *C. proxima* (Haworth) and *C. alsinella* (Zeller), *qq.v.* Genitalic examination is recommended for the separation of this species from *C. alsinella*.

Life history

Ovum. British records are from lesser stitchwort (*Stellaria graminea*) and *Cerastium* spp. (Parsons, 1995 [1996]); A. N. B. Simpson and M. W. Harper found larvae on plants of lesser stitchwort and sticky mouse-ear (*Cerastium glomeratum*) growing on anthills (Simpson, 1996). Abroad recorded from common chickweed (*Stellaria media*) (Huemer, 1988) and field mouse-ear (*Cerastium arvense*) (Benander, 1928); exceptionally on knotgrass (*Polygonum aviculare*) (Park, 1993); oviposition site not known.

Larva. Head black. Prothoracic segment brown with black plate; abdomen, anal plate and pinacula apple-

green; thoracic legs blackish brown.

The larva initially mines a leaf, later spinning together the tips of the shoots. May–early June (Simpson, *loc.cit.*).

Pupa. Apparently undescribed. Mid-June–July.

Imago. Univoltine; July–May. One of the few European gelechiids hibernating as an adult. In spring it has been taken flying in early evening sunshine (Simpson, *loc.cit.*). It rests on the trunks of oaks, pines and fir and has also been beaten out of yew-trees.

Distribution (Map 120)

Distribution uncertain because of confusion with related species. It occurs mainly in woodland and along sheltered hedgerows and lanes (Tutt, 1891); also found breeding in a meadow adjacent to woodland (Simpson, *loc.cit.*). Records are from the Midlands, Wales, north-western England and northern Scotland; not recorded from Ireland. Apparently declining (Parsons, 1995[1996]). Palaearctic, from Spain to China and Japan.

CARYOCOLUM TRICOLORELLA (Haworth)

Tinea tricolorella Haworth, 1812, *Trans.ent.Soc.Lond.* [1]1: 338.
Recurvaria contigua Haworth, 1828, *Lepid.Br.*: 552.
Type locality: England.

Description of imago (Pl.5, fig.17)

Wingspan 10.0–14.5mm. Head fuscous, frons shining whitish or pale grey; antenna with scape dark fuscous, whitish at apex, flagellum grey, ringed blackish, paler beneath; labial palpus with segment 2 fuscous, mainly white on inner side, segment 3 dark fuscous, with a few white scales, white at apex on outer side. Thorax fuscous anteriorly, posterior half of thorax and tegulae ochreous or pale ochreous, sometimes pale grey, with two or three dark fuscous spots. Forewing ochreous-orange with scattered white scales, mixed light fuscous in basal area and along dorsum; terminal area dark fuscous, without ochreous scales; white costal spot at four-fifths much wider than dorsal spot, from which it is just separated by the end of an ochreous patch or by a few fuscous scales; smaller white costal spot at five-sixths; small blackish fuscous spots in middle of wing-base, beneath costa and on dorsal side of fold at one-sixth, followed by oblique, ochreous-speckled white bar which precedes large subtriangular fuscous patch based on costa and reaching fold, completely obscuring plical and first discal spots; second discal spot sometimes obscured by small patch of fuscous

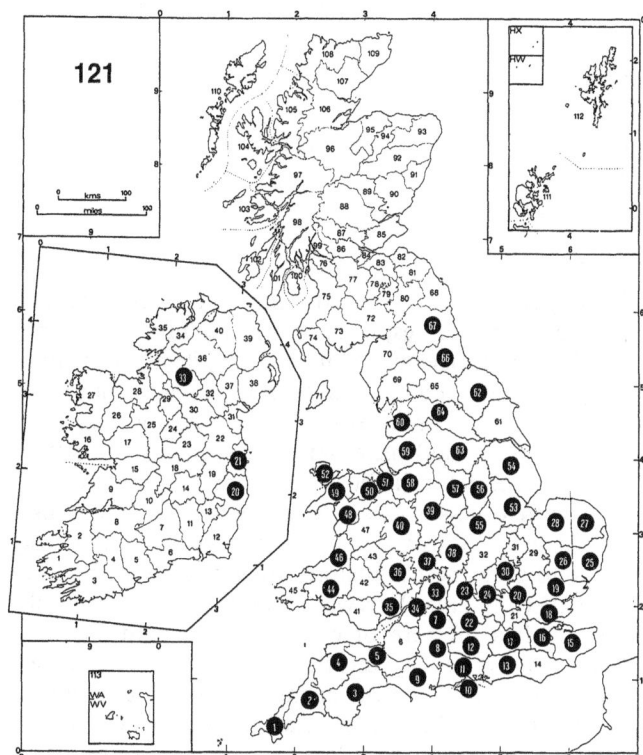

Caryocolum tricolorella

scales, but more often extended to costa and forming a second triangle, smaller than the first and partially merged with it; cilia light grey with greyish fuscous ciliary line, buff beyond line. Hindwing light grey, greyish fuscous on costa; cilia greyish buff. Legs whitish on inner sides, fuscous on outer sides, three whitish or pink bands on tibiae, tarsal segments whitish distally. Abdomen pale grey or greyish fuscous on upperside, on underside white but pale grey at margins. Genitalia, see figures 15j,39f.

Similar species. The blackish triangular costal mark separates *C. tricolorella* from all other *Caryocolum* species. *Scrobipalpa costella* (Humphreys & Westwood) has a similar costal mark, but does not have an orange-ochreous dorsal area.

Life history

Ovum. Restricted in Britain to greater stitchwort (*Stellaria holostea*); on the Continent also recorded from bog stitchwort (*S. uliginosa*); oviposition site not known.

Larva. Head and prothoracic plate black. Body dull whitish green with dull pink dorsal, subdorsal and lateral lines; pinacula small, black.

Initially, in December–January, it produces gallery-like mines in a leaf. Later it feeds on the buds and between spun terminal shoots. December–April.

Pupa. Brown; in a cocoon among detritus on the ground. May–June.

Imago. Univoltine; late June–August. It is attracted to light and sugar (Stainton, 1867a).

Distribution (Map 121)

Widespread and locally common in open woodland and along hedgerows in England and Wales; not recorded from Scotland; Cos Wicklow, Dublin and Fermanagh in Ireland (Beirne, 1941). Europe, except the Mediterranean region.

CARYOCOLUM BLANDULELLA (Tutt)

Gelechia (*Lita*) *blandulella* Tutt, 1887, *Entomologist's mon.Mag.* **24**: 105.

Type locality: England; Deal, Kent.

Description of imago (Pl.5, fig.18)

Wingspan 8.5–11.0mm. Head creamy white, with a few pale greyish fuscous scales, frons white; antenna with scape light fuscous at base, gradually whiter to apex, flagellum white, barred fuscous, white beneath; labial palpus white dorsally, segment 2 light fuscous below, segment 3 darker. Thorax creamy white, more ochreous on anterior edge, sometimes one to three fuscous dots on mesoscutellum. Forewing white, lightly dusted with pale greyish fuscous and ochreous scales, particularly at base and on dorsum; costa and dorsum more heavily mottled fuscous immediately before white, slightly angled fascia at four-fifths; terminal area heavily mottled dark fuscous; dark fuscous dots in middle of wing-base, beneath costa near base, and on dorsum at one-sixth; larger spots on costa at one-fifth, and on fold; two discal spots, the second often right-angled; costal spot at one-sixth and spot on fold obscured by oblique dark fuscous bar joining them, this bar preceded by white bar; cilia pale grey with indistinct ciliary line, apex pale buff. Hindwing light grey, scarcely darker on costa; cilia greyish buff. Legs white, tibiae with two fuscous bars, tarsal segments light fuscous, white distally. Abdomen pale grey or white, white below. Genitalia, see figures 16a,40c.

Similar species. C. blandella (Douglas), *q.v.*

Life history

Ovum. British records are from little mouse-ear (*Cerastium semidecandrum*) (D. J. L. Agassiz, pers. comm.); abroad also on dwarf mouse-ear (*C. pumilum*); oviposition site not known.

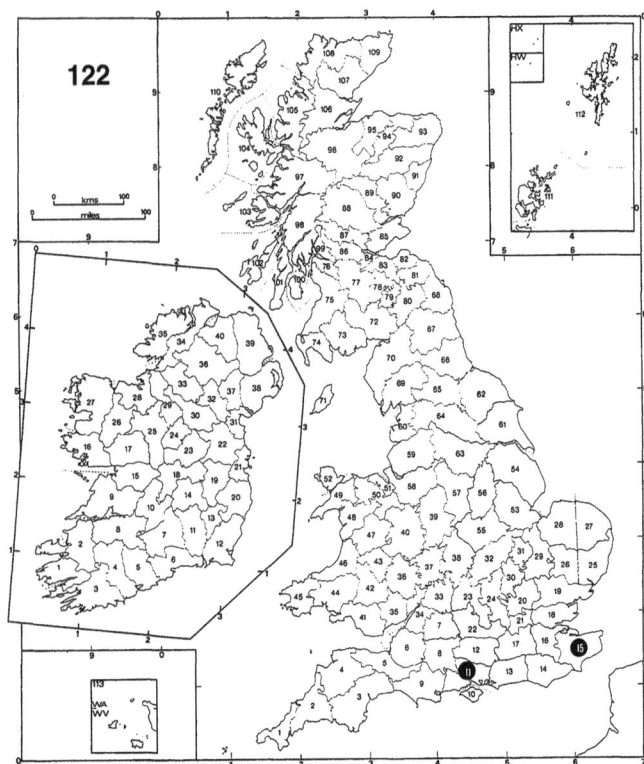

Caryocolum blandulella

Larva. Undescribed. It feeds in seed-capsules which are spun together (Benander, 1965). Early May–June.

Pupa. Apparently undescribed. Mid-June–mid-July.

Imago. Univoltine; mid-July–late August. It flies at night and is attracted to light; it can be swept from vegetation or disturbed from its resting place during the day (Karsholt, 1981).

Distribution (Map 122)

The only recorded British habitats are sand-dunes in two localities in Kent and Hampshire respectively. Since 1969 records have been restricted to Sandwich, east Kent (Parsons, 1995[1996]). Southern part of Scandinavia to central and southern Europe.

CARYOCOLUM KROESMANNIELLA (Herrich-Schäffer)

Gelechia kroesmanniella Herrich-Schäffer, 1854, *Syst. Bearb.Schmett.Eur.* **5**: 166.

Caryocolum huebneri sensu auctt., *nec* Haworth, 1828, *Lepid.Br.*: 551.

Type locality: Germany; Hanover.

Description of imago (Pl.5, fig.19)

Wingspan 12–15mm. Head white with some greyish fuscous-tipped scales, frons white; antenna with scape speckled light fuscous, white at apex, flagellum white, ringed fuscous and blackish fuscous; labial palpus with segment 2 white on inner side and above, mottled light fuscous below and on outer side, segment 3 mottled dark fuscous, whitish above and at apex. Thorax white, with more or less extensively scattered greyish fuscous-tipped scales, thus varying in appearance from almost white to grey, mesoscutellum white, usually with three dark fuscous dots. Forewing white, lightly dusted all over with a mixture of ochreous and light fuscous scales; more heavily mottled with fuscous scales at base, on costa, especially at three-fifths, on dorsum at three-fifths and in subapical area; more or less indistinct, narrow, white angulate fascia at four-fifths; dark or blackish spots on costa at one-fifth, on fold, two in disc, the first small, the second inverted L-shaped, often extending to meet indistinct spot on dorsum at three-fifths, and one at apex; small fuscous dots in wing-base towards costa, in mid-wing and on dorsum at one-sixth, dorsad of fold at one-fifth and around termen; ochreous patches in mid-wing at one-fifth, connecting costal spot at one-fifth and elongate plical spot, and from second discal spot to fascia; cilia pale greyish ochreous, ciliary line indistinct, light fuscous, apex whitish buff. Hindwing light grey; cilia pale greyish buff. Legs white on inner sides, fuscous on outer sides, tibiae with three white bands, tarsal segments white distally. Abdomen light grey above, white below. Genitalia, see figures 16c,40a.

Similar species. C. *blandelloides* Karsholt and C. *huebneri* (Haworth), *qq.v.*

Life history

Ovum. British records are from greater stitchwort (*Stellaria holostea*) or bog stitchwort (*S. uliginosa*); on the Continent also recorded on common chickweed (*S. media*); probably lesser stitchwort (*S. graminea*) is an additional hostplant. Oviposition site not known.

Larva. Head and prothoracic plate black. Body light green, first pair of thoracic legs black (Spuler, 1910). On the Continent it is reported as a leaf-miner in the

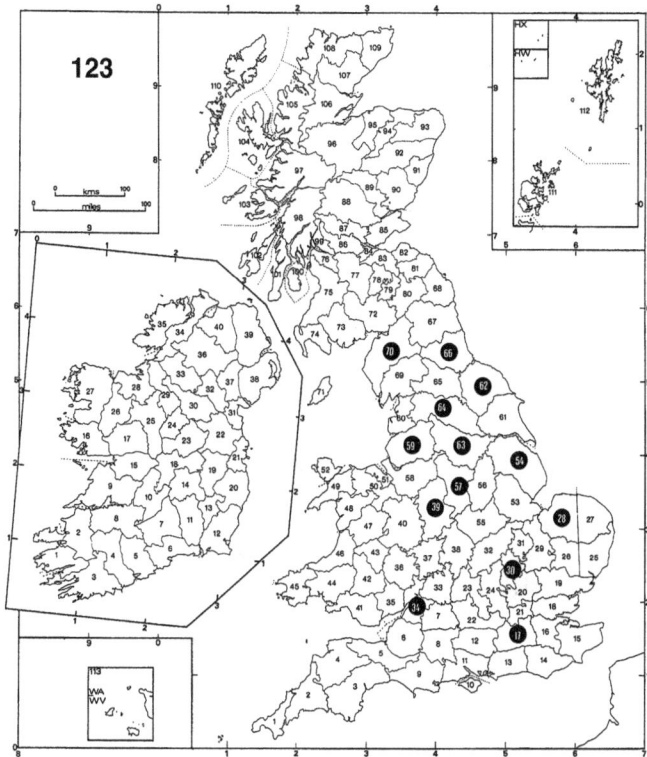

Caryocolum kroesmanniella

autumn. After hibernation it feeds in spun shoots. Autumn–May.

Pupa. Reddish brown. June.

Imago. Univoltine; early July–early September. May be found resting on trunks of oak, and at night has been recorded at light.

Distribution (Map 123)

Occurs in open woodland. Its distribution is imperfectly known because of confusion with *C. huebneri* and *C. blandella* (Douglas); the vice-county records shown on the map should therefore be accepted with caution. Among recent records, only those from Co. Durham have been confirmed (Parsons, 1995[1996]). Continental Europe, except the south.

CARYOCOLUM HUEBNERI (Haworth)

Recurvaria huebneri Haworth, 1828, *Lepid.Br.*: 551.
Gelechia hubnerella Doubleday, 1859, *Zoologist synonymic List Br.Lepid.*: 30.
Gelechia knaggsiella Stainton, 1866, *Entomologist's Annu.* **1866**: 167.

Type locality: England.

Description of imago (Pl.5, fig.20)

Wingspan 9–12mm. Head whitish, with numerous fuscous-tipped scales, frons white; antenna with scape white, mottled fuscous, white at apex, flagellum white, barred fuscous, whitish beneath; labial palpus white, segment 2 mottled fuscous, mainly on outer side and below, segment 3 mottled dark fuscous, white above and at apex. Thorax white, mottled with fuscous-tipped scales, apices of tegulae fuscous, mesoscutellum white, with three fuscous dots. Forewing white, dusted with fuscous-tipped scales, but less so in two oblique white stripes either side of dark costal spot and more or less indistinct angulate fascia at four-fifths; more heavily mottled fuscous at base, from costal spot to plical spot, on costa and dorsum at three-fifths and in apical area; blackish fuscous spots on costa at one-fifth, on fold, two in disc, the first small, the second elongate, sometimes inverted L-shaped, but not extending much towards dorsum; smaller dark fuscous spots at wing-base in middle and on costa, dorsad of fold at one-fifth and around termen; ochreous scales almost absent in some specimens, scattered mainly in central part of wing in others, and more numerous in mid-wing at one-sixth, above and below plical spot to first discal spot, and between second discal spot and white fascia; cilia light grey, ciliary line fuscous, apex pale buff. Hindwing light grey, greyish fuscous along costa; cilia greyish buff. Legs white on inner sides, fuscous on outer sides, tibiae with three white bars, tarsal segments white distally. Abdomen grey above, white below. Genitalia, see figures 16b,40b.

Similar species. *C. kroesmanniella* (Herrich-Schäffer) is larger and has no connecting bar between the costa and the plical spot. *C. blandelloides* Karsholt has the costal and dorsal white spots separated by an ochreous patch and the wing is less extensively dusted with fuscous scales. *C. blandella* (Douglas) and *C. blandulella* (Tutt) lack fuscous-tipped scales on the head and thorax and have the costal and plical spots obscured by the dark fuscous bar.

Life history

Insufficiently known, possibly also owing to confusion with *C. kroesmanniella* (Ridout, 1977).

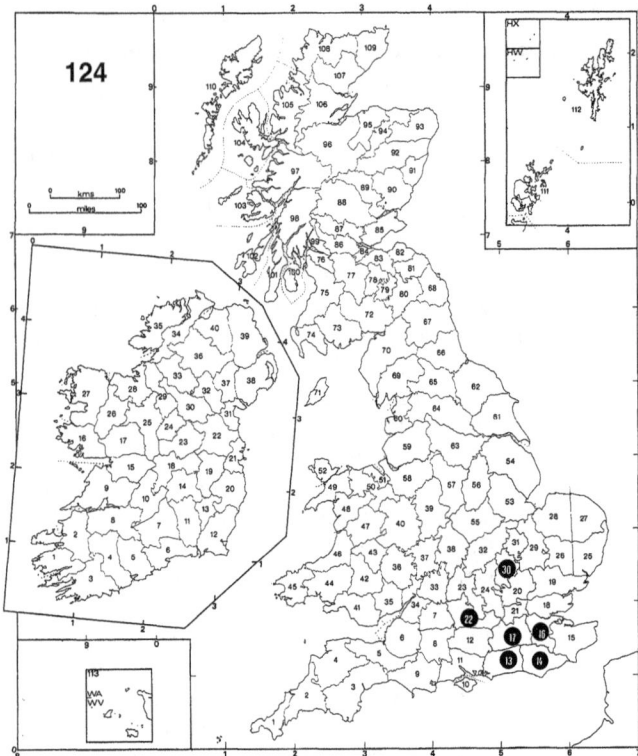

124

Caryocolum huebneri

Ovum. Restricted to greater stitchwort (*Stellaria holostea*); oviposition site not known.

Larva. Undescribed. It feeds in spun shoots. May.

Pupa. Apparently undescribed. June.

Imago. Univoltine; July–August. Flies by night and comes to light.

Distribution (Map 124)

Occurred mainly in woodland. The majority of literature records of this species appear to be based on misidentifications of *C. kroesmanniella*. Records from Sussex, Kent, Surrey, Berkshire and Bedfordshire are the most likely to be authentic, but Parsons (1995[1996]) considers that it is now extinct in Britain. It certainly occurred formerly, since the type locality is in England. Europe, except the south.

Anacampsinae

SOPHRONIA Hübner

Sophronia Hübner, [1825], *Verz. bekannt. Schmett.*: 407.

A small genus represented mainly in the Palaearctic region. Some 15 species occur in Europe but only two have been reported from Britain. Hodges (1983) placed the genus in the Gelechiinae, but Karsholt & Razowski (1996) found its systematic position unclear. Here, Leraut (1997) and Bradley (2000) are followed in placing it in the Anacampsinae.

Imago. Medium-sized to large gelechiids with segment 3 of labial palpus long, slender and projecting above head, segment 2 greatly thickened with forward-projecting scales below; antenna without pecten. Forewing without tufts of erect scales, elongate-oblong with apex produced to form an acute angle; tornus rounded. Hindwing broader than forewing, trapezoidal, apex acute and tornus rounded. Abdomen with tergites 1–3 having modified scales but concolorous with the rest of abdomen. Valva of male genitalia reduced to slender digitate costal process, sacculus similar but smaller or absent; signum of female genitalia thorn-like or absent.

Larva. Larval stages poorly known.

Key to species (imagines) of the genus *Sophronia*

–   White markings on forewing consisting only of a narrow costal stripe along basal two-thirds .... *semicostella* (p. 204)

–   White markings on forewing more extensive .... *humerella* (p. 205)

SOPHRONIA SEMICOSTELLA (Hübner)

Tinea semicostella Hübner, [1813], *Samml. eur. Schmett.* 8: 59, fig. 396.

Tinea marginella sensu Thunberg, 1794, *D. D. Diss. ent. sistens Insecta suecica* (7): 108, *nec* [Denis & Schiffermuller, 1775].

Ypsolophus parenthesellus sensu Haworth, 1828, *nec* Linnaeus, 1761.

Type locality: Europe.

Description of imago (Pl. 5, fig. 21)

Wingspan 18–19mm. Head covered with pale fuscous scales with whitish tips; antenna dark fuscous; labial palpus segment 2 strongly tufted below with pale fusc-

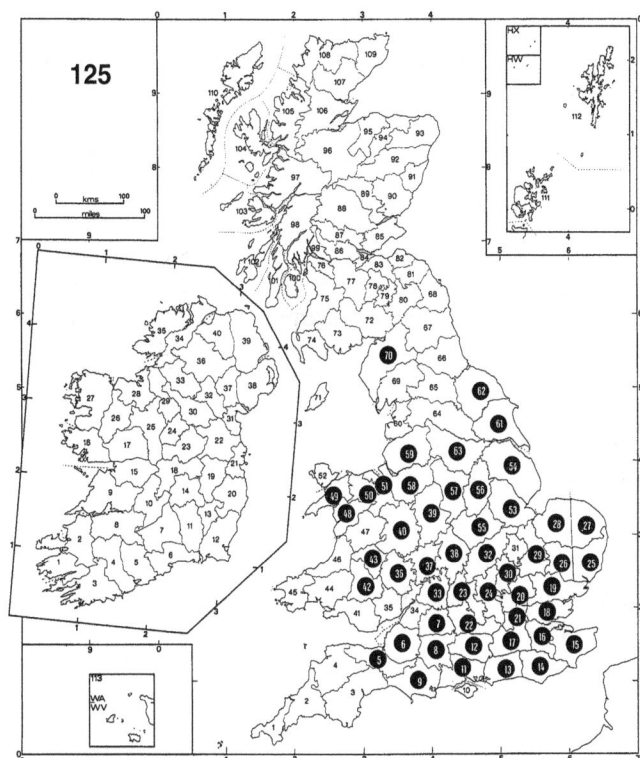

Sophronia semicostella

ous scales at least as long as segment, above whitish and pale fuscous, segment 3 smooth, white above, narrowly dark fuscous below. Thorax and tegulae pale fuscous. Forewing fuscous, suffused with chestnut scales along fold, below costa and between veins in apical one-third; broad white stripe along basal two-thirds of costa; other markings dark fuscous consisting of dash on fold just before one-half, a pair of spots, one above the other, in discal area at two-thirds and black spot at extreme apex at base of cilia; cilia on apical half of termen white with three neat lines of dark fuscous-tipped scales, on tornal half fuscous without ciliary lines; underside of forewing dark fuscous, suffused white at apex. Hindwing fuscous; cilia concolorous except at wing apex where ciliary line present, white-tipped beyond line. Legs dark fuscous, tarsi narrowly ringed white distally, tibial tuft of hindleg whitish. Abdomen fuscous. Genitalia, see figures 16e,40g.

Life history

Ovum. Laid on sweet vernal-grass (*Anthoxanthum odoratum*) according to Continental authors (Bradford & Sokoloff, 1988).

Larva. Undescribed. No details of the larval stages of this species seem to be available. Michaelis (1977)

found a larva in a rolled leaf of grass (species not mentioned) in May, but thought it might have been spinning up for pupation as no signs of feeding were seen.

Pupa. Undescribed.

Imago. Univoltine; June–July. Flies in the early afternoon and again at dusk. At night it occasionally comes to light.

Distribution (Map 125)

Found throughout England and Wales except for the far south-west of both countries; not recorded from Scotland or Ireland. Throughout Europe.

SOPHRONIA HUMERELLA ([Denis & Schiffermüller])

Tinea humerella [Denis & Schiffermüller], 1775, *Schmett. Wien.*: 137.

Type locality: [Austria]; Vienna district.

Description of imago (Pl.5, fig.22)

Wingspan 12–13 mm. Head pale ochreous-fuscous, frons white; antenna dark fuscous dorsally, white below; labial palpus with segment 2 pale ochreous-fuscous, with long projecting scales ventrally which are about three-quarters length of segment 3, segment 3 mixed whitish and fuscous. Thorax and tegulae pale ochreous-fuscous. Forewing dark fuscous with rufous and violet reflections and white, black and rufous markings; white markings consisting of broad costal streak from base to three-quarters, veering slightly away from costa at just before one-half, and a narrow acute-angled fascia at two-thirds with thin longitudinal black streak bisecting angle; other markings consisting of a band of black-tipped white scales around apical aspect of angle of fascia and extending to tornus but not reaching costa, rufous band apical to this and parallel to costal arm of fascia; beyond this a white costal patch of scales and a more apical patch of black-tipped white scales; cilia broad and dark fuscous from costa to apex, remainder pale ochreous-fuscous with double ciliary line formed from long white scales with black tips. Hindwing fuscous; cilia concolorous. Legs fuscous, apices of tarsal segments white, tibial tuft of hindleg ochreous. Abdomen fuscous. Genitalia, see figures 16f,40f.

Life history

Ovum. According to Continental authors, laid on field wormwood (*Artemisia campestris*) during July.

Larva. Head brown. Prothoracic plate yellowish green, with four dark spots; body dull green with darker green dorsal and subdorsal lines (Stainton, 1859).

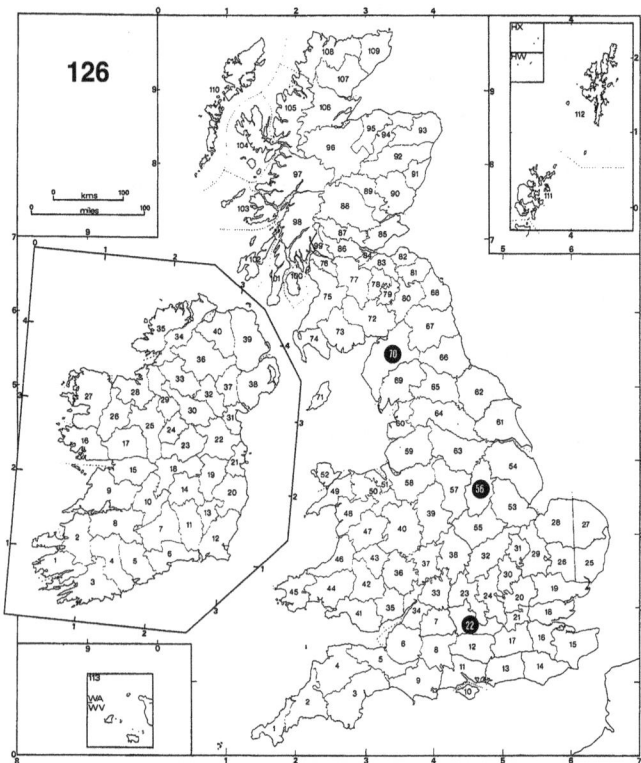

Sophronia humerella

In what stage of development winter is passed in unknown, but in May the larva feeds between spun leaves of its foodplant.

Pupa. Undescribed. Pupation site not recorded. June.
Imago. Univoltine; July.

Distribution (Map 126)

First recorded in Britain prior to 1850 at an unknown locality (Stainton, *loc.cit.*). Pre-1970 records exist for Berkshire, Nottinghamshire, Cumberland and Co. Durham, though Dunn & Parrack (1992) regard the last as unconfirmed. As the supposed foodplant has never occurred in these areas the species is probably an occasional immigrant. There has been no record of the species since 1970; thus Parsons (1995[1996]) considers that if the species had ever been a resident it is now extinct. Throughout most of Europe including Scandinavia and western Russia; Asia Minor.

APROAEREMA Durrant

Aproaerema Durrant, 1897, *Entomologist's mon.Mag.* **33**: 221.

A genus represented in the Palaearctic region and South Africa by about six species, one of which occurs in Great Britain and Ireland.

Imago. Head without ocelli; scape of antenna without pecten; labial palpus long and slender, segment 3 longer than segment 2, segment 2 slightly thickened with appressed scales. Forewing with veins R_3 (9) to M_1 (6) stalked, M_1 long. Male genitalia with cleft uncus having a number of black pegs. Female genitalia without signum in corpus bursae.

Larva. On Fabaceae (Leguminosae).

APROAEREMA ANTHYLLIDELLA (Hübner)

Tinea anthyllidella Hübner, [1813], *Samml.eur.Schmett.* **8**: pl.48, fig.330.

Gelechia nigritella sensu Douglas, 1851, *Trans.ent.Soc. Lond.* (N.S.) **1**: 108, *nec* Zeller, 1847.

Gelechia (Anacampsis) sparsiciliella Barrett, 1891, *Entomologist's mon.Mag.* **27**: 7.

Aproaerema aureliana Căpuşe, 1964, *Ent.Tidskr.* **85**: 15.

Type locality: [Germany; Augsburg].

Description of imago (Pl.5, figs 23,24)

Wingspan 10–12mm. Head slaty fuscous, frons buffish fuscous; antenna purplish fuscous, leading edge of scape lined buffish fuscous, flagellum with alternate segments buffish fuscous fading towards apex; labial palpus segment 1 purplish fuscous, segment 2 inner surface buffish white or purplish fuscous variably mixed buffish white, outer surface variably mixed purplish fuscous, segment 3 buffish white, outer edge purplish fuscous. Thorax and tegulae slaty fuscous. Forewing dark slaty fuscous, sometimes almost blackish, with scattered greyish scales; stigmata darker, sometimes obsolete; a few yellowish ochreous scales on fold, sometimes obsolete; small yellowish ochreous spot on costa at two-thirds usually outwardly oblique, and smaller one opposite on tornus, these spots sometimes obsolete; cilia fuscous, darker ciliary line fading to tornus. Hindwing light fuscous; cilia concolorous, paler basally. Legs fuscous, tarsal segments banded whitish ochreous distally. Abdomen dark fuscous. Genitalia, see figures 16g,40e.

Similar species. *Eulamprotes immaculatella* (Douglas), and *Syncopacma albipalpella* (Herrich-Schäffer), *qq.v.*

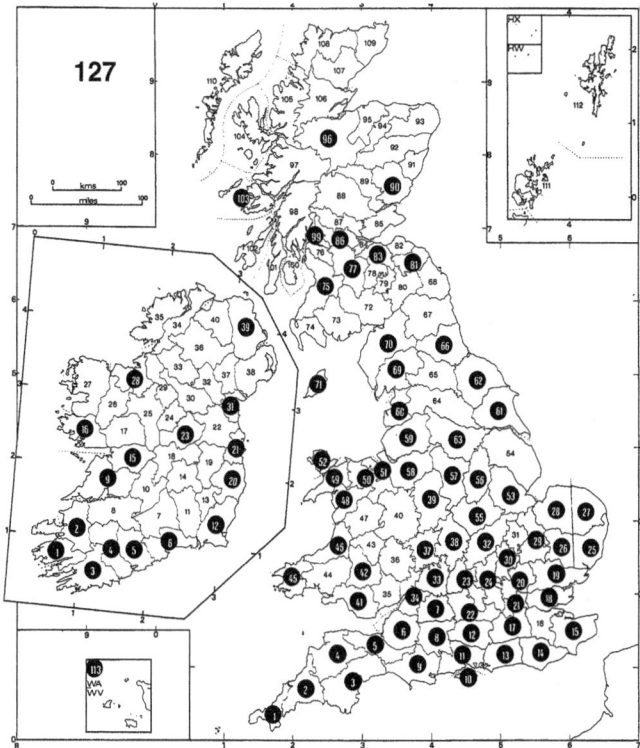

Aproaerema anthyllidella

Life history

Ovum. Undescribed. Oviposition site unknown, but presumably on the foodplant, kidney vetch (*Anthyllis vulneraria*), restharrow (*Ononis* spp.), sainfoin (*Onobrychis viciifolia*), lucerne (*Medicago sativa*) or clover (*Trifolium* spp.), but *Anthyllis* is the preferred foodplant.

Larva. Head blackish brown. Prothoracic plate black, bisected brownish; body blackish purple, lateral incisions between thoracic segments 1 to 3 white, faint grey dorsal line on these segments, dorsal incisions between thoracic segments 2 and 3 and abdominal segment 1 grey; interrupted grey lateral line fading posteriorly; pinacula small, black; anal plate black; thoracic legs black; prolegs concolorous with body.

The larva of the first generation mines the leaf, packing frass at one end, making whitish blotches, eventually causing the leaf to fold upwards to form a pod; that of the second generation on kidney vetch feeds on the flowers and seeds, that on restharrow in spun leaves (R. J. Heckford, pers.obs.); the second generation has not been recorded on sainfoin or clover. April–May, sometimes as early as January; July, August and sometimes September.

Pupa. Undescribed; in a slight cocoon. First generation in detritus; second generation on kidney vetch in the flower or seedhead, second generation on restharrow in detritus; second generation on sainfoin and clover unrecorded. May; July, August and sometimes September.

Imago. Bivoltine; May–June and August–September. Flies from late afternoon to evening and at night comes to light.

Distribution (Map 127)

Occurs on rough ground, dry pastures and in coastal areas, and is locally common throughout the British Isles as far north as Inverness-shire. Widely distributed in Europe, its range extending to central Asia.

SYNCOPACMA Meyrick

Syncopacma Meyrick, 1925, *Genera Insect.* **184**: 14 (key), 72.

Harpagus Stephens, 1834, *Ill.Br.Ent.* (Haust.) **4**: 278, *nec* Vigors, 1824.

Stomopteryx sensu auctt., *nec* Heinemann, 1870.

A mainly Palaearctic genus, but extending to North America and South Africa; represented by over 40 species, eight of which have been found in Great Britain, including one accidental importation, and three in Ireland. The species usually have blackish forewings, often with either a whitish transverse fascia, or opposing whitish costal and dorsal spots, sometimes with whitish plical spots. In some species, specimens occur with the fascia or spots obsolete. Most are difficult to determine without dissection, although several can be identified without the need for this if reared.

Imago. Head with or without ocelli; scape of antenna without pecten; labial palpus long and slender, segment 3 longer than segment 2, segment 2 slightly thickened with appressed scales. Forewing, in some species with veins R_4 (8) and R_5 (7) on a common stalk from M_1 (6), in other species with R_4 stalked with R_5 and M_1 free. Male genitalia with uncus not cleft and with a number of black pegs. Female genitalia without signum in corpus bursae.

Larva. In spun leaves of Fabaceae (Leguminosae).

Key to species (imagines) of the genus *Syncopacma*

1   Forewing dark fuscous, except base and wide fascia, both light buff .... *polychromella* (p. 215)

—   Forewing either unicolorous dark fuscous, dark fuscous with narrow yellowish white fascia or with yellowish white costal and tornal spots (rarely costal only) 2

2(1)   Underside of forewing with narrow yellowish white fascia; underside of hindwing with yellowish white costal spot .... *taeniolella* (p. 211)

—   Underside otherwise 3

3(2)   Forewing with yellowish white fascia, or clear yellowish white costal and tornal spots (rarely costal only) 4

—   Forewing unicolorous, or at most with very faint yellowish white costal and tornal spots 11

4(3)   Wingspan 8–11mm; species restricted to *Genista pilosa* and *G. anglica* 5

—   Wingspan 11–15mm; species found on other Fabaceae 6

5(4)   Wingspan 8–9mm; labial palpus segment 3 purplish, mixed cream; species restricted to *G. pilosa* ....
.... *suecicella* (p. 215)

—   Wingspan 10–11mm; labial palpus segment 3 buffish fuscous, lower edge purple-fuscous; species restricted to *G. anglica* .... *albipalpella* (p. 213)

6(4)   Forewing with yellowish white costal and tornal spots (rarely costal only) 7

—   Forewing with yellowish white fascia 9

7(6)   Forewing with yellowish white costal spot only ....
.... *cinctella* (part) (p. 210)

—   Forewing with yellowish white costal and tornal spots 8

8(7)   Wingspan generally 11–12mm; species restricted to *G. tinctoria* .... *vinella* (part) (p. 212)

—   Wingspan generally 12–15mm; species restricted to *Lotus corniculatus* .... *sangiella* (p. 208)

9(6)   Wingspan generally 11–12mm; species restricted to *G. tinctoria* .... *vinella* (part) (p. 212)

—   Wingspan generally 12–15mm; species not restricted to *G. tinctoria* 10

10(9)   Forewing appearing broader distally, with fascia straight or slightly inwardly curved ....
....*cinctella* (part) (p. 210)

—   Forewing not appearing broader distally, with fascia straight and not slightly inwardly curved ....
....*larseniella* (part) (p. 209)

11(3)   Forewing unicolorous or at most with very faint yellowish costal and tornal spots, species restricted to *G. tinctoria* .... *vinella* (part) (p. 212)

—   Forewing unicolorous without any sign of very faint yellowish costal and tornal spots; species not restricted to *G. tinctoria* 12

12(11) Forewing appearing broader distally ....
.... *cinctella* (part) (p. 210)

—   Forewing not appearing broader distally ....
.... *larseniella* (part) (p. 209)

The asterisk* denotes that the specimen must be dissected to confirm identification.

SYNCOPACMA SANGIELLA (Stainton)

Gelechia sangiella Stainton, 1863, *Entomologist's Annu.* **1863**: 149.

Lita coronillella sensu auctt. (partim), *nec* Treitschke, 1833.

Type locality: England; Darlington, Co. Durham.

Description of imago (Pl.5, fig.25)

Wingspan 11–15mm. Head including frons purple-fuscous, line of cream scales above eye curved inwardly towards frons; antenna purple-fuscous, scape with longitudinal yellowish white line along leading edge, sometimes with longitudinal faint yellowish white line along trailing edge, flagellum with alternate segments yellowish white fading towards apex; labial palpus segment 1 purple-fuscous, segment 2 inner surface buffish white, outer surface variably mixed purple-fuscous, segment 3 buffish white, outer edge purple-fuscous. Thorax and tegulae purple-black. Forewing purplish or slaty black, sometimes with scattered lighter scales; plical and discal stigmata black in fresh specimens but tending to fade to ground colour of forewing after death, a few creamy yellow scales at distal edge of plical and first discal and at proximal and distal edges of second discal, such scales being easily shed; creamy yellow spot on costa at three-quarters, sometimes outwardly oblique, and similarly coloured spot or scaling opposite on tornus; dark greyish scales from apex to tornus; cilia dark grey, black ciliary line fading to tornus. Hindwing dark fuscous; cilia concolorous. Legs purple-fuscous, tarsal segments banded whitish distally; hindleg with distinct white band at tibial spurs. Abdomen purplish fuscous. Genitalia, see figures 17b,41a.

    Douglas (1850), Stainton (1867a) and Meyrick ([1928]) included *S. coronillella* (Treitschke) as British, but a specimen under this name in the E. R. Bankes collection in the BMNH, collected at Hilly Field, Mickleham, Surrey, the only locality recorded for

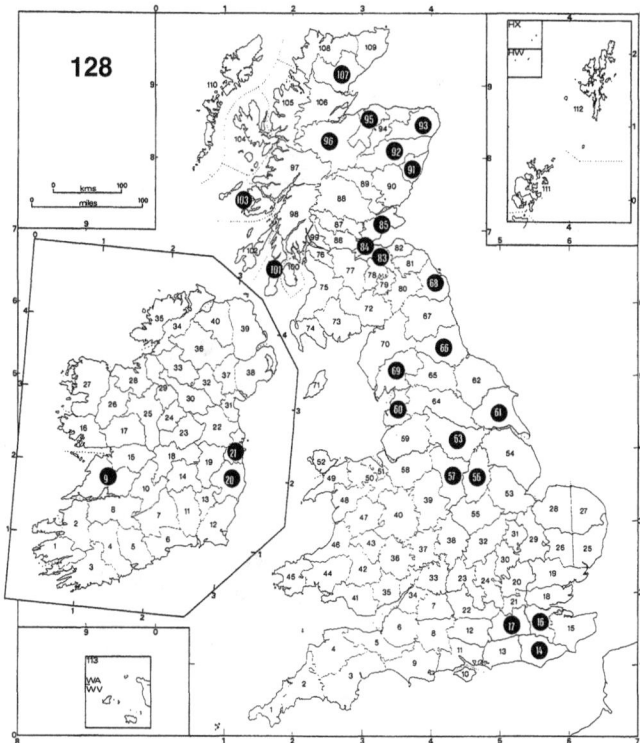

Syncopacma sangiella

coronillella, proved on dissection to be *S. sangiella* (Bradley, 1966).

Life history

Ovum. Undescribed. Oviposition site unknown, but presumably on the foodplant, common bird's-foot-trefoil (*Lotus corniculatus*).

Larva. Head honey-coloured. Prothoracic plate black with fine honey-coloured median sulcus or black with two honey-coloured semicircles anteriorly; body dark purple-fuscous; fine whitish green dorsal line on thoracic segments 2 and 3 and whitish green between thoracic segment 2 and abdominal segment 1; pinacula small, black; anal plate black; thoracic legs black; prolegs concolorous with body.

The larva feeds between spun leaves, often those at the tip of a stem, bending them downwards and forming a small rounded bunch, and appears to prefer plants with fleshy leaves. Fully fed larvae are found from May to June.

Pupa. Reddish brown to dark reddish brown; in a white silken cocoon. A fuller description is given by Bankes (1898a). June–July.

Imago. Univoltine; June–July.

Distribution (Map 128)

Frequents rough ground, often on calcareous soils, but occurs on dry heaths in Scotland. A very local species with a disjunct distribution; there are records from south-east England and scattered records, some old and mainly eastern, from the Midlands to the north of Scotland; Cos Clare, Wicklow and Dublin, Ireland. Western Europe to the former U.S.S.R.; Asia Minor.

SYNCOPACMA LARSENIELLA (Gozmány)

Stomopteryx larseniella Gozmány, 1957, *Acta zool.hung.* **3**: 116.

Stomopteryx ligulella sensu Pierce & Metcalfe, 1935, *Gen.tineid Fam.Lepid.Br.Isl*: 19, pl.11, *nec* [Denis & Schiffermüller], 1775.

Type locality: Poland; Szczecin.

Description of imago (Pl.5, figs 27,28)

Wingspan 11–14mm. Head purplish fuscous; frons buffish fuscous; antenna purplish fuscous, scape with faint longitudinal yellowish white line along leading edge, flagellum with alternate segments almost completely ringed yellowish white, fading distally; labial palpus segment 1 buffish fuscous, segment 2 buffish white, basally and on inner edge mixed fuscous, segment 3 buffish white, outer edge purplish fuscous. Thorax and tegulae purplish fuscous. Forewing black, with scattered silvery grey scaling; yellowish white straight fascia at two-thirds, edged with scattered ochreous scales, rarely obsolete; cilia fuscous, darker but indistinct ciliary line fading to tornus. Hindwing dark fuscous; cilia concolorous. Legs and abdomen similar to *S. taeniolella* (Zeller), *q.v.* Genitalia, see figures 17a,41b.

Similar species. *S. cinctella* (Clerck) and *S. taeniolella*, *qq.v.*

Life history

Ovum. Undescribed. Oviposition site unknown, but presumably on the foodplant, greater bird's-foot-trefoil (*Lotus pedunculatus* (=*uliginosus*)), but the moth has also been reared from common bird's-foot-trefoil (*L. corniculatus*) (A. N. B. Simpson, pers.comm.).

Larva. Similar to *S. taeniolella* except that segments 1 and 2 of the thoracic legs are blackish.

The larva feeds between spun leaves. Fully fed larvae are found from May to June.

Pupa. Reddish brown; in a silken cocoon. June.

Imago. Univoltine; June–July. Can be disturbed in the late afternoon, and at night comes to light.

Distribution (Map 129; see also Map 130)

Frequents rough ground and damp areas, and is not uncommon where the foodplant occurs. Widely distributed in the British Isles as far north as Co. Durham. It is more widespread than *S. cinctella* and there are unconfirmed reports from Dorset, Hertfordshire, Warwickshire and Shropshire; unlike that species, however, it has not yet been recorded from Ireland. Records of these two species are confused because the adults cannot be separated reliably without dissection and, until the publication of Pierce & Metcalfe (1935a), they were mostly considered to be one species. Throughout most of Europe.

SYNCOPACMA CINCTELLA (Clerck)

[*Phalaena*] *cinctella* Clerck, 1759, *Icones Insect.rar.* **1**: pl.11, fig.2.

Phalaena vorticella Scopoli, 1763, *Ent.Carn.*: 252.

Phalaena (Tinea) ligulella [Denis & Schiffermüller], 1775, *Schmett.Wien.*: 319.

?*Harpagus albistrigella* Stephens, 1834, *Ill.Br.Ent.* (Haust.) **4**: 279.

?*Gelechia sircomella* Stainton, 1854, *Ins.Br.Lepid.*: 132.

Type locality: [Sweden].

Description of imago (Pl.5, fig.29)

Wingspan 11–14mm. Indistinguishable from *S. larseniella*, *q.v.* In some specimens the yellowish white fascia is slightly inwardly curved and the forewing appears broader distally, but certain identification is possible only by examination of the genitalia. Rarely, the fascia may be obsolete or represented only by a small costal spot. Genitalia, see figures 17e,41c.

Similar species. S. larseniella (Gozmány) and *S. taeniolella* (Zeller), *qq.v.*

Life history

Ovum. Undescribed. Oviposition site unknown, but presumably on the foodplant, common bird's-foot-trefoil (*Lotus corniculatus*) or, possibly, *Genista* spp.

Larva. Described by Stainton (1867b) from Continental larvae, feeding between united leaves of *Lotus corniculatus*, as 'reddish, mixed with pale green; the latter colour prevailing on the anterior half of the larva; head pale-yellowish brown; the second segment [thoracic segment 1] with four black semi-lunules above, the inside of these lunules being filled up with yellowish-brown'. *S. larseniella* had not been described when Stainton wrote this. The description of the prothoracic plate does not agree with that of *S. larseniella* and so Stainton's description must, presumably, be of *S. cinctella*. Meyrick's larval description ([1928]) is simi-

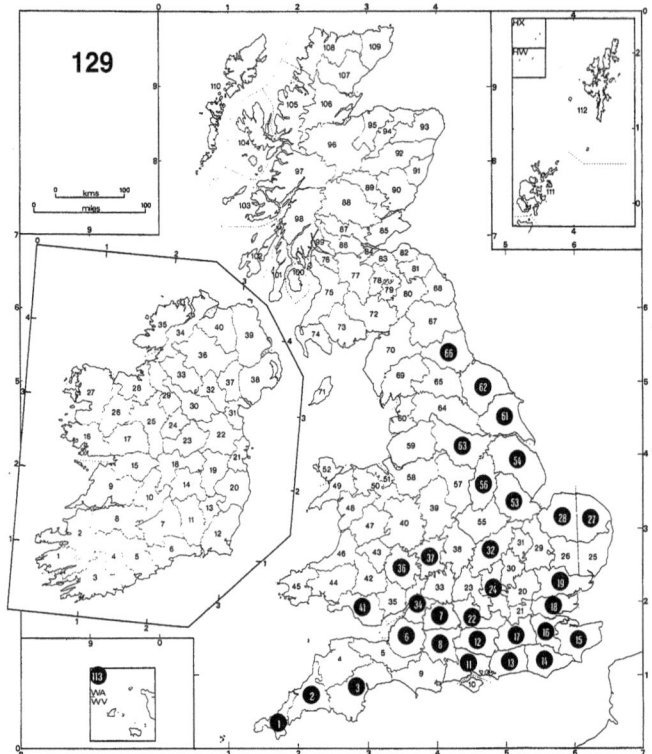

Syncopacma larseniella

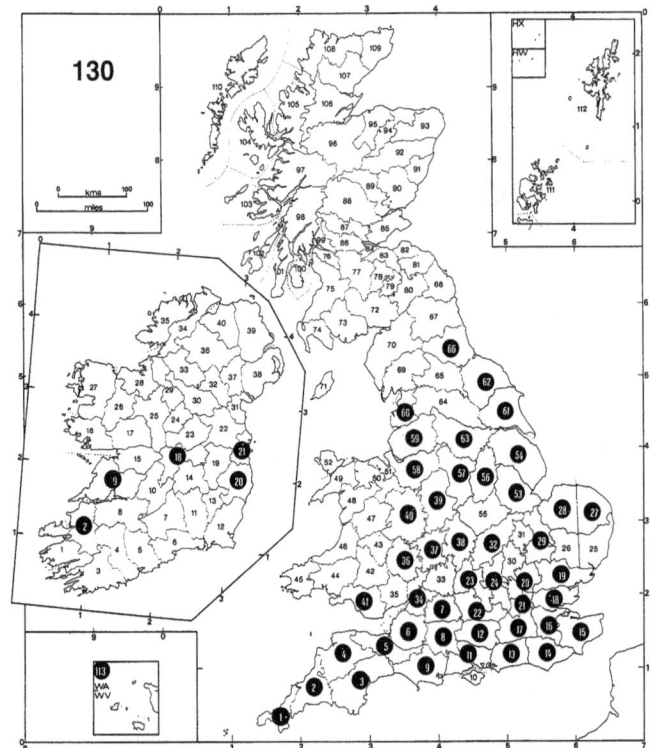

Syncopacma larseniella / *S. cinctella* (includes records not determined)

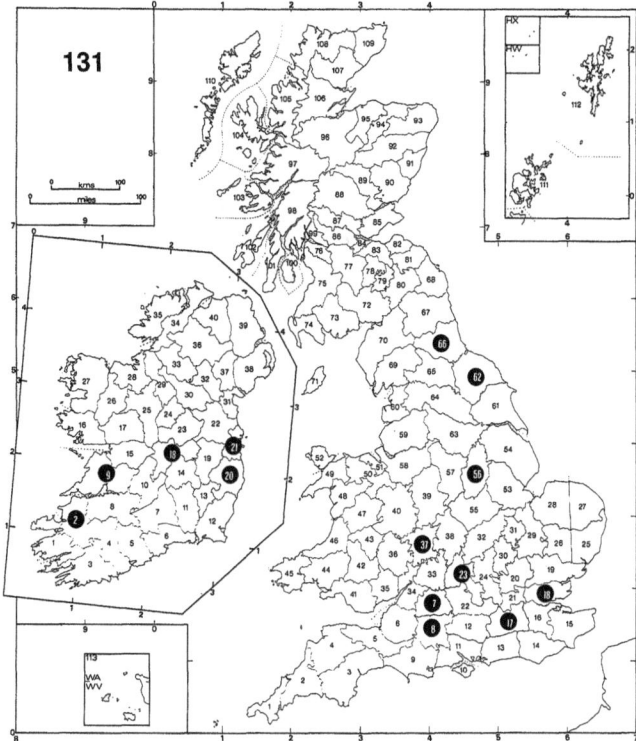

131

Syncopacma cinctella

lar. He states that it feeds between joined leaves of *Lotus* and *Genista*. Parsons (1995[1996]) reported incorrectly that it had been reared from strawberry clover (*Trifolium fragiferum*) but, on re-examination, the moth was shown to have been *Aproaerema anthyllidella* (Hübner). May–June.

Pupa. Undescribed. June.

Imago. Univoltine; July.

Distribution (Map 131; see also Map 130)

Frequents rough dry ground and may have a preference for calcareous soils. *S. cinctella* seems to be far less common and widespread than *S. larseniella* having a southern and central distribution extending north to Co. Durham; also recorded across central Ireland. Since 1960 it appears to have been reliably recorded from only half a dozen localities in England although there are unconfirmed records from north and south Somerset, south Wiltshire, Dorset, Berkshire, Buckinghamshire, Middlesex and south Lincolnshire. Widely distributed in the Palaearctic region, its range extending to North Africa and Asia Minor.

SYNCOPACMA TAENIOLELLA (Zeller)

Gelechia taeniolella Zeller, 1839, *Isis, Leipzig* **1839**: 201.

Type localities: Switzerland, Hungary and Germany; Dresden.

Description of imago (Pl.5, figs 30,31)

Wingspan 11–14mm. Head purplish fuscous, often with a line of yellowish white scales above eye curved inwardly towards frons, frons creamy to greyish fuscous; antenna purplish fuscous, scape with longitudinal yellowish white line along leading edge, flagellum with alternate segments almost completely ringed yellowish white, rings fading distally and often less complete from one-third; labial palpus segment 1 greyish fuscous, segment 2 greyish fuscous banded whitish distally, lower edge dark fuscous, segment 3 white, lower and upper edges lined dark fuscous. Thorax and tegulae purplish fuscous. Forewing black, basally purplish fuscous, with scattered silvery grey scaling, towards termen with coarser light grey, black-tipped scales; yellowish white fascia at two-thirds, edged with scattered ochreous scales, straight or, more often, slightly inwardly curved; cilia fuscous, darker ciliary line fading to tornus. Hindwing dark fuscous; cilia concolorous. Underside: forewing dark fuscous with distinct yellowish white fascia at two-thirds, wider than on upperside; hindwing pale fuscous, darker towards costa with distinct yellowish white spot on costa forming continuation of forewing fascia. Rarely fascia on upperside obsolete and fascia on underside less distinct or reduced to spots, with costal spot on hindwing reduced or obsolete. Legs purplish fuscous, tarsal segments banded yellowish white distally; hindleg with distinct white band at tibial spurs. Abdomen purplish fuscous. Genitalia, see figures 17d,41g.

Similar species. *S. larseniella* (Gozmány) and *S. cinctella* (Clerck), neither of which has a fascia on the underside of the forewing nor a costal spot on the underside of the hindwing.

Life history

Ovum. Undescribed. Oviposition site unknown, but presumably on the foodplant, common bird's-foot-trefoil (*Lotus corniculatus*) or, occasionally, greater bird's-foot-trefoil (*L. pedunculatus* (=*uliginosus*)), clover (*Trifolium* spp.) or medick (*Medicago* spp.) (Emmet, 1988b).

Larva. Head honey-coloured. Prothoracic plate lighter, posteriorly a pair of black dorsal lines extending three-quarters length of plate, posterolaterally a black elongate subtriangular mark; body light reddish brown, extensively marked translucent green on thoracic segments 2 and 3 including dorsal line and between thoracic seg-

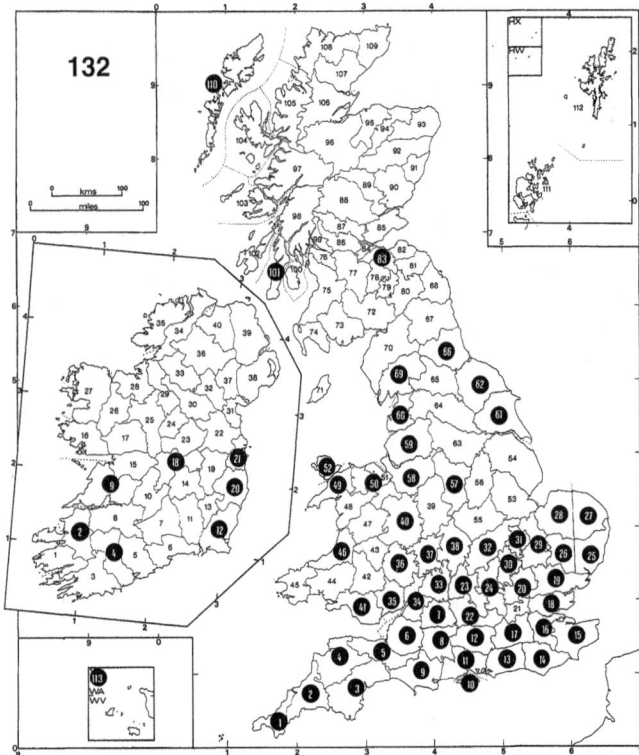

Syncopacma taeniolella

ment 2 and abdominal segment 6, with translucent green lateral line fading posteriorly; pinacula small, black; anal plate brownish honey-coloured; thoracic legs, segment 1 blackish, others translucent; prolegs concolorous with body.

The larva feeds between spun leaves. Fully fed larvae are found from May to June.

Pupa. Reddish brown; in a slight cocoon in detritus. June.

Imago. Univoltine; July. Can be disturbed from late afternoon, and at night comes to light.

Distribution (Map 132)

Frequents rough ground. Widely distributed and not uncommon in England south of a line from Shropshire to The Wash, becoming less common northwards; scattered records from Wales and Ireland, and so far recorded from only three localities in Scotland. Throughout most of Europe to the former U.S.S.R. and Asia Minor.

SYNCOPACMA VINELLA (Bankes)

Aproaerema vinella Bankes, 1898, *Entomologist's mon. Mag.* **34**: 242.

Lita coronillella sensu auctt. (partim), *nec* Treitschke, 1833.

Anacampsis cincticulella sensu auctt., *nec* Bruand, 1850.

Type locality: England; Brighton, Sussex.

Description of imago (Pl.5, figs 32,33)

Wingspan 11–12mm. Head purple-fuscous, frons buffish fuscous; antenna purple-fuscous, scape with longitudinal yellowish white line along leading edge, flagellum with alternate segments anteriorly lined yellowish white to three-quarters; labial palpus segment 1 purple-fuscous, segment 2 inner surface buffish white or purple-fuscous variably mixed buffish white, outer surface variably mixed purple-fuscous, segment 3 buffish white, outer edge purplish fuscous. Thorax and tegulae purple-fuscous. Forewing purple-fuscous, scattered plumbeous scaling along costa to apex and along termen to tornus; occasionally very small yellowish white costal and tornal spots, sometimes very faint, or, even more rarely, a yellowish white fascia at two-thirds (Bankes (1898b) called the latter f. *fasciata*); cilia fuscous with indistinct ciliary line. Hindwing dark fuscous; cilia concolorous. Legs purple-fuscous, tarsal segments indistinctly banded yellowish white distally. Abdomen purple-fuscous. The figure of the female genitalia in Pierce & Metcalfe (1935a) depicts *S. albipalpella* (Herrich-Schäffer) (Heckford, 1991). Genitalia, see figures 16h,41f.

Life history

Ovum. Undescribed. Oviposition site unknown, but presumably on the foodplant, dyer's greenweed (*Genista tinctoria*).

Larva. Head varying from honey-coloured to yellowish brown, sometimes blackish posterolaterally. Prothoracic plate narrower than prothoracic segment, varying from orange-ochreous somewhat marked black, to grey with large blackish lateral areas and six submedian spots, to black or black with a honey-coloured median triangular area with three black spots, one anterior to the others, and finely bisected whitish; body varying from dull whitish, subdorsal lateral and spiracular lines dull reddish, anteriorly macular, to dull reddish purple with faint grey dorsal line fading to anal plate, or to dull purplish brown; first thoracic segment often greyish white; pinacula small, black; anal plate shining black; thoracic legs shining black; prolegs translucent with black markings.

The larva feeds between spun leaves, often mining

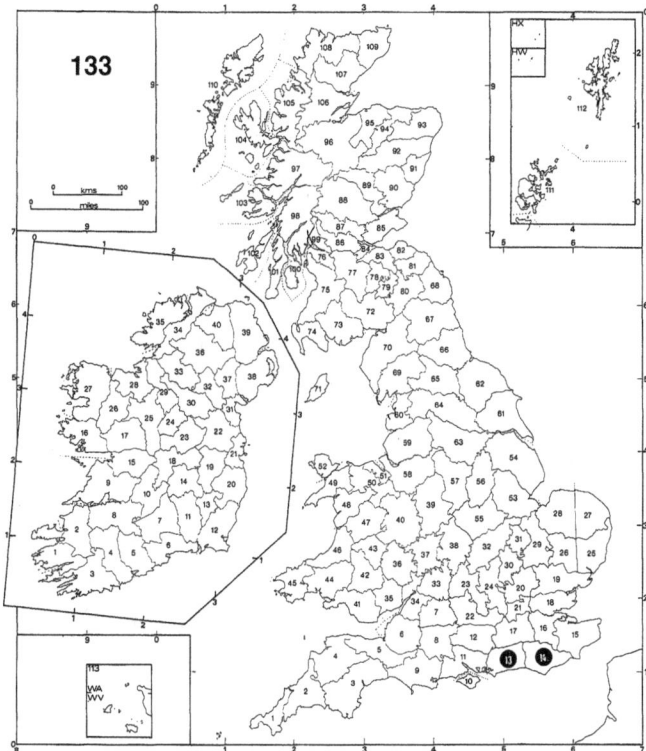

Syncopacma vinella

them. Bradford (*in* Emmet [1979]) and Emmet (1988b; 1991) also give crown vetch (*Coronilla varia*) as a foodplant. This is almost certainly incorrect. *Coronilla varia* is a very rare introduction in southern England and recorded from only a few sites, none of them near the known localities of *S. vinella*. Further, there is no record of its feeding on this plant. Meyrick ([1928]) and Ford (1949b) do not give it as a foodplant of *S. vinella* but as the foodplant of *S. coronillella* (Treitschke), a species which was subsequently shown by Bradley (1966) not to occur in the British Isles. All supposed British specimens of that species were either *S. vinella* or *S. sangiella* (Stainton). Larvae have been found in late April (Wakely, 1962), late May and June (R. J. Heckford, pers.obs.) and October and November (Bankes, 1898b). Bankes (1899b) stated that larvae of the autumnal brood hibernated full-fed and refused to touch fresh shoots of the foodplant in the spring.

Pupa. Shiny dark brown; in a slight cocoon in spun leaves of the foodplant or in detritus. A fuller description is given by Bankes (*loc.cit.*). May, June and July.

Imago. Adults have been found in the wild in June, July, August and September. Bankes (*loc.cit.*), Meyrick (*loc.cit.*) and Ford (*loc.cit.*) consider the species to be bivoltine; Bradford (*loc.cit.*) and Emmet (1991) to be

univoltine. In view of the dates when larvae and adults have been found in the wild the present author considers it to be bivoltine.

History and distribution (Map 133)

Known from only four or possibly five localities, all in Sussex and all in rough fields on or very near Wealden clay. It was originally discovered in 1886 by A. C. Vine, who gave the locality as the neighbourhood of Brighton. It was evidently fairly common, as a long series, both of reared and captured moths, was later taken by W. H. B. Fletcher at the same site (Bankes, 1898b). It is generally believed that this was Ditchling Common, from which the species has proved to be best known. Other localities are a rough field near Lewes (Griffith, 1932), Tilgate Forest, 1934–36 (specimens in BMNH) and two fields near Plaistow (Wakely, *loc.cit.*; Fairclough, 1962). The most recent record was of larvae at Ditchling Common in 1990 (R. J. Heckford & J. R. Langmaid, pers.obs.). Subsequent searches have proved unsuccessful. At all its former localities the foodplant has declined or been exterminated, mainly through agricultural practice. Status RDB2 was accorded to the species in Shirt (1987) but Parsons (1995 [1996]) suggests that it should now be given status RDB1. However, the moth may already be extinct in Britain (Heckford, 1999b). The Netherlands, Denmark, Switzerland, Germany, Czech Republic, Slovakia, Austria and Hungary.

SYNCOPACMA ALBIPALPELLA (Herrich-Schäffer)

Anacampsis albipalpella Herrich-Schäffer, 1854, *Syst. Bearb.Schmett.Eur.* 5: 195.

leucopalpella Herrich-Schäffer, 1853, *Samml.eur.Schmett.* 5: pl. 70, fig. 523 (unavailable, non-binominal).

Type localities: Austria; Vienna, and Germany; Regensburg.

Description of imago (Pl.5, fig.26)

Wingspan 10–11mm. Head purple-fuscous, with line of yellowish white scales above eye; antenna purple-fuscous, scape with longitudinal cream line along leading edge, flagellum with alternate segments almost completely ringed cream to two-thirds; labial palpus segment 1 purple-fuscous, segment 2 purple-fuscous extensively mixed whitish and ringed whitish at apex, segment 3 buffish fuscous, lower edge purple-fuscous. Thorax and tegulae purple-fuscous. Forewing purple-fuscous, towards termen scales coarser; white costal and tornal spots at two-thirds, rarely forming a fascia, sometimes costal spot much reduced and tornal spot obsolete; cilia fuscous, darker ciliary line fading to tornus. Hindwing fuscous; cilia concolorous. Legs purple-fuscous, tarsal segments banded whitish distally; hindleg with indistinct whitish band at tibial spurs. Abdomen purple-fuscous. The figure of the female genitalia of *S. vinella* (Bankes) in Pierce & Metcalfe (1935a) is in fact of this species (Heckford, 1991). Genitalia, see figures 17c,41d.

Similar species. Aproaerema anthyllidella (Hübner) usually differs in having a few yellowish ochreous scales on the fold and dark stigmata, but these markings are sometimes obsolete, and segment 2 of the labial palpus is usually not ringed whitish at the apex. The species are easily separable on the genitalia.

Life history

Ovum. Undescribed. Oviposition site unknown, but presumably on the foodplant, petty whin (*Genista anglica*).

Larva. Head pale yellowish-brown. Prothoracic plate similar, posteriorly a pair of black dorsal lines extending half length of plate, posterolaterally a black round mark; body dull crimson, except thoracic segments 2 and 3 and abdominal segments 1 and 2 which are pale ochreous-yellow; pinacula minute, black; anal plate black; thoracic legs black, whitish at joints; prolegs translucent, anal pair blackish anteriorly.

The larva feeds between spun leaves, mining most of them by entering through irregular, suboval holes, usually eating all the parenchyma without depositing frass within the mine; the leaves turn primrose-yellow. September–June.

Pupa. Reddish brown. June.

Imago. Univoltine; July.

History and distribution (Map 134)

Frequents damp heathland where the foodplant occurs; the foodplant is declining nationally. The species was first reported in the British Isles by Stainton (1858), who found larvae near Horsell, Surrey, on 20 June 1857; he also recorded (*loc.cit.*) that a Mr Scott collected larvae that autumn near York. Dunn & Parrack (1992) state that J. Sang found larvae at Darlington, Co. Durham, but give no date, though it was probably in the 1890s and, according to Meyrick ([1928]), it also occurred in Surrey, Devon, Essex and York. The present author cannot trace the source of the Devon record, which must be considered unconfirmed. Emmet (1981) believed it to be extinct in Essex, the last record being in 1883. Sutton & Beaumont (1989) knew of no other Yorkshire record than that of the larvae found near York in 1857, referred to by Stainton. There are unconfirmed and improbable records from

Syncopacma albipalpella

Formby, Lancashire, in 1919 (Mansbridge, 1934) and Pakefield, Suffolk, in 1950 (Morley, 1957). It has also been recorded from Southampton, Hook Common and Silchester, Hampshire, and Ashdown Forest and Ditchling Common, Sussex. Until 1999, when R. W. J. Uffen found vacated feedings and one larva in Hertfordshire from which a moth was reared, it had not been recorded in the British Isles since 1973 when larvae were found at Silchester (Goater, 1974). Searches at the Hampshire locality and others were unsuccessful, and an enquiry about recent records (Heckford, 1992) drew no response. It is possible that Hertfordshire now has the only British locality. Parsons (1995 [1996]) proposes that it be given RDB1 status. On the Continent, recorded from The Netherlands, Belgium, Germany, Poland, Czech Republic, Austria, Spain, France, Italy, Sicily and Greece.

SYNCOPACMA SUECICELLA (Wolff)

Stomopteryx suecicella Wolff, 1958, *Ent.Meddr* **28**: 268.
Type locality: Sweden; Bonarpshed, Skåne.

Description of imago (Pl.5, figs 34,35)

Wingspan 8–9mm. Head purplish black, frons dirty cream; antenna black, scape with longitudinal cream line along leading edge, flagellum with alternate segments anteriorly lined cream to one-half; labial palpus purplish black, mixed cream. Thorax and tegulae purplish black. Forewing purplish black, becoming black after death; small yellowish white dot consisting of four to six scales at two-fifths on fold, sometimes obsolete; in male a yellowish white, narrow, straight dash on costa at two-thirds and a similar mark opposite on tornus, in female a complete yellowish white fascia at two-thirds; in both sexes beyond these marks coarser scaling with scales light grey, tipped black; cilia dark fuscous, distinct black ciliary line fading to tornus. Hindwing dark fuscous; cilia concolorous, lighter at bases. Legs purplish fuscous, tarsal segments banded whitish distally; hindleg with distinct whitish band at tibial spurs. Abdomen purplish fuscous. Genitalia, see figures 17f,41h.

Life history

Ovum. Undescribed. Oviposition site unknown, but presumably on the foodplant, hairy greenweed (*Genista pilosa*).

Larva. Head honey-coloured with black posterolateral markings. Prothoracic plate black, finely bisected yellowish white; body reddish purple, marbled grey, the marbling stronger on thoracic segments 2 and 3, with grey dorsal line becoming fainter posteriorly and with a broken grey lateral line; pinacula very small, black, encircled dark grey; anal plate black; thoracic legs black; prolegs concolorous with body.

The larva feeds between spun leaves, often mining them, usually at or near the tip of a stem. Fully fed larvae are found from May to June. On 16 July 1993 one small larva was found which may suggest that the species has a long larval period and overwinters in this stage (R. J. Heckford, pers.obs.).

Pupa. Shining black, within spun leaves. June.

Imago. Univoltine; June–July.

Distribution (Map 135)

Known from only one small area on the Lizard peninsula, Cornwall, where it was discovered in 1984 (Heckford, 1986). It has been searched for without success in other Cornish localities where *Genista pilosa* occurs. Parsons (1995[1996]) proposes that it should be given RDB1 status and should not be collected. Recorded

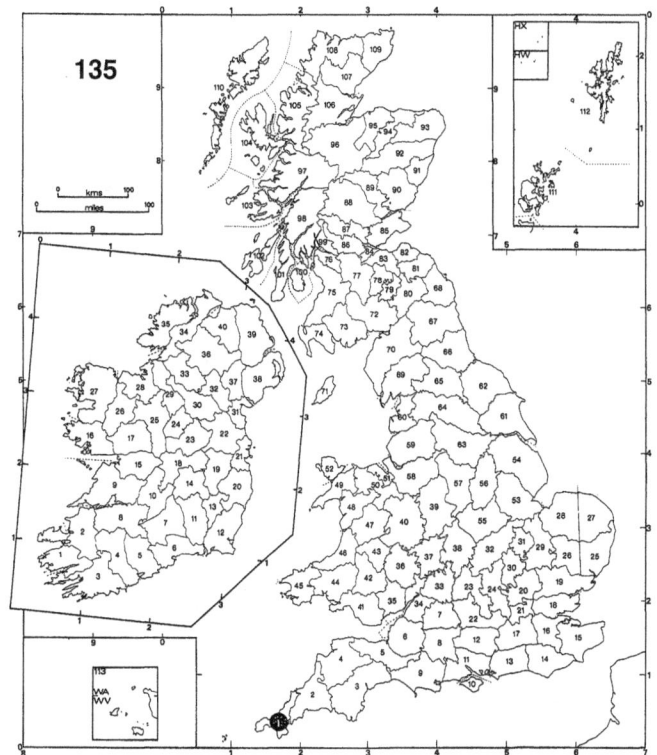

Syncopacma suecicella

from Sweden, Denmark, Germany, Austria, Czech Republic, Slovakia, Spain, France, Corsica, Sardinia, Italy and Greece.

SYNCOPACMA POLYCHROMELLA (Rebel)

Anacampsis polychromella Rebel, 1902, *Dt.ent.Z.Iris* **15**: 109.

Type locality: Israel; Haifa.

Description of imago (Pl.5, fig.36)

Wingspan 7–10mm. Head varying from whitish ochreous to pale brown, frons whitish; antenna dark reddish brown, upperside with interrupted white line along each side; labial palpus white, segment 3 lined dark fuscous beneath. Thorax and tegulae light brownish ochreous. Forewing with basal one-half light brownish ochreous darkening to dark golden brown or fuscous before white fascia at two-thirds; base of dorsum and basal one-quarter of costa indistinctly edged white; fascia wider at costa than at dorsum, its inner edge straight; beyond this, apical one-third of wing densely overlaid with dark fuscous-tipped scales; cilia greyish ochreous with one dark fuscous ciliary line. Hindwing light ochreous-fuscous, cilia concolorous. Legs mixed

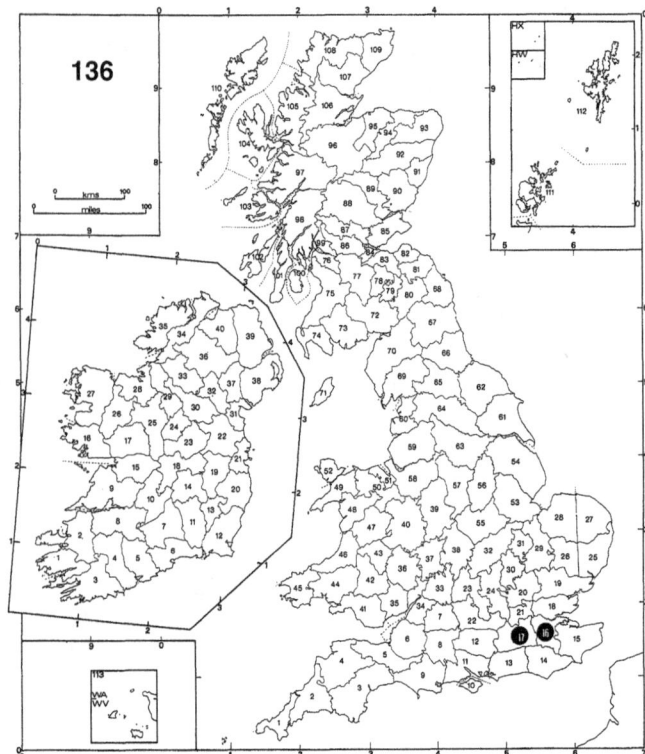

Syncopacma polychromella

whitish and greyish fuscous; fore- and midleg tarsi dark fuscous, ringed whitish at joints. Abdomen ochreous-fuscous; anal tuft paler. Genitalia, see figures 17g,41a.

Life history

Unknown. The only recorded foodplant that the present author can trace is *Lotus sessilifolius* (Klimesch, 1983) which does not occur in the British Isles.

History and distribution (Map 136)

Adventive. The first British specimen was taken by L. T. Ford on 20 February 1952 at rest on a fence near Bexley railway station, Kent. He considered that it had probably been imported accidentally with esparto grass as a pupa (Ford, 1953), although it is not clear why he took this view. A second specimen was found on 1 February 1999 at rest on the inside of a ground floor corridor window of offices at Hook, Chessington, Surrey, by J. Porter (pers.comm.). Portugal, Spain, Italy, Sicily, Malta, the former Yugoslavia; Africa; Israel, Jordan, Syria; India.

ANACAMPSIS Curtis

Anacampsis Curtis, 1827, *Br.Ent.* **4**: 189.

Tachyptilia Heinemann, 1870, *Schmett.Dtl.Schweiz* (2)**2**(1): 321.

A worldwide genus with its stronghold in the Nearctic and Neotropical regions. Eight species are represented in Europe, of which only three occur in Britain.

Imago. Small to large gelechiids with labial palpus long, slender and sometimes with segment 2 thickened; antenna without pecten. Forewing elongate, subtrapezoidal with termen oblique, without tufts of erect scales. Hindwing broader than forewing, elongate-trapezoidal, apex and tornus rounded, no emargination of termen. Abdomen with tergites 1–3 buff. Valva of male genitalia simple and linear, aedeagus large and often curved; signum of female genitalia a small plate with single lateral spine rudimentary or absent.

Larva. In a rolled or folded leaf or spun shoot on a range of deciduous trees and shrubs.

Key to species (imagines) of the genus *Anacampsis*

1	Forewing blackish with velvet-black markings .... *temerella* (p. 216)
–	Forewings grey or blackish with pale-coloured markings 2
2(1)	Larva on *Betula*; male genitalia with uncus flat on top and aedeagus strongly curved; female genitalia with ostial plate ovate with obvious apical pimple .... *blattariella* (p. 218)
–	Larva on *Salix* or *Populus*; uncus of male conical and aedeagus weakly curved; ostial plate of female subquadrate with only a suggestion of an apical pimple .... *populella* (p. 217)

ANACAMPSIS TEMERELLA (Lienig & Zeller)

Gelechia temerella Lienig & Zeller, 1846, *Isis, Leipzig* **1846**: 284.

Gelechia pernigrella Douglas, 1850, *Trans.ent.Soc.Lond.* (N.S.) **1**: 64.

Type locality: Latvia; Rambdau.

Description of imago (Pl.6, fig.1)

Wingspan 11–13mm. Head black becoming brown or ochreous on frons; antenna black, paler beneath; labial palpus pale ochreous above, ochreous beneath, tip of segment 3 black. Thorax and tegulae black. Forewing dark fuscous-black; black markings consisting of ill-

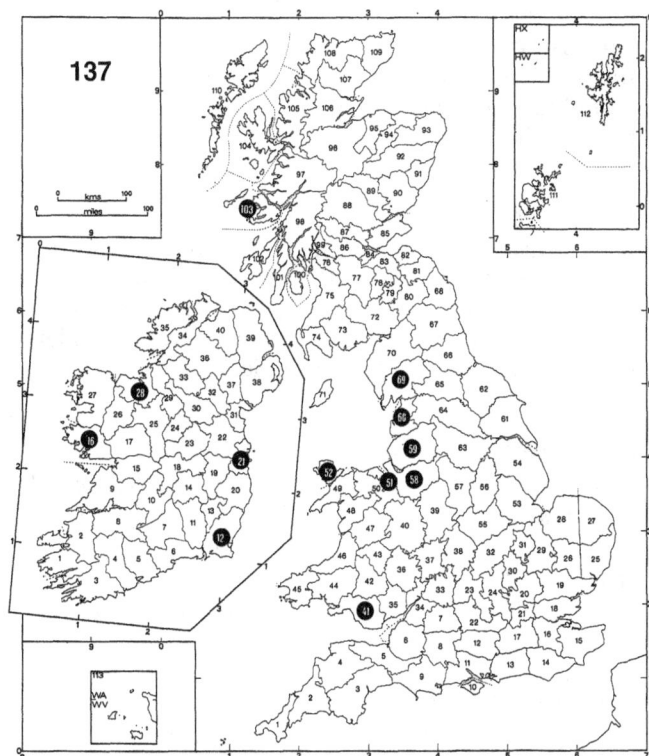

Anacampsis temerella

defined transverse fascia at two-thirds, two spots in discal area, one at one-third and other just before fascia, and a row of spots at base of cilia; cilia dark fuscous. Hindwing ochreous-fuscous; cilia concolorous, becoming fuscous at apex. Legs fuscous-brown, tarsi browner. Abdomen dark fuscous. Genitalia, see figures 17h,42a.

Life history

Ovum. Laid on a twig of creeping willow (*Salix repens*). July.

Larva. Final instar length 8mm. Head and prothoracic plate black. Body dull whitish; pinacula small and black; thoracic legs blackish; prolegs dull whitish (Stainton, 1865b).

The larva spins the leaves of the terminal 5cm of a twig of the foodplant together to form a tube around the twig. The larva lives in this tube, eating the leaves from within. At what season the eggs hatch is unknown but feeding larvae can be found throughout June, becoming full-grown by early July.

Pupa. Length 6 mm. Short, robust, brown, covered with fine short hair; terminal cremaster of about eight long hooked bristles and many more on underside of last two segments. Within the larval habitation. July.

Imago. Univoltine; July and early August. Can best be obtained by rearing or sweeping its foodplant in suitable localities.

Distribution (Map 137)

A very local species occurring in widely separated coastal localities throughout Britain and Ireland. Its main strongholds are the coastal sandy areas bordering the Irish Sea from Anglesey to Cumbria. In Scotland, known only from the Isle of Coll (Bland *et al.*, 1987a). Unconfirmed reports of its occurrence on the sandy heaths of Berkshire and Buckinghamshire suggest that it could be more widespread than currently thought. The local nature of this species has led Parsons (1995[1996]) to ascribe it a Nationally Notable (Scarce) category A conservation status. Widespread throughout northern and central Europe; less widespread in southern Europe and apparently absent from the Iberian Peninsula.

ANACAMPSIS POPULELLA (Clerck)

Phalaena populella Clerck, 1759, *Icones Insect.rar.* **1**: pl.11, fig.5.

Recurvaria populi Haworth, 1828, *Lepid.Br.*: 548.

Recurvaria blattariae Haworth, 1828, *Lepid.Br.*: 553 (partim).

Anacampsis laticinctella Stephens, 1834, *Ill.Br.Ent.* (Haust.). **4**: 205.

Anacampsis tremulella Duponchel, [1839], *Hist nat.Lép. Fr.* **11**(1838): 272, pl.296, fig.4.

Tachyptilia populella fuscatella Bentinck, 1934, *Tijdschr.Ent.* **77**: xxii.

Tachyptilia populella ambronella Meder, 1934, *Schr. Naturw.Ver.f.Schlesw.-Holst.* **20**: 362.

Type locality: [Sweden].

NOMENCLATURE. The subspecies *fuscatella* Bentinck was relegated to f. *fuscatella* by Bradley & Fletcher (1979). The name applied to specimens which were bred from creeping willow (*Salix repens*) and differed from the typical form (wingspan 17–19mm) only in their smaller size (wingspan 14–15mm). As typically sized specimens have also been bred from creeping willow, the description below serves for both forms.

Description of imago (Pl.6, fig.2,3)

Wingspan 14–19mm. Sexes similar except that abdominal tergites 2–4 of female are yellowish ochreous. Very variable with range of coloration and basic pattern same as for the following species from which it can be reliably separated only on genitalia differences (see figures 17i,42b) and larval foodplant.

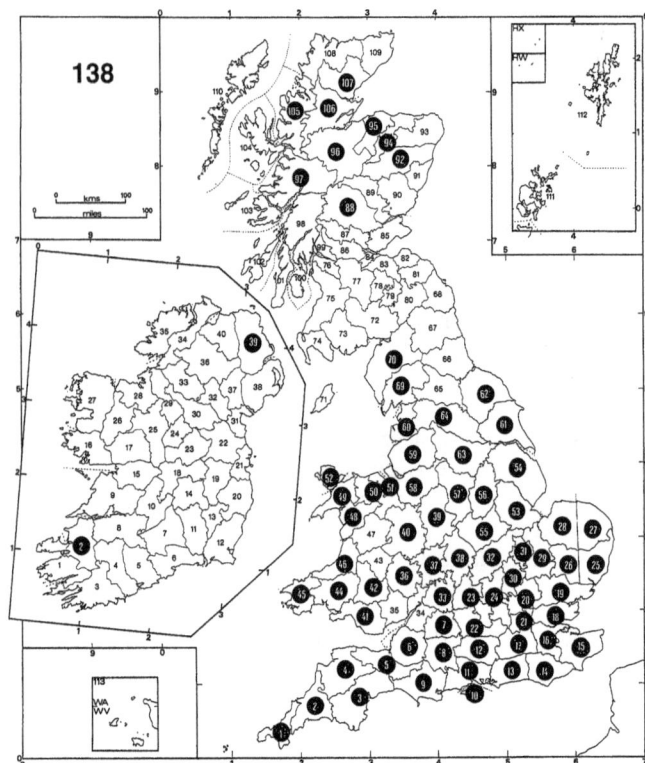

Anacampsis populella

Similar species. A. *populella* is usually slightly larger, and the transverse fascia (when discernible) is less abruptly angular than in A. *blattariella* (Hübner). The contention that A. *populella* is grey and fuscous and A. *blattariella* is black and white (Vári, 1941) is not reliable, although the latter is frequently more strongly marked than the former. In the male genitalia of A. *populella* the uncus is more conical, the valva shorter and the aedeagus less curved; in the female genitalia the apophyses are shorter, the ostium plate less ovate and the apical projection less prominent than in A. *blattariella*.

Life history

Ovum. Laid on various species of poplar and aspen (*Populus* spp.) or willows (*Salix* spp.) including creeping willow (*S. repens*).

Larva. Head and divided prothoracic plate black. Body greyish green, with conspicuous round black pinacula (Doets, 1941); anal plate brownish.

The larva feeds within a rolled leaf of its foodplant during May and June. When fully fed it pupates within a cocoon in the rolled leaf.

Pupa. Length 7–9mm. Pale brown, somewhat flattened; short and broad, tapering strongly caudad from two-thirds and covered with fine hairs; cremaster bluntly pointed, with six to eight hooked ochreous hairs. June–July.

Imago. Univoltine; July and August. On willows, it is best obtained by beating the foodplant, when it readily takes to the wing, whereas on large poplars it rests on the trunk by day.

Distribution (Map 138)

Widely distributed throughout mainland Britain, but appears to be absent from the whole of southern Scotland and is reported from only two widely separated parts of Ireland. It occurs throughout Europe; accidentally introduced into North America.

ANACAMPSIS BLATTARIELLA (Hübner)

Tinea blattariella Hübner, 1796, *Samml.eur.Schmett.* **8**: p.22, fig.148.
Recurvaria blattariae Haworth, 1828, *Lepid.Br:* 553 (partim).
Anacampsis betulinella Vári, 1941, *Tijdschr.Ent.* **84**: 351.
Type locality: Europe.

Description of imago (Pl.6, figs 4,5)

Wingspan 16–19mm. Sexes similar except that abdominal tergites 2 to 4 of female are clothed with small tightly-packed yellowish ochreous scales. Head ochreous-white to light fuscous; antenna dark fuscous, barred paler or sometimes white; labial palpus with basal four-fifths of segment 2 fuscous, remainder pale ochreous or sometimes white. Thorax and tegulae varying from greyish fuscous to mixture of white and dark fuscous. Forewing very variable from almost unmarked to having contrasting black and white markings; in the unmarked form the wing is greyish fuscous with ill-defined dark fuscous spots at one-quarter and one-half along fold, others in discal area just before and also just beyond halfway along wing and as many as six regularly spaced along base of apical and terminal cilia; cilia pale ochreous beyond two ciliary lines of fuscous-tipped scales; in the well-marked forms the forewing is blackish fuscous heavily mixed with white, with a prominent strongly angled narrow transverse white fascia at three-quarters and with the spots black and well-defined; cilia white beyond double ciliary line. Hindwing fuscous; cilia concolorous, two poorly defined ciliary lines. Legs fuscous, distal part of tarsi edged whitish; hindleg tibial tuft ochreous-fuscous. Abdomen greyish fuscous to fuscous. Genitalia, see figures 17j,42c.

Similar species. A. *blattariella* can be differentiated with certainty from A. *populella* (Clerck) only by the structure of the genitalia and the larval foodplant, although

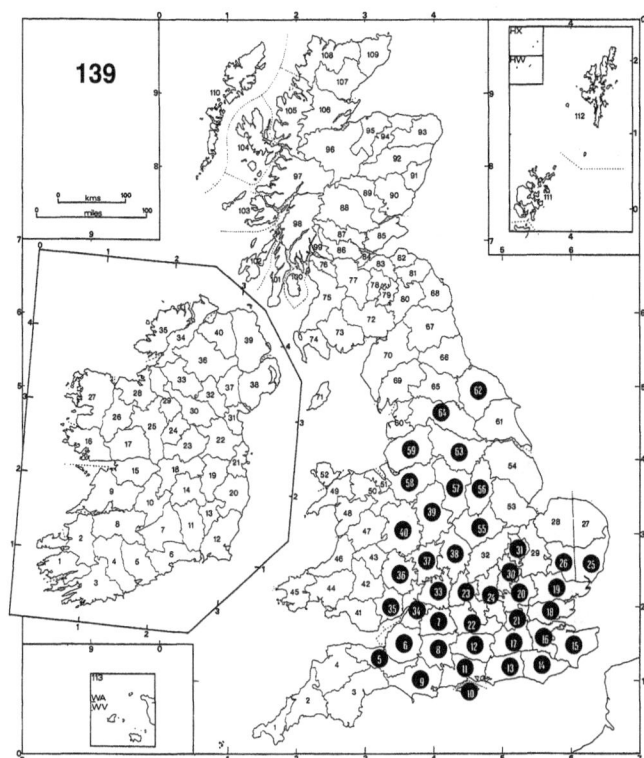

Anacampsis blattariella

the more strongly angled fascia (when visible) allows tentative identification. In the male genitalia of *A. blattariella* the uncus is more flattened apically, the valvae longer and the aedeagus more strongly curved; in the female genitalia the apophyses are longer and the ostial plate more ovate and the apical pimple more prominent than in *A. populella*. *A. blattariella* and *A. populella* are closely related species which appear to be clearly separated on the Continent. This separation may not be so complete in Britain. Specimens reared by D. A. B. Macnicol from larvae 'in rolled *Betula* leaves' from Surrey in 1952 are large, weakly marked and have female genitalia approximating more to *A. populella* than to *A. blattariella*.

Life history

Ovum. Laid on birch (*Betula* spp.).

Larva. Head, prothoracic plate and thoracic legs black. Body greyish green when young but yellowish green in final instar; pinacula small and black; anal plate brownish.

Feeds during May and June within a leaf of birch that has been rolled longitudinally along the midrib. The ends of the rolled leaf are crimped closed (Doets, 1941). Larvae of *Epinotia solandriana* (Linnaeus) (Tortricidae) feed in the same way but can be distinguished

by their yellowish brown heads mottled with dark brown.

Pupa. Length 7–8mm. Yellow-brown, darker dorsally, covered with fine ochreous yellow pubescence; cremaster of 12–20 hooked bristles and others more ventrally; within the rolled leaf or between two opposing leaves. June–July.

Imago. Univoltine; July–September. Flies at night, sometimes coming to light. Rests during the day on the trunks of large birch trees.

Distribution (Map 139)

First taken in Britain by S. N. A. Jacobs in the New Forest in 1947 (Jacobs, 1948). Since then found throughout most of England except the extreme north and south-west. Absent from Scotland, Ireland and the greater part of Wales. Throughout Europe, except the Iberian Peninsula.

MESOPHLEPS Hübner

Mesophleps Hübner, [1825], *Verz.bekannt.Schmett.*: 406.

A genus represented by about 35 species, most of which are Australian; four occur in Europe, one of which used to be found in England.

Imago. Head with appressed scales, slightly tufted at sides; ocelli present; antenna three-quarters length of forewing, apical one third serrate; labial palpus with segment 2 longer than segment 3. Forewing elongate, rather pointed. Hindwing broader than forewing, trapezoidal; apex moderately produced; termen somewhat sinuate; tornus rounded; cilia length equal to breadth of wing.

Larva. Feeds on Cistaceae.

MESOPHLEPS SILACELLA (Hübner)

Tinea silacella Hübner, 1796, *Samml.eur.Schmett.* **8**: 37, pl.17, fig.117.
Recurvaria silacea Haworth, 1828, *Lepid.Br.*: 555
Type locality: Europe.

Description of imago (Pl.6, fig.6)

Wingspan 13–18mm. Head whitish; antenna greyish brown or ochreous-brown, indistinctly annulate darker; labial palpus with segment 2 thickened with dense appressed scales beneath and loosely clad with long scales above, reddish brown at sides and beneath, whitish above, segment 3 whitish. Thorax and tegulae whitish or pale yellowish. Forewing pale yellowish ochreous; stigmata reddish brown, first discal, some-

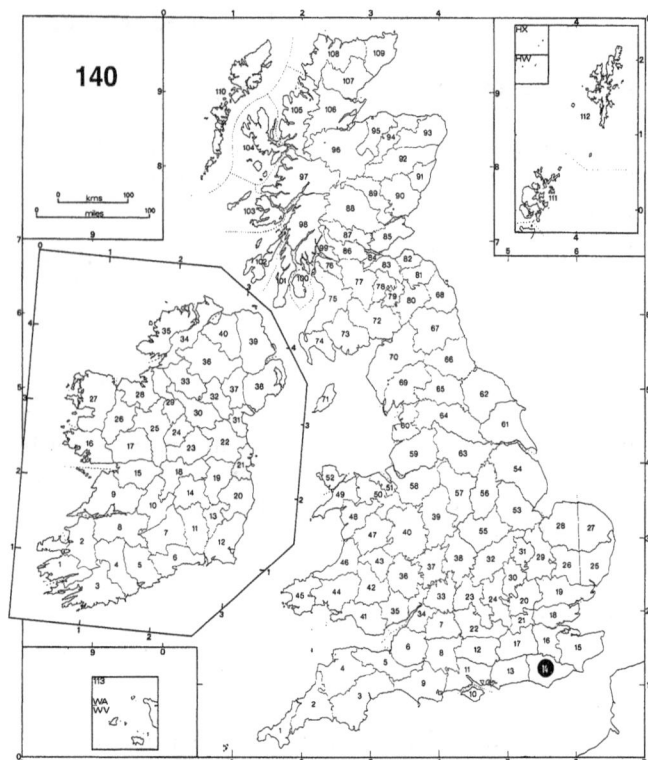

Mesophleps silacella

times obsolete, well beyond plical; costa shortly edged reddish brown near base; reddish brown costal streak from just beyond middle to apex; irregular reddish brown mark below second discal stigma; terminal area suffused reddish fuscous; cilia fuscous, darker ciliary line fading toward tornus. Hindwing grey; cilia concolorous. Legs reddish fuscous; hind tibia and tarsus whitish ochreous. Abdomen ochreous-grey, segments 1 and 2 dorsally clad with specialized spiny scales, yellow. Genitalia, see figures 16d,42d.

Life history
Ovum. Undescribed. Laid on common rock-rose (*Helianthemum nummularium*).
Larva. Undescribed. It feeds in seed-capsules. August.
Pupa. Undescribed.
Imago. June–July.

Distribution (Map 140)
The most recent record is from Moulsecoomb, East Sussex, in 1906, before which it was found only in the Brighton area (Parsons, 1995[1996]). South-eastern Sweden, central and southern Europe; Asia Minor.

Chelariinae

NEOFACULTA Gozmány
Neofaculta Gozmány, 1955, *Ann.hist.-nat.Mus.natn. hung.* (S.N.) **6**: 308–309 (keys), 312.

A Holarctic genus represented by two species in Europe, one of which is also found in North America. One species occurs in the British Isles.

Imago. Head with appressed scales; ocelli present; antenna three-quarters length of forewing, serrate and shortly ciliate in male, simple in female, scape without pecten; labial palpus with segment 3 a little shorter than segment 2. Forewing elongate. Hindwing subtrapezoidal, broader than forewing; apex slightly produced; termen slightly sinuate, merging gently into dorsum, with tornus indefinable; cilia about one-half breadth of wing; veins Rs (7) and M_1 (6) stalked, M_3 (4) and CuA_1 (3) connate or stalked.
Larva. The British species feeds on heather (*Calluna vulgaris*) and *Erica* spp.

NEOFACULTA ERICETELLA (Geyer)
Tinea ericetella Geyer, [1832], *in* Hübner, *Samml.eur. Schmett.* **8**: pl.70, fig.470.
Recurvaria betulea sensu Haworth, 1828, *Lepid.Br.*: 549 [unjustified emendation and misidentification of *Tinea betulinella* Hübner].
Anacampsis lanceolella Stephens, 1834, *Ill.Br.Ent.* (Haust.) **4**: 211.
Type locality: [Europe].

Description of imago (Pl.6, fig.7)
Wingspan 13–18 mm. Head greyish fuscous, frons paler; antenna fuscous; labial palpus greyish fuscous, darker at apex, inner side of segment 2 and upperside of segment 3 extensively whitish. Thorax and tegulae greyish fuscous. Forewing pale to dark greyish fuscous, veins more or less irrorate with white and black scales; indistinct blackish spot in disc at one-fifth; two blackish spots at one-third, that in disc beyond that on fold, the former sometimes elongate forming blackish streak extending to second blackish discal spot just beyond one-half; indistinct dark greyish fuscous tornal spot sometimes extending to second discal spot; indistinct blackish terminal spots forming broken line from costa at two-thirds around termen to tornus, sometimes with larger blackish spot at apex; cilia greyish fuscous with broad dark grey ciliary line, cilia paler at tips. Hindwing and cilia greyish, both paler at base. Legs greyish

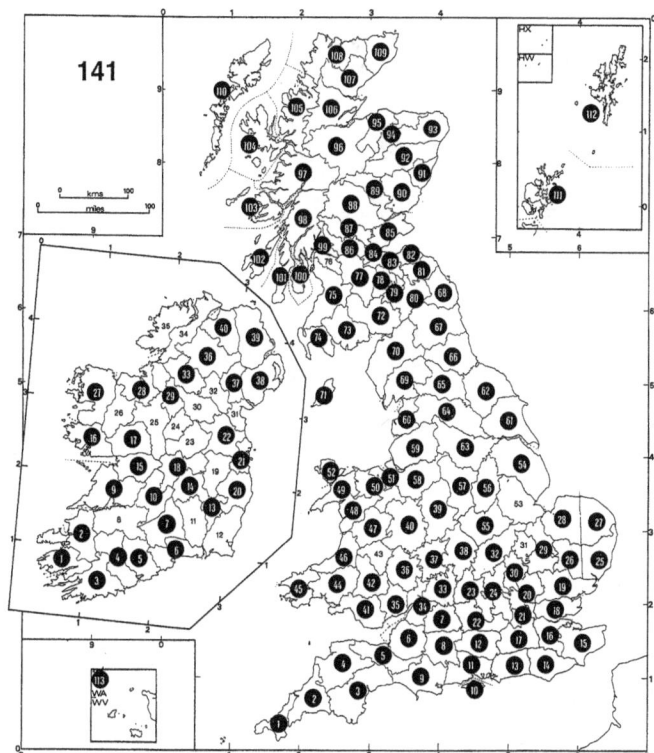

Neofaculta ericetella

fuscous with inner sides pale ochreous, tarsi narrowly ringed pale ochreous. Abdomen greyish fuscous; anal tuft of male dull whitish. Genitalia, see figures 18a,42e.

Life history

Ovum. Laid on heather (*Calluna vulgaris*), bell heather (*Erica cinerea*) or cross-leaved heath (*E. tetralix*). Abroad it has also been noted on *Rhododendron hirsutum* (Lhomme, [1946–49]).

Larva. Head pale brown with sparse reddish brown spots. Prothoracic plate grey with several small black spots; body pale greenish grey with dull reddish dorsal and broader subdorsal lines, darkening posteriorly; pinacula small and black; thoracic segments 2 and 3 with dark reddish subdorsal blotches; anal plate dull pale brownish grey with scattered darker brownish marks.

When small, the larva feeds on flowers of the foodplant, hiding in the interior; later it feeds in spun shoots. September–February.

Pupa. Within the larval spinning or in a cocoon in the soil. March–April.

Imago. Univoltine; late April–June. Easily disturbed from heather during the day and later comes to light.

Distribution (Map 141)

Occurs on heaths, moorland and in gardens where heathers are grown. Common throughout the British Isles as far north as Orkney and Shetland. Europe; Asia Minor; North Africa.

NOTHRIS Hübner

Nothris Hübner, [1825], *Verz.bekannt.Schmett.*: 411.

A Palaearctic genus represented by about 20 species, three of which occur in Europe. Two species have been recorded in Britain, but one may now be extinct.

Imago. Labial palpus segment 3 longer than segment 2; segment 2 with dense triangular tuft of scales at apex beneath. Forewing elongate; veins R_5 (7) and R_4 (8) stalked, R_5 to costa. Hindwing broader than forewing, trapezoidal, apex weakly produced, cilia two-thirds width of wing; veins CuA_1 (3) and M_3 (4) connate, M_2 (5) nearly parallel, M_1 (6) and Rs (7) stalked.

Larva. British species feed on the leaves of Scrophulariaceae.

Key to species (imagines) of the genus *Nothris*

– Forewing with black longitudinal lines on fold and in disc ....*congressariella* (p. 223)
– Forewing without such lines .... *verbascella* (p. 221)

NOTHRIS VERBASCELLA ([Denis & Schiffermüller])

Tinea verbascella [Denis & Schiffermüller], 1775, *Schmett.Wien.*: 136.

Type locality: [Austria]; Vienna district.

Description of imago (Pl.6, fig.8)

Wingspan 17–22mm, moths of presumed second generation smaller. Head glossy pale ochreous; antenna ochreous, scape pale brown; labial palpus with segment 3 longer than segment 2, segment 2 with prominent triangular tuft beneath, black on outer side and beneath, pale buff on inner side and above, segment 3 slender, gently curved, arising almost at right angle from segment 2, pale buff, posterior side irrorate fuscous; haustellum short, coiled, pale buff. Thorax pale ochreous. Forewing pale buff, sparsely irrorate black; black spot at base of costa; costa finely edged black to three-quarters; stigmata black, first discal well beyond plical, plical and first discal very small and sometimes absent; termen with interneural black spots; cilia con-colorous with wing at apex, becoming paler at tornus, with indistinct darker ciliary line. Hindwing slightly broader than forewing, trapezoidal, tornus rounded, apex weakly produced; pale grey; cilia pale buff, darker on costa and at apex. Foreleg pale fuscous, annulate buff at intersegmental joints; midleg with femur and tibia blackish fuscous, tarsus ochreous, outer side irro-rate fuscous; hindleg buff, irrorate pale fuscous on outer side. Abdomen ochreous-fuscous above, fuscous below with ochreous median stripe. Genitalia, see fig-ures 18b,43b.

Life history

Ovum. Laid on hoary mullein (*Verbascum pulverulen-tum*).

Larva. Head and prothoracic plate black. Body brown (Meyrick, [1928]).

The rate of development of larvae is variable; most hatch in September, but it is possible that some ova overwinter and do not hatch until spring (Barrett, 1869). The larvae that hatch in autumn feed, often gre-gariously, on the undeveloped leaves at the heart of the plant, continuing to do so throughout the winter and growing very slowly. In spring they feed up rapidly under a web of silk mixed with meal from the plant, almost always on the underside of the larger lower leaves. At this time of year small, probably newly hatched, larvae feed on the younger leaves or bore into leaf-stalks or stems. On 30 May 1967 larvae at all stages of development were observed, together with pupae and an imago (Chalmers-Hunt, 1967; A. M. Emmet, pers.obs.). Larvae occur throughout the summer until September, when young and fully grown larvae are to be found simultaneously, probably indi-cating a small second brood.

Pupa. Under a web, almost always on the underside of a leaf, generally at an angle of veins, but sometimes under a turned-down edge of a leaf. Throughout the summer, but mainly May–June and September.

Imago. Probably bivoltine; late May–September, with peaks in June and September, the September moths being smaller than those that emerge in June. The adult has rarely been observed in the wild.

History and distribution (Map 142)

The first British specimens were taken near Norwich, Norfolk, by W. Wing (1853). Thereafter it continued to be found in Norfolk at irregular intervals but some-times commonly until the end of the century. It was also found near Bury St Edmunds in Suffolk from 1895–99 (Morley, 1937). Thereafter it passed un-noticed until 1967, when it was found in reasonable

Nothris verbascella

plenty in a gravel quarry at Snettisham in west Nor-folk, which is the current headquarters of the food-plant. It was still present in 1971 (A. M. Emmet, pers.obs.), but apparently was not looked for again until the late 1980s, when it was not found. Several subsequent searches have likewise failed and it is possi-ble that the species is now extinct in Britain. The own-ers of the quarry, which is an SSSI, are well aware of the rarity and importance of hoary mullein and are careful to conserve it; currently it occurs in reasonable abundance in and around the quarry. In 1967 it was noted that no parasitoids were reared from the larvae and pupae that were collected (Chalmers-Hunt, *loc.cit.*). Overcollecting cannot be to blame; in 1971, when it was last observed, E. C. Pelham-Clinton restricted himself to two larvae in view of its rarity and A. M. Emmet took none. It is likely, therefore, that adverse weather conditions have eliminated a species that has always had a precarious foothold in East Anglia. Old records from Oxfordshire (Walker & Hobby, 1939) are unconfirmed and improbable. Throughout Europe; Turkmenistan; North Africa.

NOTHRIS CONGRESSARIELLA (Bruand)

Ypsolopha congressariella Bruand, 1858, *Annls Soc.ent. Fr.* (3)**6**: 471, pl.11, fig.7.

Nothris declaratella Staudinger, 1859, *Stettin.ent.Ztg* **20**: 238.

Type locality: France; Montpellier.

Description of imago (Pl.6, fig.9)

Wingspan 15–20mm. Head buffish ochreous; antenna fuscous becoming buffish fuscous on scape; labial palpus segment 1 dark brown, segment 2 outer edge with very large brush of dark brown scales, apically buffish ochreous, segment 3 buffish ochreous. Thorax dark buff; tegulae buffish ochreous. Forewing dark brown, becoming buffish ochreous towards costa with a few scattered very small subcostal black spots; black lines on edge of costa at base and on fold; black longitudinal line in disc interrupted by two small ochreous dots; black scaling on veins towards termen and black dashes at termen often preceded by ochreous dashes; cilia ochreous with fuscous ciliary line. Hindwing light fuscous; cilia lighter. Legs light to dark brown, tarsal segments banded ochreous distally; hindleg with ochreous band at tibial spurs. Abdomen dark buffish ochreous. Genitalia, see figures 18c,43a.

Life history

Ovum. Undescribed. Oviposition site unknown, but presumably on the foodplant, balm-leaved figwort (*Scrophularia scorodonia*). June; September–October.

Larva. Head shiny dark brown, blackish bar medially widening anteriorly. Prothoracic plate shining black; body dull olive-green; pinacula small, black, encircled light olive-green; anal plate shining light yellowish brown; thoracic legs black; prolegs concolorous with body. An illustration is given by Key (1995).

In early instars the larva usually feeds between two spun leaves, in later instars under a folded leaf. July–September; October–May.

Pupa. Dark brown; in a cocoon. May; September.

Imago. Bivoltine; May–early July and September–October.

Distribution (Map 143)

Occurs on waste ground, sandhills and low cliffs. Added to the British list by Richardson & Mere (1958) who found it in 1957 on Tresco, Isles of Scilly. It has since been found on Bryher, Gugh, St Martin's and St Mary's, Isles of Scilly, at three or four localities on the north coast of Cornwall and on Lundy off the north Devon coast. It is also known from Guernsey and Herm, the Channel Islands. The foodplant is restricted to the Isles of Scilly, parts of Cornwall, Lundy, parts of

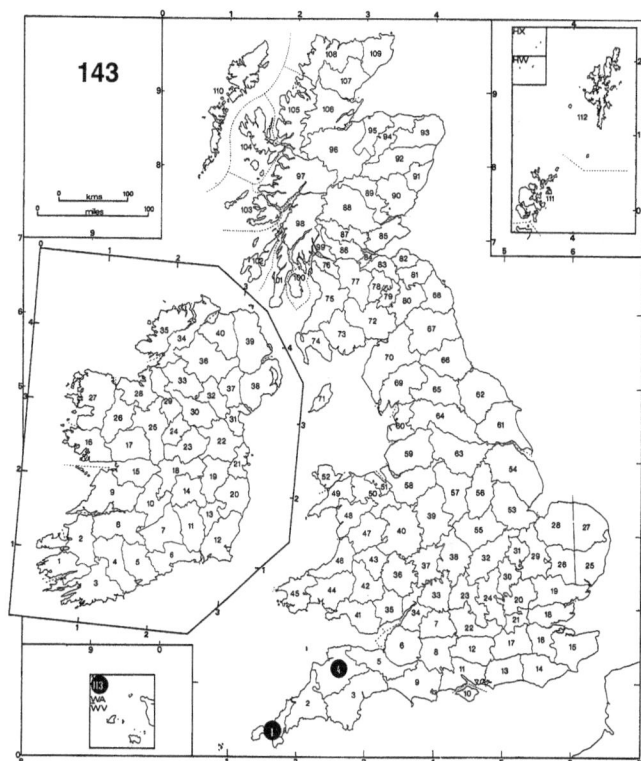

Nothris congressariella

south Devon, an island off the Pembrokeshire coast and Jersey, Guernsey and Herm, the Channel Islands. Parsons (1995[1996]) proposes RDB3 status. Spain, France, Switzerland, Sardinia, Sicily, Italy, the former Yugoslavia and Greece.

ANARSIA Zeller

Anarsia Zeller, 1839, *Isis, Leipzig* **1839**: 190.

A genus of over 50 species occurring mainly in Africa and Malaysia, with a few represented in Australia. Four species are known from Europe, two of which are found in Britain, one as a resident, the other as an adventive. According to Karsholt & Razowski (1996), the European species are in need of revision.

Imago. Labial palpus porrect, segment 2 beneath with a dense tuft of projecting scales, segment 3 in male very short and almost concealed in tuft (figure 60a), in female longer than segment 2 and exposed (figure

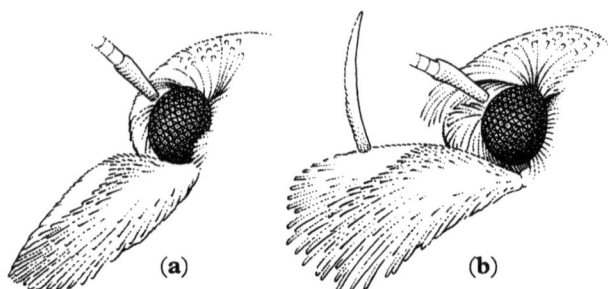

Figure 60 *Anarsia spartiella* (Schrank), side view of head of (a) male; (b) female

60b). Forewing elongate-oval; veins R_5 (7) and R_4 (8) stalked, R_4 to costa, M_1 (6) sometimes out of R_4 near base. Hindwing as broad as forewing, trapezoidal, termen weakly sinuate, apex not produced; veins Cu_1A (3) and M_3 (4) connate or approximate, M_2 (5) also approximate, M_1 (6) and R_s (7) stalked. Male genitalia with valvae asymmetrical and with specialized peg-like scales. Female genitalia with ductus bursae membranous; signum present in corpus bursae.

Larva. In spun leaves or in the fruit of Fabaceae (Leguminosae).

Key to species (imagines) of the genus *Anarsia*

- Forewing with blackish subquadrate costal spot at one-half .... *lineatella* (p. 225)
- Forewing without this spot, though sometimes with fuscous costal streaks .... *spartiella* (p. 224)

ANARSIA SPARTIELLA (Schrank)

Tinea spartiella Schrank, 1802, *Fauna Boica* **2**: 104.
Aplota robertsonella Curtis, 1837, *Br.Ent.* **14**: 655.
Anarsia genistae Stainton, 1854, *Ins.Br.Lepid.*: 144.
Anarsia genistella Doubleday, 1859, *Zoologist synonymic List Br.Lepid.*: 31.
Type locality: [Czech Republic; Bohemia].

Description of imago (Pl.6, figs 10,11)
Wingspan 12–15mm. Head whitish grey to grey; antenna two-thirds length of forewing, white, annulate fuscous, apical segments in male weakly serrate; labial palpus segment 2 with dense tuft of scales, smooth, slightly arched and concolorous with head above, below with long, forward-directed fuscous to dark fuscous hair-scales, segment 3 in male very short and generally completely hidden in tuft, in female as long as segment 2, slender, gently curved, porrect. Thorax and tegulae grey to dark grey. Forewing grey to dark grey; often diffuse paler subcostal streak and similar streak on fold; sometimes interneural pale streaks or irroration in distal half; costa sometimes with three short, outwardly oblique fuscous streaks, that at midwing largest and most often present; stigmata, if visible, consisting of thin lines of black scales, first discal beyond plical; occasionally a series of two or three subterminal fuscous spots, that below apex largest and most often present; cilia concolorous, with two darker ciliary lines, the inner line often diffuse, especially towards tornus. Hindwing grey, apical area in male sometimes darker; cilia whitish grey, but fuscous at apex in males with darker apical area. All markings are variable in expression and both pale grey and dark forms may have the forewing unmarked except for whitish irroration. Foreleg with tibia fuscous, tarsus grey with whitish annulation at joints; mid- and hindlegs grey with obscure whitish annulation at tarsal joints, hind tibia with dense whitish hair-scales. Abdomen grey; anal tuft in male paler. Genitalia in male (see figure 18e) markedly asymmetrical; valvae shaped differently and each with a long, slender, curved process, but of different length and arising from a different point on each valva. Genitalia, see figures 18e,42f.

Life history
Ovum. Laid on gorse (*Ulex europaeus*), broom (*Cytisus scoparius*) or dyer's greenweed (*Genista tinctoria*).
Larva. Length 11–12mm. Somewhat spindle-shaped. Head and prothoracic plate black. Body dark reddish brown; pinacula concolorous; anal plate and thoracic legs black.

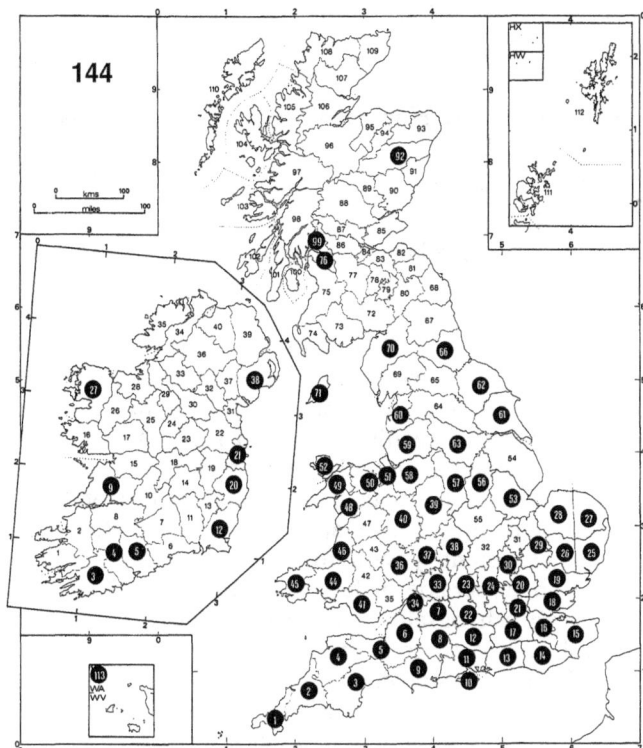

Anarsia spartiella

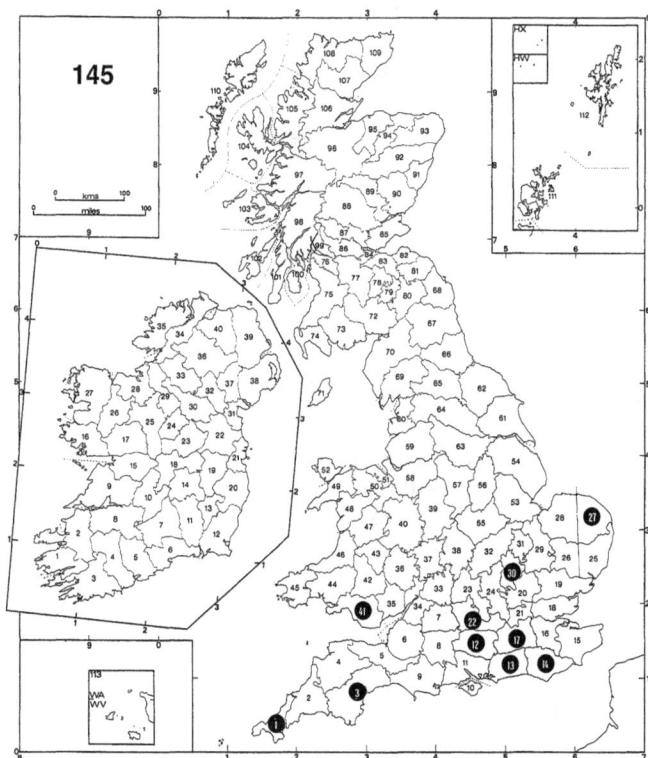

Anarsia lineatella

The larva feeds in spun shoots in May and June. It is not known whether the winter is passed as an ovum or larva.

Pupa. In a cocoon in the larval feeding place.

Imago. Univoltine; June–August. It may be disturbed from its foodplant by day and comes to light after dark.

Distribution (Map 144)

Occurs on heathland, downland and waste ground; although rather local, it is usually common where its foodplants are well established. Throughout England north to Cumberland; the Isle of Man; in Scotland recorded only from Renfrewshire, Dunbartonshire and Aberdeenshire; coastal counties in Wales and Ireland, but probably under-recorded in the latter; the Channel Islands. Central and southern Europe; Sweden.

ANARSIA LINEATELLA Zeller
Peach Twig Borer

Anarsia lineatella Zeller, 1839, *Isis, Leipzig* **1839**: 190.

Anarsia pruniella Clemens, 1860, *Proc.Acad.nat.Sci. Philad.* **1860**: 169.

Type locality: Austria; Vienna.

Description of imago (Pl.6, fig.12)

Wingspan 11–14mm. Head with dark grey or grey scales tipped paler grey, giving a speckled appearance, frons sometimes pale grey, some dark grey scales anterior to eye; antenna dark grey, flagellum mixed lighter grey; labial palpus segment 1 dark grey, segment 2 outer edge with very large brush of scales speckled dark and light grey, segment 3 vestigial in male, in female light grey, speckled darker, subapically dark grey, speckled light grey. Thorax and tegulae concolorous with head. Forewing grey or dark grey, sometimes suffused light grey especially towards costa; several short oblique dark grey or black costal marks with wider black subquadrate costal mark at one-half; several longitudinal black dashes across wing from base to termen; whitish dash at outer edge of disc; a few black dots or dashes along termen; cilia dark grey, darker ciliary line with a few black dashes. Hindwing light grey; cilia concolorous but darker basally. Legs dark grey, speckled light grey, tarsal segments banded light grey distally; hindleg light grey, banded lighter grey at tibial spurs. Abdomen dark grey. Genitalia, see figures 18f,42g.

Life history

Ovum. Oval, shiny, white when newly laid, becoming yellowish to orange. For foodplants, see *History and distribution* below.

Larva. Larvae have not been found in the wild in the British Isles although the moth appears to have established itself in one locality. Head and prothoracic plate black. Body dark honey-brown, intersegmental divisions whitish; pinacula small, black; anal plate black; thoracic legs black; prolegs concolorous with body.

Pupa. Dark brown; in a cocoon.

Imago. June–August.

History and distribution (Map 145)

Added to the British list by Uffen (1959) who reared it in 1957 from an imported apricot (*Prunus armeniaca*). Between 1971 and 1999 there are eight further records of moths being reared from or of larvae found in imported plums (*Prunus* spp.) peaches (*P. persica*) and nectarines (*P. persica* var. *nectarina*). It appears to have established itself in the wild at Walberton, Arundel, Sussex, as single specimens were taken at light on 2 August 1991; 1 July 1993; 28 June and 5, 8 and 13 July 1995 (J. T. Radford, pers.comm.); and one came to light at Cardiff, south Glamorgan, on 16 August 1997 (D. J. Slade, pers.comm.). Abroad it occurs mainly on almond, apricot and peach, less frequently on cherry and plum, and is considered a serious pest, in particular of apricot and peach, causing economic damage.

Throughout temperate and southern Europe; North Africa; Asia Minor; Iran; India; introduced into U.S.A, Australia, Japan and China.

HYPATIMA Hübner

Hypatima Hübner, [1825], *Verz. bekannt. Schmett.*: 415.
Chelaria Haworth, 1828, *Lepid. Br.*: 526.

A genus with over 100 species, represented mainly in Australia, South-east Asia and Africa. A single species occurs in Europe including the British Isles.

Imago. Head with scales tufted at sides; ocelli present; antenna three-quarters length of forewing in male, five-sixths in female, apical one-fifth serrate, scape without pecten; labial palpus with projecting scales on segment 2 in male, and on segments 2 and 3 in female; forewing elongate, rather pointed; vein M_1 (6) from joint stalk of veins R_4 (8) and R_5 (7). Hindwing a little broader than forewing, elongate-subtrapezoidal, apex hardly produced, termen sinuate, merging inconspicuously into dorsum, with tornus indefinite.

Larva. The British species feeds on Betulaceae.

HYPATIMA RHOMBOIDELLA (Linnaeus)

Phalaena rhomboidella Linnaeus, 1758, *Syst. Nat.* (Edn 10): 538.
Tinea conscriptella Hübner, [1805], *Samml. eur. Schmett.* **8**: pl.41, fig.283.
Phalaena (Tinea) hubnerella Donovan, 1806, *Br. Ins.* **11**: pl.382, fig.2.
Chelaria conscripta Haworth, 1828, *Lepid. Br.*: 526.
Type locality: Europe.

Description of imago (Pl.6, fig.13)

Wingspan 15–19mm. Female larger than male. Head greyish brown; antenna whitish, faintly annulate greyish brown and irregularly marked fuscous in apical one-fifth; labial palpus with segment 3 longer than segment 2, segment 2 with large triangular tuft of long scales beneath, whitish apically, brownish fuscous basally, segment 3 with forward-projecting tuft of scales on upperside in female, hardly discernible in male, whitish basally, brownish fuscous in middle, whitish mixed fuscous apically. Thorax and tegulae greyish brown. Forewing with ground colour whitish, suffused and mottled greyish brown, sometimes with darker lines along some veins; short dark fuscous dash subcostally at base; large dark brownish fuscous subtriangular mid-costal blotch, preceded by one or two small costal marks of same colour; apical one-half of costa maculate brown; short longitudinal blackish line to termen below apex; cilia concolorous with wing, darker ciliary line around apex and along termen, fading toward tornus. Hindwing light fuscous, slightly darker toward apex; cilia concolorous, a little paler toward base. Fore-and midlegs dark fuscous above, tarsi marked whitish at joints, paler beneath; hindleg greyish ochreous. Abdomen grey dorsally, whitish ventrally; anal tuft of male greyish ochreous. Genitalia, see figures 18d,43c.

Life history

Ovum. Laid on birch (*Betula* spp.) or hazel (*Corylus avellana*), probably on a twig. July–September.

Larva. Head and prothoracic plate black. Body bright green with pinacula and anal plate concolorous. Shortly before pupation the body becomes suffused pinkish brown (Heckford & Langmaid, 1991).

Feeds in a spun shoot or rolled leaf. May–June.

Pupa. In a silken cocoon amongst detritus on the ground. June–July.

Imago. Univoltine; July–September. Flies at night and comes to light.

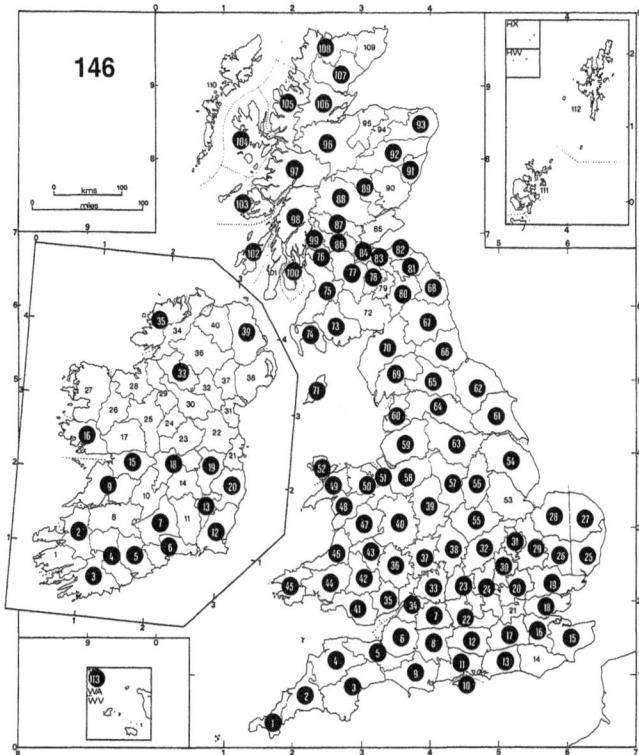

146

Hypatima rhomboidella

Distribution (Map 146)

Inhabits woodland, heathland and scrub. Common throughout the British Isles as far north as Sutherland, but not recorded from the Outer Hebrides, Orkney or Shetland. Central and northern Europe.

Dichomeridinae

ACOMPSIA Hübner

Acompsia Hübner, [1825], *Verz. bekannt. Schmett.*: 409.
Brachycrossata Heinemann, 1870, *Schmett. Dtl. Schweiz* (2)**2**(1): 323.
Telephila Meyrick, 1923, *Exot. Microlep.* **2**: 626.

A small genus represented by 20 species, eight of which occur in Europe, the remainder in America and Africa. Two species occur in Britain.

Imago. Medium-sized to large gelechiids with labial palpus long and slender; antenna without pecten. Forewing without scale-tufts, subtrapezoidal, slightly tapering, termen slightly emarginate. Hindwing much broader than forewing, trapezoidal, apex acute, termen slightly emarginate just before apex. Abdominal tergites concolorous. Valva of male genitalia spatulate, aedeagus short and broad; female genitalia diverse.

Larva. Early stages on low plants.

Key to species (imagines) of the genus *Acompsia*

-     Forewing ochreous-brown and unmarked .... *cinerella* (p. 227)
-     Forewing ochreous-orange with small black spot halfway along fold .... *schmidtiellus* (p. 228)

ACOMPSIA CINERELLA (Clerck)

Phalaena cinerella Clerck, 1759, *Icones Insect. rar.* **1**: pl. 11, fig. 6.
Recurvaria cinerea Haworth, 1828, *Lepid. Br.*: 547.
Type locality: [Sweden].

Description of imago (Pl. 6, fig. 14)

Wingspan 16–19mm. Head and antenna ochreous-golden; labial palpus ochreous above, ochreous-brown beneath. Thorax and tegulae ochreous-brown. Forewing and cilia ochreous-brown. Hindwing paler ochreous-brown; cilia concolorous. Fore- and midlegs ochreous-brown; hindleg ochreous. Abdomen ochreous-brown with ochreous genital tuft. Genitalia, see figures 19a, 43e.

Life history

Ovum. Laid on moss.

Larva. Undescribed. Feeds in moss at the base of a tree during May and June. Probably also occurs on ground-dwelling mosses as the moth is not restricted to areas where there are trees (Bradford & Sokoloff, 1988).

Pupa. Undescribed. June.

Imago. Univoltine; June–July. Occasionally comes to light.

Distribution (Map 147)

Widespread throughout the British Isles except for the Outer Hebrides, Orkney and Shetland. Throughout Europe.

ACOMPSIA SCHMIDTIELLUS (Heyden)

Ypsolophus schmidtiellus Heyden, 1848, *in* Koch, *Isis, Leipzig* **1848**: 954.

Ypsolophus durdhamellus Stainton, 1849, *Syst.Cat.Br. Tineidae & Pterophoridae*: 12.

Type locality: Germany; Wetterau [Frankfurt am Main].

Description of imago (Pl.6, fig.15)

Wingspan 15–17mm. Head ochreous-yellow, vertex sometimes tinged greyish, frons a little paler; antenna ochreous-grey; labial palpus with segment 3 longer than segment 2, segment 2 with long dense brush of forward-projecting scales beneath, ochreous-yellow, suffused fuscous on outer side, segment 3 very slender, ascending, ochreous-yellow with thin fuscous median line beneath. Thorax and tegulae concolorous with forewing. Forewing ochreous-yellow, suffused reddish brown in outer one-third; leading edge of costa lined black at base; stigmata black, first discal, sometimes obsolete, above larger plical, second discal small and also sometimes obsolete; blackish or dark reddish brown dot on dorsum before tornus, sometimes obsolete; slender irregular fascia of ground colour in suffused area subterminally; thin black line along termen, fading toward tornus; cilia concolorous with wing. Hindwing light grey, paler toward base; cilia pale ochreous. Foreleg fuscous, tarsus paler, ringed whitish at joints, tibia blackish above; midleg fuscous; hindleg paler, sometimes greyish ochreous. Abdomen light grey, segments 1 and 2 tinged ochreous dorsally, venter paler. Genitalia, see figures 19b,43d.

Life history

Ovum. Laid on marjoram (*Origanum vulgare*).

Larva. Head black. Body yellowish white, dorsal and subdorsal lines deep purplish brown; prothoracic plate brown or blackish, meso- and metathorax swollen, dark purplish brown or blackish (Meyrick, [1928]).

Feeds in a folded or rolled leaf which sometimes

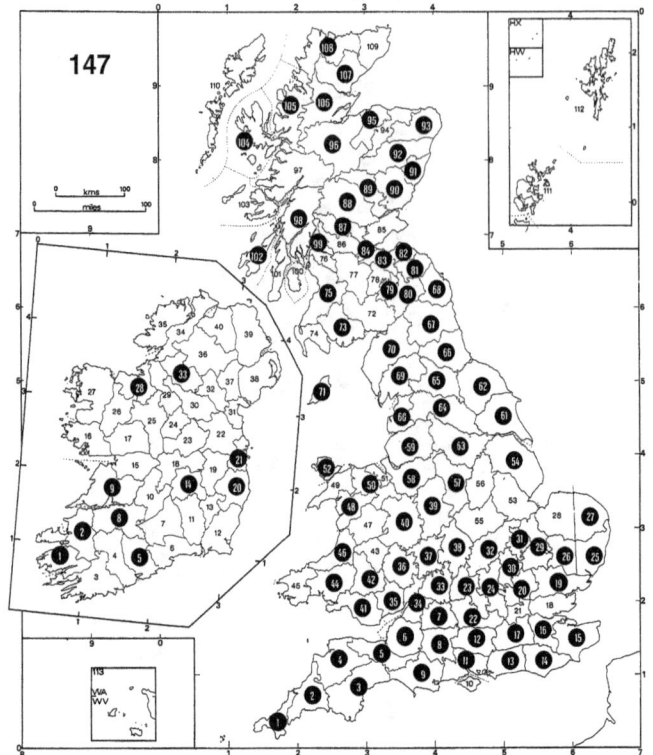

Acompsia cinerella

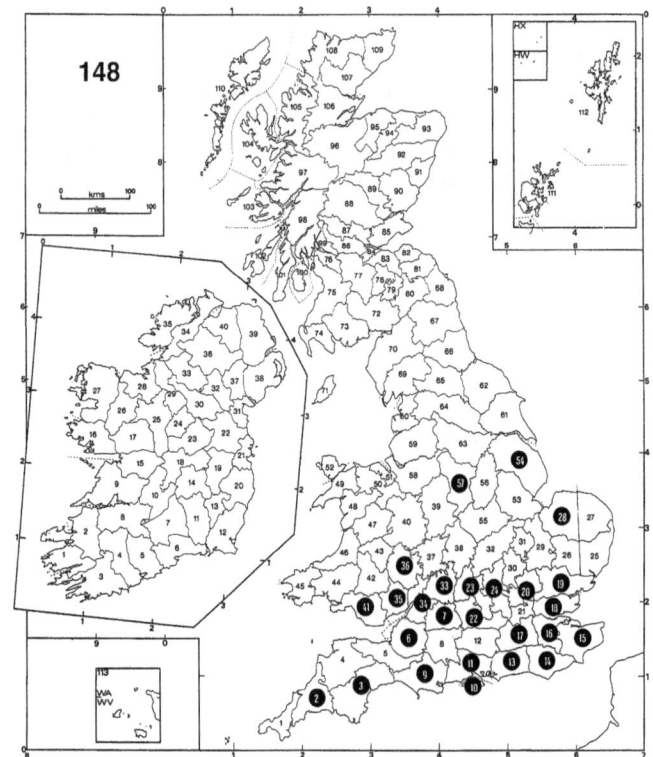

Acompsia schmidtiellus

turns scarlet (Tutt, 1905). May–June.

Pupa. Amongst detritus on the ground or in a folded leaf on the foodplant. June–July.

Imago. Univoltine; July–August. Flies at night and sometimes comes to light.

Distribution (Map 148)

Inhabits chalk or limestone downland and roadside verges where its foodplant occurs. Local in southern England and South Wales, and there are old records from Derbyshire (Meyrick, [1928]) and Lincolnshire (Mason, 1910). Unrecorded from Scotland or Ireland. South-west and central Europe from Spain and Italy northwards to Denmark and Estonia.

DICHOMERIS Hübner

Dichomeris Hübner, 1818, *Zutr.Samml.exot.Schmett.* 1: 25.

Acanthophila Heinemann, 1870, *Schmett.Dtl.Schweiz* (2)2(1): 320.

A very large genus containing over 500 species with an almost worldwide distribution, but absent from New Zealand. Eighteen species are known in Europe, of which four occur, and one other has occurred, in Great Britain.

Imago. Head with rather loosely tufted scales at sides, with appressed scales in *D. alacella* (Duponchel); ocelli present; antenna with apical two-thirds serrate, ciliate in male except *D. alacella*; labial palpus usually with segment 3 as long as, or longer than segment 2, but shorter in *D. alacella*, segment 2 usually with large brush of scales beneath, except in *D. alacella* where it is thickened with appressed scales, scape without pecten. Forewing elongate; veins R_4 (8) and R_5 (7) stalked in most species with R_5 to costa, but in *D. alacella* with R_5 absent and R_3 (9) and R_4 stalked. Hindwing usually broader than forewing except in *D. alacella* where they are of equal width, trapezoidal or subtrapezoidal, apex hardly produced, but more so and pointed in *D. alacella*; termen slightly sinuate; tornal area gently rounded; veins Rs (7) and M_1 (6) connate or stalked, M_3 (4) and CuA_1 (3) connate (figure 61).

Larva. Most species feed on the leaves of trees or shrubs; *D. alacella* on lichens.

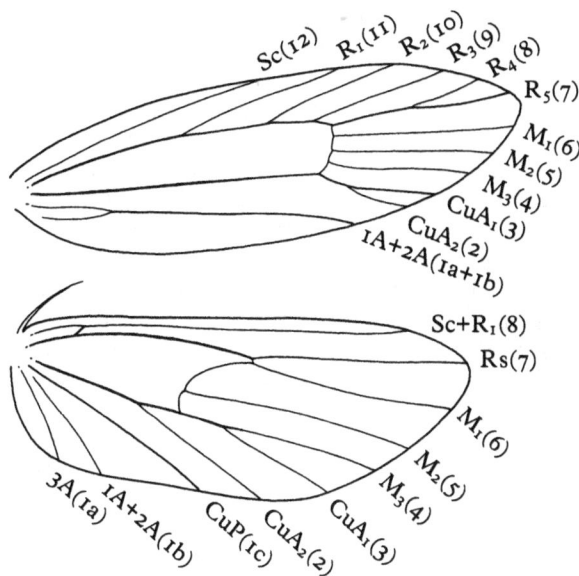

Figure 61 *Dichomeris marginella* (Fabricius), wing venation

Key to species (imagines) of the genus *Dichomeris*

1 Forewing dark fuscous; segment 2 of labial palpus thickened with appressed scales .... *alacella* (p. 233)

– Forewing not dark fuscous; segment 2 of labial palpus with large brush of scales beneath 2

2(1) Forewing with white streaks along costa and dorsum .... *marginella* (p. 229)

– Forewing without white streaks 3

3(2) Forewing grey .... *juniperella* (p. 231)

– Forewing light brown or dark reddish brown 4

4(3) Forewing with ground colour light brown .... *derasella* (p. 232)

– Forewing with ground colour dark reddish brown .... *ustalella* (p. 231)

DICHOMERIS MARGINELLA (Fabricius)
Juniper Webber

Alucita marginella Fabricius, 1781, *Spec.Insect.* 2: 307.

Type locality: England.

Description of imago (Pl.6, fig.16)

Wingspan 15–17mm. Head white; antenna dark fuscous, indistinctly annulate blackish; labial palpus with segment 3 a little longer than segment 2, segment 2 with large, dense, slightly forwardly-projecting brush of long scales beneath, and a smaller brush above,

brownish fuscous at sides and beneath, white anteriorly and above, segment 3 very slender, white above,
fuscous at sides and beneath. Thorax white; tegulae
concolorous with forewing. Forewing rather shining
dark brown; costal streak from base to near apex and
dorsal streak from base to middle of termen, white;
costa edged dark brown near base; thin brown line
along termen and around tornus; cilia greyish white,
outer margin and two ciliary lines brownish fuscous.
Hindwing whitish grey; cilia concolorous. Fore- and
midlegs brownish fuscous above, paler beneath; hindleg paler. Abdomen greyish white above, ochreous-
brown beneath. Genitalia, see figures 19c,44e.

Life history

Ovum. Laid on junipers (*Juniperus* spp.), including
common juniper (*J. communis*) and garden species and
cultivars.

Larva. Head dark reddish brown. Prothoracic plate
paler brown, bordered dark brown posteriorly; body
pinkish white, slender dorsal and broad dorsolateral
lines dark red; pinacula black; anal plate concolorous
with body; thoracic legs blackish.

The larva commences feeding in the autumn by
mining a needle, and then overwinters in the mine
(Emmet, 1988b). In the spring it makes a considerable
amount of silken webbing amongst the needles and
feeds on the leaves therefrom; sometimes there are several larvae in the same web. It can be a pest on cultivated forms of juniper. September–June.

Pupa. In a cocoon in the larval web. June–July.

Imago. Univoltine; July–August. Rests by day in
juniper bushes whence it can be beaten out. Flies at
night and comes to light.

Distribution (Map 149)

Inhabits chalk and limestone downland where juniper
occurs, and suburban gardens. Widespread and locally
common in England to south Yorkshire, but absent
from the south-west; Westmorland (R. G. Warren,
pers.comm.); there is an old record from Newcastle-
on-Tyne (Stainton, 1859; Morris, 1891; Meyrick
[1928]), but this was rejected by Dunn & Parrack
(1992); South Wales; the Burren, Co. Clare, Ireland,
where it feeds on *J. communis* ssp. *alpina*; the Channel
Islands. Not recorded from Scotland. Throughout
most of continental Europe, but absent from much of
Scandinavia and the Mediterranean islands; Siberia.

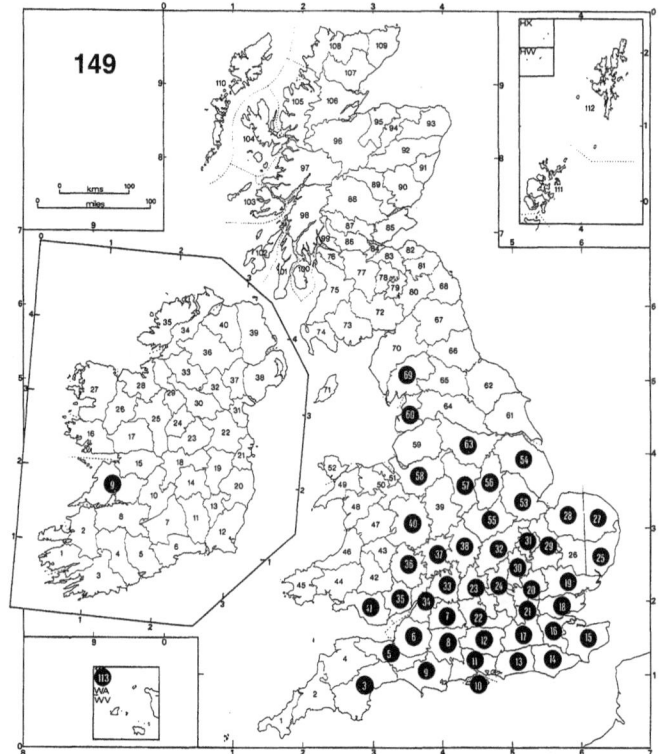

Dichomeris marginella

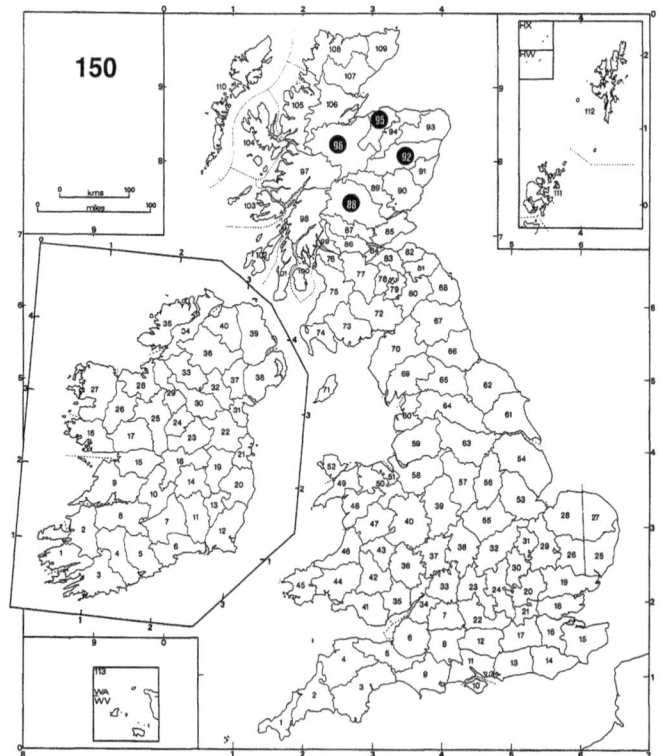

Dichomeris juniperella

DICHOMERIS JUNIPERELLA (Linnaeus)

Phalaena (Tinea) juniperella Linnaeus, 1761, *Fauna Suecica* (Edn 2): 370.
Type locality: Sweden.

Description of imago (Pl.6, fig.17)

Wingspan 18–22mm. Head light brownish grey or grey; antenna light grey, indistinctly annulate darker; labial palpus with segments 2 and 3 of approximately equal length, segment 2 with large brush of long scales beneath, brown at sides and beneath, white above, segment 3 very slender, whitish, lined fuscous beneath. Thorax and tegulae light grey. Forewing light grey, sprinkled with dark fuscous scales; stigmata dark fuscous, first discal slightly beyond plical; indistinct pale fascia at three-quarters indent below costa; a row of dark fuscous dots around apex and along termen; cilia light grey, with two fuscous ciliary lines. Hindwing light grey; cilia concolorous, paler toward base. Legs brownish fuscous, tarsi ringed paler at joints; hind tibia paler above. Abdomen grey above, darker brownish grey beneath. Genitalia, see figures 19e,44b.

Life history

Based largely on personal observations made by R. J. Heckford and the author in 2001.

Ovum. Laid on common juniper (*Juniperus communis*).
Larva. Full-fed *c*.12–13mm long. Head and prothoracic plate marbled black and reddish brown. Body whitish ochreous; dorsal, broad dorsolateral and interrupted spiracular lines purplish brown; pinacula black; anal plate pale brown, extensively marked blackish; thoracic legs black; prolegs concolorous with body.

It feeds in a silken spinning amongst the needles, apparently preferring bushes, with very dense foliage, growing in sheltered, south-facing, open situations on sparsely birch-wooded slopes. Many spinnings may be found on a single bush, while neighbouring ones have none. April–early July. It is not known whether overwintering takes place as an egg or a young larva.

Pupa. Chestnut-brown, becoming darker before eclosion; without a cocoon, the cremaster attached to silk within the larval spinning. June–July.

Imago. Univoltine; July. Can be beaten from the foodplant. Flies at night and comes to light.

Distribution (Map 150)

Occurs where its foodplant grows in open woodland on the lower slopes of mountains and in glens. Found only in the central and eastern Highlands of Scotland, where it is apparently local and scarce, but in 2001 larvae were found in some numbers very locally at two localities on Deeside, Aberdeenshire (R. J. Heckford & J. R. Langmaid, pers.obs.). Inhabits montane regions in continental Europe from Spain northwards to Scandinavia and eastwards to the Balkans; Asia Minor.

DICHOMERIS USTALELLA (Fabricius)

Tinea ustalella Fabricius, 1794, *Ent.syst.* **3**(2): 302.
Type locality: Italy.

Description of imago (Pl.6, fig.18)

Wingspan 15.5–22.0mm. Female larger than male. Head reddish brown, tinged purplish; antenna reddish brown, annulate black; labial palpus with segment 3 nearly twice length of segment 2, segment 2 with large, dense, triangular tuft of long scales beneath, projecting downwards and forwards, smaller tuft above projecting upwards and forwards, reddish brown on outer side and anteriorly, ochreous on inner side with reddish brown anterior margin, segment 3 very slender, recurved, whitish, lined reddish brown beneath. Thorax and tegulae light to dark reddish brown. Forewing dark reddish brown, darker toward termen; broad golden-ochreous suffusion in costal half of discal area, more marked in female, sometimes barely perceptible in male; cilia reddish or reddish brown proximally, ochreous distally. Hindwing dark reddish fuscous; cilia whitish, fuscous at apex, dark basal ciliary line. Legs reddish brown, tarsi ringed white at joints; hind tibia ochreous above. Abdomen reddish fuscous, venter densely sprinked with whitish scales. Genitalia, see figures 19d,44c.

Life history

Based on Simpson (1989).

Ovum. Laid on small-leaved lime (*Tilia cordata*). On the Continent hornbeam (*Carpinus betulus*), hazel (*Corylus avellana*), field maple (*Acer campestre*), beech (*Fagus sylvatica*), *Salix* and *Prunus* are also mentioned as foodplants (Lhomme, [1946–49]).

Larva. Head dark brown. Prothoracic plate brown anteriorly, with two large black marks occupying posterior three-quarters; meso- and metathorax each with anterior one-third pale blue-grey and posterior two-thirds black; body rather translucent greyish green; pinacula large and black, with whitish hairs; anal plate marked black; hind prolegs black with tips chestnut-brown; first pair of thoracic legs black. Head and thoracic segments noticeably broader than abdomen.

The larva feeds in September and early October between flatly spun leaves, making holes between the veins. From October it overwinters, fully fed, until April in a folded leaf which falls to the ground.

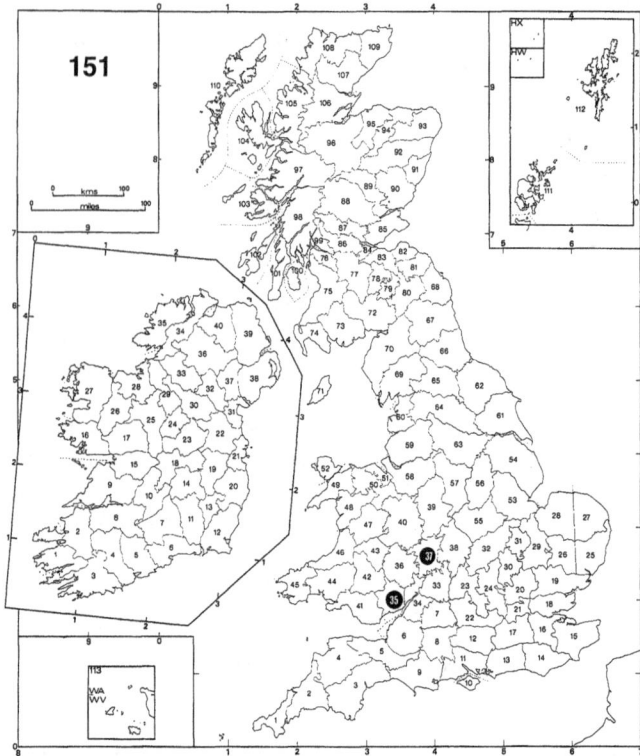

Dichomeris ustalella

Pupa. In the spinning in which the larva overwintered. April–May.

Imago. Univoltine; May–June. Though apparently of retiring habits, it has been observed 'sunning itself' on leaves of lime (Horton, 1867).

Distribution (Map 151)

First found in England in Worcestershire in 1861 (Horton, *loc.cit.*), where it was rediscovered in 1987, having not been seen for many years (Simpson, *loc.cit.*). This remained its only known locality in the British Isles until 1999 when two specimens were taken on 25 June near Tintern, Monmouthshire (D. O'Keeffe, pers. comm.). Central and eastern Europe from France eastwards, and as far north as Denmark and the Baltic states; eastern Siberia.

DICHOMERIS DERASELLA ([Denis & Schiffermüller])

Tinea derasella [Denis & Schiffermüller], 1775, *Schmett. Wien.*: 140.

Tinea fasciella Hübner, 1796, *Samml.eur.Schmett.* **8**: pl.16, fig.111.

Type locality: [Austria]; Vienna district.

Description of imago (Pl.6, fig.19)

Wingspan 21–22mm. Head light brown, darker on vertex; antenna light brown, indistinctly annulate darker; labial palpus with segment 3 a little longer than segment 2, segment 2 with dense brush of scales beneath and smaller brush above, both downward- and forward-projecting, reddish brown with a few scattered white scales on outer side, paler inwardly, segment 3 very slender, whitish, lined reddish brown beneath. Thorax and tegulae light brown. Forewing light brown; stigmata indistinct and contained in darker reddish brown suffused area, first discal slightly beyond plical; broad reddish brown angulate fascia at three-quarters; cilia concolorous with wing, inconspicuous darker ciliary line. Hindwing grey, paler toward base, with purplish reflections. Fore- and midlegs brownish fuscous above, paler beneath, tarsi ringed whitish at joints; hind tibia whitish above. Abdomen light reddish brown above, slightly paler beneath. Genitalia, see figures 19f,44d.

Life history

Ovum. Laid on blackthorn (*Prunus spinosa*).

Larva. Head reddish ochreous. Prothoracic plate pale ochreous with four black spots; body pale yellowish grey, dorsal and fainter subdorsal lines greenish grey; pinacula black (Meyrick, [1928]).

It feeds in an upwardly rolled or folded leaf, open at each end, and is best obtained by beating (Tutt, 1905). September–October.

Pupa. Amongst detritus on the ground. September–May.

Imago. May–June.

Distribution (Map 152)

Formerly found locally in the south-eastern sector of England. The last known record was in 1933 from Chiddingfold, Surrey, and the species is now thought to be extinct in Britain, though it is possibly overlooked owing to its retiring habits (Parsons, 1995[1996]). Southern and central Europe as far north as Denmark and the Baltic States, absent from Iberia and the Mediterranean islands; Asia Minor.

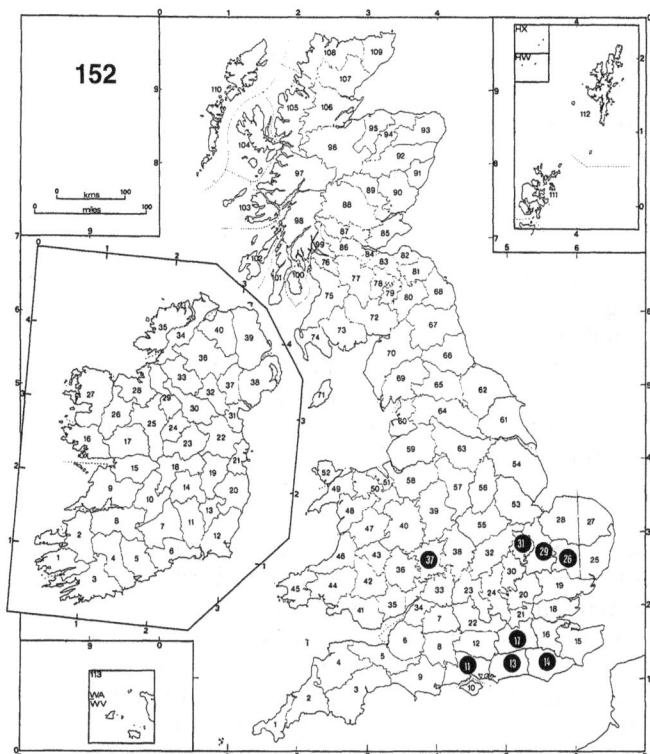

Dichomeris derasella

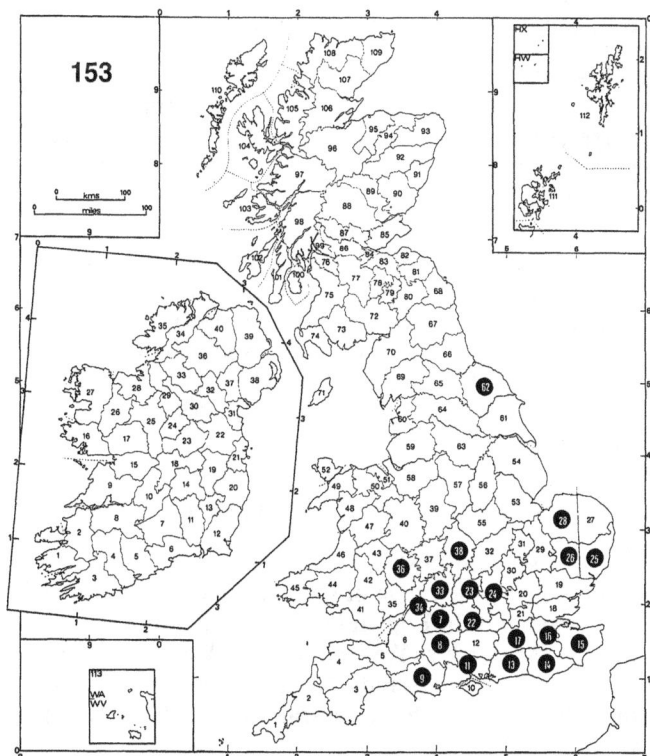

Dichomeris alacella

DICHOMERIS ALACELLA (Zeller)

Gelechia alacella Zeller, 1839, *Isis, Leipzig* **1839**: 199.

Type localities: Germany; Berlin, Frankfurt am Main and Glogau (now Poland; Glogów).

Description of imago (Pl.6, fig.20)

Wingspan 13–14mm. Head, neck-tufts and sides of frons dark brown, frons brownish grey to white centrally; antenna rather rough-scaled and dark brown, barred lighter; labial palpus dark brown, segment 3 tipped ochreous to greyish. Thorax and tegulae dark brown. Forewing chocolate-brown; subtriangular whitish spot on costa at three-quarters and much smaller one, often obsolete, on dorsum at two-thirds; other markings black, consisting of large spot at two-thirds along fold, another in discal area directly above it and another, usually smaller, in discal area at two-thirds, all usually edged with scattered white scales; cilia dark brown with regular patches of pale buff and scattered clumps of black scales basally. Hindwing brown; cilia concolorous. Legs brown, tarsi edged whitish distally; tibial tuft of hindleg buff. Abdomen pale brown; anal tuft cream and rufous brown. Genitalia, see figures 19g,44a.

Life history

Ovum. Laid on lichens on tree-trunks in August. Time of hatching unknown.

Larva. Head and divided prothoracic plate black. Body whitish grey (Meyrick, [1928]).

Feeds in June on lichens growing on tree-trunks.

Pupa. Undescribed. June–July.

Imago. Univoltine; July–August. Rests on tree-trunks by day and flies by night; comes to light and sugar (Parsons, 1995[1996]).

Distribution (Map 153)

Widespread, but local, in southern England as far north as Herefordshire, Warwickshire and north-east Yorkshire. Not recorded from Wales, Scotland or Ireland. Records in the last 30 years suggest that the distribution of this species is contracting. This has led Parsons (*loc.cit.*) to ascribe the species a Nationally Notable (Scarce) conservation status. Widespread throughout Europe; also known from eastern Siberia and Iran.

BRACHMIA Hübner

Brachmia Hübner, [1825], *Verz.bekannt.Schmett.*: 419.

A genus containing four species in Europe (Karsholt & Razowski, 1996), two of which are represented in Britain.

Imago. Head thickly clothed with scales, covering frons; antenna four-fifths length of forewing; labial palpus long, curved, ascending, segment 2 thickened with appressed scales, and shorter or as long as segment 3. Forewing and hindwing of equal breadth. Termen of forewing distinctly sinuate; forewing with veins CuA_2 (2) and CuA_1 (3) stalked, R_5 (7) and R_4 (8) stalked, with R_5 to apex and R_3 (9) often out of R_5. Hindwing trapezoidal, termen sinuate, with cilia as long as breadth of wing; veins CuA_1 (3) and M_3 (4) connate or stalked, M_2 (5) somewhat approximated, and M_1 (6) and Rs (7) stalked.

Early stages. Not described. Larva lives and feeds on or within plant tissues.

Key to species (imagines) of the genus *Brachmia*

– Forewing with dark brown fasciae ....
.... *blandella* (p. 234)
– Forewing without darker fasciae ....
.... *inornatella* (p. 235)

BRACHMIA BLANDELLA (Fabricius)

Tinea blandella Fabricius, 1798, *Suppl.Ent.syst.*: 499.
Gelechia gerronella Zeller, 1850, *Stettin.ent.Ztg* **11**: 155.
Type locality: Germany; Kiel.

Description of imago (Pl.6, fig.21)
Wingspan 9–12mm. Head creamy whitish to ochreous; antenna whitish ochreous, ringed dark brown, rings sometimes fading towards apex; labial palpus whitish, segment 3 with dark brown band. Thorax ochreous; tegulae ochreous, mixed dark brown. Forewing with apex weakly falcate, termen sinuate; ochreous, irrorate dark brown; dark brown, indistinct, angulate fascia at one-third, and another, more distinct and oblique, from costa at three-fifths to dorsum at four-fifths, the two fasciae sometimes linked with dark brown; dark brown subterminal fascia bifurcate near costa to enclose triangular buff marking; stigmata blackish, often formed of raised scales, first discal beyond, and often smaller than, plical, second discal variable in shape, formed of one or two dots, or vertical line; cilia fuscous, paler towards termen, darker at apex, with indistinct dark brown line in middle. Hindwing grey; cilia grey. Foreleg whitish, mixed fuscous, preterminal tarsal segment with dark brown band; mid- and hindlegs whitish, mixed fuscous. Abdomen grey above and below, with pale ochreous or whitish scales on anal tuft. Genitalia, see figures 20b,45a.

Life history
Ovum. Presumed to be laid on the pabulum, including gorse (*Ulex europaeus*), insect galls on grand fir (*Abies grandis*), marsh thistle (*Cirsium palustre*), and probably other plants.

Larva. The larva is apparently undescribed. It is stated by Meyrick [1928] to feed on gorse in June, and has been reared inadvertently from this foodplant (Stainton, 1878; J. R. Langmaid, pers.comm.). Similarly, it has been reared from within an insect gall on grand fir (Langmaid, 1978a), and from seedheads of marsh thistle collected in winter (D. H. Sterling, pers.comm.).

Pupa. Undescribed.

Imago. Univoltine; July and early August. Can be disturbed by day, and is attracted to mercury vapour light.

Distribution (Map 154)
Occurs in a wide range of habitats including grassland and woodland. In England widespread and fairly common as far north as Yorkshire; South Wales; Cardiganshire; the Channel Islands. Not recorded from Scotland or Ireland. Throughout central and southern Europe to the Middle East.

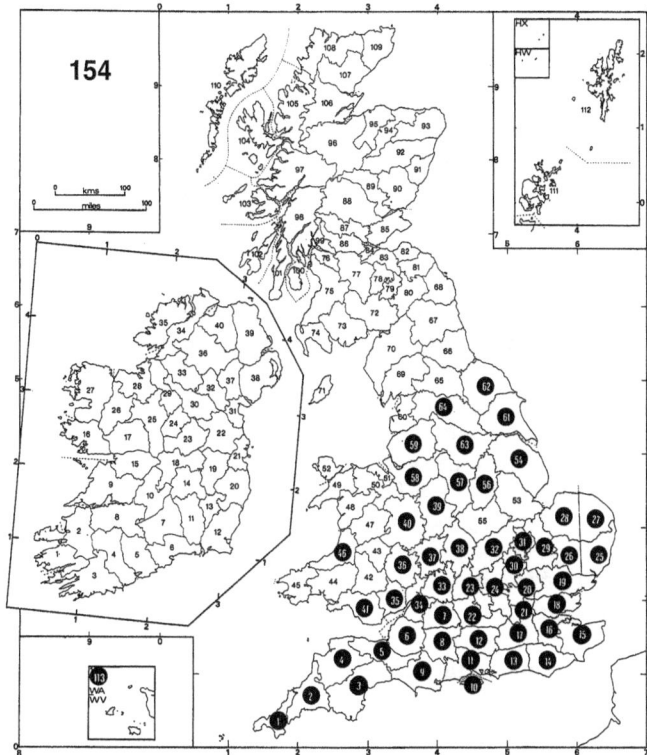

154

Brachmia blandella

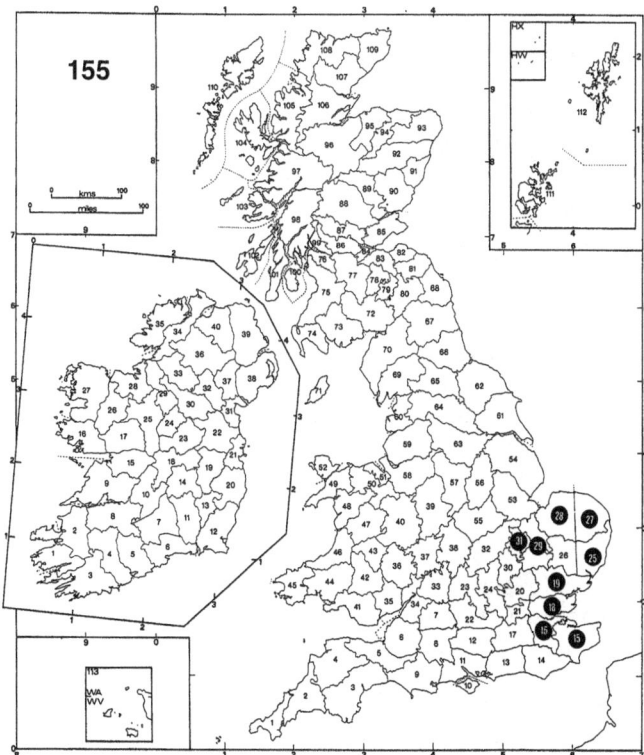

155

Brachmia inornatella

BRACHMIA INORNATELLA (Douglas)

Gelechia inornatella Douglas, 1850, *Trans. ent. Soc. Lond.* (N.S.) **1**: 65.

Type locality: England; Whittlesea Mere, Cambridgeshire.

Description of imago (Pl.6, fig.22)

Wingspan 11–15mm. Head pale grey-brown, frons whitish; antenna pale grey, ringed dark grey-brown, indistinctly from base to one-quarter and near apex; labial palpus whitish above with scattered dark brown scales beneath and on inner margin of segment 3. Thorax pale greyish ochreous with dark brown median line; tegulae pale greyish ochreous. Forewing with apex pointed, termen sinuate; pale greyish ochreous; paler grey fascia at four-fifths sometimes indicated; termen edged with dark brown scales, especially in middle; stigmata black, variable in shape and size and sometimes formed of raised scales, first discal beyond plical, second discal sometimes edged with whitish scales; one or more black scales usually present in middle of wing before termen; cilia pale grey, darker at tornus and apex, with dark brown line in middle. Hindwing grey; cilia pale grey with indistinct darker ciliary lines. Foreleg whitish, mixed dark brown above; midleg and hindleg whitish, mixed dark brown, with shorter of paired tibial spurs dark brown. Abdomen grey above and below, with some pale ochreous or whitish scales on anal tuft. Genitalia, see figures 20a,45b.

Similar species. Could be confused with *Helcystogramma lutatella* (Herrich-Schäffer), but that species does not have the termen of the forewing distinctly sinuate.

Life history

Ovum. Undescribed.

Larva. The larva is apparently undescribed. It is stated by Meyrick ([1928]) to feed in the stems of common reed (*Phragmites australis*) between September and April.

Pupa. Undescribed.

Imago. Univoltine; June and July. Can be found flying freely at dusk and later is attracted to mercury vapour light.

Distribution (Map 155)

The moth frequents fenland, but it is a scarce species which has been recorded only from south-east England and East Anglia. Scandinavia, Germany, The Netherlands and western Russia.

HELCYSTOGRAMMA Zeller

Helcystogramma Zeller, 1877, *Horae Soc.ent.ross.* **13**: 371–373.

A genus containing six species that occur in Europe (Karsholt & Razowski, 1996), two of which are represented in Britain.

Imago. Head thickly clothed with scales, covering frons; antenna four-fifths length of forewing, shortly ciliate in male; labial palpus long, curved, ascending, segment 2 thickened with appressed scales, and shorter or as long as segment 3. Forewing and hindwing of equal breadth. Termen of forewing straight or hardly sinuate; forewing with veins CuA_2 (2) and CuA_1 (3) stalked, R_5 (7) and R_4 (8) stalked, with R_5 to apex and R_3 (9) often out of or connate with R_5. Hindwing trapezoidal, termen sinuate, with hindwing cilia as long as breadth of wing; veins CuA_1 (3) and M_3 (4) connate or stalked, M_2 (5) somewhat approximated, and M_1 (6) and Rs (7) stalked.

Larva. A highly mobile, usually distinctly patterned larva. Feeds within a grass blade spun to form a linear roll; vacates the spinning rapidly if disturbed.

Pupa. In a light silken cocoon within a grass blade spun to form a roll but not that in which the larva has fed.

Imago. Univoltine. Both species can be disturbed by day and are attracted to light.

Key to species (imagines) of the genus *Helcystogramma*

–   Forewing ochreous, or veins of forewing ochreous with brown scaling between; outer margin of labial palpus with brown scales .... *rufescens* (p. 236)
–   Forewing grey-brown or brown, without ochreous veins; labial palpus white above and below and without brown scales .... *lutatella* (p. 237)

HELCYSTOGRAMMA RUFESCENS (Haworth)

Recurvaria rufescens Haworth, 1828, *Lepid.Br.*: 555.
Gelechia rufescentella Doubleday, 1859, *Zoologist synonymic List Br.Lepid.*: 29.
Type locality: England; London.

Description of imago (Pl.6, fig.23)

Wingspan 14–18mm. Head ochreous; antenna whitish ochreous with brown markings or rings on first, second and alternate segments to near apex; labial palpus ochreous or whitish, outer margin mixed with brown scales. Thorax and tegulae ochreous. Forewing with termen straight or hardly sinuate; ochreous and often unicolorous, but sometimes brown or dark brown between veins, the veins appearing as pale lines; darker ochreous, angulate fascia at two-thirds, and another, paler ochreous, beyond, sometimes present; very occasionally, row of dark brown dots along termen; stigmata often completely absent, but may be up to three dark brown or black, first discal above plical, second discal beyond; cilia ochreous with two indistinct dark brown lines. Hindwing of male grey, often ochreous towards apex; cilia pale ochreous, sometimes darker; hindwing of female whitish, pale ochreous towards apex; cilia cream. Foreleg pale ochreous, inner margin of femur and tibia clothed with dark brown scales; mid- and hindlegs pale ochreous. Abdomen grey-brown above, grey-brown with ochreous or whitish brown median and lateral lines below; pale ochreous or whitish scales on anal tuft. Genitalia and progenital abdomen, see figures 20c,45c.

Similar species. The ochreous coloration of at least the forewing veins, markings on the labial palpus, and the whitish hindwing (in female only) distinguish *H. rufescens* from fresh examples of *H. lutatella* (Herrich-Schäffer). However, worn, pale specimens of *H. lutatella* may appear ochreous and may be reliably separated only on the labial palpus, the shape of abdominal tergites 1–2 and genitalia characters in the female only. The key character given in Meyrick ([1928]), the number of dots in the forewing, does not distinguish the two species. See also Heckford & Sterling (1999) for further details on the separation of this species and *H. lutatella*.

Life history

Ovum. Presumed to be laid on a wide range of grasses (Poaceae (=Gramineae)), including false oat-grass (*Arrhenatherum elatius*), meadow grasses (*Poa* spp.), cock's-foot grass (*Dactylis glomerata*), false brome (*Brachypodium sylvaticum*), tor-grass (*B. pinnatum*) and tall fescue (*Festuca arundinacea*). The ovum is laid in summer, but may not hatch until the following spring.

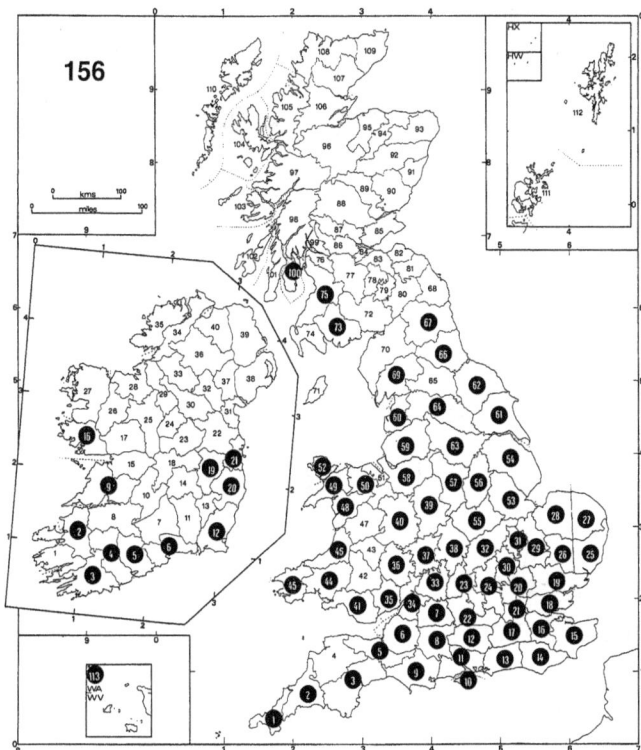

Helcystogramma rufescens

Larva. Head black, mouth-parts brown. Prothoracic plate black without median sulcus, mesothoracic and metathoracic segments black; intersegmental membranes of thorax heavily creased, white, but with small dark area along dorsum between prothoracic and mesothoracic segments; abdominal segments 1 and 2 black, with dark olive sclerotized plate anteriorly on each, very small on segment 1, consisting of a narrow, linear band, hardly widening on dorsum; remainder of body colour white, broad subdorsal line blackish with oblique extensions from subdorso-anterior to latero-posterior parts of each segment; anal plate concolorous with pattern of body colour; thoracic legs black; prolegs whitish.

The colour and shape of the sclerotized plates of the abdominal segments provide reliable characters to distinguish larvae of this species and *H. lutatella* in the field. Other differences between larvae of the two species in the presence and shape of plates on the abdominal segments were recognized by Stainton (1865a), but these have not been found to hold true in material examined for the preparation of this text.

The larva feeds from April to July. It spins a grass blade into a spiral to make a roll, open either end, in which it feeds and hides, changing rolls several times before pupation. The larva fenestrates the leaf from within the roll, and the feeding may cause blanching of the leaf-tip.

Pupa. Black. Tubercles of the cremaster arranged in two rows of four; upper row formed as digitate projections, with a wide gap between middle pair. The arrangement and shape of the tubercles provide reliable characters to distinguish pupae of this species and *H. lutatella* (Heckford & Sterling, *loc.cit.*).

The pupal stage lasts two to three weeks in July or August. The pupa is contained within a light silken cocoon inside a roll not used for feeding. The larva makes a smaller, tighter spinning, often incorporating and bending over the tip of the leaf blade so that the roll has only one exit. On the emergence of the moth the pupal exuviae remain within the roll.

Imago. Univoltine; July and August. It can be disturbed by day, flies freely at dusk and is attracted to light.

Distribution (Map 156)

A species of rank, often ungrazed or lightly grazed grassland and found in a variety of habitats including downland, coastal areas and woodland rides. Widespread and common throughout England and Wales; south-western Scotland; the southern half of Ireland, where it is mainly coastal; the Channel Islands. Abroad it has been recorded throughout Europe.

HELCYSTOGRAMMA LUTATELLA (Herrich-Schäffer)

Anacampsis lutatella Herrich-Schäffer, 1854, *Syst.Bearb. Schmett.Eur.* **5**: 191 (key), 201; 1853, *ibid.* **5**: pl.63, fig.467 (non-binominal).

Type locality: Germany; Frankfurt am Main.

Description of imago (Pl.6, fig.24)

Wingspan 12–18mm. Head grey-brown, frons whitish brown; antenna whitish grey or grey-brown with darker grey rings on first, second and alternate segments to near apex; labial palpus pale greyish ochreous, whitish above and below. Thorax grey-brown; tegulae pale grey-brown. Forewing with termen hardly sinuate; pale grey-brown to brown, sometimes mixed ochreous; paler, doubly angulate fascia usually present at four-fifths; beyond fascia often brown to termen; row of dark brown dots or dashes along termen, sometimes forming complete line; up to three dark brown or black stigmata usually present, first discal above plical, second discal beyond; cilia coloured as forewing, with two dark brown ciliary lines. Hindwing pale to dark grey, darker towards apex; cilia grey, with creamy whitish line adjacent to termen and darker grey band beyond. Legs creamy brown. Abdomen grey-brown above, grey-brown with ochreous or whitish brown median and lateral lines below; pale ochreous or whitish scales on anal tuft. Genitalia and pregenital abdomen, see figures 20d,45d.

Similar species. H. rufescens (Haworth) and *Brachmia inornatella* (Douglas), *qq.v.*

Life history

Ovum. Presumed to be laid on grasses (Poaceae (=Gramineae)), including cock's-foot grass (*Dactylis glomerata*) and false brome (*Brachypodium sylvaticum*). On the Continent the larva apparently feeds only on wood small-reed (*Calamagrostis epigejos*) (Stainton, 1865a). The ovum is laid in summer, but may not hatch until the following spring.

Larva. Head black, mouth-parts brown. Prothoracic plate black with median sulcus, mesothoracic and metathoracic segments black; intersegmental membranes of thorax heavily creased, white, but with small dark area along dorsum between prothoracic and mesothoracic segments; abdominal segments 1 and 2 black, with reddish brown sclerotized plate anteriorly on each, smaller on segment 1, confined to dorsum, rounded with small lateral extensions; remainder of body colour white; broad subdorsal line blackish with oblique extensions from subdorso-anterior to latero-posterior parts of each segment; anal plate concolorous with pattern of body colour; thoracic legs black; prolegs whitish. Differences between the larvae of this species and *H. rufescens* are given under the latter, *q.v.*

The larva is known to feed slowly from April to July (Waters, 1925b). It spins a grass blade into a spiral to make a roll, open each end, in which it feeds and hides, changing rolls several times before pupation. The larva fenestrates the rolled part of the leaf. Exact habitat requirements in Britain are not fully understood; Waters (*loc.cit.*) found larvae on cock's-foot grass, especially on stunted plants growing on the cliff face; many larvae have been found in false brome and cock's-foot grass at the top of a cliff and a single larva on false brome amongst broken turf within a recently land-slipped area (P. H. Sterling, pers.obs.). These observations suggest that vegetated sea-cliff and other maritime grassland communities may be important.

Pupa. Brown or blackish brown, exuviae sometimes appearing brown and translucent once the adult has hatched. Glands of the cremaster arranged in two rows of four; upper row formed as small warts, evenly spaced in the row. Differences between the pupae of this species and *H. rufescens* are given under the latter, *q.v.*

The pupal stage lasts two to three weeks in July or August. The pupa is contained within a light silken

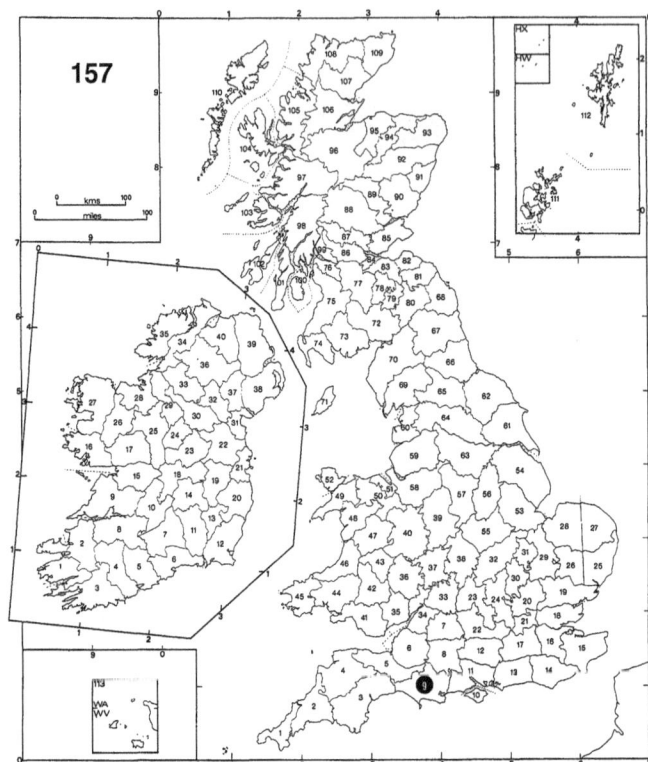

Helcystogramma lutatella

cocoon inside a roll not used for feeding. The larva makes a smaller, tighter spinning, often incorporating and bending over the tip of the leaf blade so that the roll has only one exit. On emergence of the moth the pupal exuviae remain within the roll.

Imago. Univoltine; July–early September. Adults have been disturbed by day from vegetation growing close to the sea, and at night are attracted to light.

Distribution (Map 157)

A rare species in Britain, thus far recorded only from the Dorset coast between Portland and Kimmeridge. Few specimens from Britain can be authenticated owing to confusion with *H. rufescens*, and it is almost certainly overlooked. The species was first found on Portland in 1884, but the specimen remained unidentified within a series of *H. rufescens* for many years. It was added to the British list by Waters (1925a), who had found adults at Portland in 1921 and Lulworth in 1922. On Portland it occurs on limestone, at Lulworth on chalk, and at Gad Cliff on clay (P. H. Sterling, pers.obs.); in view of the rather unspecific foodplant requirements in Britain it is possible that this species will be found to be more widespread amongst maritime cliff and slope vegetation. It occurs widely throughout Europe to western Russia.

Pexicopiinae

PLATYEDRA Meyrick

Platyedra Meyrick, 1895, *Handbk.Br.Lepid.*: 605.

A very small Palaearctic genus represented in Europe by a single species, which also occurs in Britain.

Imago. Medium-sized gelechiids with labial palpus long and somewhat thickened, especially segment 2; antenna with a reduced pecten (figure 62). Forewing

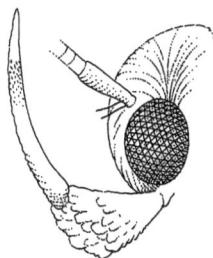

Figure 62 *Platyedra subcinerea* (Haworth), side view of head showing labial palpus

without scale-tufts, ovate and elongate. Hindwing broader than forewing, trapezoidal; termen not emarginate. All abdominal tergites concolorous. Valva of male genitalia spatulate with a strongly sclerotized pointed central ridge; signum of female genitalia a pair of small spiny plates.

Larva. On the seeds of Malvaceae.

PLATYEDRA SUBCINEREA (Haworth)

Recurvaria subcinerea Haworth, 1828, *Lepid.Br.*: 548.
Gelechia vilella Zeller, 1847, *Isis, Leipzig* **1847**: 845.
Type locality: Britain.

Description of imago (Pl.6, fig.25)

Wingspan 16–18mm. Head ochreous; antenna brownish fuscous, ringed ochreous; labial palpus ochreous, becoming dark ochreous-fuscous towards apex except extreme tip. Thorax and tegulae ochreous, mixed brownish anteriorly. Forewing ochreous, mixed brownish and brownish fuscous; markings fuscous, consisting of spot at base of dorsum, spot halfway across wing at one-third, another at two-thirds which is sometimes absent, and frequently a suffusion along basal part of costa; cilia ochreous with intermixed brownish scales, some of which form an indistinct ciliary line. Underside of forewing brownish fuscous. Hindwing light ochreous-fuscous, somewhat darker towards termen; cilia ochreous-fuscous basally, ochreous apically with a suggestion of a ciliary line round apex. Fore- and

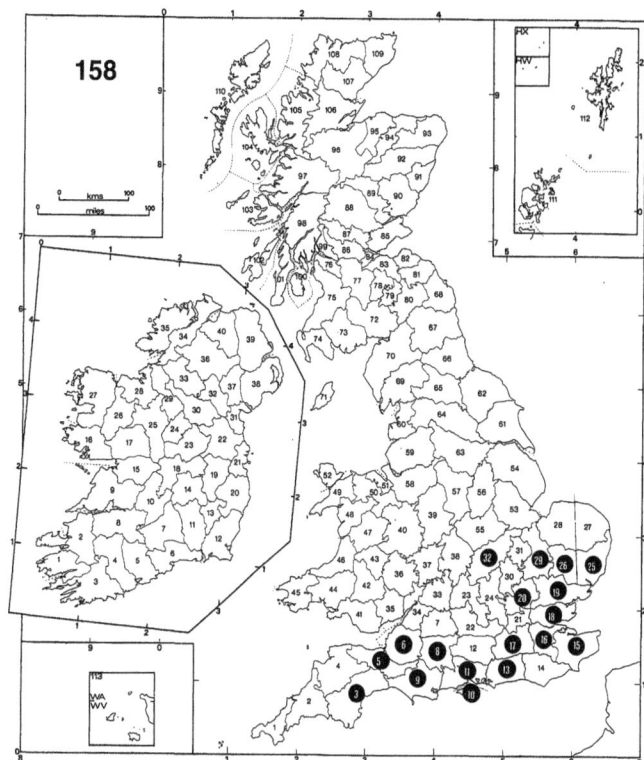

Platyedra subcinerea

midlegs dark fuscous-brown, tarsi edged paler distally, hindleg pale ochreous-brown, tarsus ringed darker proximally. Abdomen ochreous-brown. Genitalia, see figures 20e,46b.

Life history

Ovum. Laid on common mallow (*Malva sylvestris*) or occasionally on garden hollyhock (*Alcea rosea*). May.

Larva. Head and divided prothoracic plate black, but sometimes pale to dark brown. According to Meyrick (1895), body pinkish with large brownish spots; however a preserved larva in the BMNH would be better described as having body creamy white, with each segment bearing a broad anterior and narrower posterior interrupted transverse ring of carmine pink.

The larva feeds during June and July in the flowers and seeds of its foodplant. A seed containing a larva has the sepals sealed down so as to cover the seed completely. (Bradford, [1979]).

Pupa. Length 8mm. Pale brown and covered with fine pale hairs; rather stout, almost parallel-sided then tapering abruptly to a caudal point; caudal tip forming a dorsally curving sharp spine with six hooked ochreous hairs; also a scattered patch of similar hooked hairs more ventrally. July–August.

Imago. August–June. Thatch is often cited as the favoured hibernation site. It flies at night and comes to light, especially in the spring and early summer after hibernation.

Distribution (Map 158)

In Britain confined to southern England where its range seems to be contracting. In view of this Parsons (1995[1996]) ascribed the species a Nationally Notable (Scarce) conservation status. Throughout Europe; Asia Minor; central Asia; North Africa. Accidentally introduced into North America.

PEXICOPIA Common

Pexicopia Common, 1958, *Aust. J. Zool.* **6**: 271,178.

A small genus centred in the Australian region, represented by one species in Europe including Britain.

Imago. Medium-sized to large gelechiids with labial palpus long, slender, segment 2 thickened ventrally; antenna with a pecten. Forewing without scale-tufts, subtrapezoidal with apex and tornus rounded. Hindwing much broader than forewing, trapezoidal, tornus gently rounded. Abdominal tergites concolorous. Valva of male genitalia subtriangular, uncus bipinnate; signum of female genitalia a pair of spiny plates.

Larva. In seeds of Malvaceae.

PEXICOPIA MALVELLA (Hübner)
Hollyhock Seed Moth

Tinea malvella Hübner, [1805], *Samml. eur. Schmett.* **8**: pl.41, fig.281.

Recurvaria lutarea Haworth, 1828, *Lepid. Br.*: 549.

Type locality: Europe.

Description of imago (Pl.6, fig.26)

Wingspan 18–20mm. Head ochreous-brown; antenna ochreous-brown, banded darker and finely ciliate on underside in male; labial palpus ochreous-brown, sometimes with narrow fuscous ring just below apex. Thorax and tegulae ochreous-brown, tegulae darker anteriorly. Forewing ochreous-brown, suffused fuscous especially in apical one-third and along costa; ill-defined markings reddish fuscous, consisting of diffuse spot halfway along fold, spot halfway across wing just beyond it and another in discal area just before two-thirds; cilia ochreous with double row of fuscous-tipped scales; underside of forewing darkish fuscous, tinged ochreous. Hindwing pale ochreous-fuscous; cilia ochreous with double row of fuscous-tipped scales around apex. Fore- and midlegs dark fuscous, distal

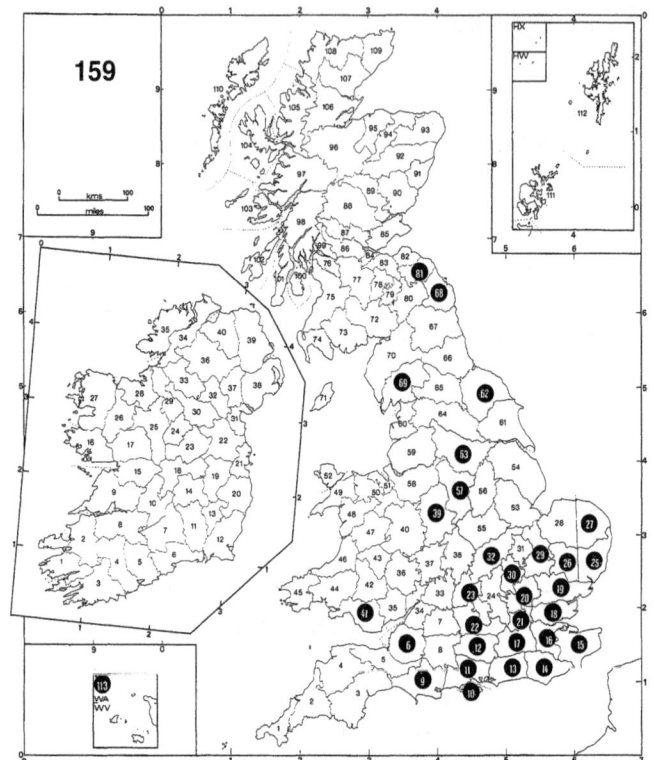

Pexicopia malvella

ends of tarsi ochreous; hindleg paler, tibial tuft pale ochreous-fuscous. Abdomen brownish fuscous with tergites 2 and 3 smooth-scaled and slightly less fuscous. Genitalia, see figures 20g,46a.

Life history

Ovum. Laid in July on marsh-mallow (*Althaea officinalis*) or cultivated hollyhock (*Alcea rosea*).

Larva. Head blackish. Prothoracic plate blackish, trapezoidal and divided down the midline. Body whitish ochreous with reddish spots; pinacula small and black (Fischer von Röslerstamm, 1834).

The larva feeds in the seeds of its foodplant from late August to October. It then spins a cocoon in the larval workings and passes the winter as a fully fed larva within this cocoon until late April.

Pupa. Pale yellowish, short, stout and covered with fine short hairs; cremaster usually consisting of six hooked hairs (Fischer von Röslerstamm, *loc. cit.*). May–June.

Imago. Univoltine; June–August. At night occasionally comes to light.

Distribution (Map 159)

Most frequent in south-eastern England, where it sometimes ranks as a garden pest, but although once widespread throughout England and Wales as far north as the eastern border counties of Scotland (Bolam, 1932) it now appears to be contracting its range. This has led Parsons (1995[1996]) to ascribe the species a Nationally Notable (Scarce) category B conservation status. Throughout Europe; Asia Minor; central Asia; North Africa.

SITOTROGA Heinemann

Sitotroga Heinemann, 1870, *Schmett. Dtl. Schweiz* (2)**2**(2): 287.

A small, mainly tropical genus, represented in Europe by two species, one of which occurs in Britain and also in North America.

Imago. Antennal scape long, with pecten; labial palpus with segment 3 longer than segment 2, segment 2 tufted beneath; ocelli present. Forewing elongate and narrow, apex pointed; veins R_5 (7) and R_4 (8) stalked with M_1 (6), R_5 to costa. Hindwing narrower than forewing, trapezoidal, termen sinuate, apical projection very long; veins CuA_1 (3)–M_2 (5) remote and parallel, M_1 (6) and R_s (7) stalked. Male genitalia completely ringed by abdominal segment 8; uncus with two lobes.

The genus was formerly placed in the Anomologinae on the basis of similarity of wing shape. It has now been reallocated to the Pexicopiinae because of the presence of a pecten on the antennal scape and the structure of the male genitalia.

Early stages. See under *S. cerealella* (Olivier).

SITOTROGA CEREALELLA (Olivier)
Angoumois Grain Moth

Alucita cerealella Olivier, 1789, *Encyclopédie méthodique Hist. nat. (Insectes)* **4**: 121.

Tinea hordei Beckman, 1815, *in* Kirby & Spence, *An Introduction to Entomology: or Elements of the natural History of Insects*: **1**: 175.

Type locality: central Europe.

Description of imago (Pl.6, fig.27)
Wingspan 11–19mm, size varying between cultures. Head glossy white to pale ochreous; antenna slightly more than half length of forewing, ochreous, ringed fuscous, scape with short pecten; labial palpus with segment 3 longer than segment 2, ochreous, segment 2 with tuft of yellowish scales beneath, not furrowed, segment 3 banded fuscous, apex white. Thorax ochreous. Forewing ochreous, paler in disc, variably irrorate fuscous; costal edge black, especially at base; subdorsal black dot at base of wing; often black streak on fold

extending to plical stigma; first discal stigma obsolete, second a black spot at three-quarters, sometimes elongate; cluster of blackish scales at apex and row of terminal black spots round apex to tornus; cilia ochreous, darker at tornus, with black basal line and ciliary line beyond, black on costal cilia, grey and indistinct on termen. Hindwing slightly narrower than forewing, trapezoidal, tornus rounded, termen oblique, apical projection abrupt, twice as long as termen; grey; cilia one and one-half times breadth of wing, pale ochreous. Fore- and midlegs fuscous; hindleg fuscous with inner surface and tarsus ochreous. Abdomen ochreous-fuscous, ventral surface whitish ochreous. Genitalia, see figures 20h,46c.

Similar species. S. cerealella could be mistaken for a *Metzneria* Zeller sp., but can be distinguished by the pecten present on the antennal scape.

Two laboratory cultures were studied; the moths of one were larger and paler, those of the other smaller and with stronger black markings. We are grateful to the Central Science Laboratory, Sand Hutton, York, for the second of these cultures. The larger moths, of Mexican origin, were supplied by the Pest Infestation Laboratory, Slough, in 1969.

Life history

Ovum. In the open, laid on a ripening grain of wheat (*Triticum* spp.), barley (*Hordeum* spp.) or rye (*Secale* spp.), hatching in six to seven days. Indoors a wider range of foodstuffs is used (see below).

Larva. Full-fed *c*.5mm long, short and stout. Head pale honey-brown, mouth-parts dark brown. Body pale yellowish white; gut greyish, sometimes widened on abdominal segments 4–6 to form a grey blotch; thoracic legs concolorous with body.

The larva enters the grain through a minute hole and can complete its growth in a single grain and pupate therein without external evidence of its presence; it is claimed that it can even survive threshing. However, for climatic reasons it can seldom complete its life cycle out of doors in Britain, where it exists mainly indoors as a pest of stored products. Additional foodstuffs indoors include oats, maize, rice, sorghum, buckwheat and bamboo (Sokoloff & Bradford, 1993). It can cause damage in museums by eating corn dollies (Cooter, 1985), or old stooks of wheat (Irwin, 1997). It has been found feeding on dried flowers, presumably their seeds, at a garden centre (Sokoloff, 1995b). These indoor infestations sometimes occur in very large numbers.

Pupa. Length 5mm. Yellowish brown; cremaster two short, stout, widely separated, outwardly pointing processes. Appendages darken before the emergence of

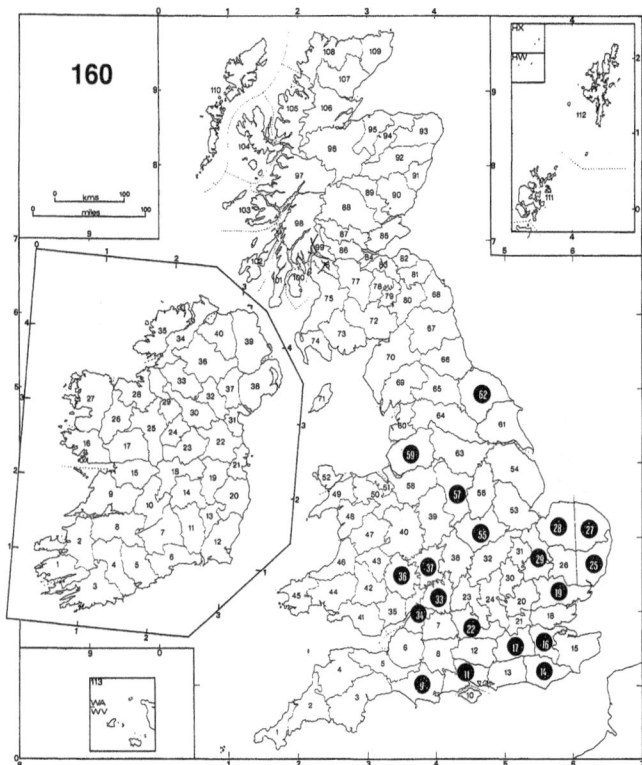

Sitotroga cerealella

the adult. Inside a grain or without a cocoon amongst the foodstuff.

Imago. Continuously brooded in heated warehouses amongst crops that are harvested once a year. The adult tends to hide by day in nooks such as cracks in the brickwork. It has occasionally been taken in the open at light (Langmaid, 1996; A. M. Emmet, pers.obs.).

Distribution (Map 160)
In the wild the species occurs in arable areas where its foodplants are cultivated. The British Isles are at the northern limit of its range; rare and local as far north as Herefordshire and the south Midlands (J. D. Bradley, pers.comm.). As a pest of stored food products it has a disjunct distribution in Britain that reaches Aberdeenshire, having been sporadically introduced by commercial activity. Where it effects an infestation, numbers can build up rapidly. Not recorded from Wales, Ireland, the Channel Islands or the Isle of Man. Throughout continental Europe, but in the north restricted to artificial habitats and occurring naturally most commonly in the Mediterranean region; worldwide in tropical and warmer temperate regions.

THIOTRICHA Meyrick

Thiotricha Meyrick, 1886, *Trans. N.Z. Inst.* **18**: 162,164.

Thistricha Meyrick,1885, *N.Z.Jl Sci.Dunedin* **2**: 590.

Reuttia Hofmann, 1898, *Dt.ent.Z.Iris* **10**: 228.

NOMENCLATURE. *Thistricha* Meyrick, 1885, can be considered as an incorrect original spelling of *Thiotricha* Meyrick, 1886, or as an unused senior synonym. However, it has never been used. In the interests of nomenclatural stability and following usage, *Thiotricha* Meyrick, 1886, is retained here.

A worldwide genus centred in the Australasian region. Only three species are represented in Europe, one of which has been recorded from Britain.

Imago. Small to medium-sized gelechiids with labial palpus long and slender; antenna without pecten. Forewing without scale-tufts, elongate lanceolate. Hindwing slightly less broad than forewing, elongate trapezoidal, tornus rounded, apex produced to a sharp point. Abdominal tergites concolorous. Valvae of male genitalia asymmetrical, both being spatulate and elongate but right valva with additional digitate process from ventral margin; sternite 8 bifurcate; signum of female genitalia a small plate with a central ridge.

Larva. Unusually amongst the Gelechiidae, all known larvae of this genus are case-bearing, feeding on flowers and seeds from a portable case.

THIOTRICHA SUBOCELLEA (Stephens)

Anacampsis subocellea Stephens, 1834, *Ill.Br.Ent.* (Haust.) **4**: 214.

Gelechia internella Lienig & Zeller, 1846, *Isis, Leipzig* **1846**: 291.

Gelechia subocellella Doubleday, 1859, *Zoologist synonymic List Br.Lepid.*: 31.

Epithectis lathyri sensu Pierce & Metcalfe, 1935, *Gen. Tineid Fam.Lepid.Br.Isl.*: 14, *nec* Stainton, 1865.

Type locality: England; New Forest, Hampshire.

Description of imago (Pl.6, fig.28)
Wingspan 10–11mm. Head white, becoming greyer on frons; antenna pale greyish fuscous, ringed paler or whitish; labial palpus greyish white, darker below. Thorax and tegulae white, mixed greyish white. Forewing dull white, sparsely sprinkled with fuscous scales; markings fuscous and black, the fuscous ones consisting of patch occupying most of basal one-quarter of wing, and streak along dorsal half of wing, almost interrupted at one-half and narrowing markedly in apical quarter but continuing around apex

to costa – all these may be replaced with pale grey, reduced or even absent; black markings consisting of diffuse spot halfway across wing at one-half, small spot beyond it near costa and thin line along base of costal cilia; cilia whitish, becoming ochreous-fuscous around tornus, three ciliary lines of dark-tipped scales at apex but only one extending along termen; underside of forewing dark fuscous. Hindwing fuscous; cilia ochreous-fuscous becoming paler at apex. Fore- and midlegs dark fuscous; hindleg more ochreous, tibial tuft ochreous. Abdomen greyish fuscous. Genitalia, see figures 20f,46d.

Life history

Ovum. Laid on a flowerhead of marjoram (*Origanum vulgare*) or, more rarely, water mint (*Mentha aquatica*) (Fletcher, 1884) or thyme (*Thymus polytrichus*) (Bland, 1996); July–August.

Larva: Final instar length 5mm. Head and prothoracic plate light brown. Body dull ochreous with small dark brown pinacula; anal plate brown (Stainton, 1867a).

The larva feeds on the seeds of its foodplant from August to late October, concealed in a portable case. The case is constructed sequentially, as the larva grows, from four to five empty calyxes stacked inside each other (figure 63). Pupation occurs within the larval case attached to a stem or to debris on the ground.

Figure 63 *Thiotricha subocellea* (Stephens), larval case (×5)

Pupa. Length 4.5mm. Pale yellowish brown; rather slender, parallel-sided, with segments 8 and 9 tapering abruptly; no cremaster or hooked hairs, but tergite 8 with a dense transverse row of short teeth anteriorly and sternite 8 with ventrolateral short lengths of similar teeth; tergite 9 with a short dorsolateral pointed projection on each side.

Imago. Univoltine; July–August. Best obtained by sweeping the foodplant.

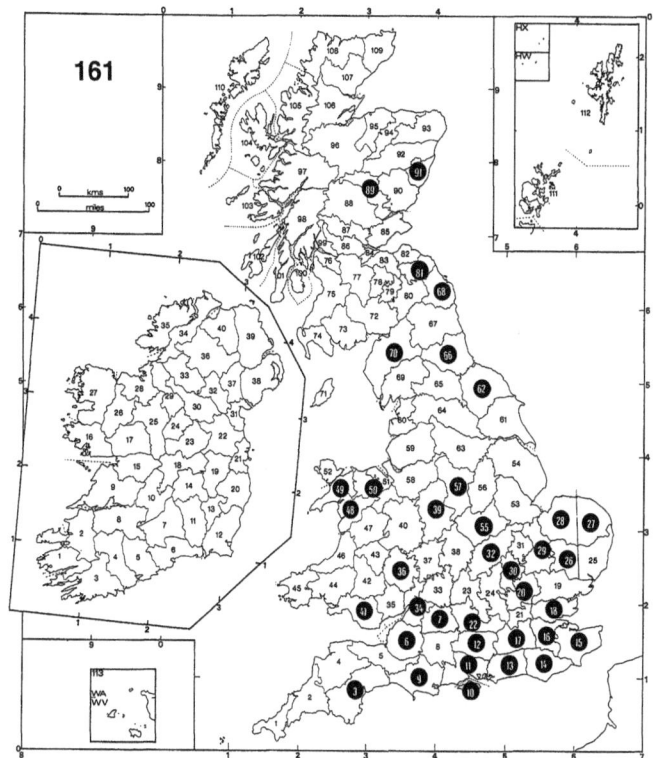

Thiotricha subocellea

Distribution (Map 161)

Widespread in Wales and southern England and occurring in scattered localities as far north as Kincardineshire. Absent from Ireland. Throughout Europe.

References

Agassiz, D. J. L., 1978. *Gelechia sabinella* (Zeller) (Lepidoptera: Gelechiidae), a species new to Britain. *Entomologist's Gaz.* **29**: 136–138.

——, 1984. Microlepidoptera – a review of the year 1983. *Entomologist's Rec.J.Var.* **96**: 245–258.

——, 1986. *Scrobipalpa klimeschi* Povolný (Lepidoptera: Gelechiidae) new to Britain. *Entomologist's Gaz.* **37**: 33–35.

——, 1988. Microlepidoptera – a review of the year 1986. *Entomologist's Rec.J.Var.* **100**: 118–130.

——, 1989. *Gelechia senticetella* Staudinger (Lepidoptera: Gelechiidae) new to the British Isles. *Entomologist's Gaz.* **40**: 189–192.

——, 1990. Microlepidoptera – a review of the year 1988. *Entomologist's Rec.J.Var.* **102**: 129–141.

——, Heckford, R. J. & Langmaid, J. R., 1998. Microlepidoptera review of 1996. *Ibid.* **110**: 97–114.

Anonymous, 1852. [No title]. *Proc.ent.Soc.Lond.* (2) **1852**: 3–6.

Arnold, V. W., Baker, C. R. B., Manning, D. V. & Woiwod, I. P., 1997. *The butterflies and moths of Bedfordshire*, [iv], 408 pp., text-figs, maps, 104 col.pls. Bedford.

Baker, B. R., 1994. *The butterflies and moths of Berkshire*, xxxi, 368 pp., 3 pls, 2 maps. Uffington, Oxfordshire.

Bankes, E. R., 1893. *Gelechia (Lita) strelitziella* not a British insect. *Entomologist's mon.Mag.* **29**: 213.

——, 1894. *Lita instabilella*, Dgl., and its nearest British allies. *Ibid.* **30**: 80–83, 125–128, 188–194.

——, 1898a. Descriptions of the larva and pupa of *Aproaerema sangiella*, Stn. *Ibid.* **34**: 2–3.

——, 1898b. On a new species of the genus *Aproaerema*, Drnt. (=*Anacampsis* auct., nec Crt.) from England. *Ibid.* **34**: 242–244.

——, 1899a. *Aristotelia unicolorella*, Dp., identified as a British species. *Ibid.* **35**: 33–36.

——, 1899b. Descriptions of the larva and pupa of *Aproaerema vinella*, Bnks. *Ibid.* **35**: 202–205.

——, 1907. *Gelechia streliciella*, H.-S., in Britain. *Ibid.* **43**: 236–237.

Barrett, C. G., 1865. Notes on Microlepidoptera occurring near Haslemere. *Ibid.* **2**: 42–44.

——, 1869. Notes on the early stages of *Nothris verbascella*. *Ibid.* **6**: 163–164.

——, 1893. Occurrence of *Gelechia (Bryotropha) figulella* Staud. in England. *Ibid.* **29**: 158.

——, 1901. Lepidoptera, pp. 135–162. *In* Salaman, L. G. (Ed.), *The Victoria history of the county of Norfolk* I.

Beirne, B. P., 1941. A list of the Microlepidoptera of Ireland. *Proc.R.Ir.Acad.* **47**(B): 53–147.

Benander, P., 1928. *Svensk Insektfauna*, 10. Lepidoptera. 2. Microlepidoptera 1. Familjen Gelechiidae. 97 pp. Stockholm.

——, 1965. Notes on larvae of Swedish Microlepidoptera 2. *Opusc.ent.* **30**: 1–23.

Blackburn, T. & Blackburn, J. B., 1867. Captures of Lepidoptera at Rannoch. *Entomologist's mon.Mag.* **4**: 138–140.

Bland, K. P., 1987. Is *Scrobipalpa costella* (H. & W.) (Lepidoptera: Gelechiidae) double brooded? *Entomologist's Rec.J.Var.* **99**: 43.

——, 1990. Exhibit, Annual Exhibition, 1989. *Br.J.Ent.nat.Hist.* **3**: 70.

——, 1992. An instance of *Scrobipalpa acuminatella* (Sircom, 1850) (Lepidoptera: Gelechiidae) mining *Tussilago farfara* L. in Scotland. *Entomologist's Gaz.* **43**: 101–102.

——, 1996. Exhibit, Annual Exhibition, 1995. *Br.J.Ent.nat.Hist.* **9**: 218.

——, Christie, I. C. & Wormell, P., 1987a. The Lepidoptera of The Isle of Coll, Inner Hebrides. *Glasg.Nat.* **21**: 309–330.

——, Knill-Jones, R., Palmer, R. M. & Young, M. R., 1987b. An exceptional evening in east Perthshire. *Entomologist's Gaz.* **38**: 114.

Bolam, G., 1932. The Lepidoptera of Northumberland and the Eastern Borders. *Hist.Berwicksh.Nat.Club* **27**: 221–265.

Bond, K. G. M., 1984. Invertebrates of Irish Midlands raised bogs. Part III. Lepidoptera. *Bull.Ir.biogeog.Soc.* **8**: 103–110.

——, 1991. *Eulamprotes wilkella* ab. *tarquiniella* Stainton (Lepidoptera: Gelechiidae) rediscovered in eastern Ireland. *Entomologist's Gaz.* **42**: 71–74.

——, 1995. Irish Microlepidoptera check-list. *Bull.Ir.biogeog.Soc.* **18**(2): 176–262.

——, 1996. Previously unpublished records of Microlepidoptera to be added to the Irish list. *Ir.Nat.J.* **25**: 194–207.

——, 1999. *Metzneria aestivella* (Zeller, 1839) (Lepidoptera: Gelechiidae): first confirmed Irish record, with a note on the locality of its capture. *Entomologist's Gaz.* **50**: 115.

Boyd, T., 1858. Notes on an entomological tour in Cornwall. *Entomologist's wkly Intell.* **4**: 142–144.

Bradford, E. S., 1978. *Gelechia scotinella* Herrich-Schäffer (Lep.: Gelechiidae) in Herts. *Entomologist's Rec.J.Var.* **90**: 262.

——, [1979]. Gelechiidae, pp.115–132. *In* Emmet, A. M. (Ed.), *A field guide to the smaller British Lepidoptera*, 271 pp. London.

—— & Sokoloff, P. A., 1988. Gelechiidae, pp. 123–141. *In* Emmet, A. M. (Ed.), *A field guide to the smaller British Lepidoptera* (Edn 2), 288 pp. London.

Bradley, J. D., 1964. Lepidoptera in Ireland, May–June, 1962. Part 3: Descriptions of *Aphelia unitana* (Hübner) (Tortricidae) and *Scrobipalpa murinella* (H.-S.) (Gelechiidae), species new to the British list, and notes on variation in *Pyrausta funebris* (Ström) (Pyralidae). *Entomologist's Gaz.* **15**: 74–82.

——, 1966. Some changes in the nomenclature of British Lepidoptera. Part 4. *Ibid.* **17**: 213–235.

——, 1998. *Checklist of Lepidoptera recorded from the British Isles*, vi, 106 pp. Fordingbridge. (Edn 2 (revised), 2000).

——, 2000. *Log book of British Lepidoptera*, iv, 108 pp. Fordingbridge.

—— & Fletcher, D. S., 1979. *A recorder's log book of label list of British butterflies and moths*, [vi], 136 pp. London

—— & Mere, R. M., 1964. Lepidoptera. *In* Natural history of the garden of Buckingham Palace. *Proc. Trans. S. Lond. ent. nat. Hist. Soc.* **1963**: 55–74.

—— & Pelham-Clinton, E. C., 1967. The Lepidoptera of the Burren, Co. Clare, W. Ireland. *Entomologist's Gaz.* **18**: 115–153.

Buhl, O., Falck, P., Karsholt, O., Larsen, K. & Schnack, K., 1989. Fund af småsommerfugle fra Danmark i 1987. *Ent. Meddr* **57**: 133–135.

——, ——, Jørgensen, B., Karsholt, O., Larsen, K. & Schnack, K., 1992. Fund af småsommerfugle fra Danmark i 1990. *Ibid.* **60**: 1–12.

——, ——, ——, ——, ——, —— & Vilhelmsen, F., 1996. Fund af småsommerfugle fra Danmark i 1995. *Ibid.* **64**: 277–288.

Burmann, K., 1950. *Nothris obscuripennis* Frey in Nordtirol. *Z. Lepid.* **1**: 31–34.

——, 1990. Beiträge zur Microlepidopteren-Fauna Tirols 14, *Caryocolum* Gregor & Povolný, 1954 (Insecta: Lepidoptera, Gelechiidae). *Ber. naturw.-med. Ver. Innsbruck* **77**: 171–184.

Carter, D., 1984. Pest Lepidoptera of Europe with special reference to the British Isles. *Series ent.* **31**, 431 pp. Dordrecht.

Chalmers-Hunt, J. M., 1967. *Nothris verbascella* Hübn. (Lep.: Tinaeina) rediscovered. *Entomologist's Rec. J. Var.* **79**: 216–219.

——, 1970. *Phthorimaea operculella* (Zeller) (Lep. Tineina) reared from a tomato. *Ibid.* **82**: 216.

——, 1985. *Monochroa niphognatha* Gozmány, 1953 and *Athrips rancidella* Herrich-Schäffer, 1984 (Lepidoptera: Gelechiidae) new to the British fauna. *Ibid.* **97**: 20–24.

—— & Wakely, S., 1964. Holiday at Thorpeness, Suffolk, 1964. *Ibid.* **76**: 271–275.

Cooter, J., 1985. *Sitotroga cerealella* Ol. (Lep.: Gelechiidae) in Herefordshire. *Ibid.* **97**: 108.

Corley, M. F. V., 1988. Early reports of *Metzneria aprilella* (Herrich-Schäffer) (Lepidoptera: Gelechiidae) from Berkshire. *Entomologist's Gaz.* **39**: 198.

——, 1995. *Caryocolum proximum* (Haworth, 1828) (Lepidoptera: Gelechiidae) in Oxfordshire. *Ibid.* **46**: 231–232.

Cruttwell, C. T., 1907. Occurrence of *Gelechia streliciella*, H.-S., in the Highlands. *Entomologist's mon. Mag.* **43**: 235–236.

Doets, C., 1941. Biology of *Anacampsis betulinella* Vári and *populella* Cl. *Tijdschr. Ent.* **84**: 354–355.

Douglas, J. W., 1850. On the British species of the genus *Gelechia* of Zeller. *Trans. ent. Soc. Lond.* (N.S.) **1**: 60–68.

——, 1879. Notes on *Gelechia nanella*, Hübn. *Entomologist's mon. Mag.* **15**: 207.

Dowling, D. N., 1979. *Lita virgella* Thunberg (Lep.: Gelechiidae). A species new to Ireland. *Entomologist's Rec. J. Var.* **91**: 73.

Dunn, T. C. & Parrack, J. D., 1992. The moths and butterflies of Northumberland and Durham. Part 2: Microlepidoptera. *Vasculum* – Suppl. **3**, 378 pp., distr. maps.

Ellerton, J., 1970. Presidential Address: Microlepidoptera added to the British List since L. T. Ford's review. *Proc. Trans. Br. ent. nat. Hist. Soc.* **3**: 33–41.

Elsner, G., Huemer, P. & Tokár, Z., 1999. *Die Palpenmotten (Lepidoptera, Gelechiidae) Mitteleuropas*, 208 pp., 113 pls (28 col.). Bratislava.

Emmet, A. M., 1968. Lepidoptera in West Galway. *Entomologist's Gaz.* **19**: 45–58.

—— (Ed.), [1979]. *A field guide to the smaller British Lepidoptera*, 271 pp. London.

——, 1981. *The smaller moths of Essex*, 158 pp., maps. London.

——, 1988a. *Teleiodes wagae* (Nowicki) (Lepidoptera: Gelechiidae) on *Betula*. *Entomologist's Gaz.* **39**: 76.

—— (Ed.), 1988b. *A field guide to the smaller British Lepidoptera* (Edn 2), 288 pp. London.

——, 1991. Chart showing the life history and habits of the British Lepidoptera, pp. 61–303. *In* Emmet, A. M. & Heath, J. (Eds.), *The moths and butterflies of Great Britain and Ireland* **7**(2), 400 pp., 8 text figs, 28 maps, 4 col. pls. Colchester.

——, 1997. Exhibit, Annual Exhibition, 1996. *Br. J. Ent. nat. Hist.* **10**: 156.

——, 1998. Exhibit, Annual Exhibition, 1997. *Ibid.* **11**: 90.

Fairclough, A. J. & Fairclough, R., 1980. Exhibit, Annual Exhibition, 1979. *Proc. Trans. Br. ent. nat. Hist. Soc.* **13**: 7.

Fairclough, R., 1962. Collecting Lepidoptera in 1961. *Entomologist's Rec. J. Var.* **74**: 93–98.

Ffennell, D. W. H., 1974. Rediscovery of *Psamathocrita argentella* Pierce & Metcalfe (Lepidoptera: Gelechiidae) as a British species. *Entomologist's Gaz.* **25**: 302–304.

Fischer von Röslerstamm, J. E., 1834–[1843]. *Abbildungen zur Berichtigung und Ergänzung der Schmetterlingskunde, besonders der Microlepidopterologie*. 304 pp., 100 pls. Leipzig.

Fletcher, T. B. & Clutterbuck, G. C., 1939. Microlepidoptera of Gloucestershire. *Proc. Cotteswold Nat. Fld Club* **27**: 34–53.

Fletcher, W. H. B., 1884. Note on the food plant of *Gelechia subocellea*. *Entomologist's mon. Mag.* **21**: 22.

Ford, L. T., 1949a. Presidential Address, 1948. *Proc. Trans. S. Lond. ent. nat. Hist. Soc.* **1947–1948**: 48–58.

———, 1949b. *A guide to the smaller British Lepidoptera*, 230 pp. London.

———, 1953. *Stomopteryx polychromella* Rebel – a gelechiid moth new to Britain. *Entomologist's Gaz.* **4**: 37.

Freeman, T. N., 1967. Annotated keys to some Nearctic leaf-mining Lepidoptera on conifers. *Can. Ent.* **99**: 410–435.

Fryer, J. C. F. & Edelsten, H. M., 1938. Lepidoptera, pp. 139–161. *In* Salzmann, L. F. (Ed.), *The Victoria history of the County of Cambridgeshire and the Isle of Ely* **1**.

Goater, B., 1974. *The butterflies and moths of Hampshire and the Isle of Wight*, xiv, 439 pp. Faringdon.

Gozmány, L. A., 1963. The family Symmocidae and the description of new taxa mainly from the Near East (Lepidoptera). *Acta zool. Acad. Sci. hung.* **9**: 67–134.

Griffith, A. F., 1932. Captures in 1931: Lepidoptera. *Entomologist* **65**: 163.

Hartig, F., 1964. Microlepidotteri della Venezia Tridentina e delle regioni adiacenti. Parte 3 (Fam. Gelechiidae–Micropterygidae). *Studi trent. Sci. nat.* **41**(3–4): 1–292.

Heckford, R. J., 1986. *Syncopacma suecicella* (Wolff) (Lepidoptera: Gelechiidae) new to the British Isles. *Entomologist's Gaz.* **37**: 87–89, 1 pl.

———, 1991. The female genitalia of *Syncopacma vinella* (Bankes) and *S. albipalpella* (Herrich-Schäffer) (Lepidoptera: Gelechiidae), two species confused by Pierce & Metcalfe. *Ibid.* **42**: 227–230, 2 figs.

———, 1992. Is *Syncopacma albipalpella* (Herrich-Schäffer) (Lep.: Gelechiidae) extinct in the British Isles? *Entomologist's Rec. J. Var.* **104**: 62.

———, 1995a. *Scrobipalpa artemisiella* (Treitschke) (Lep.: Gelechiidae), a larval description. *Ibid.* **107**: 38.

———, 1995b. Misidentification by Pierce & Metcalfe of the female of *Gnorimoschema streliciella* (Herrich-Schäffer) (Lepidoptera: Gelechiidae). *Entomologist's Gaz.* **46**: 283.

———, 1997. Dead stems of *Aster tripolium* L. – pupation sites for *Coleophora atriplicis* (Meyrick) and *Scrobipalpa salinella* (Zeller) (Lepidoptera: Coleophoridae and Gelechiidae). *Ibid.* **48**: 106.

———, 1998. Notes on larvae of four species of Microlepidoptera not previously described in the British literature. *Ibid.* **49**: 155–160.

———, 1999a. The British larval foodplants of *Prolita solutella* (Zeller) (Lepidoptera: Gelechiidae). *Ibid.* **50**: 106–107.

———, 1999b. The decline and possible extinction of *Syncopacma vinella* (Bankes, 1898) (Lepidoptera: Gelechiidae) in the British Isles. *Ibid.* **50**: 161–162.

———, 1999c. Notes on the larvae of seven species of Microlepidoptera (Oecophoridae, Gelechiidae and Pyralidae) not previously described in the British literature, together with the redescription of one and a further description of another. *Ibid.* **50**: 223–237, 7 col. figs.

———, 2000. *Caryocolum marmoreum* (Haworth) (Lepidoptera: Gelechiidae): some apparently unrecorded observations on the early larval stages. *Ibid.* **51**: 194.

———, 2001. *Coleophora pappiferella* Hofmann, 1869 (Lepidoptera: Coleophoridae) new to Britain and *Scrobipalpa murinella* (Duponchel, 1843) (Lepidoptera: Gelechiidae) at the same locality. *Ibid.* **52**: 101–104.

——— & Langmaid, J. R., 1988. *Eulamprotes phaeella* sp. n. (Lepidoptera: Gelechiidae) in the British Isles. *Ibid.* **39**: 1–11, 1 pl., 6 figs.

——— & ———, 1991. Lepidoptera in Ireland, 1989. *Ibid.* **42**: 15–29.

——— & ———, 1999. Discovery of the larva of *Monochroa hornigi* (Staudinger) (Lepidoptera: Gelechiidae) in the British Isles. *Ibid.* **50**: 57–58.

——— & Sattler, K., 1999. *Scrobipalpula diffluella* (Frey, 1870) in Britain (Lepidoptera: Gelechiidae). *Ibid.* **50**: 255–260, 2 figs.

——— & ———, 2002. *Bryotropha dryadella* (Zeller, 1850) a newly recognised species, and the removal of *B. figulella* (Staudinger, 1859) from the British list (Lepidoptera: Gelechiidae). *Ibid.* **53** (in press).

———, Sattler, K. & York, P. V., 1999. The taxonomic status of *Eulamprotes immaculatella* (Douglas, 1850) (Lepidoptera: Gelechiidae). *Ibid.* **50**: 155–160, 2 figs.

——— & Sterling, P. H., 1999. Separation of *Helcystogramma rufescens* (Haworth, 1828) and *H. lutatella* (Herrich-Schäffer, 1854) (Lepidoptera: Gelechiidae) on mainly previously unrecorded characters. *Ibid.* **50**: 239–250, 16 figs (incl. 2 col.).

—— & ——, 2002. The discovery of the larva of *Bryotropha dryadella* (Zeller, 1850) and larval descriptions of this, *B. basaltinella* (Zeller, 1839) and *B. senectella* (Zeller, 1839) (Lepidoptera: Gelechiidae). *Ibid.* **53** (in press).

Hering, E. M., 1957. *Bestimmungstabellen der Blattminen von Europa* 1–3, 1185, 221 pp., 725 figs, 's-Gravenhage.

Hoare, R. J. B., Langmaid, J. R., Simpson, A. N. B. & Young, M. R., 1999. *Caryocolum blandelloides* Karsholt, 1981 (Lepidoptera: Gelechiidae) in the British Isles. *Entomologist's Gaz.* **50**: 149–154, 5 figs.

Hodges, R. W. (Ed.), 1983. *Check list of the Lepidoptera of America north of Mexico*, 284 pp. London.

Horton, E., 1867, Occurrence of *Ypsolophus ustellus*, Fab., a lepidopteron new to Britain. *Entomologist's mon.Mag.* **4**: 152.

Huemer, P., 1987. *Caryocolum blandelloides* Karsholt, 1981, neu für Österreich und Mitteleuropa (Insecta: Lepidoptera: Gelechiidae). *Ber.naturw.-med. Ver.Innsbruck* **74**: 207–209.

——, 1988. A taxonomic revision of *Caryocolum* (Lepidoptera: Gelechiidae). *Bull.Br.Mus.nat.Hist.* (Ent.) **57**: 439–571.

——, 1989. Eine neue, gallenerzeugende *Caryocolum*-Art aus Mitteleuropa (Lepidoptera: Gelechiidae). *Nota lepid.* **12**: 21–28.

——, 1993. The British species of *Caryocolum* Gregor & Povolný. *Br.J.Ent.nat.Hist.* **6**: 145–157, pl. 5.

—— & Karsholt, O., 1998. A review of Old World *Scrobipalpula* (Gelechiidae), with special reference to central and northern Europe. *Nota lepid.* **21**: 37–65, 49 figs.

—— & ——, 1999. Gelechiidae I (Gelechiinae: Teleiodini, Gelechiini). *Microlepidoptera of Europe* **3**: 1–356.

—— & Sattler, K., 1995. A taxonomic revision of Palaearctic *Chionodes* (Lepidoptera: Gelechiidae). *Beitr.Ent.* **45**: 3–108, 198 figs.

Irwin, T., 1997. Angoumois grain moth – *Sitotroga cerealella*. *Norfolk moth survey Newsletter* **52**: 7.

Jacobs, S. N. A., 1936. *In* Abstract of proceedings – indoor meetings. *Proc.Trans.S.Lond.ent.nat.Hist.Soc.* **1935–1936**: 27.

——, 1948. Exhibit, ordinary meeting, 14 January 1948. *Ibid.* **1947–1948**: 46.

——, 1958. Editor's note to '*Phthorimaea operculella* Zell. in Kent'. *Entomologist's Rec.J.Var.* **70**: 57.

Kaitila, J.-P., 1996. Suomen jäytäjäkoiden (Gelechiidae) elintavat. *Baptria* **21**: 81–105.

Karsholt, O., 1981. Northern European species of the genus *Caryocolum* Gregor & Povolný, 1954, feeding on *Cerastium*

and *Stellaria*, with the description of a new species (Lepidoptera: Gelechiidae). *Entomologica scand.* **12**: 251–270.

—— & Razowski, J., (Eds), 1996. *The Lepidoptera of Europe. A distributional checklist.* 380 pp. Stenstrup.

Kennard, A., 1965. *Argolamprotes micella* Schiff. (Lep. Gelechiidae) taken in Britain. *Proc.Trans.S.Lond.ent.nat. Hist.Soc.* **1965** (2): 42–43, 1 fig.

Key, R. S., 1995. *Nothris congressariella* (Bruand, 1858) (Lepidoptera: Gelechiidae) reared from Lundy. *Entomologist's Rec.J.Var.* **107**: 273–275.

Klimesch, J., 1951. Über Microlepidopteren des Traunsteingebietes in Oberösterreich. *Z.wien.ent.Ges.* **36**: 101–117.

——, 1958. Beiträge zur Kenntnis der blattminierenden Insektenlarven des Linzer Gebietes und Oberösterreichs. III. Gelechiidae, Acrolepiidae. *Naturk.Jb.Stadt Linz* **1958**: 265–279.

——, 1961. Ordnung Lepidoptera 1. Teil: Pyralidina, Tortricina, Tineina, Eriocraniina, und Micropterygina. *In* Franz, H., *Die Nordost-Alpen im Spiegel ihrer Landtierwelt* **2**: 431–789.

——, 1983. Beiträge zur Kenntnis der Microlepidopteren Fauna des Kanarischen Archipels: 6. Beitrag: Gelechiidae. *Vieraea* **13**: 145–182.

Krogerus, H. & von Schantz, M., 1970. Einige Ergebnisse der nordischen microlepidopterologischen Symposien. *Notul.ent.* **50**: 117–121.

——, Opheim, M., Schanz, M. von, Svensson, I. & Wolff, N. L., 1971. *Catalogus Lepidopterorum Fenniae et Scandinaviae.* Microlepidoptera, 40 pp. Helsinki.

Langmaid, J. R., 1978a. *Abies grandis* Lindl. (Giant or Grand Fir) – A lepidopterous pabulum. *Entomologist's Rec.J.Var.* **90**: 67–68.

——, 1978b. A previously unrecorded foodplant of *Teleiodes proximella* (Hübner). *Ibid.* **90**: 307.

——, 1980. The biology of *Teleiodes wagae* (Nowicki) (Lepidoptera: Gelechiidae). *Entomologist's Gaz.* **31**: 253–254.

——, 1992. *Monochroa moyses* Uffen (Lepidoptera: Gelechiidae) in Sussex and Hampshire, with a note on the probable hibernation habit. *Ibid.* **43**: 144.

——, 1994. A third British record of *Gelechia senticetella* (Staudinger) (Lepidoptera: Gelechiidae). *Ibid.* **45**: 36.

——, 1995. An alternative feeding habit by the larva of *Scrobipalpa obsoletella* (Fischer von Röslerstamm) (Lepidoptera: Gelechiidae). *Ibid.* **46**: 277.

——, 1996. *Sitotroga cerealella* (Olivier) (Lepidoptera: Gelechiidae) and *Cryptophlebia leucotreta* (Meyrick) (Lepidoptera:Tortricidae) at m.v. light in Hampshire. *Ibid.* **47**: 50.

——, 1997. Further observations on the biology of *Monochroa moyses* Uffen (Lepidoptera: Gelechiidae). *Ibid.* **48**: 208.

—— & Sattler, K., 1999. *Teleiodes flavimaculella* (Herrich-Schäffer, 1854) new to the British fauna. *Ibid.* **50**: 5–10, 4 figs.

Larsen, C. S., 1927. Tillæg til fortegnelse over Danmarks Microlepidoptera. *Ent.Meddr* **17**: 7–212.

Leraut, P. J. A., 1997. *Liste systématique et synonymique des Lépidoptères de France, Belgique et Corse* (Edn 2), 526 pp. Paris.

Lhomme, L., [1946–49]. Gelechiidae, pp. 530–677. *In* Lhomme, L., 1935-[1963]. *Catalogue des Lépidoptères de France et de Belgique* **2**, 1253 pp. Le Carriol.

MacAlpine, E. A. M., 1979. The Lepidoptera of the Cairngorms National Nature Reserve. *Entomologist's Rec.J.Var.* **91**: 1–6, 65–70, 213–216, 242–244, map.

Mansbridge, W., 1934. Recent records of Lepidoptera new to Lancashire and Cheshire. *Rep.Proc.Lancs.Chesh.ent.Soc.* **1931–33**: 30–32.

Mariani, M., 1943. Fauna Lepidopterorum Italiae. Parte 1. Catalogo ragionato dei Lepidotteri d'Italia. *G.Sci.nat.econ.Palermo* **42**(3): 81–237, 1 pl.

Martini, W., 1916. Verzeichnis Thüringer Falter aus den Familien Pyralidae–Micropterygidae. *Dt.ent.Z.* **30**: 110–144.

Mason, G. W., 1910. The Lepidoptera of Lincolnshire 4. *Trans.Lincs.Nat.Un.* **1910**: 176–219.

McLeod, J. M., 1966. Notes on the biology of a Spruce Needle-miner, *Pulicalvaria piceaella* (Kearfott) (Lepidoptera: Gelechiidae). *Can.Ent.* **98**: 225–236.

Meyrick, E., 1895. *A handbook of British Lepidoptera*, 843 pp. London.

Meyrick, E., [1928]. *A revised Handbook of British Lepidoptera*, vi, 914 pp. London.

Michaelis, H. N., 1951. Notes on Tortricina and Tineina found in Cheshire. *Entomologist's Rec.J.Var.* **63**: 107–108.

——, 1977. Records of Gelechiidae (Lepidoptera) from North Wales. *Entomologist's Gaz.* **28**: 217–222.

——, 1979. Exhibit, Annual Exhibition, 1978. *Proc.Trans.Br.ent.nat.Hist.Soc.* **12**: 9.

Morley, C. (Ed.), 1937. Final catalogue of the Lepidoptera of Suffolk. *Suffolk Nat.Soc.* **1937**: 1–214.

——, 1957. Tineid new to Suffolk. *Trans.Suffolk Nat.Soc.* **7**(2): 90.

Morris, F. O., 1891. *A natural history of British moths.* **4**: 115.

Müller-Rutz, J., 1913–14. *Die Schmetterlinge der Schweiz* **2**, 727 pp., 1 pl. Bern.

Newton, N., 1985. Supplement to Clutterbuck and Bainbrigge-Fletcher's Microlepidoptera of Gloucestershire. *Gloucs.Nat.* **2**: 3–29.

Nickerl, O., 1908. Die Motten Böhmens (Tineen). *Beiträge zur Insekten-Fauna Böhmens* **6**, 161 pp. Prague.

O'Keeffe, D., 1993. *Gelechia senticetella* (Staud.) (Lep.: Gelechiidae), a second British record. *Entomologist's Rec.J.Var.* **105**: 176.

Opheim, M., 1978. *The Lepidoptera of Norway, check-list*, Part 3. Gelechioidea (first part), 30 pp. Trondheim.

Park, K. T., 1993. Two species of the genus *Caryocolum* (Gelechiidae, Lepidoptera) new to Korea. *Korean J.Ent.* **23**: 17–21.

Parry, J. A., 1980. *Neofriseria singula* Stdgr (Lep., Gelechiidae) in rabbit burrows. *Entomologist's Rec.J.Var.* **92**: 257.

Parsons, M. S., 1995. Exhibit, Annual Exhibition, 1994. *Br.J.Ent.nat.Hist.* **8**: 191.

——, 1995[1996]. A review of the scarce and threatened ethmiine, stathmopodine and gelechiid moths of Great Britain. *U.K. Nature Conservation* No. **16** (1995), 130 pp. Peterborough.

Patočka, J., 1989. Über die Puppen der mitteleuropäischen Gelechiidae (Lepidoptera). 5. Teil, Tribus Gnorimoschemini. *Vest.csl.Spol.zool.* **53**: 123–140.

Pelham-Clinton, E. C., 1964. Lepidoptera in Ireland, May–June 1962. Part IV – list of Lepidoptera recorded. *Entomologist's Gaz.* **15**: 83–92.

——, 1971. *Scrobipalpa clintoni* Povolný (Lepidoptera, Gelechiidae): a note on the type series and some further records. *Ibid.* **22**: 244.

——, 1989. *Scrobipalpa tussilaginis* (Frey) (Lepidoptera: Gelechiidae) new to the British Isles. *Ibid.* **40**: 103–108, 1 pl., 2 figs.

Pierce, F. N. & Metcalfe, J. W., 1935a. *The genitalia of the tineid families of the Lepidoptera of the British Islands.* xxii, 116 pp., 34 pls. Oundle.

—— & ——, 1935b. *Phthorimaea obsoletella* F.v.R. and an allied species. *Entomologist* **68**: 97–99.

—— & ——, 1942. *Psamathocrita argentella* sp. nov. (Lep., Gelechiidae) an addition to the British Fauna. *Ibid.* **75**: 255–256.

Piskunov, V. I., 1981. Family Gelechiidae, pp. 659–748. *In* Medvedev, G. S. (Ed), *Keys to the Insects of the European Part of the USSR* **4**: Lepidoptera, part 2, 1092 pp., 747 figs. Leningrad [in Russian]. English translation, Leiden, 1990.

Pitkin, L. M., 1984. Gelechiid moths of the genus *Mirificarma*. *Bull.Br.Mus.nat.Hist.* (Ent.) **48**: 1–70.

——, 1986. A technique for the preparation of complex male genitalia in Microlepidoptera. *Entomologist's Gaz.* **37**: 173–179.

Povolný, D., 1964. Gnorimoschemini trib. nov. – eine neue Tribus der Familie Gelechiidae nebst Bemerkungen zu ihrer Taxonomie (Lepidoptera). *Cas.ceské Spol.ent.* **61**: 330–359.

——, 1967. Ein kritischer Beitrag zur taxonomischen Klärung einiger palaearktischer Arten der Gattung *Scrobipalpa* (Lepidoptera, Gelechiidae). *Prirodov.Pr.Cesk.Akad. Véd.* (N.S.) **1**: 209–250.

——, 1968. *Scrobipalpa clintoni* sp. nov. (Lep., Gelechiidae) a surprising new discovery from Scotland. *Entomologist's Gaz.* **19**: 113–118.

——, 1979. Eine Ausbeute der Tribus Gnorimoschemini aus Tunis (Lepidoptera: Gelechiidae). *Folia ent.hung.* (S.N.) **32**: 111–119.

——, 1980. Die bisher bekannten Futterpflanzen der Tribus Gnorimoschemini (Lepidoptera, Gelechiidae) und deren Bedeutung für taxonomisch-ökologische Erwägungen. *Acta Univ.Agric.Brno* (A) **28**(1): 189–210.

——, 1990. Zur heutigen Kenntnis von Nahrungspflanzen der Tribus Gnorimoschemini (Lepidoptera, Gelechiidae). *Acta Univ.Agric.Brno* (A) **38**(3–4): 191–204.

——, 1996. Gnorimoschemini, pp. 113–118. *In* Karsholt, O. & Razowski, J. (Eds), *The Lepidoptera of Europe. A distributional checklist*, 380 pp. Stenstrup.

—— & Bradley, J. D., 1965. *Scrobipalpa psilella* (Herrich-Schäffer) (Lep., Gelechiidae) – a species hitherto overlooked in the British Isles. *Entomologist's Gaz.* **16**: 9–12, 2 figs.

Poynton, D., 1996. New and updated Microlepidoptera records for Cheshire (VC58), Shropshire (VC40) and Denbighshire (VC50). *Entomologist's Rec.J.Var.* **108**: 25–26.

Rebel, H., 1901. Famil. Pyralidae–Micropterygidae. *In* Staudinger, O. & Rebel, H., *Catalog der Lepidopteren des palaearktischen Faunengebiets* **2**, xxxii, 368 pp. Berlin.

Richardson, A. & Mere, R. M., 1958. Some preliminary observations on the Lepidoptera of the Isles of Scilly with particular reference to Tresco. *Entomologist's Gaz.* **9**: 115–147, 2 pls.

Richardson, N. M., 1893. On some members of the *instabilella* group of the genus *Lita* (*Gelechia*, partim), with descriptions of *L. suaedella*, n. sp. and *L. instabilella*, Douglas. *Entomologist's mon.Mag.* **29**: 241–248.

Ridout, B. V., 1977. Two new synonymies of Microlepidoptera (Gelechiidae: Gnorimoschemini and Oecophoridae: Depressariinae). *Entomologist's Gaz.* **28**: 38–42.

Riley, A. M., 1991. *A natural history of the butterflies and moths of Shropshire*, 205 pp., 17 figs, 32 col.pls. Shrewsbury.

Russell, S. G. C. & Turner, H. J., 1941. Records and full descriptions of varieties and aberrations. *Entomologist's Rec.J.Var.* **53**: (1)–(8).

Sattler, K., 1980. *Teleiodes wagae* (Nowicki, 1860) new to the British list. *Ibid.* **31**: 235–245, 2 pls, figs.

——, 1981. *Metzneria aprilella* (Herrich-Schäffer, 1854) new to the British fauna (Lepidoptera Gelechiidae). *Ibid.* **32**: 83–90.

——, 1986. Exhibit, Annual Exhibition,1985. *Proc.Trans.Br. ent.nat.Hist.Soc.* **19**: 60.

——, 1987a. Exhibit, Annual Exhibition, 1986. *Ibid.* **20**: 53.

——, 1987b. Die an Compositen gebundenen *Scrobipalpa*-Arten des östlichen Österreichs (Lepidoptera: Gelechiidae). *Annln naturh.Mus.Wien* **88/89**B: 435–456.

——, 1989. The taxonomic status of *Scrobipalpa klimeschi* Povolný, 1967, and *Lita pauperella* Heinemann, 1870 (Lepidoptera: Gelechiidae). *Entomologist's Gaz.* **40**: 7–12.

Schmid, A., 1886. Die Lepidopteren-Fauna der Regensburger Umgebung mit Kelheim und Wörth. *KorrespBl. naturw.Ver.Regensburg* **40**: 101–224.

Shirt, D. B. (Ed.), 1987. *British red data books*: 2. Insects, xliv, 402 pp. Peterborough.

Simpson, A. N. B., 1979. Exhibit, Annual Exhibition, 1978. *Proc.Trans.Br.ent.nat.Hist.Soc.* **12**: 10.

——, 1989. *Dichomeris ustalella* Fab. (Lep.: Gelechiidae) rediscovered in Britain. *Entomologist's Rec.J.Var.* **101**: 17–18.

——, 1996. *Caryocolum junctella* (Douglas) (Lep.: Gelechiidae) in Worcestershire (VC37) in 1994. *Ibid.* **108**: 145–146.

Smith, F. N. H., 1997. *The moths and butterflies of Cornwall and the Isles of Scilly*, xiv, 434 pp., 32 col.pls. Wallingford.

Snellen, P. C. T., 1882. *De vlinders van Nederland. Microlepidoptera, systematisch beschreven* **2**, 1196 pp. Leiden.

Sokoloff, P. A., 1979. The voltinism of *Teleiopsis diffinis* (Haw.) (Gelechiidae). *Entomologist's Rec.J.Var.* **91**: 329–330.

——, 1983. *Teleiodes vulgella* (Hübner) on juniper. *Ibid.* **95**: 116.

——, 1995a. Appendix to Presidential Address. The British species of *Teleiodes* and *Teleiopsis* (Lep.: Gelechiidae). *Proc.Trans.Br.ent.nat.Hist.Soc.* **18**: 103–105, 1 col.pl.

——, 1995b. A flurry of *Sitotroga cerealella* (Olivier) (Lep.: Gelechiidae) in Kent. *Entomologist's Rec.J.Var.* **107**: 14.

—— & Bradford, E. S., 1993. The British species of *Monochroa*, *Chrysoesthia*, *Ptocheuusa* and *Sitotroga* (Lepidoptera: Gelechiidae). *Br.J.Ent.nat.Hist.* **6**: 37–44, 1 col.pl.

—— & Chalmers-Hunt, J. M., 1987. Notes on the biology of *Athrips rancidella* H.-S. (Lep.: Gelechiidae). *Entomologist's Rec.J.Var.* **99**: 253–254.

Sønderup, H. P. S., 1949. Fortegnelse over de Danske Miner (Hyponomer). *Spolia zool.Mus.haun.* **10**: 1–256.

Sorhagen, L., 1886. *Die Kleinschmetterlinge der Mark Brandenburg und einiger angrenzenden Landschaften mit besonderer Berücksichtigung der Berliner Arten*, x, 368 pp. Berlin.

Spuler, A., 1910. *Die Schmetterlinge Europas* **2**, [vi], 524 pp., 23 text figs; **3**, [vi], [95], 95 col.pls. Stuttgart.

Stainton, H. T., 1854. *Insecta Britannica*. Lepidoptera: Tineina. viii, 313 pp, 10 pls. London.

——, 1858. Lepidoptera – New British species in 1857. *Entomologist's Annu.* **1858**: 90.

——, 1859. *A manual of British butterflies and moths* **2**, xi, 480 pp. London.

——, 1862. Observations on British Tineina. *Entomologist's Annu.* **1862**: 129–131.

——, 1864. Observations on Tineina. *Ibid.* **1864**: 164–165.

——, 1865a. Observations on Tineina. *Ibid.* **1865**: 132–142.

——, 1865b. *The natural history of the Tineina* **9**, [iv], 276 pp., 8 col.pls. London.

——, 1866. Observations on Tineina. *Entomologist's mon. Mag.* **3**: 54–57.

——, 1867a. *The natural history of the Tineina* **10**, ix, 304 pp., 8 col.pls. London.

——, 1867b. Observations on Tineina. *Entomologist's Annu.* **1867**: 17–30.

——, 1869a. *The Tineina of southern Europe*, [viii], 370 pp. London.

——, 1869b. Strange pupation of the larva of *Gelechia atrella*, Haw. *Entomologist's mon.Mag.* **6**: 36–37.

——, 1871. New British Tineina in 1870. *Entomologist's Annu.* **1871**: 96–100.

——, 1872. New British Tineina in 1871. *Ibid.* **1872**: 122–123.

——, 1878. Habits of *Gelechia gerronella*. *Entomologist's mon. Mag.* **15**: 89.

——, 1883. On two of the species of *Gelechia* which frequent our salt-marshes. *Ibid.* **19**: 251–263.

Sterling, P. H. & Heckford, R. J., 2001. The discovery of the larva of *Bryotropha desertella* (Douglas, 1850) in Britain and a further note on the larva of *B. terrella* ([D. & S.], 1775) (Lepidoptera: Gelechiidae). *Entomologist's Gaz.* **52**: 143–148.

Stüning, D., 1988. Biologisch-ökologische Untersuchungen an Lepidopteren des Supralitorals der Nordseeküste. *Faun.-Ökol.Mitt.* Suppl. **7**: 1–116.

Styles, J. H., 1959. Notes on some microlepidoptera. *Entomologist's Gaz.* **10**: 43–44.

Sutton, S. L. & Beaumont, H. E., 1989. *Butterflies and moths of Yorkshire*, xi, 367 pp., figs. Doncaster.

Svensson, I., 1976. Anmärkningsvärda fynd av Microlepidoptera i Sverige 1975. *Ent.Tidskr.* **97**: 124–134.

——, 1992. *Monochroa inflexella* n. sp. (Lepidoptera, Gelechiidae). *Ibid.* **113**: 47–51.

Threlfall, J. H., 1878. Notes on Tineina bred in 1877 and 1878. *Entomologist's mon.Mag.* **15**: 89–90.

Treitschke, F., 1833. *Die Schmetterlinge von Europa*. **9**(2), 294 pp. Leipzig.

Turner, A. H., 1955. *Lepidoptera of Somerset*, 188 pp., map. Taunton.

Tutt, J.W., 1891. Notes on *Lita* (*Gelechia*) *junctella*. *Entomologist's Rec.J.Var.* **1**: 7–9.

——, 1905. *Practical hints for the field lepidopterist* **3**, 166 pp. London.

Uffen, R. W. J., 1959. *Anarsia lineatella* Zeller (Lep., Gelechiidae) in Britain. *Entomologist's Gaz.* **10**: 57–58.

——, 1991. *Monochroa moyses* n. sp., a new gelechiid moth mining the leaves of *Scirpus maritimus* L. *Br.J.Ent.nat.Hist.* **4**: 1–7.

Vári, L., 1941. *Anacampsis betulinella*, a new species of the Gelechiidae. *Tijdschr.Ent.* **84**: 351–353.

Wakely, S., 1936. Notes on Lepidoptera collected during 1935. *Entomologist* **69**: 197–199.

——, 1951. *Gelechia hippophaeella* Schrank in Norfolk. *Entomologist's Rec.J.Var.* **63**: 248.

——, 1956. Records from Camber and district, including the capture of the rare *Aristotelia micrometra* Meyr. *Ibid.* **68**: 34–36.

——, 1957. Notes on *Isophrictis tanacetella* Schrank. *Ibid.* **69**: 257–258.

——, 1962. Collecting notes, 1961. *Ibid.* **74**: 165–168.

Walker, J. J. & Hobby, B. M., 1939. Lepidoptera, pp. 82–106. *In* Salzman, L. F. (Ed.), *The Victoria history of the county of Oxfordshire*. **1**. London.

Waters, E. G. R., 1925a. Three additions to the British list of Tineina. *Entomologist's mon.Mag.* **61**: 82–89.

——, 1925b. A supplementary note on *Brachmia lutatella* H.-S. *Ibid.* **61**: 227.

——, 1929. A list of the Microlepidoptera of the Oxford district. *Proc.Rep.Ashmol.nat.Hist.Soc.* Suppl. **1928**, 78 pp.

Wing, W., 1853. *Ypsolopha verbascella*, a new British species. *Proc.ent.Soc.Lond.* **1853**: 127.

Wolff, N. L., 1958. Further notes on the *Stomopteryx* group. *Ent.Meddr* **28**: 224–281.

——, 1971. *The zoology of Iceland*. Lepidoptera. **3**(45), 193 pp., 33 figs., 15 pls., 4 tables. Copenhagen & Reykjavik.

Wormell, P., 1982. The entomology of the Isle of Rhum National Nature Reserve. *Biol. J. Linn. Soc.* **18**: 291–401.

Zeller, P. C., 1868. Lepidopterologische Ergebnisse einer Reise in Oberkärnthen. *Stettin. ent. Ztg* **29**: 121–149.

Addenda and Notes

Addenda and Notes

Addenda and Notes

Addenda and Notes

Addenda and Notes

Addenda and Notes

Addenda and Notes

THE PLATES

Plate 1: Gelechiidae Anomologinae

Figs 1–33, × 4

Plate 1: *Gelechiidae Anomologinae*   Figs 1–33, × 4

Plate 2: Gelechiidae Anomologinae, Gelechiinae

Figs 1–36, × 4

Plate 2: *Gelechiidae Anomologinae, Gelechiinae*

Figs 1–36, × 4

Plate 3: Gelechiidae Gelechiinae

Figs 1–30, × 4

1 *Pseudotelphusa scalella* (Scopoli) ♀ *Page 129*

2 *Pseudotelphusa paripunctella* (Thunberg) ♂ *Page 130*

3 *Pseudotelphusa paripunctella* (Thunberg) ♀ *Page 130*

4 *Altenia scriptella* (Hübner) ♀ *Page 131*

5 *Teleiodes vulgella* ([Denis & Schiffermüller]) ♀ *Page 132*

6 *Teleiodes wagae* (Nowicki) ♂ *Page 133*

7 *Teleiodes wagae* (Nowicki) ♀ *Page 133*

8 *Teleiodes luculella* (Hübner) ♀ *Page 134*

9 *Teleiodes flavimaculella* (Herrich-Schäffer) ♀ *Page 135*

10 *Teleiodes sequax* (Haworth) ♀ *Page 136*

11 *Carpatolechia decorella* (Haworth) ♀ *Page 137*

12 *Carpatolechia decorella* (Haworth) ♀ *Page 137*

13 *Carpatolechia decorella* (Haworth) ♀ *Page 137*

14 *Carpatolechia notatella* (Hübner) ♀ *Page 138*

15 *Carpatolechia fugitivella* (Zeller) ♀ *Page 141*

16 *Carpatolechia fugitivella* (Zeller) ♂ *Page 141*

17 *Carpatolechia proximella* (Hübner) ♂ *Page 139*

18 *Carpatolechia alburnella* (Zeller) (Scottish form) ♂ *Page 140*

19 *Carpatolechia alburnella* (Zeller) ♂ *Page 139*

20 *Teleiopsis diffinis* (Haworth) ♂ *Page 141*

21 *Chionodes fumatella* (Douglas) ♀ *Page 143*

22 *Chionodes distinctella* (Zeller) ♀ *Page 143*

23 *Psoricoptera gibbosella* (Zeller) ♀ *Page 144*

24 *Mirificarma mulinella* (Zeller) ♀ *Page 145*

25 *Mirificarma mulinella* (Zeller) ♀ *Page 145*

26 *Mirificarma lentiginosella* (Zeller) ♀ *Page 146*

27 *Prolita sexpunctella* (Fabricius) ♀ *Page 148*

28 *Prolita solutella* (Zeller) (northern form) ♂ *Page 148*

29 *Prolita solutella* (Zeller) ♀ *Page 148*

30 *Aroga velocella* (Zeller) ♂ *Page 150*

Plate 4: Gelechiidae Gelechiinae

Figs 1–31, × 4

Plate 4: Gelechiidae Gelechiinae

Figs 1–31, × 4

Plate 5: Gelechiidae Gelechiinae, Anacampsinae

Figs 1–36, × 4

1 *Scrobipalpa costella* (Humphreys & Westwood) ♂ *Page 182*

2 *Scrobipalpa costella* (Humphreys & Westwood) ♀ *Page 182*

3 *Scrobipalpula diffluella* (Frey) ♂ *Page 184*

4 *Scrobipalpula tussilaginis* (Stainton) ♂ *Page 186*

5 *Gnorimoschema streliciella* (Herrich-Schäffer) ♀ *Page 187*

6 *Phthorimaea operculella* (Zeller) ♂ *Page 188*

7 *Caryocolum vicinella* (Douglas) ♀ *Page 191*

8 *Caryocolum alsinella* (Zeller) ♀ *Page 192*

9 *Caryocolum viscariella* (Stainton) ♀ *Page 193*

10 *Caryocolum marmorea* (Haworth) ♂ *Page 194*

11 *Caryocolum marmorea* (Haworth) ♀ *Page 194*

12 *Caryocolum fraternella* (Douglas) ♀ *Page 195*

13 *Caryocolum proxima* (Haworth) ♀ *Page 196*

14 *Caryocolum blandella* (Douglas) ♂ *Page 197*

15 *Caryocolum blandelloides* Karsholt ♀ *Page 198*

16 *Caryocolum junctella* (Douglas) ♂ *Page 199*

17 *Caryocolum tricolorella* (Haworth) ♂ *Page 200*

18 *Caryocolum blandulella* (Tutt) ♂ *Page 201*

19 *Caryocolum kroesmanniella* (Herrich-Schäffer) ♀ *Page 202*

20 *Caryocolum huebneri* (Haworth) ♀ *Page 203*

21 *Sophronia semicostella* (Hübner) ♂ *Page 204*

22 *Sophronia humerella* ([Denis & Schiffermüller]) ♀ *Page 205*

23 *Aproaerema anthyllidella* (Hübner) ♂ *Page 206*

24 *Aproaerema anthyllidella* (Hübner) ♀ *Page 206*

25 *Syncopacma sangiella* (Stainton) ♀ *Page 208*

26 *Syncopacma albipalpella* (Herrich-Schäffer) ♀ *Page 213*

27 *Syncopacma larseniella* (Gozmány) ♀ *Page 209*

28 *Syncopacma larseniella* (Gozmány) ♀ *Page 209*

29 *Syncopacma cinctella* (Clerck) ♂ *Page 210*

30 *Syncopacma taeniolella* (Zeller) ♀ *Page 211*

31 *Syncopacma taeniolella* (Zeller) ♀ *Page 211*

32 *Syncopacma vinella* (Bankes) ♀ *Page 212*

33 *Syncopacma vinella* (Bankes) ♂ *Page 212*

34 *Syncopacma suecicella* (Wolff) ♂ *Page 215*

35 *Syncopacma suecicella* (Wolff) ♀ *Page 215*

36 *Syncopacma polychromella* (Rebel) ♀ *Page 215*

Plate 6: Gelechiidae Anacampsinae, Chelariinae, Dichomeridinae, Pexicopiinae

Figs 1–28, × 4

Plate 6: Gelechiidae Anacampsinae, Chelariinae, Dichomeridinae, Pexicopiinae

Figs 1–28, × 4

General Index

Principal entries are given in **bold type**. Plate references are shown as **69** (1:3). The index also includes references to figures in the text as 34 (text fig.14d) and keys.

Index of Host Plants

and other food substances and attractants

www.ingramcontent.com/pod-product-compliance
Lightning Source LLC
Chambersburg PA
CBHW061352210326
41598CB00035B/5960